Advanced Sciences and Technologies for Security Applications

Indexed by SCOPUS

The series Advanced Sciences and Technologies for Security Applications comprises interdisciplinary research covering the theory, foundations and domain-specific topics pertaining to security. Publications within the series are peer-reviewed monographs and edited works in the areas of:

- biological and chemical threat recognition and detection (e.g., biosensors, aerosols, forensics)
- crisis and disaster management
- terrorism
- cyber security and secure information systems (e.g., encryption, optical and photonic systems)
- traditional and non-traditional security
- energy, food and resource security
- economic security and securitization (including associated infrastructures)
- transnational crime
- human security and health security
- social, political and psychological aspects of security
- recognition and identification (e.g., optical imaging, biometrics, authentication and verification)
- smart surveillance systems
- applications of theoretical frameworks and methodologies (e.g., grounded theory, complexity, network sciences, modelling and simulation)

Together, the high-quality contributions to this series provide a cross-disciplinary overview of forefront research endeavours aiming to make the world a safer place.

The editors encourage prospective authors to correspond with them in advance of submitting a manuscript. Submission of manuscripts should be made to the Editor-in-Chief or one of the Editors.

Hamid Jahankhani · Biju Issac
Editors

Cybersecurity and Human Capabilities Through Symbiotic Artificial Intelligence

Proceedings of the 16th International Conference on Global Security, Safety and Sustainability, London, November 2024

Springer

Editors
Hamid Jahankhani
Northumbria University London
London, UK

Biju Issac
Department of Computer and Information Sciences, Academic Centre of Excellence in Cyber Security Research
Northumbria University London
Newcastle upon Tyne, UK

ISSN 1613-5113 ISSN 2363-9466 (electronic)
Advanced Sciences and Technologies for Security Applications
ISBN 978-3-031-82030-4 ISBN 978-3-031-82031-1 (eBook)
https://doi.org/10.1007/978-3-031-82031-1

This Springer imprint is published by the registered company Springer Nature Switzerland AG
The registered company address is: Gewerbestrasse 11, 6330 Cham, Switzerland

Committee

General Chair

Prof. Hamid Jahankhani
hamid.jahankhani@northumbria.ac.uk

Co-chair

Dr. Biju Issac (Director, ACE-CSR)
biju.issac@northumbria.ac.uk

Steering Committee

Prof. Amin Hosseinian-Far, University of Hertfordshire, UK
Prof. Mohammad Dastbaz, Deputy Vice-Chancellor, University of Suffolk, UK
Prof. Magalhaes, Sergio Tenreiro de, Champlain College Online, USA
Dr. Sina Pournouri, Sheffield Hallam University, UK
Dr. Reza Montasari, University of Swansea, UK
Dr. Rose Cheuk Wai, QAHE and Northumbria University London, UK
Prof. Babak Akhgar, Director of CENTRIC, Sheffield Hallam University, UK
Prof. Gianluigi Me, Luiss University, Italy
Dr. Sara El-Deeb, German University in Cairo, Egypt
Prof. Giovanni Bottazzi, Luiss University, Italy
Prof. Bobby Tait, University of South Africa, South Africa
Dr. John McCarthy, Director, Oxford Systems, UK
Vivek Jain, ISACA London Chapter, UK

Osama Akram Amin Metwally, Northumbria University, UK
Dr. Muhammad Muzammal, Northumbria University, UK
Dr. Jordan Schoenherr, Carleton University, Canada
Dr. Mo Adda, Portsmouth University, UK
Dr. Uchenna Ani, University of Keele, UK
Dr. Nigel Houlden, Manchester Metropolitan University, UK
Dr. Abdur Rakib, Coventry University, UK

About the Conference

Over the last two decades, technology relating to cyber space (satellite, drones, UAVs), cybersecurity, artificial intelligence and Generative AI has evolved rapidly. Today, criminals have identified rewards from online frauds; therefore, the risks and threats of cyber attacks have increased too.

The aim of 16th ICGS3-24 is to understand the full impact of cybersecurity, AI, deepfake and quantum computing on humanity. We are now at an age of technological advances such as satellite communications, drones and UAVs in conjunction with AI and GenAI that has never seen before. The ability to manage cybersecurity threats which requires identification, assessment and understanding of the threat together with clear accountability will be the foundation of the our debates and future strategies.

Detection of the threat is another strand to the strategy and will require dynamic risk management techniques, strong and up to date information governance standards and frameworks with AI responsive approaches in order to successfully monitor and co-ordinate efforts between the parties.

Thus, the ability to minimize the threats from cyber is an important requirement. This will be a mission critical aspect of the strategy with development of the right cybersecurity skills, knowledge and culture that are imperative for the implementation of the cyber strategies. Especially when this strategy is created within the artificial intelligence concept. As a result, the requirement for how AI demand will influence business change and thus influence organizations and governments is becoming important.

In an era of unprecedented volatile, political and economic environment across the world, computer-based systems face ever more increasing challenges, disputes and responsibilities, and while the Internet has created a global platform for the exchange of ideas, goods and services, however, it has also created boundless opportunities for cyber-crime.

The ethical and legal implications of connecting the physical and digital worlds, and presenting the reality of a truly interconnected society, present the realization of the concept of smart societies.

Drawing on 15 years of successful events, the 16th ICGS3-24 conference aims to provide attendees with an information-packed agenda with representatives from across the industry and the globe.

This Annual International Conference is an established platform in which security, safety, and sustainability issues can be examined from several global perspectives through dialogue between academics, students, government representatives, chief executives, security professionals, and research scientists from the United Kingdom and from around the globe.

The THREE days Virtual conference will focus on the challenges of complexity, the rapid pace of change, and risk/opportunity issues associated with the 21st-century lifestyle, systems, and infrastructures.

This conference is organised by Northumbria University, UK, which is an Academic Center for Excellence in Cyber Security Research (ACE-CSR) recognised by the National Cyber Security Centre (NCSC) and Engineering and Physical Sciences Research Council (EPSRC), along with other partners.

October 2024

Prof. Hamid Jahankhani
General Chair
hamid.jahankhani@northumbria.ac.uk

Dr. Biju Issac (Director, ACE-CSR)
Co-chair
biju.issac@northumbria.ac.uk

Contents

Critical Space Infrastructure and Cybersecurity Challenges in Outer Space: A Review of India's Space Security

Adithya Variath

Abstract Outer space tech architecture including satellites and various other functionalities is critical to a State's complex cyber infrastructures. While there is an overarching concern about state-sponsored cyber-attacks in the South Asian region, the states in the bloc have not developed regional treaties or domestic policies that adequately address these concerns. With rapidly growing technology and booming population, it is becoming difficult for the South Asian states to suspect the vulnerabilities of cyber politics. The volatile region is dominated by outer space security threats that affect the geopolitical stability of the region. In this context, it becomes critical for states to develop strong policies that can counter these growing concerns. India, which is slowly becoming a major space-faring nation in the region introduced its new Indian Space Policy (hereafter 'ISP') in 2023. India does not have a national space legislation which further exaggerates the problem of not having systems in place to deal with critical issues. The 2023 ISP has several inherent defects in its structure, terminologies and relation with international law obligations. However, one of the most concerning parts is the complete absence of 'cybersecurity' in the policy document. This paper attempts to critically review the Indian Space Policy through a comparative analysis of various national security policies and develop a holistic framework for collecting sensitive data related to satellite payloads, remote sensing, and imaging capabilities. The paper also delves into how global governance efforts can develop technological architecture against cyber threats and vulnerabilities in space.

Keywords India · Cybersecurity · Critical infrastructures · National security · Indian space policy 2023

A. Variath (✉)
Oxford Department of International Development, Oxford India Centre for Sustainable Development, University of Oxford, Oxford, UK
e-mail: adithya.variath@some.ox.ac.uk

H. Jahankhani and B. Issac (eds.), *Cybersecurity and Human Capabilities Through Symbiotic Artificial Intelligence*, Advanced Sciences and Technologies for Security Applications, https://doi.org/10.1007/978-3-031-82031-1_1

1 Introduction

Last year, circa. 2023 has been a vibrant year for policymaking oriented towards protecting cyberspace and space security. The protection of national security and cybersecurity has become a concerning factor for states. The ever-evolving interconnected character of cyber security has developed close ties with the aspect of outer space security. Maintaining safe channels of communication, precise control, and accurate navigation in space all depend on cybersecurity infrastructure. The fragility of this domain has opened a new set of vulnerabilities. In April 2023, a Cyberspace Solarium Commission (CSC) report made the case for identifying outer space as the 17th critical infrastructure sector, with the goal of adopting enhanced cybersecurity measures among satellite operators.[1] In November 2023, the European Space Agency (ESA) came out with a policy document, ESA Security for Space: Shaping the Future, Protecting the Present, with the goal of protecting the ESA's critical space infrastructure.[2] In December 2023, the U.S. National Aeronautics and Space Administration (NASA) announced Space Security: Best Practices Guide in order "to bolster mission cybersecurity efforts for both public sector and private sector space activities."[3] While that was also a response to the growing concern of protecting critical infrastructure, this was also an attempt to prevent the ever-growing threat of infiltration by foreign spy agencies.[4] India introduced its new Indian Space Policy in April 2023, however, the policy mentions the words, "national security", only twice.[5]

[1] Time to Designate Space Systems as Critical Infrastructure, CSC 2.0 Report (14 April 2023), https://www.fdd.org/analysis/2023/04/14/time-to-designate-space-systems-as-critical-infrastructure/.

[2] ESA Security for Space: Shaping the Future, Protecting the Present, ESA, https://esamultimedia.esa.int/docs/corporate/ESA_Cyber_Security_Resilience_Achievement.pdf.

[3] Space Security: Best Practices Guide, NASA, https://swehb.nasa.gov/display/SWEHBVD/7.22+-+Space+Security%3A+Best+Practices+Guide.

[4] The Federal Bureau of Investigation (FBI), the National Counterintelligence and Security Center (NCSC) and the Air Force Office of Special Investigations (AFOSI) issued a two-page bulletin, highlighting unspecified foreign entities were using cyberattacks and techniques such as strategic investment through joint ventures and acquisitions to gain access to the U.S. space industry. According to the Bulletin: "The US financial sector estimates, the global space economy is projected to grow from $469 billion in 2021 to more than $1 trillion by 2030. The United States is the main driver of this growth through its role as a global leader in space investment, research, innovation, and production. Space is fundamental to every aspect of our society, including emergency services, energy, financial services, telecommunications, transportation, and food and agriculture. All rely on space services to operate. Foreign intelligence entities (FIEs) recognize the importance of the commercial space industry to the US economy and national security, including the growing dependence of critical infrastructure on space-based assets. They see US space-related innovation and assets as potential threats as well as valuable opportunities to acquire vital technologies and expertise. FIEs use cyberattacks, strategic investment (including joint ventures and acquisitions), the targeting of key supply chain nodes, and other techniques to gain access to the US space industry", https://www.dni.gov/files/NCSC/documents/SafeguardingOurFuture/FINAL%20FINAL%20Safeguarding%20the%20US%20Space%20Industry%20-%20Digital.pdf.

[5] Indian Space Policy—2023, ISRO, https://www.isro.gov.in/media_isro/pdf/IndianSpacePolicy2023.pdf.

While this is upsetting, what is more delusional is that India despite being one of the fastest-growing space-faring nations has yet to make its own National Space Law or develop its National Space Security Policy. This saddening state of affairs will be critiqued and delved into detail in Part III of this paper.

According to the ESA document "the evolution of the threat landscape affects not only IT networks but the overall space ecosystem, including the space segment (satellites, launchers, etc.), ground segment (ground stations, ground networks, payload data ground segment, etc.), and signal in space (SiS)."[6] It adds that along with the evolution of technologies, there has been almost a simultaneous growth of the threats as well. The document noted that "even small, organized hacker groups or hacktivists are in the position to identify space systems vulnerabilities and launch hybrid attacks, usually driven by a willingness to experiment new hacking strategies or by public recognition and visibility." Geopolitics also play a major role in reshaping the power dynamics of security governance.[7] With the advent of space commercialisation, private players are also involved in military activities indirectly. With the increasing interconnectedness between civilian, commercial, and military use of space, any disruptions or unmitigable attacks can fundamentally disrupt the entire ecosystem.[8]

With the use of space technology in multiple sectors ranging from medicine, and finance to disaster management, any attack can destabilise society. While the financial cost of these disruptions cannot be computable in totality, there has to be more investments in developing security architecture to prevent the threats. Part II of this paper explores these issues from a national security perspective. The issue becomes critical since these disruptions in space technology can also hamper military activities as their ability to gather intelligence, surveillance, reconnaissance (ISR) data and position, navigation, and timing (PNT) data can be disrupted. Part II of this paper will also explore the importance of ensuring the security of cyber systems in this context. Part III of the paper reviews India's critical space infrastructure and suggests some recommendations to rethink and rework the way the Government looks at cybersecurity. Jamming and spoofing leading to data destruction have become quite attractive options to damage space architecture.[9] The most recent example of this can be seen in the Russian-Ukraine or Israel-Palestine conflict. Part IV of this paper also explores the issue of cybersecurity and its relationship with space security and suggests best practices and norms to prevent interference in satellite operations. This part narrows its focus on India's cyber/space security capabilities and reviews the national legislative framework around it.

---

[6] ESA Security for Space: Shaping the Future, Protecting the Present, ESA, https://esamultimedia.esa.int/docs/corporate/ESA_Cyber_Security_Resilience_Achievement.pdf.

[7] Squire, Vicki. "Reshaping Critical Geopolitics? The Materialist Challenge." *Review of International Studies* 41, no. 1 (2015): 139–59. https://doi.org/10.1017/S0260210514000102.

[8] Trachtman, Joel P. "Cyberspace and Cybersecurity." Chapter. In *The Future of International Law: Global Government*, 85–117. ASIL Studies in International Legal Theory. Cambridge: Cambridge University Press, 2013.

[9] Hunt, Jennifer S. "Cybersecurity: Emerging Threats." Chapter. In *Global Security in an Age of Crisis*, edited by Aiden Warren, 290–313. Edinburgh University Press, 2023.

2 Changing Character of National Security and Cybersecurity: What's Next for India?

In international relations, data privacy and protection are crucial concerns. The international community has been compelled to create global standards due to the growth of multinational technology corporations and the expansion of cyber activities beyond national borders. Threats in the modern world include ransomware attacks, identity theft, data breaches, cyberattacks, and cyber fraud such as terrorist financing and money laundering.[10] Cyber hackers are mercenary elements that pose a severe threat to international security since they frequently operate with impunity under the pretence of nation-states. Lower-middle-income and economically vulnerable nations are particularly vulnerable to cyberattacks, which might have catastrophic effects on their fragile economy. Every year, cybercrime damages the global economy by billions of dollars.[11] Cybercrime costs the world economy billions of dollars annually. More time, money, and resources are currently being allocated by governments worldwide to improving cyber-security systems and erecting barriers to lower the risks. By strengthening defences against cyberattacks, new technologies are being used to enhance national asset protection and worldwide security. The guiding concepts for draughting legislation pertaining to cyberspace activities in the context of international law are sovereignty and non-interference. It also discusses how to shield private data from hackers, fraudsters, and other bad actors in order to prevent identity theft and data breaches.

The absence of a unified legal framework and inadequate national data protection infrastructure across nations worldwide are the largest obstacle to cyber security diplomacy. The main reasons India lags are a lack of essential infrastructure, poor performance, and a lack of strong, legally binding cybersecurity legislation. India has so far declined to ratify the 2001 Budapest Convention on Cybercrime, arguing that it would be a breach of its national sovereignty to share data with international law enforcement.[12] The G20 acknowledged cyber security as a vital element of the digital economy during India's G20 Presidency.[13] As a member of the Quad cybersecurity alliance, India has also contributed to projects pertaining to supply chain security, resilience, and cybersecurity, as well as secure software. Recently, the worldwide cybersecurity foundation ISAC and SIA India formed a cooperative agreement.

---

[10] See also Mitsilegas, Valsamis, and Bill Gilmore. "The Eu Legislative Framework Against Money Laundering and Terrorist Finance: A Critical Analysis in the Light of Evolving Global Standards." *International and Comparative Law Quarterly* 56, no. 1 (2007): 119–40. https://doi.org/10.1093/iclq/lei152.

[11] How Much Money Is Lost to Cybercrime?, Statista (31 May 2024), https://www.statista.com/chart/32341/worldwide-reported-losses-connected-to-cybercrime/#:~:text=According%20to%20the%20Bureau's%20analysis,compared%20to%20four%20years%20prior.

[12] India and the Budapest Convention: Why not?, ORF, https://www.orfonline.org/expert-speak/india-and-the-budapest-convention-why-not.

[13] G20 Digital Ministers Recognize Digital Public Infrastructure as an Accelerator of the Global Goals, UNDP, https://www.undp.org/press-releases/g20-digital-ministers-recognize-digital-public-infrastructure-accelerator-global-goals.

The rivalry between the main powers and their incapacity to reach an agreement are two factors contributing to the rarity of international cooperation. International collaborations are also essential for creating a framework against cyber threats and vulnerabilities in space, particularly when they involve like-minded partners. While orbital, kinetic, and electronic warfare are among the attack vectors that ambush space systems, cyberattacks pose a number of problems that target vital infrastructure. Satellites are riskier than most people realise. A cyberattack is defined by the National Institute of Standards and Technology at the U.S. Department of Commerce as "one that targets an organisation by compromising the integrity of data, stealing sensitive information, or destroying or hostilely administering a computing infrastructure".[14] The increased digitisation and digital piloting of satellites have increased the surface area of invasion. These attacks range from nation-state-level assaults, such as supply-chain intrusions or space-based strikes, to script kiddie attacks, which are primarily carried out by individuals on the ground.[15]

Information systems and networks are the foundation of space systems, just like they are for other high-tech systems. This is true not only during the design and manufacturing phase but also from the time of launch until the space vehicle is eventually left in orbit and permitted to burn out. Its fundamental systems, including onboard controls, data transmission channels, and telemetry interfaces to ground stations, are all very susceptible to malicious acts intended to prevent, impair, or interfere with their functionality. The possible consequences of deflecting a compromised satellite from its orbit might be catastrophic. Spoofing sensor data, manipulating sensor systems to produce erroneous responses, jamming command links, infecting software that controls communications or spy satellites in orbit with malware, and even launching denial of service attacks for ransomware are all cases of hostile cyber activity.

Some states have taken proactive steps to mitigate many of these threats. For example, the US government released its "Memorandum on Space Policy Directive-5: Cybersecurity Principles for Space Systems" in 2020.[16] It has designated space a critical infrastructure and has emphasised the importance of giving it cybersecurity protection. Space will also play a key role in nuclear command and global positioning systems. On the other side, India has not drafted a national policy on space security. However, the growing geopolitical tensions between India and China is forcing India to rethink its cybersecurity architecture. In 2010, India established a Space Security Coordination group and in 2019 a centre dedicated to the development of military

---

[14] Understanding the NIST cybersecurity framework, Federal Trade Commission, https://www.ftc.gov/business-guidance/small-businesses/cybersecurity/nist-framework.

[15] Lahmann, Henning. "Unilateral Remedies to Cybersecurity Incidents." Part. In *Unilateral Remedies to Cyber Operations: Self-Defence, Countermeasures, Necessity, and the Question of Attribution*, 43–258. Cambridge: Cambridge University Press, 2020.

[16] Memorandum on Space Policy Directive-5—Cybersecurity Principles for Space Systems, U.S. Department of Transportation, https://www.transportation.gov/pnt/memorandum-space-policy-directive-5-cybersecurity-principles-space-systems.

space capacities within the DRDO was set in place.[17] India also plans to develop satellites for India's defence forces to enable communication between their other forces. It makes cybersecurity more crucial to protect the value of these investments. India is strengthening its weaponisation components when it comes to space security with the testing of ASAT weapons in 2019.[18] India became only the 4th country to conduct the ASAT test. India has aspirations to further develop its military capacity. To enable this, there has been rapid involvement of private actors entering into partnerships to assist the State. In the coming years, entrepreneurship and national security will become the two key components of India's space capacity. The entry of private actors can also increase vulnerability as critical information becomes volatile with several new actors coming into the picture. It calls for a holistic and systematic policy that protects space assets from non-kinetic or kinetic attacks and cyber concerns. Emerging technologies like artificial intelligence and robotics add a new layer to policymaking these technologies also possess the power to jeopardise the stability and safety of assets.

The changing nature of cyberspace attacks has also enabled terrorist organisations and non-state actors to pursue activities to disable national assets and critical infrastructure like power grids, transport network logistics and nuclear infrastructure.[19] These trends put the information and data of citizens against cyber theft and cyber-intrusions. It would require international cooperation to hold many of these actors accountable, as the problem often transcends borders. Many states have also criticised the role of some states that often protect these non-state actors and use these actors to carry out covert operations on behalf of states.[20] Many times these states could be states having the capacity to even veto consensus building. This makes it

---

[17] Rajeswari Pillai, India's Space Priorities Are Shifting Toward National Security, Carnegie, https://carnegieendowment.org/posts/2022/09/indias-space-priorities-are-shifting-toward-national-security?lang=en.

[18] The RAND report on Responsible Space Behavior for the New Space Era: Preserving the Province of Humanity stated that "in March 2019, an Indian ASAT test generated approximately 400 fragments that will decay in weeks or perhaps a few months because the target was orbiting at less than 300 km. As of the end of 2019, less than 20 pieces of the debris created from the Indian ASAT test earlier that year remained on orbit", Responsible Space Behavior for the New Space Era, https://www.rand.org/pubs/perspectives/PEA887-2.html#:~:text=Humans%20have%20explored%20and%20exploited,risk%20of%20collisions%20and%20conflict.

[19] Peng, Shin-yi. "Data Network as Critical Infrastructure: National Security and the Digital Economy." Chapter. In *International Economic Law in the Era of Datafication*, 48–76. Cambridge International Trade and Economic Law. Cambridge: Cambridge University Press, 2024; Taubman, Cara, Alexander Hart, Attila Hertelendy, Derrick Tin, Ryan Hata, and Gregory R. Ciottone. "Reviewing the Health Care Impacts of Attacks on Critical Infrastructure." *Prehospital and Disaster Medicine* 38, no. 5 (2023): 660–67. https://doi.org/10.1017/S1049023X23006167; Brem, Stefan. "Critical Infrastructure Protection from a National Perspective." *European Journal of Risk Regulation* 6, no. 2 (2015): 191–99. https://doi.org/10.1017/S1867299X00004499; Kunneke, Rolf, Claude Ménard, and John Groenewegen. *Network Infrastructures: Technology meets Institutions.* Cambridge: Cambridge University Press, 2021.

[20] Lahmann, Henning. "Cybersecurity Incidents and International Law." Part. In *Unilateral Remedies to Cyber Operations: Self-Defence, Countermeasures, Necessity, and the Question of Attribution*, 1–42. Cambridge: Cambridge University Press, 2020.

difficult to build a formidable consensus to frame binding and effective legal frameworks at the global level. While international experiences have shown a dismaying future it is also critical to analyse the domestic experiences and how national laws are also shaping how to combat cybercrime.

The critical need for advanced cybersecurity tools has grown with India's recent advancements in its space capabilities. These advancements are also possible with leadership, collaboration and mentorship that would require participation from private and government sectors. Something certain is that when the applications expand, threats also expand. ISRO's internal capabilities in software development, mentioning the success of tools like FEAST (Finite Element Analysis of Structures) are now used by structural engineers and designers across industries, including aerospace and architecture.[21] The ISRO Chairman has publicly said that "Most of the software ISRO uses for our space domain whether it's rocket dynamics, satellite control, or structural designed and developed in-house such as the Computational Fluid Dynamics (CFD)"[22] as an attempt to reinforce ISRO's self-reliant approach to space technologies. ISRO's Analysis of Collision Possibilities tool, designed to assess the risk of satellites colliding in space has also proved to be one of the best in the world in terms of its software capabilities. ISRO's research in areas like thermal design, orbital mechanics, and the long-term evolution of space objects is proving to be extremely cost-effective.

India's space program started early as compared to its Asian counterparts way back in 1962. With the old Indian National Committee for Space Research (INCOSPAR) evolving into the Indian Space Research Organisation in 1969, it also gave mettle to the organisation to influence policy making. Later, the Government established the Space Commission and the Department of Space (DoS). The space program is monitored by the Prime Minister's Office and managed through the Space Commission and the DoS. India, with the support of the USSR, launched its first satellite in 1975. In the later years, India developed its GSAT-7 and GSAT-6, India's military satellites, to enhance space-based maritime communications. In 2018, India established the Defence Space Agency (DSA), and in 2019, ISRO also developed an electromagnetic intelligence-gathering satellite to intercept enemy radars. India has also signed several space treaties.[23] At the domestic level, India revised its Remote Sensing Data Policy, which lays down "the ground rules for distributing and using satellite remote sensing data by private users, both from Indian and foreign space platforms" and in

---

[21] S. Unnikrishnan Nair, S. Somanath, Introduction to Finite Element Analysis: A Textbook for Engineering Students (Springer, 2024).

[22] Need for cybersecurity tools evident as India's space capabilities grow: ISRO chairman, Hindustan Times (September 5, 2024), https://www.hindustantimes.com/india-news/need-for-cybersecurity-tools-evident-as-india-s-space-capabilities-grow-isro-chairman-101725520984821.html.

[23] These treaties include the Outer Space Treaty, signed in 1967 and ratified in 1982; the Convention on International Liability for Damage Caused by Space Objects, recognised in 1979; the Agreement on the Rescue of Astronauts, the Return of Astronauts and the Return of Objects Launched into Outer Space, endorsed in 1979; and the Convention on Registration of Objects Launched into Outer Space, agreed to in 1982.

2018 formulated its National Frequency Allocation Plan.[24] The Defence Research and Development Organisation (DRDO) is responsible for designing and making the weapon systems, while the Defence Space Research Organisation (DSRO) works on developing models of space warfare. With the new Indian Space Policy 2023, the Indian National Space Promotion and Authorisation Centre (IN-SPACe) has been designated as the independent nodal agency to engage with space activities and usage of DoS-owned facilities by non-government private entities. However, the Indian Space Policy has been silent on many crucial aspects including the contentious domain of cybersecurity.

I have argued elsewhere that: "India expects to play a key role in the future in the drafting of international law on the prevention of an arms race in outer space while remaining committed to protecting the country's national security interests. India also supported the 2014 UNGA resolution, 69/32, on 'No First Placement of Weapons on Outer Space'. India has reiterated that it has no intention of entering an arms race in outer space but regional instability forces it to develop its outer space military capabilities. But while India is investing in and building on its space assets, the element of securing the space assets will always trickle down to the issue of militarisation. In 2023, the Indian government unveiled its 'Indian Space Policy', which has used the words, "national security", only twice. Despite being one of the fastest-growing space economies, India has not developed its own National Space Law or National Space Security Policy. Drafting them will help India develop military space capabilities and build space security-oriented architecture."[25]

With its aggressive pursuit of space weaponization by neighbouring China there is always a threat regarding the safety of critical data acquired or received by its satellites. China's substantial intelligence, surveillance and reconnaissance capacity looms over India's lack of strong policy mechanisms. India's national security space infrastructure is also important to tackle the shimmers between India and Pakistan at the border and to tackle state-sponsored terrorism.[26]

There have been reports saying that China-based cyber operations have targeted India's critical sectors. DoS and the ISRO also conducted a space situational assessment in 2021 and warned the "growing threats of space objects, including orbital debris colliding with operational space assets, have become a perennial problem for outer space's safe and sustainable use."[27] In this context, it is worth studying how India's space policy performs to mitigate these risks. The question is—'is it enough?'

---

[24] Remote Sensing Data Policy (RSDP), ISRO, https://www.nrsc.gov.in/EOP_irsdata_Policy/page_1?language_content_entity=en.

[25] Adithya Variath, Time for an Indian National Space Security Policy?, CASL (June 1, 2024), https://nujs.edu/casl/time-for-an-indian-national-space-security-policy/.

[26] Adithya Variath. (2024). Space economy and the geopolitics of South Asia's space security trilemma. *The Round Table, 113*(2), 202–203. https://doi.org/10.1080/00358533.2024.2333201.

[27] Space Situational Assessment 2021, ISRO, https://www.isro.gov.in/SSA.html.

3 India's Space Cyber Security Architecture: Reviewing India's Space Policy 2023

In 2020, the Government of India introduced sectoral reforms in the space sector. On 24 October 2024, "the Union Cabinet approved setting up of Rs.1000 crore Venture Capital Fund dedicated to the space sector, under aegis of IN-SPACe.These reforms were designed to encourage and empower new generation enterprises to actively participate in a wide range of space activities, from start to finish."[28] The objective was to establish a just and welcoming atmosphere that offers each participant equal chances. Building on these reforms, the government is currently concentrating on creating trustworthy and transparent legislation for space operations carried out by different parties. This will create a robust and dynamic space environment that will allow for the growth of creativity and teamwork. The Cabinet-approved reform agenda is embodied in the Indian Space Policy 2023, or ISP, which is a comprehensive and adaptable framework. It lays the groundwork for India's space industry to flourish and advance dynamically, laying a strong basis for the future. Thus, the Indian Space Policy 2023 has been developed as a comprehensive, multifaceted, and flexible framework to carry out the Cabinet-approved reform objective.

The vision of the ISP is to "augment space capabilities; enable, encourage and develop a flourishing commercial presence in space; use space as a driver of technology development and derived benefits in allied areas; pursue international relations, and create an ecosystem for effective implementation of space applications among all stakeholders; for, the nation's socio-economic development and security, protection of environment and lives, pursuing peaceful exploration of outer space, stimulation of public awareness and scientific quest."[29] The ISP seeks to promote the commercialisation of the space industry and encourage increased private sector involvement in space operations. Increasing the involvement of the private sector by offering a stable regulatory environment, a single point of contact for all space activity authorisation, and clearly defined roles for the many agencies involved in space operations in India.

---

[28] "The Government of India, as part of its 2020 space sector reforms, established IN-SPACe to promote and oversee private sector participation in space activities. IN-SPACe has proposed a Rs.1000 crore Venture Capital Fund to support the growth of India's space, economy, currently valued at S8.4 billion, with a target to reach $44 billion by 2033. The fund aims to address the critical need for risk capital, as traditional lenders are hesitant to fund startups in this high-tech sector. With nearly 250 space startups emerging across the value chain, timely financial support is crucial to ensure their growth and prevent talent loss overseas. The proposed government-backed fund will boost investor confidence, attract private capital, and signal the government's commitment to advancing space reforms. It will serve as an Alternative investment Fund under SEBI regulations, providing early-stage equity to startups and enabling them to scale for further private equity investments." *Union Cabinet approves establishment of Rs.1000 crore Venture Capital Fund for Space Sector under aegis of IN-SPACe*, Press Information Bureau (24 October 2024), https://pib.gov.in/PressReleaseIframePage.aspx?PRID=2067667.

[29] Indian Space Policy—2023, ISRO, https://www.isro.gov.in/media_isro/pdf/IndianSpacePolicy2023.pdf.

The ISP laid down the institutional framework and has laid down that the NewSpace India Limited (NSIL) as the primary agency for the commercialisation of space technology and platforms of ISRO. The ISP states that the "Indian National Space Promotion & Authorisation Centre or 'IN-SPACe' shall function as an autonomous Government organization, mandated to promote, hand-hold, guide and authorize space activities in the country". For this purpose, IN-SPACe shall periodically issue guidelines and procedures, that would among other things promote ease of doing business. Prof. Sandeepa Bhat critiquing the policy writes "the function of promoting space activities, which is entrusted with IN-SPACe, is within the policy domain of the executive, however, granting authorization to private players and conferring power to frame guidelines and regulations are the essential functions of the legislature."[30]

IN-SPACe is responsible for promoting manufacturing hubs, incubation centres, and industry clusters or zones related to the space industry. According to the strategy, "IN-SPACe shall also promote identified space activities and establish India as a preferred service provider for global requirements of products/services in the space sector by working with industry both domestically and internationally."[31] In order to expand the space ecosystem and facilitate industry-academics connections, IN-SPACe will collaborate with academia. Through public-private partnerships, the program seeks to increase India's economic potential and creativity as a spacefaring country. The ISP highlights that the ISRO's primary priority will be on developing new space technology and applications and advancing human knowledge of space.

The Department of Space and the NSIL cover nearly every aspect of the new ISP, IN-SPACe, from promotion and governance to regulation, and they mirror the new, optimistic actions of the Indian government. Private startups would undoubtedly be empowered and encouraged to develop space-tech solutions under the new legislation. ISP can be viewed as a comprehensive, forward-looking document that could influence the Indian Space Program's discourse and establish India as a world leader in space administration. Although academics have expressed issues over the ISP's definition, interpreting the policy in light of international treaty responsibilities will be the real test of the policy. The policy can be considered as the first step, but it needs a Parliamentary enactment to make the ideas in the policy more effective in nature. However, the most concerning inconsistency in the policy is the absence of cybersecurity aspects. The policy is completely silent on this concerning issue.

The 2023 Open-ended Working Group on Reducing Space Threats via Norms, Rules, and Principles of Responsible Behaviours gave the space security discussion a boost on a worldwide scale. Resolution 76/231 of the General Assembly called for it to be held with the theme of "Reducing space threats through norms, rules, and principles of responsible behaviours."[32] The Working Group was given

---

[30] Sandeepa Bhat, Outlining Inconsistencies in the Indian Space Policy 2023, CASL (May 1, 2023), https://nujs.edu/casl/outlining-inconsistencies-in-the-indian-space-policy-2023/.

[31] Indian Space Policy—2023, ISRO, https://www.isro.gov.in/media_isro/pdf/IndianSpacePolicy2023.pdf.

[32] Open-ended working group on reducing space threats through norms, rules and principles of responsible behaviours, UN Office of Disarmament Affairs, https://meetings.unoda.org/open-ended-working-group-on-reducing-space-threats-2022.

the task of recommending appropriate actions regarding the hazards that states pose to space systems, particularly how to stop an arms race in space. There was never any agreement on the report. States' belligerent attitude and resistance to any kind of self-regulation are reflected in the report's outcome. While the international community was silent on this, India had a chance to develop a framework on space security concerns by evolving a holistic space security framework within its new policy.

Although academics contend that the policy has a number of intrinsic flaws and contradictions, it can be viewed as the state's initial move to capitalise on commercial space capabilities and optimise the multibillion-dollar space industry. The term "national security" has only been used twice by the 2023 ISP. Space-based intelligence, lunar and cislunar activities, space situational awareness, and military communication in positioning, navigation, and timing capabilities need to be developed in a national security structure in the near future, even though a commercial policy developed by a civilian space agency does not have to address these issues.

The next priority should be developing a national space security policy and a space defence strategy, even if national space legislation is still required. The policy should be based on developing guidelines against hostile space acts and establishing institutional frameworks that can enhance space situational awareness and counter-space capabilities. Additionally, given the nuclear security conundrum that China, Pakistan, and India are facing as well as the tensions in India's immediate neighbourhood, it becomes imperative. Given the geopolitical and astropolitical tensions in South Asia, China's growing space domination and space diplomacy have implications for India's security.

Space technology is enhancing India's ability to manage the military and civilian aspects of space security, security readiness, and military operations. Because space assets are also vulnerable to a growing array of orbital, kinetic, and electronic types of warfare, a cybersecurity component is introduced to space technologies. To resist the risks of space activities, India needs a national space security policy that places a strong focus on utilising technological developments and bolstering space-based capabilities. Given the global and cross-border implications of cyberspace policies, India may have created a framework to persuade other participating countries to encourage cooperation, coordination, and mutual involvement.

According to the India Cyber Threat Report 2023 in 2024 "some of the major cyberspace threats would include- Ransomware and Digital Extortion, MFA Fatigue Attacks, Event-Based Attacks, Phishing/Vishing attacks & Dating App Scams, Exploitation of Supply Chains, AI-powered malware and use of Deep Fake for Deceptive Social Engineering."[33] With the publication of the Joint Doctrine for Cyberspace Operations in 2024, India has made its first step in the advancement of its national security strategy. This doctrine not only demonstrates India's dedication to protecting its cyberspace but also highlights the wider ramifications for its national security and defence as cyber threats grow more complex and widespread. However, this is yet to

[33] India Cyber Threat Report 2023, DSCi, https://www.dsci.in/files/content/knowledge-centre/2023/India-Cyber-Threat-Report-2023_0.pdf.

be formulated in the space sector. So, the answer to the question I raised in the last part is that it ‘is not enough’.

4 Concluding Suggestions

There is an ever-growing need to develop a framework for safeguarding the nation’s digital frontiers. Promoting resilience and enhancing collaboration to counteract the sophisticated tactics employed by cyber adversaries can be central points for space diplomacy. There is a critical need to embed cyberspace operations within the broader framework of national security. Integration of cybersecurity within India’s national framework will be critical for effectively addressing the complex security landscapes. While major space powers like Russia, the United States and Israel have developed their military space cybersecurity capabilities, China’s Strategic Support Force has centralised space, cyber, electronic, and psychological warfare capabilities.

Space technology is essential to industries like manufacturing, defence, power, oil and gas, and their supply networks. In this regard, India must create a thorough national space strategy that may explore crucial cybersecurity protection measures at the national infrastructure level in order to create a watchful and robust cybersecurity framework. Increasing the budgetary allotment for space from 0.04% of GDP is the first step. To establish credible deterrents, India must spend on infrastructure, technology, and human resource training. Developing a space security agenda involves not only the scientific community but also space lawyers and diplomats. Additionally, it is necessary to update current national laws and think about enacting new space laws. The ESA has created a multi-platform for cybersecurity resilience that consists of a geographically dispersed set of systems with common security features and capabilities. From design to operation to disposal, it facilitates the engineering stage of secure systems. Additionally, it certifies and verifies secure functions, performs cybersecurity operational monitoring on Earth and in space, and aids in threat and vulnerability assessments, threat modelling, and intelligence collecting. India should investigate whether it can create a comparable system in South Asia. Guidelines for responding to any hostile activity in space must also be updated, and a specialised cyber security mechanism must be created.

Outsourcing in NATO Cyberspace: A Warfighting Imperative

Manisha Parmar and Andy Miles

Abstract Since the establishment of Cyberspace as a NATO warfighting domain, an endorsement taken by NATO Allies at the Warsaw Summit in 2016 (North Atlantic Treaty Organization in Warsaw summit key decisions. Factsheet. https://www.nato.int/nato_static_fl2014/assets/pdf/pdf_2017_02/20170206_1702-factsheet-warsaw-summit-key-en.pdf, 2017 [1]), NATO has seen a rapid buildup of requirements relating to key Cyberspace capabilities such as Situational Awareness and Command and Control (C2) for the new operational domain. In addition, NATO, in its 2030 Agenda (North Atlantic Treaty Organization in NATO 2030. Factsheet. https://www.nato.int/nato_static_fl2014/assets/pdf/2021/6/pdf/2106-factsheet-nato2030-en.pdf, 2023 [2]), has identified an ambitious set of proposals, which bring implications and growing need for the bolstering of NATO Cyber defences and delivery of NATO Cyber Security services. This brings continuous demand for Cyber-related expertise and foreshadows the reality that NATO in-house capacity is significantly outweighed by the demand. NATO must rely now, more than ever before, on industry partnerships to help deploy and deliver Cyberspace capabilities, including the strengthening of Cyber Security services, to meet the demand of NATO stakeholders. NATO, as a defensive, consensus driven body comprised of 32 Nations, is not alone in this paradigm where Cyber demand outweighs available capacity. NATO Nations, especially those Nations with mature Cyberspace practices such as Canada and the United States, have already faced this challenge. This paper highlights the approach taken by these two nations, regarding nationally adopted outsourcing models in support of Cyber. It reviews related activities and phases within the capability development lifecycle where industry partnerships have been leveraged in support of developing Cyber capabilities and services. This information (approaches, organizational construct, etc.) is assessed to identify

M. Parmar (✉)
NATO Communications and Information Agency, Oude Waalsdorperweg 61, Den Haag, 2597AK The Hague, Netherlands
e-mail: manisha.parmar@gmail.com

A. Miles
Made to Measure Mentoring, Thorncroft Drive, Leatherhead KT22 8JB, Surrey, UK

H. Jahankhani and B. Issac (eds.), *Cybersecurity and Human Capabilities Through Symbiotic Artificial Intelligence*, Advanced Sciences and Technologies for Security Applications, https://doi.org/10.1007/978-3-031-82031-1_2

lessons identified and is used to derive a set of recommendations which can be applied to the NATO context.

Keywords NATO · Outsourcing · Cyberspace · Cyber security · Warfighting

1 Introduction

At the Warsaw Summit in 2016, NATO Nations endorsed Cyberspace as an operational domain [1], meaning NATO must be able to effectively defend itself in Cyberspace with the same proficiency as it defends itself on Land, in Air, at Sea and in Space (which was subsequently recognized as a domain in 2019 [3]).

Nations have also set sight on a very ambitious agenda referred to as the NATO 2030 Agenda [2], where nine lines of effort have been proposed, ranging from deeper political coordination, improved resilience and the preservation of the technological edge, to continued strategy growth and NATO investments. Due to the cross-cutting nature of Cyber Security, where Cyber Security is a key enabler to several aspects of the Agenda, it is highly likely that every proposal within the Agenda brings Cyber related considerations. This will increase requirements for NATO warfare development, capability development and subsequent service uplifts to be undertaken in Cyberspace and Cyber Security. NATO must rely now, more than ever before, on industry partnerships to help deploy and deliver Cyberspace capabilities, including the strengthening of Cyber Security services, to meet the demand of NATO stakeholders. By looking at two NATO Nations with mature Cyberspace practices such as Canada and the US, NATO can assess approaches taken for outsourcing which can be self-applied. Given NATO is, by its nature, a defensive organization, the scope of analysis/assessment is primarily limited to those activities contributing to maintaining a defensive posture.

2 Demand and Outsourcing Within NATO Today

Cyberspace as an operational domain implies the Commander must be provided tools to execute missions either taking place in Cyberspace itself or through Cyberspace (where Cyber is an enabler to missions in other domains). The possible courses of action selection and ability to direct warfighters are underpinned, respectively, by gaining Situational Awareness in operations and achieving Command and Control (C2) in the Cyberspace domain [4]. The ability to integrate Cyberspace into the mission lifecycle also derives additional requirements aligning to the "DOTMLPFI" paradigm [5], where capabilities are expanded using Doctrine, Organization, Training, Materiel, Leadership, Personnel, Facility, and Interoperability considerations.

"A capability's lifecycle begins with its founding, when a group of individuals organizes around an objective. Subsequently, in the development stage, a capability is built, refined and improved. Finally, a capability enters the maturity stage, where it ceases development and is maintained through exercise." [6]. Simplified here in the following phases (1) Research and Development (R&D), (2) Development and Deployment (D&D) and (3) Service Operations and Maintenance (O&M), standardly1 defined as follows:

- R&D "…comprise creative and systematic work undertaken in order to increase the stock of knowledge - including knowledge of humankind, culture and society - and to devise new applications of available knowledge." [7]
- D&D, adapted from [8], is the design and implementation of requirements through to the completed enhancement available in production.
- Service O&M is "the processes, functions, organization, and tools used to underpin the ongoing activities required to manage and support services [9].

Considering the DOTMLPFI paradigm, outsourcing focuses on "M", and possibly "T" and/or "F", whereby the remaining items are typically undertaken in house, specifically noting that Doctrine development or Leadership changes should not be outsourced.

NATO already pursues outsourcing activities in all three of the described phases, however, there is a need to increase outsourcing while streamlining the methodologies applied. Outsourcing can either take place via "Adopt" (from a NATO Nation), "Buy" or "Create" options. This construct, referred to as "ABC", is the first evaluation to take place when considering procurement of capabilities and is presented in order of preference by the NATO Nations [10].

Before an ABC decision can be taken, however, NATO conducts R&D activities via several entities. Within NATO, the NATO Allied Command Transformation (ACT) is the primary scientific body mandated with NATO's transformation via warfare development activities. ACT works alongside several other bodies, both within the NATO community (such as the NCI Agency or NATO Science and Technology Organization [STO]) and within industry and/or academia, to research concepts and conduct Proof of Concepts (POCs). ACT's efforts are further augmented by scientific activities undertaken by other NATO bodies, namely the Allied Command Operations (ACO) and the NATO Office of the Chief Information Officer (OCIO).

Following R&D, if a "Buy" decision is reached, it is likely (although not always) the NCI Agency, as the "Cyber arm" of NATO, to engage. In cases where an "Adopt" option is pursued, the Agency will support, but in later parts of the lifecycle, to support integration of national tooling into existing systems and providing O&M.

NCIA, by effect of the common funding regiment, where 32 Nations contribute to the cost of NCIA operations, has a requirement to outsource as much support as possible. D&D activities corresponding to "Buy" options rely heavily on industry support with Agency contracts exceeding 226 MEUR in 2022 [11].

O&M is also outsourced in varying models, depending on the capability/service rendered. The three, varying models include: (1) NATO Owned, Contractor Operated

(NOCO), (2) Contractor Owned, Contractor Operated (COCO) and (3) Contractor Owned and NATO Operated (CONO) noting that the last possible permutation, NATO Owned, NATO Operated (NONO) refers to a completely insourced approach.

NOCO models rely on industry to provide a level of effort for Operations and Maintenance (O&M) support whereby NATO has opted to own the resources associated with the service. These resources can include hardware and/or software (licenses). Industry partners that work alongside NATO Subject Matter Experts (SMEs) may or may not work onsite with the Agency as it would depend on the service delivery model and the classification being handled.

In contrast, COCO also relies on a level of effort support to O&M but allows the third-party ownership of resources providing NATO services, which implies the equipment can be on or off premises, depending on the classification of data and offsets the risk of equipment ownership (equipment failure, for example) to the contractor instead of to NATO. In the COCO model, NATO personnel are involved to varying degrees, as the service may be almost fully outsourced (for example, reverse malware engineering) or partially outsourced (for example, Incident Handling and Management). It is vital to note that, as a rule of thumb, a customer facing service is never completely outsourced and that NCIA is always involved and accountable to the customer.

Lastly, the CONO model exists but is least likely to be leveraged compared to NOCO or COCO, simply because there are limited reasons why NATO would need to explicitly operate a service but would allow third party ownership of the supporting resources.

For all phases (R&D, D&D and O&M), solicitation for RFPs to acquire industry support across the lifecycle are advertised by each individual NATO body; there is no centralized website across all NATO bodies for acquisition activities. NATO does, however, have the NATO Industrial Advisory Group (NIAG) to facilitate NATO to industry partnerships, although use of the group is not formalized or prescribed procedurally.

Subsequently, third parties may or may not require clearances, depending on the topic at hand and phase of the capability lifecycle. For example, D&D activities typically require clearance due to the classification of facilities and access requirements to both physical premises and CIS. R&D may not, however, require support from personnel with clearances, depending on content. A generic assessment on quantum threat, for example, could be developed by anyone, but a paper reviewing options for centralization of incident management on NS networks may, however, have sensitivities requiring clearance. The O&M considerations are dependent on the outsourcing model being used. For example, in NOCO instances where personnel are working alongside NATO personnel, clearances would be required for facility and CIS access.

Security clearances are provided by the NATO Nation and incur both cost and lead time—in some cases, this lead time can be significant, upwards to one year [12]. This requirement can either limit opportunities to outsource or constrain the accessible talent pool.

3 Outsourcing in Canada

In 2013, the Canadian Armed Forces (CAF) began operationalization of the Cyber domain in Canada [13], starting with creation of the Cyber Force Commander function, which is responsible for the full spectrum of force management and deployment activities in support of Canadian and/or joint missions. Prior to this, the CAF already had an active stance in Computer Network Operations (CNO) and have regularly worked with industry through various partnerships to supplement force capacity in the areas of research, design/development, deployment, and operations of CNO specific and related capabilities.

The Canadian Department of National Defence (DND) is supported scientifically by the Defence Research Development Canada (DRDC), as a "… science, technology and innovation leader, trusted advisor, collaborative partner, and knowledge integrator for defence and security" [14].

Partnerships range from both internal Government of Canada (GoC) entities providing Information Technology and Communication (ICT) services to the CAF and other GoC departments via Shared Services Canada (SSC), as well as external, Canadian companies working in the defence industry.

SSC, established in 2011, is a GoC entity established to centralize common ICT services across several GoC entities and agencies to reduce duplication of services, gain efficiencies through collation of customer requirements and experience the benefit of reduced costs, both investment (hardware and/or software costs) and operations (costs to support and maintain a service). Offerings range from Infrastructure as a Service (IaaS) in the form of cloud offerings or hosting private infrastructure (in cases where privacy and security prevail), to automated, repeatable services such as time accounting or leave recording for staff.

For external Canadian companies, the CAF issues Request For Proposals (RFPs) seeking both level of effort support where companies provide Subject Matter Expertise (SME) on a specific topic to augment internal teams, or product based activities, where companies are hired to prepare/produce a specific artefact or deliver a service, via the GoC website "Buy and Sell GC" [15]. Based on the needs of the CAF, respondents are evaluated for compliance against both mandatory and optional requirements.

Thus, the outsourcing paradigm within the CAF is codified, however, how and where the process is applied is a vital consideration. Referring to the design, development, deployment and operations stages, one can apply both level of effort and product/service-based support at all stages. From information publicly available, it seems the design, development and deployment of CAF Information Management (IM)/Information Technology (IT) is GoC internally outsourced to DIMEI—the Director Information Management Engineering and Integration.

DIMEI, a GoC entity with relationships to both the CAF and SSC, has a mandate to "develop and oversee the implementation of technical and security standards for all architecture, engineering, configuration and integration of IT and CIS [Communication and Information System] that support corporate, C2 [Command and Control], and ISR [Intelligence, Surveillance and Reconnaissance] systems in the DND/CAF" [16]. In this function, DIMEI consults with SSC to determine the most effective manner to provide a shared service to the CAF and is responsible to the CAF to ensure their requirements are satisfied. DIMEI is divided into six program areas, three of which are directly related to Cyberspace: (1) Cyber Operations, (2) Ready Cyber Forces and (3) Cyber and Command Control. Other areas pertain generically to acquisition, IM/IT services management and internal IT services, which indirectly support Cyberspace.

In a report from 2020 [16], DIMEI was evaluated for effectiveness in this role and the findings highlighted that, in the Cyber domain, some projects are suffering schedule delays and that Cyber projects are cross cutting to other IM/IT projects which makes the assessment of Cyber programs in isolation less valuable.

Generic findings have assessed a lack of prioritization of customer requirements, inconsistent service delivery despite defined and agreed Key Performance Indicators (KPIs) and that delays in Shared Services Canada IT support have further compounded delays in DIMEI's support to CAF operations. Findings with respect to economy and efficiency have been deemed difficult to assess, due to cost tracking challenges and lack of defined model between DIMEI and SSC. The report did state, however, that the services provided by SSC are 20% higher than services provided by industry (but did not go on to further assess how industry is leveraged or whether this increased cost has resulted in increased benefits).

It is known, however, from public RFP responses such as the one shown in [17], that in addition to the relationship between DIMEI and SSC, DIMEI also engages with industry directly for both level of effort and product/services support, during the design, development and deployment of capabilities.

Operationally, the CAF has a history of outsourcing stemming from operations in Afghanistan, specifically with respect to logistic requirements. In a Canadian House of Commons report on Cyber Defence in Canada, from 2023 [18], a recommendation is included for Canada to "... expand its collaboration with Canadian security and defence industries to bolster Canada's offensive and defensive cyber infrastructure amidst the growing assertiveness of malign foreign states." Thus, while specific examples where CAF has leveraged outsourcing relationships in support of Cyber are not publicly available, this recommendation illudes to the fact that industry support is required in the domain, including in support of offensive requirements. Further reports [19] have identified the shortage in personnel support to the CAF in Cyberspace, only bolstering this assumption.

4 Outsourcing in the United States

The US Cyberspace Command was established in June 2009 under decree from, then, Secretary of Defence, Robert Gates. Recognizing the growing importance and vulnerability of computer-based assets in the US and around the world. Later, in 2004, the US recognized Cyberspace as an operational domain [20] and the Joint Task

Force (JTF) on CNO was divided into defensive and offensive entities, respectively the JTF-Global Network Operations (GNO) and JTF-Network Warfare (NW).

The outsourcing of shared services does not seem to be formally centralized and appears quite diversified in the US, although, the US General Services Administration (GSA) was established in 1949 as a resource for government entities to turn to, if desired, for some shared service support. The GSA can be leveraged by any government entity to provide IaaS via cloud hosting or other infrastructure support, to services for invoices, payroll, travel planning, etc. It is difficult to tell from publicly available information how strongly the US Department of Defence (DoD) may use or rely on services provided by the GSA noting the US DoD entities may optionally leverage GSA support.

Each branch of the DoD has a scientific arm where preparation of new innovations is conducted and, in addition, there is a central entity specifically for military research called DARPA, the Defense Advanced Research Projects Agency (DARPA). DARPA has a long-standing history as the first creators of the Internet [21] and seeks partnerships with industry.

Additionally, the United States military has strong ties with major non-profit organizations to support its Research and Development (R&D) initiatives, such as the creation of Mitre for the United States Air Force, or the RAND corporation which is also a non-profit primarily funded by the Department of Defence (DoD). These not-for-profit corporations are provided regular funding via the US government and, in return, provide both on-site contractor support and product-based support (through research and the creation of prototypes, papers, etc.). These organizations are also permitted to do work for other entities, for example, NATO has contracted both Mitre and RAND to perform studies and conduct assessments.

External industry providers are also engaged by the various branches of the DoD (Army, Navy, etc.) via the publicly available "Sam.gov" site which is the US government's official website for contract opportunities, amongst other information, where the DoD posts RFPs and solicits responses to major defence initiatives within areas such as Cyberspace. DARPA also advertise their RFPs via Sam.gov. Looking at the robustness and diversity in the network leveraged by the US DoD in support of national security and defence, it is not surprising that the US DoD spending on defence contracts is already anticipated in early 2024 to exceed 105 billion US Dollars (USD) [22] (in 2023, total value of issued contracts was 466 billion USD) [23].

Similarly to Canada, the outsourcing by the US DoD in support of design, development and deployment phases are not only anticipated but are the norm. This brings focus back to operations where we again ask ourselves about the relationship between US DoD military operations and relationships with third party industry support. As described in [24], in kinetic domains, a soldier who has never fired a weapon would never exercise Command and Control (C2) over troops, however, in the Cyberspace domain, there is more flexibility, where civilian operators could be performing C2, directing activities within the Cyber realm where kinetic effect is experienced. This disconnect, as the authors further explain, is due to a significant skill and capacity shortage.

The authors further suggest that the augmentation of existing Cyber work force need not be solved only by adding capacity—the decentralization of Cyber trained personnel across the various parts of the US DoD—Army, Navy, Air Force, Space and Cyber require an element of centralization to allow for coordination.

Thus, while it is not clear from publicly available information whether operational arms of the US DoD working in Cyberspace augment the personnel/skills gap with outsourced personnel, it is highly likely given the requirements and challenges in meeting operational need.

5 Summary

There are differences and commonalities across both countries which can be summarized using a generic lifecycle of capability development as described in [6] and summarized earlier in this paper.

Within the R&D phase, the use of industry and commercial expertise is leveraged to underpin innovation and ensure best of breed solutions are identified. All countries have dedicated entities to support this scientific and academic work.

Beyond these government entities, both countries offer RFPs and opportunities for industry to deliver both products and services, through a consolidated portal. Both countries augment their in-house personnel via outsourcing in support of R&D objectives.

One difference, however, is that the US further outsources DoD support via not-for-profit organizations and these organizations further outsource, as desired. Canada does not seem to expand its outsourcing approach via not-for-profit organizations or does not discriminate between industry and not for profit companies.

At the D&D phase, there is continued cooperation between Canada and the US, with industry, and for both countries, major contracts appear in this phase, where industry is hired to undertake significant deployment activities including the installation and configuration of hardware/software in support of new capabilities.

It is the third and last phase, O&M, in cases where a common service is required, such as cloud hosting or printer deployment, a centralized government service provisioner can be leveraged (SSC in Canada and GSA in the US). When it comes to specific Cyber based capabilities, however, the outsourcing model (if leveraged) becomes unclear. Given the known capacity shortage experienced in both Canada and the US, it is reasonable to assume that outsourcing is leveraged to augment the existing workforce, but when, where and how is not obvious from publicly available information.

Other approaches to close this gap include the earlier recommendation for consolidation of some common Cyber requirements which may exist across multiple government entities. By centralizing certain capabilities and establishing a single service provider, certain duplication in skill and personnel can be avoided, thereby reducing the support footprint.

There are, however, limitations such as offensive capabilities and legal considerations such as C2 where non-vetted personnel may be prohibited (both out of scope of this paper).

6 Recommendations for NATO

In a comparison and contrast of approaches for outsourcing taken by Canada and the US, the following recommendations for NATO are proposed.

- Consider all Cyberspace and Cyber Security priorities to ensure operational urgency is addressed first, to position the workforce appropriately.
- Centralize acquisition and procurement activities to allow for a reduced operating footprint and increased situational awareness on like activities which may increase throughput of knowledge and content. Activities that may complement one another could allow for greater usage and application of developed products across all of NATO, rather than focusing on individual entities.
- Ensure relevant and appropriate KPIs exist to allow measurement of services rendered and in support of continuous improvement. In cases where it is difficult to generate a KPI (for example, it is not possible to measure how many cyber-attacks have been prevented) ensure there is a focus on obtaining qualitative measurements and that processes and/or procedures are being followed.
- Ensure appropriate cost models for the delivery of Cyber services are in place to support effective decision-making, balancing cost, and risk. For example, where shared infrastructure is used to support multiple customers, derive consistent approaches to determine how customers are charged for services and ensure this information is used when assessing outsourcing economy.
- Increase reliance and partnerships with not-for-profit organizations within NATO Nations, which may prove to be an economical method to rapidly develop knowledge in R&D. This may also introduce means to access nationally funded products or services with economies of scale.
- Increase cooperation with the NIAG to gain National partnerships and build industry engagements. This would further enable Nations to track NATO activities which, in turn, could influence the commercial landscape.
- Increase opportunities for third party engagement without clearance requirements (especially in R&D and D&D). Clearance requirements limit the talent pool and introduce significant lead time to procurement.
- Consolidate Cyberspace O&M activities to the greatest extent possible to reduce duplication of effort, maximize economy and increase effectiveness. Ensure clear roles and responsibilities exist across all NATO entities to allow individual entities mastership of a specific function while relying on others to perform their functions (there is no need to create something new, per say). For example, the division between Cyber Security and Cyberspace Operations, where Cyberspace operators

can rely on Cyber Security teams to perform the underlying functions to secure and defend CIS.
- Develop a strategy to define appropriate NOCO, CONO and COCO to ensure robust service delivery with the necessary security and resiliency.

7 Conclusions

Using NATO Nations such as Canada and the US as examples, due to their recognized maturity in the Cyberspace domain, NATO can take lessons identified and implement certain practices to increase outsourcing, streamline adopted approaches and ensure partnerships with industry are put in place to maximize both efficiency and economy while maintaining a technological edge and security.

Considering the rapid buildup of capabilities expected via the NATO 2030 Agenda and the growing need for NATO to maintain a high operational state of readiness, the workload across NATO bodies will significantly increase, making outsourcing not just a nice to have practice but an imperative for the Commander to maintain mission assurance through diversification and operations support.

Disclosure of Interests The authors have no competing interests to declare that are relevant to the content of this article.

References

1. North Atlantic Treaty Organization (2017, Feb) Warsaw summit key decisions. Fact-sheet. https://www.nato.int/nato_static_fl2014/assets/pdf/pdf_2017_02/20170206_1702-factsheet-warsaw-summit-key-en.pdf
2. North Atlantic Treaty Organization (2023, Mar 11) NATO 2030. Factsheet. https://www.nato.int/nato_static_fl2014/assets/pdf/2021/6/pdf/2106-factsheet-nato2030-en.pdf
3. North Atlantic Treaty Organization (2024, Mar 14) NATO's approach to space. What we do. https://www.nato.int/cps/en/natohq/topics_175419.htm
4. Morgan S, Stone S (2019) Command and control for cyberspace operations—a call for research. Mil Cyber Aff 4(1), Article 4
5. European Defence Agency (2023) Enhancing EU military capabilities beyond 2040, Brussels, BEL. https://eda.europa.eu/docs/default-source/eda-publications/enhancing-eu-military-capabilities-beyond-2040.pdf
6. Zuzul T, Helfat C (2016) Capability lifecycle. In: The Palgrave encyclopedia of strategic management. Palgrave Macmillan, London, UK
7. Statistics Canada (2023) Annual survey of research and development in Canadian industry, Ottawa, ON. https://www23.statcan.gc.ca/imdb/p2SV.pl?Function=getSurvey&SDDS=4201
8. Chaudhary A (2023) Development and deployment process. Sales Force. https://www.apexhours.com/development-and-deployment-process/
9. Danby S (2023) What is ITIL service operation? Functions, processes, and best practices. InvGate. https://blog.invgate.com/itil-service-opera-tion#:~:text=ITIL%20service%20operation%20is%20the,root%20cause%20of%20preventable%20incidents
10. NATO Consultation, Command and Control Board (C3B) (2021, Aug 30) C3 taxonomy baseline 5.0. AC/322-D(2021)0017, Brussels, BEL

11. NATO Communications and Information Agency (2022) NCI agency annual report 2022 summary, Brussels, BEL. https://www.ncia.nato.int/resources/site1/general/about_us/publications/nci_agency_annual_report_2022.pdf
12. North Atlantic Treaty Organization (2024) General information—questions and answers, Brussels, BEL. https://www.nato.int/cps/en/natohq/86935.htm
13. Department of National Defence (DND) (2017) Joint doctrine note on cyber operations. Canadian Forces Joint Warfare Centre, Ottawa, ON, pp 1–3
14. Government of Canada. Defence research and development Canada, Ottawa, ON. https://www.canada.ca/en/defence-research-development.html
15. Government of Canada. Tender opportunities, Ottawa, ON. https://canadabuys.canada.ca/en/tender-opportunities?status%5B0%5D=1920&status%5B1%5D=87
16. Government of Canada (2020) Evaluation of the defence information management/information technology programme, Ottawa, ON. https://www.canada.ca/en/department-national-defence/corporate/reports-publications/audit-evaluation/evaluation-defence-imit-programme.html
17. Government of Canada. Buy and sell, Ottawa, https://buyandsell.gc.ca/cds/public/2021/07/07/5818ddfe3effd7b53dbaa53e88332e41/w6369-22-p5oa_-_rfp_en.pdf
18. Canada House of Commons (2020) The cyber defence of Canada: report of the standing committee on national defence, Ottawa, ON. https://www.ourcommons.ca/Content/Committee/441/NDDN/Reports/RP12548256/nddnrp05/nddnrp05-e.pdf
19. The Ottawa Citizen (2022) DND cyber force hindered by lack of staff and training, assessment team warns, Ottawa, ON. https://ottawacitizen.com/news/national/defence-watch/dnd-cyber-force-hindered-by-lack-of-staff-and-training-assessment-team-warns
20. United States Cyber Command. About: US cyber command, Washington, DC. https://www.cybercom.mil/About/
21. Defense Advanced Research Projects Agency. Arpanet, Arlington, VA. https://www.darpa.mil/about-us/timeline/arpanet
22. USA Spending (2024) Department of defence, Washington, DC. https://www.usaspending.gov/agency/department-of-defense?fy=2024
23. USA Spending (2024) Department of defence, Washington, DC. https://www.usaspending.gov/agency/department-of-defense?fy=2023
24. Lonergan E, Montgomery M (2025) United States cyber force: a defense imperative. Foundation for Defense of Democracies. https://www.fdd.org/analy-sis/2024/03/25/united-states-cyber-force/

Near Space Instability. Geopolitical Tensions, Debris Crisis, and Cyberattacks

Stefania Paladini

Abstract Never before Near Space, that portion of outer space closest to Earth and crucial for human activities, has been under threat like in present days. There are several factors responsible for the current status of things, but three of them (geopolitics, debris and cyberthreats) are emerging as the most critical in terms of impact and long-term implications. What is more important, those factors are now colliding, with state-sponsored cyberattacks in outer space that risks exacerbating the debris crisis. The aim of this study is therefore to investigate how geopolitical tensions are pushing existing criticalities such as the debris crisis in the Earth Orbit and the growing threat of cyberattacks to satellite and ground space infrastructure toward an unprecedent level of tensions. Building on the analysis of existing datasets, it will present some cases for discussion and attempts a scenario analysis for the short-medium term.

Keywords Debris · Geopolitics · ASAT · Satellite · HSN · QKD · Space governance

1 The Shifting Landscape of Space

Nothing expresses better the state of the space sector than the recent EU policy document in which space is defined as a "*contested and challenged environment*" [1], a statement widely endorsed [2, 3]. With good reasons: new players, both public and private, are emerging globally, driven in part by the decreasing costs of satellite development and launch, driving partnership and cooperation initiatives but also heightening competition and, increasingly, debates about what should be the 'new space order' to come. Geopolitical tensions are only one side of this 'contested' environment; the legal framework (or the lack of a modern one) is a substantial component of the issues the sector is now facing.

S. Paladini (✉)
Queen Margaret University, Edinburgh, Scotland
e-mail: spaladini@qmu.ac.uk

H. Jahankhani and B. Issac (eds.), *Cybersecurity and Human Capabilities Through Symbiotic Artificial Intelligence*, Advanced Sciences and Technologies for Security Applications, https://doi.org/10.1007/978-3-031-82031-1_3

General challenges do abound, ranging from space debris and cyber threats to the effects of space weather and the growing unsustainability of the space industry as a whole, and it would be short-sighted to consider them separately one from the other, given how interlinked they are.

A good example in this sense is the following, which shows how geopolitics, power plays and the overall sustainability of the sector are connected. In November 2021, a Russian ASAT (anti-satellite) missile test created over 1500 pieces of trackable debris and forced astronauts aboard the ISS to take shelter, in one that was only the latest in a series of wargame tests conducted in Low Earth Orbit (LEO), conducted by multiple states, USA, China and India among them.

While ASAT tests are by no means not the only cause of the growing amount of debris in Earth's orbit, they are still one of the most severe (see further for this) and each event significantly worsens what has become a major threat to space operations, let alone a military security concern. Moreover, the increasing number of spacefaring nations only heightens the urgency to address this issue, as countries continue to test their own ASAT missiles, as evidenced by China and India's own ASAT tests.

An important component here is the current legal framework governing the use of weapons in space, which explicitly bans weapons of mass destruction (WMDs) but does not prohibit ASATs. Legal framework is also a contributing factor to a series of challenges the sector faces at the moment, including the complexity it adds to ADR (active debris removal), insurance, and operations in LEO and beyond. Until a substantial revision of the laws and treaties that govern the space milieu is not undertaken at an international level it is doubtful that sustainable growth for the sector can be achieved.

But geopolitical tensions and what we can define ongoing weaponisation of space also means that cyber threats are on the rise, too. Satellites are now becoming a target for cyberattacks, and, given the fact that have been traditionally protected by security by obscurity, their vulnerabilities can successfully be exploited. Technological advances such as the use of widespread use of LLMs in an offensive postures and post-quantum computing capabilities will only make these threats more severe and will demand mitigation measures to avoid dangerous incidents.

This article intends to offer a snapshot of the current security issues in space, with a focus on the Earth's orbits, and outlines possible scenarios going forward.

For the purpose of this discussion, Near Space consists of the various Earth's orbits that can host satellites, as shown in Fig. 1.

Different orbits have different characteristics. LEO ranges from 160 to 2000 km above Earth, ideal for communication and Earth observation. It is where the majority of satellites have been launched, especially Cubesats, and it is also the place for constellations like Starlinks. The International Space Station (ISS) hovers there, too. Going farther away from the planet, the coverage expands. MEO is located between 2000 and 35,786 km altitude. It's commonly used for navigation systems like GPS, Galileo, GLONASS, and Beidou. GEO is located 35,786 km above the equator, where satellites to match Earth's rotation. The fixed position over the same point is ideal for weather monitoring, broadcasting, and communication satellites. In addition

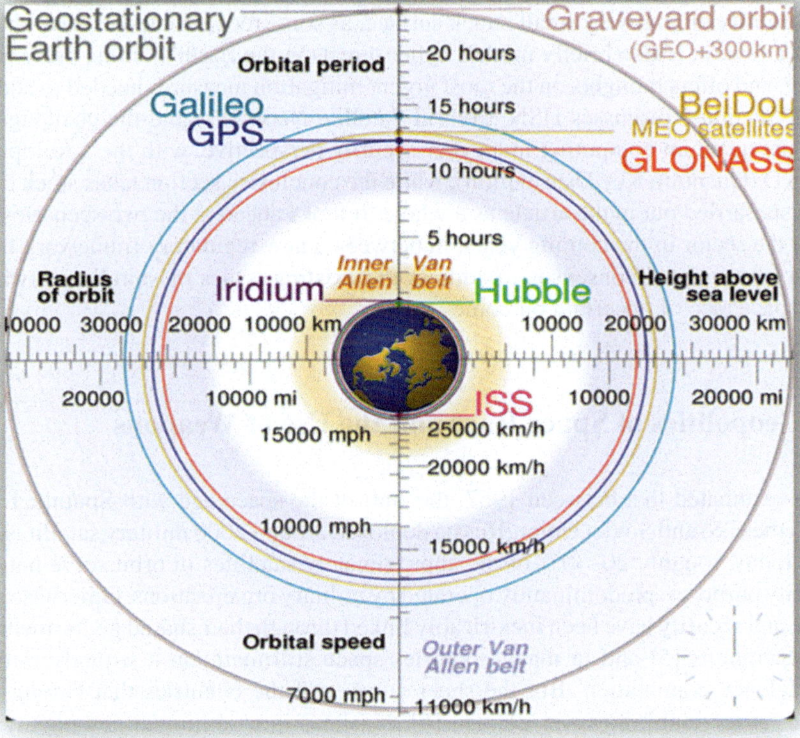

Fig. 1 The Earth's orbits. *Source* Creative Commons, 2024

to these three orbital levels, there are more—HEO (Highly Elliptical Orbit), Polar Orbit, etc.—with more specialised uses.

For their communication (generally indicated as SATCOM) satellites adopt an ITU's allocated frequency band in between 1 and 40 GHz [4], such as L-band (1–2 GHz) of GPS services and Inmarsat communication or X-band (8–12 GHz) primarily used by the military.

In this article, while making reference to some specific orbits, there is no differentiated analysis, as the points made are equally applicable to satellites in general, regardless the orbits where they are lodged.

With this in mind, Sect. 2 covers the ongoing securitisation of space, with a discussion of what war in space means in geopolitical terms and what the current weaponisation of space bring in terms of risks for the sector. An analysis of the datasets of debris producing incidents demonstrates the importance of reaching an agreement on ASAT tests. Debris crisis is the subject of Sect. 3, which offers a quantitative assessment of the phenomenon and the challenges, both technical and legal, that mitigation measures currently face. Section 4 discusses instead how cyber threats are now a new, important feature of security in Near Space, no longer protected by obscurity or relatively small attack surface, as some recent incidents have shown. After a review, it also briefly models cyber threats in the Earth's orbits, present and future, and offers thoughts on the most urgent mitigation measures needed to address them. Section 5 discusses HSN (Hybrid Satellite Networks) and the challenges of the post-quantum computing in a cyber-security perspective, with the development of QKD (Quantum Key Distribution), while the conclusive section takes stock of the analysis carried out in the article as a whole. It looks ahead at the two scenarios that await the sector in the coming years, in between a new regulatory framework based on an international consensus to address the existing issues or spiralling toward a new space race of uncertain outcome.

2 Geopolitics of Space Race and the Use of Weapons

It was estimated that between 1957, the start of the space age with Sputnik 1, and 2015, the U.S. and Soviet Union/Russia deployed around 2000 military satellites, and even today, roughly 20–30% of the approximately satellites in orbit serve national security purposes, predominantly operated by military organizations. Otherwise said, space and security have been inextricably linked through their shared goals, methods, and operations [5] and in many countries space still maintain a strongly military and defence connotation. But the sheer number of the countries that belong now to the 'space club', however defined [6] translates into additional complexity and geopolitical scenarios absent or of scarce relevance in previous decades.

Therefore, if it is certain that we are witnessing to a securitisation of the space milieu –as defined according to the Copenhagen School's definition [7, 8]—it is also true that a dimension of traditional security in space has been a feature since the beginning, and any discussion of space (geo)politics cannot disregard this fundamental points. Not only any "*geopolitical theory developed for the Earth and its geographical setting can be transferred to outer space*" [9]. Any war in space would necessarily be a war on Earth first and foremost, as the widely cited statement from Gen. John Hyten of the US Strategic Command, "*there is no such thing as a war in space, there is just war, and it's with an adversary and if it extends into space, we have to figure out how to fight it*" clarifies [10]. Good examples are what is now called the first space war (the 1991 First Gulf War) and the Russian-Ukrainian conflict, both making extensive use of satellite capabilities to direct and/or assist military operations on Earth. Whatever happens in space, it is for consequences on the ground [11–14].

With the expansion of the number of countries with space access of some sort, also the very concept of space security had to evolve from what was "*a two-dimensional model of military and environmental dimensions of space security*" [15] to a more nuanced definition that not only encompasses non-state actors, such as private operators, but also includes non-traditional threats emerging from space (such as debris, asteroids, space weather) or directed to space (e.g., contamination, cyber attacks; [16, 17]).

Which brings to the fundamental points of contention when discussing space security: what space weapons are and how they be treated, from a legal perspective first. m. Regulations are often part of the problem rather than the solution, as there is ongoing confusion and differing opinions about what qualifies as a "space weapon" (Paladini 2016; Hebert 2014; Vermeer 2008) Despite long-standing debates on space security, no official definition exists, and the common assumption that nuclear weapons are equivalent to space weapons is incorrect for multiple reasons. What is certain is that the ambiguities in the definition of space weapons and gaps in existing laws have allowed spacefaring nations to integrate space with terrestrial military capabilities, advancing their strategic interests without clear restrictions [18].

Moreover, many spacecrafts and probes are dual-use, meaning they can serve both commercial and military purposes. Notable examples include China's Aolong-1, launched in 2016 to address space debris, but there were claims it could also target satellites (Vasani 2017). Similarly, "inspection satellites" are equipped with capabilities to track and potentially disable other satellites, raising concerns about their dual-use nature and implications for space security (Dawson 2018).

But nothing is as controversial as the use of ASATs, the infamous anti-satellite systems.

Originally developed by the USSR in the 1960s and becoming operational in the 1990s, ASATs are basically probes designed to enter the same orbit of a target satellite and equipped with offensive weapons (kinetic, mechanical, laser, etc.) to destroy or disable it [19, 20] During a hypothetical conflict, these systems would pose a significant threat, since, once launched, it would be difficult for the targeted satellite to avoid collision.

Importantly, at a purely regulatory level and their danger nonetheless, international treaties do allow ASAT deployment, as the existing treaties only forbid weapons of mass destruction (WMD) but allows the deployment of everything else in orbit.

From a legal perspective, here the landscape has remained largely unchanged since 1967 (year of the OST, Outer Space Treaty, cornerstone of the entire 'space law'; [21]), despite significant advancements in space technology and their attack capabilities.

If the ASATs were formally recognized as space weapons, rather than simply 'weapons used in space', that would require important changes in how international space law and treaties are framed. At the current state of affairs, this seems particularly unlikely, given that, although the US advocate a ban on ASAT testing, China and Russia are vocals in wanting to keep them open [22, 23].

This is more than just a legal dispute or a dangerous trigger of the well-known security dilemma dynamics: ASATs are the single most important cause of debris creation, although not the only one (the datasets report at least 200 instances of satellite explosions due to malfunctioning, collisions, discussed in section or aborted launches; [24]).

But a quick look at Fig. 2 shows that ASATs are by far and large the most damaging debris-producing events, with the first one being the Chinese ASAT test targeting the

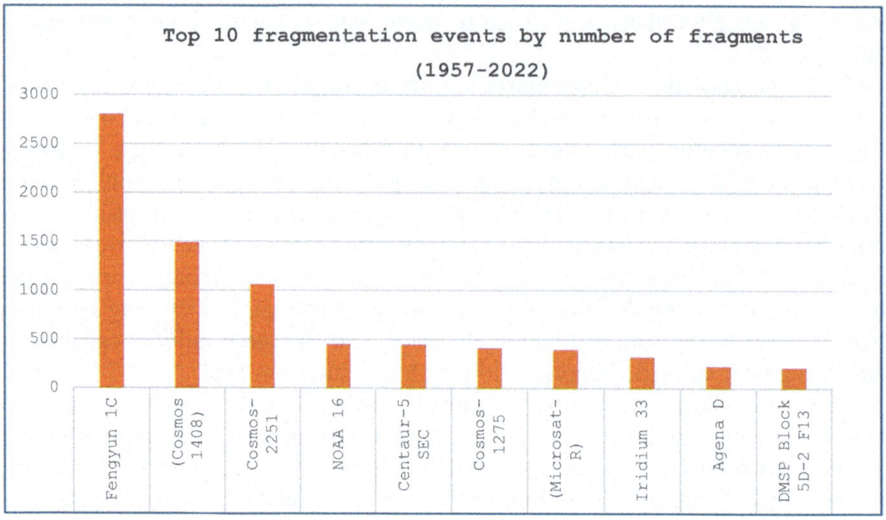

Fig. 2 Top 10 fragmentation events by number of fragments (1957–2022). *Source* Elaboration on Paladini [17]. Original data from ESA and JAXA data, 2022

Chinese satellite Fengyun-1C 'Wind Cloud' (FY-1C) and producing a debris field of 2809 objects [25, 26] and the second is the November 2021 Russian test with 1500 trackable debris, which also forced the ISS to perform an orbital correction to avoid a collision [27].

Which leads us to one of most relevant threats the sector is facing now.

3 Debris Crisis and Mitigation Effort in Between Technical Challenges and Legal Hurdles

If there is clear example of environmental [28], non-traditional security issue on the way of securitisation this is clearly the debris crisis. Which is per se not surprising considering the dimension of the problem.

Even Fig. 3, while impressive, does not fully convey the gravity of the issue itself.

A point not always clearly understood is that, if space is infinite by definition, usable space is not, and the orbits even less. And if ITU has warned the international communities states about the overpopulation of GEO since the 1970s, it is LEO the source of main concern (ITU 1992; UN 1997) in terms of availability to nest satellites in safe conditions.

Overcrowding routinely leads to incidents (the jamming of the Indonesian satellite PT Pasifik in 1996 being only one among many examples) and even incidents, like the infamous case of the 2009 collision in Iridium 33-Cosmos 2251, the third and eighth most calamitous case in terms of debris production, shows. That case has

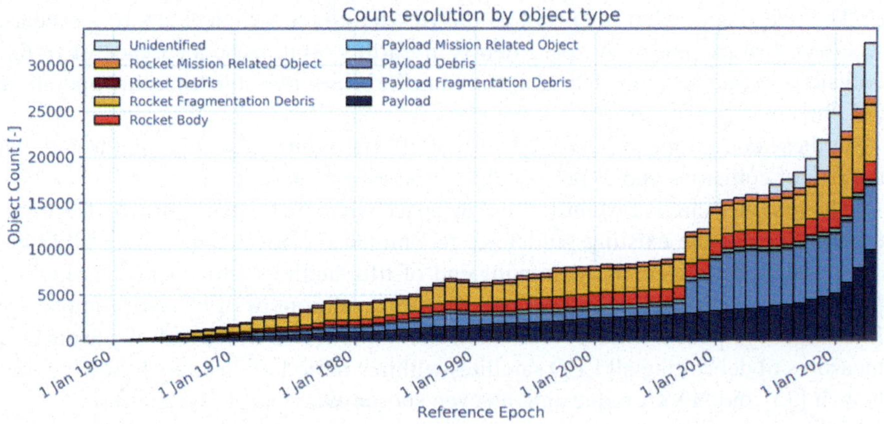

Fig. 3 Objects in orbit by reference epoch. *Source* ESA, 2024

been also studied for the crucial legal implications and for having "unchained a new chapter in the field of space debris" [29] in terms of attributions of responsibilities and even more for the compensation, considering that 1972 Liability Convention, the treaty theoretically applicable to this case, did not (and still does not include) private actors such as the company Iridium.

The legal framework will be considered in section 5, as it is one of the fundamental factors to be discussed for the future of the sector. Here it is worth noticing that the Iridium 33-Cosmos 2251 combined with the Chinese ASAT test of two years earlier, resulting in a production of 5500 debris particles, 90% of which still in orbit nowadays and accounting for 36% of all LEO objects monitored by the space agencies.

The 2024b Environmental Report offers grim statistics, summarised in Table 1.

Table 1 Space debris by the numbers

Statistics	Value
Number of rocket launches since the start of the space age in 1957	About 6740
Number of satellites these rocket launches have placed into earth orbit	About 19,590
Number of these still in space	About 13,230
Number of these still functioning	About 10,200
Number of space objects regularly tracked by space surveillance networks	About 36,860
Estimated number of break-ups, explosions, collisions, or anomalous events	> 650
Total mass of all space objects in earth orbit	> 13,000 tonnes
Space debris objects greater than 10 cm	40,500
Space debris objects from greater than 1–10 cm	1,100,000
Space debris objects from greater than 1 mm to 1 cm	130 million

Source ESA [30]

The real risk is what is known as the Kessler effect, which refers to a scenario where collisions caused by space debris in Earth's orbit cause a cascade of further collisions, create yet more debris and making the orbits unusable for future operations [31, 32].

In these conditions, it is normal a lot of efforts is currently being put to mitigate the risk of collisions and to prevent the formation of more debris.

The first measure is certainly 'free' up space when no longer required, deorbiting the satellites as the existing guidelines recommend. Until recently, only GEO had dismissal procedures in place, moving end-of-life satellites into a "graveyard orbit", just above the GEO's altitude; however, the dire conditions of LEO have prompted the Inter-Agency Space Debris Coordination Committee (IADC) for end-of-life disposal measures of deorbiting all LEO satellites within a period of 25 years following their launch [33] and NASA requesting an even shorter window of 5 years [34].

The second looks more to remediate the existing situation by actively remove debris (ADR), a technically challenging endeavour for which there are no standardised procedures and that is largely experimental or at prototype level, with no clear timeline in terms of implementation [35] (Martin-Lawson et al. 2023; Bongers and Torres 2023). Market leaders here are the Swiss-based Clearspace [36] and the UK company Astroscale [37], while, in technical terms, solutions under study include electromagnetic tethers [38], iron beams [39], nets (ESA 2014), ground-based lasers [40–43] and the use of robotic arms [44, 45].

But technical challenges are only one side of the issues surrounding the implementation of ADR. In addition of concerns over dual-use technologies—some of the solutions can be frowned upon for military reasons—the key issue in space debris management, as highlighted by the Iridium case, remains legal liability. Solutions are legally complex due to the unique nature of space law. Unlike maritime law, where abandoned vessels in international waters can be handled by anybody, no state can interfere with another state's spacecraft in space, complicating efforts for international cooperation and assistance.

Again, the 1960–70s regulatory framework appears outdated to handle today's issues.

4 Security by Obscurity? Cyberwars in LEO

Finally, the cyber posture of satellite activities.

Interestingly, this specific aspect has not historically been as covered and analysed as much as it could—or indeed should—have been. A long-standing belief in the industry is what can be summarised by the well-known 'security through obscurity' [46]. As the satellites themselves were generally running on proprietary, scarcely available for study software, and were, by definition, not directly accessible (being nested in some orbits), these factors alone would provide protection from attacks [47] (NASA 2023).

Recent and less recent events—and well-documented incidents—have proved this a severely mistaken approach [48, 49]. The attack to the German ROSAT X-ray satellite in 1998, the first of its kind, opened a new era in anti-satellite threats. While there are some aspects still not clarified [50], a cyber intrusion into the Goddard Space Flight Center servers has been linked to the incident, as causing the satellite to turn its solar panels directly at the sun, destroying its batteries and making it no longer usable (the satellite itself crashed back to Earth in 2011; as it plunged into the Indian ocean, no direct inspection was possible; [51]).

We have witnessed in the last decades an increasing number of cyberattacks toward satellites, for the majority driven by geopolitical motives—espionage, military disruption, and sabotage, from the Iranian Hackers Target Satellite Companies (2019) by the Iranian hacker group, MuddyWater (APT34) to Turla Group Hijacking Satellite Internet for Espionage in 2020 to a series of cyber-attack related to the Russian-Ukrainian conflicts [52, 53, 54, 55, 56]. The perpetrators come under the generic definition of APTs, often state-backed actors from countries such as Russia (i.e. Fancy Bear), China, and Iran, and probably others, and have targeted satellite communication systems, GPS networks, and military satellites. While different in modalities and attack procedures, from full-fledged attacks to signal jamming these incidents highlight the growing threat of cyberattacks to space infrastructure in a moment where satellites an integral feature of industry, society and government operation.

While this is not the place for a full-fledged analysis of the cyber security dimension of the satellite activities, a few points need to be noted, as they not only add another layer of instability to Earth's orbits; given to the specific nature of information security, it has the potential to affect debris mitigation efforts and directly contribute to become a geopolitical factor of prime relevance.

There are a few reasons behind this change.

First of all, the sheer growth of the sector, not specifically discussed in this article but under everybody's eyes, which has also meant a growing variety of satellites sent to space. Small satellites, particularly those using Commercial off-the-shelf (COTS) components, nano satellites (such as Cubesats and their siblings), and Starlink-like constellations have led this growth. And while attacks have not specifically or exclusively targeted small satellites, the presence of them has added other sets of possible objectives, widening the attack surface.

A quick look at satellite architectures and operations in Fig. 4 model would better clarify both vulnerabilities and possible threats.

In their most essential configuration, satellite operations can be represented as made of three main components: the ground segment, the space segment, and the user segment (NASA 2023)—although there are more sophisticated configurations, without even mentioning HSN (hybrid satellite networks), which will be briefly covered in the next section.

The ground segment, located on Earth, communicates with satellites to send commands and receive data, using a variety of space protocols (TC/TM).

The space segment includes the satellites deployed in orbit. Again, the difference of typology here is huge, from massive GEO satellites to nanostructure such as

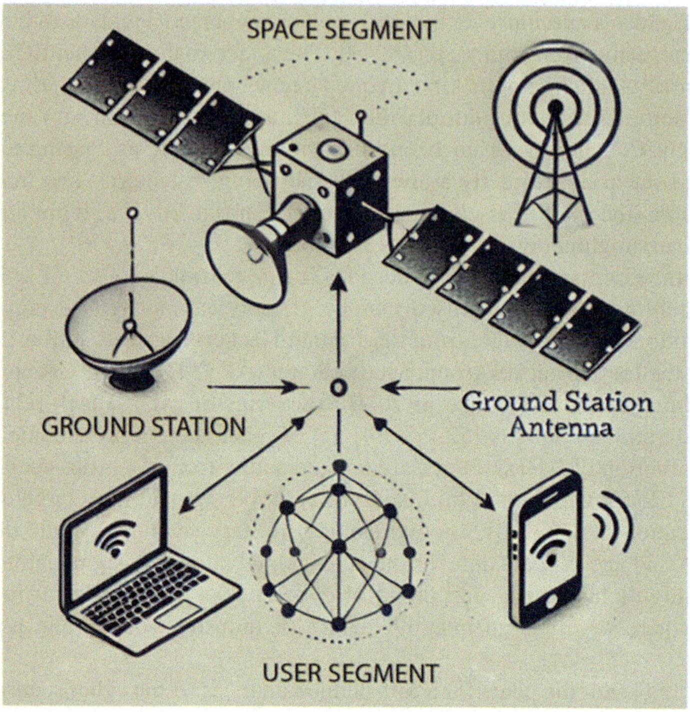

Fig. 4 Satellite operation diagram. *Source* Image generated through LLMs on author's prompts. Captions by the author, 2024

the Cubesats. Still, generally speaking, all share a basic architecture, which sees them consisting of two sections, the payload and the bus. It is worth noting that the satellite payload includes the mission-specific instruments, while the bus provides the infrastructure needed for the payload's operations, such as the Command and Data Handling System—CDHS—the 'brain' of the satellites, the Communication Module (COM), the Attitude Determination and Control System (ADCS), and the EPS, the satellite's power supply.

Finally, the user segment receives the satellite-provided services, from GPS-type location signals to communication etc.

It is the case to mention that each of the three have bilateral and at times trilateral communication exchanges (generally indicated as SATCOM) adopting a variety of protocols at different frequency bands depending on their orbits and as allocated by ITU [57].

Any and each of this specific component presents an attack surface and specific vulnerabilities that can be exploited (as they have been in the past) by malicious actor, without mentioning threats that are outside human control and that can only be mitigated but not prevented [58].

Table 2 gives a summary of the most common threats (both malicious and not) and the primarily (it is important to note that side effects are common) affected components.

It is important to notice that these threats are not equivalent in terms of severity and the danger to pose to satellite operations, for once because their likelihood varies widely and second because even the occurrence of the event does not affect all the system in the same way. Again, this is due to what previously mentioned in terms of variety of existing platforms, components, systems, and protocols, which makes the analysis of incidents in space, especially cyber-incidents, an always complex story.

And while it would be impossible to address even just the fundamental issues in this contribution, it is worth discussing at least one specific point, which is likely going to become even more crucial in the near future: communication and, closely related, encryption standards for satellites.

Table 2 Threats to satellite operations

Threat	System impacted	Type (malicious)
Space weather	Satellite components	
Physical attacks	Ground station facilities	x
Natural disasters	Ground station facilities	
Interception/Data breaches	Communication/Network[a]	x
Jamming	Communication/Network	x
MitM	All components	x
DoS	Communication/Network	x
Firmware RE	Satellite components	x
SC vulnerability exploitation	Satellite components	x
IAM/Encryption issues	Communication/Network	x
Insider threats	Ground station facilities	x
Lack of redundancy	All components	
Radiation	Satellite components	
Space objects (debris)	Satellite components	
ASAT and other weapons	All components	x

Source Author's elaboration, 2024

[a]This refers specifically to TC/TT and data link exchanges between the satellite and the ground station, conceptually distinct from the two components

5 Looking Ahead. HSNs and Post-quantum Encryption

Quantum computing is going to take all the industry by storm, and the space sector is going to be no exception. But, together with the opportunities it will certainly open, it is going to make more severe an already challenging environment for satellite communication, which has still to fully integrate 5G/6G and the rise of HSNs mentioned in the previous section. While HSNs are not a monolithic structure, their emergence now means that the space sector is '*transitioning away from traditional vertically integrated entities and towards an aggregation of independently owned and operated segments*' [59], with the addition of several terrestrial components in the networks, carrying new vulnerabilities together with the capabilities [60, 61].

Figure 5 shows an essential HSN architecture, making rather intuitive how the different components can all represent possible entry points for potential attacks.

Implementations of HSNs are going to be far more complex than this skeleton architecture outlined by NIST. A more realistic example is the HSN in the making Inmarsat ORCHESTRA [62], which aims at bringing together Inmarsat existing

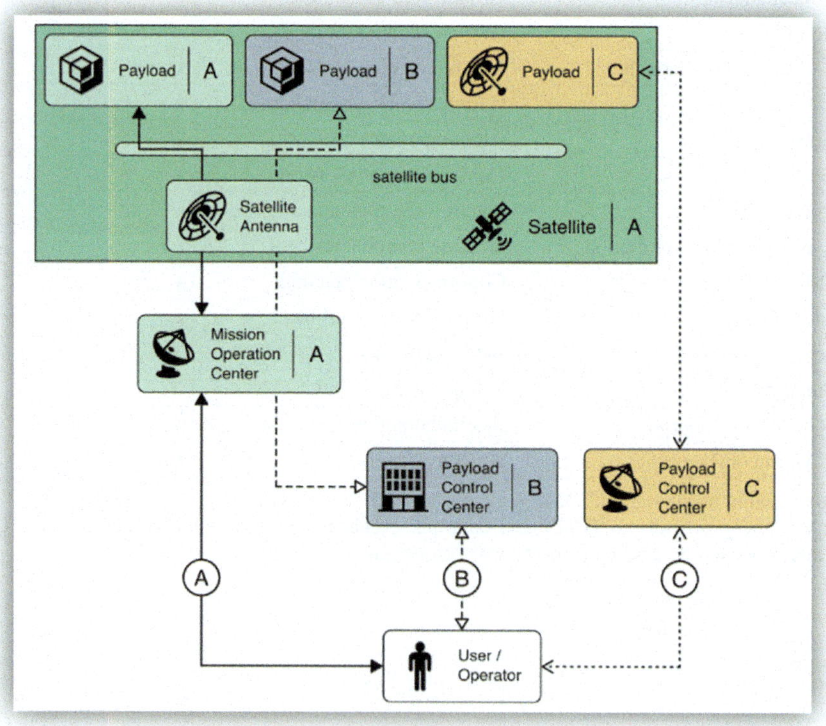

Fig. 5 A simple HSN architecture. *Source* NIST [59]

geosynchronous (GEO) satellites with new LEO satellites—in the expected number of 150–175 units, and terrestrial 5G into an integrated, high-performance solution.

And it is true that HSNs are still a minority in comparative terms—Starlink might well evolve to a HSN in future, but it is not there yet—they are going to dramatically affect the security posture of the sector. HSNs will not only widen the attack surface (for the lack of direct control over external organisations in the network, for a start) but also contribute to make the implementation of new security protocols and procedures extremely complex and anything but straightforward.

One of the most crucial aspects will be, of course, encryption standards, already a problematic point in satellite operations due to the existence of legacy systems, unsecure protocols, and even a rapidly changing technology (e.g., in the case of satellite imaging; [63]) that makes the use of traditional AES symmetric encryption no longer sufficient.

This is where post-quantum computing is going to play an important role.

There is at the moment, considerable focus from both industry and government [64]. In the US, institutions and agencies such as NIST, CISA, and NSA are scaling up efforts to prepare against future threats posed by cryptanalytically relevant quantum computer (CRQC) capabilities. There are good reasons for that: many cryptographic systems, products, and services in use today, especially in infrastructure and industrial systems, which rely on public key algorithms (such as RSA, Elliptic Curve Diffie-Hellman, and Elliptic Curve Digital Signature Algorithm) will need to be updated, replaced, or modified to mitigate the risks posed by future quantum threats [65, 66, 67]. NIST has been hosting a competition since 2017 to identify quantum resistant algorithms for implementation, with good promise [68, 69].

The space sector, however, is looking to another direction, and Quantum Key Distribution (QKD) is the one that is considered the most. QKD uses the principles of quantum mechanics to securely distribute encryption keys between satellites and ground stations, making it theoretically impossible for an eavesdropper to intercept the key without detection.

There have already been a few experiments in long-distance communication using quantum technologies in the last ten years. Japan was the first to send a very compact payload for QKD on a microsatellite SOCRATES in 2015, though results from the voyage were published in July 2017. On the other hand, Italy was the first to report satellite-to-ground QKD communication using satellite retroreflectors in 2015, while China-led satellite-based QKD by launching the Chinese Academy of Science Micius in 2016 [70].

More investments and efforts are coming, from all the corners of the globe.

China intends to keep working on QKD development. A first step will see three or five small satellites focused on QKD—generating entangled particles for use as quantum keys—with masses under 100 kg. These will be sent into sun-synchronous orbits (SSO) at altitudes of 800 km at some moment in 2026. Moreover, the China National Space Administration expects to launch a quantum satellite constellation in MEO before the end of the decade. India has signalled plans in this sense, successfully trialling free QKD (precursor to satellite based communication) in 2021 [71, 72, 73].

Quantum technology was defined as a strategic priority in Agenda 2025 of ESA, as a part of the Scylight programme, which focuses on advancing secure, laser-based satellite communication technologies, and aimed at aims at enhancing resilience and security of communication systems. ESA itself is one of the most important stakeholders of the European Quantum Communication Infrastructure (EuroQCI) initiative, designed to establish a secure quantum communication network across the EU, integrating quantum encryption technologies with existing communication infrastructures to protect data against cyber threats.

In the space sector, ESA will launch Eagle-1 between late 2025 to early 2026, the first satellite with QKD system [74]. It will then complete three years of in-orbit validation, gathering fundamental data for future implantations.

Even with a fully functioning QKD, however, here are still several technical limitations, both in the applicability of the solution to the space sector and the risks it involves—from the requirement of trusted relays to avoiding different instances of DoS. Otherwise said, in cybersecurity terms, it is not a silver bullet.

6 Conclusions. A New Regulatory Framework or a New Space Arm Race?

More satellites were launched in 2023 than in any year before, and all the indicators point to an ever-increasing level of activities in the short-medium term.

If anything, the plans of occupying the cislunar space to facilitate the planned missions to the Moon surface only hint to an expansion of the various issues and flashpoints analysed here.

From more than one side, there is alarm at the perspective of fuelling the 'space arm race' [75, 76] already ongoing, taking along also the risk of a non-consensual appropriation of resources. While space mining seems still far away from becoming a reality, from reasons more technical than strategic or political, it is important to have regulatory framework is stably in place when it happens. Which it is not the case at present. The first space mining laws in the sector are only national, not international [77], and more needs to be done to pre-empt conflicts.

Defusing geopolitical tensions is also essential to reduce the frequency and the severity of full-fledged cyber wars with state-sponsored or at least facilitated APTs, which have been steadily growing in these last years.

More in general, a global space governance based on consensus approach is essential to finally put hand to an outdated regulatory framework [78, 79] which in a sense keeps the sector back in the 1970s, with the United Nations Five Treaties regulating global space activities and a system where only (a handful of) states and no private companies are the objects of the provisions. With more than eighty countries with satellites in space and a few major private players, the orbits are more crowded than ever, with every disruption, intended or not, that can potentially prove disastrous.

A revision of the existing treaties with clear and applicable regulations for space activities present and future, based on international consensus, is a first step to de-escalate and de-securitise at least some of the issues discussed in this article, starting from the handling of end-of-life satellites and addressing the debris crisis.

A consensus-based approach and applicable legal framework will not only help to prevent the formation of new debris but also facilitate the implementation of active measures in debris removal to prevent catastrophic events such as Kessler syndrome which would make the Earth's orbits, LEO in particular, unusable for the next generations.

References

1. The EU (2016) Communication from the Commission to the European Parliament, the Council, the European Economic and Social Committee and the Committee of the Regions. Space Strategy for Europe (COM 2016/705 Final), Brussels
2. DeFrieze DC (2014) Defining and regulating the weaponization of space. Joint Force Q 74. Available at: https://ndupress.ndu.edu/Media/News/Article/577537/defining-and-regulating-the-weaponization-of-space/. Accessed 1 Oct 2024
3. Harrison RG (2013) Unpacking the three C's: congested, competitive, and contested space. Astropolit Int J Space Polit Policy 11(1–2):34–47
4. ESA (2024) ESA space environment report 2024. Available at: https://www.esa.int/Space_Safety/Space_Debris/ESA_Space_Environment_Report_2024. Accessed 4 Oct 2024
5. Jakhu RS, Pelton JN (2017) Global space governance: an international study. Springer International, Cham
6. Paladini S, Barbieri D (2024) Orbit bound: predictors of 'readiness to space' and government support. Space Policy 101628
7. Buzan B, Wæver O, de Wilde J (1998) Security: a new framework for analysis. Lynne Rienner Publishers, Boulder, CO
8. Lisitsyna K (2021) The securitization of space and the international political economy of new space. Int Polit Sociol 15(1):109–123
9. Dolman E (2002) Astropolitik: classic geopolitics in the space age. Frank Cass, London, p 15
10. Villasanta AD (2017) US will fight and win a space war—and is preparing for it. Chinatopix, 27 Feb. Available at: http://www.chinatopix.com/articles/111937/20170227/will-fight-win-space-war-preparing.htm. Accessed 18 Apr 2024
11. Anson P, Cummings D (1991) The first space war: the contribution of satellites to the Gulf War. RUSI J 136:45–53
12. Bryen S (2022) Musk's tech put to deadly weapon effect in Ukraine: starlink satellite system and GIS Arta software enabling near immediate and highly accurate artillery target identification. Asia Times, 1 July. Available at: https://asiatimes.com/2022/07/musks-tech-put-to-deadly-weapon-effect-in-ukraine/. Accessed 1 Oct 2024
13. Burbach DT (2022) Airpower after Ukraine: early lessons from the Russia-Ukraine war as a space conflict. Atlantic Council, 30 Aug. Available at: https://www.atlanticcouncil.org/content-series/airpower-after-ukraine/early-lessons-from-the-russia-ukraine-war-as-a-space-conflict/. Accessed 1 Oct 2024
14. Hostbeck L (2015) Space weapons' concepts and their international security implications. In: Schrogl K, Hays PL, Robinson J, Moura D, Giannopapa C (eds) Handbook of space security. Springer Media, New York, pp 970–995
15. Sheehan M (2015) Defining space security. In: Schrogl K-U et al (eds) Handbook of space security. Springer, New York, pp 1–15

16. Hammack K (2021) International relations in space: the role of miscalculation, militarization, and weaponization. Astropolit Int J Space Polit Policy 19(3):1–25
17. Paladini S (2023) Unsustainable wars? The use of weapons in lower earth orbit. In: Cayli Messina B (ed) Environmental injustice and catastrophe: how global insecurities threaten the future of humanity. De Gruyter, Berlin, pp 91–106
18. Khalid M (2021) Space legal regimes, militarization, and weaponization of outer space. Astropolit Int J Space Polit Policy 19(1–2):128–144
19. Ortega A (2021) Placement of weapons in outer space: the dichotomy between word and deed. Lawfare, 28 Jan. Available at: https://www.lawfareblog.com/placement-weapons-outer-space-dichotomy-between-word-and-deed. Accessed 4 Oct 2024
20. Vergun D (2023) Official details space-based threats and U.S. countermeasures. DOD News, 26 Apr. Available at: https://www.defense.gov/News/News-Stories/Article/Article/3375577/. Accessed 2 Oct 2024
21. Johnson-Freese J, Burbach D (2019) The outer space treaty and the weaponization of space. Bull At Scientists 75(4):137–141
22. Hitchens T (2022) Biden's space policy nominee backs ban on destructive ASAT testing, pushes norms. Breaking Defense, 13 Jan. Available at: https://breakingdefense.com/2022/01/bidens-space-policy-nominee-backs-ban-on-destructive-asat-testing-pushes-norms/. Accessed 4 Oct 2024
23. Sankaran J (2022) Russia's anti-satellite weapons: an asymmetric response to U.S. aerospace superiority. Arms Control Association, March. Available at: https://www.armscontrol.org/act/2022-03/features/russias-anti-satellite-weapons-asymmetric-response-us-aerospace-superiority. Accessed 1 Oct 2024
24. Lemaître A (2019) Space debris: from LEO to GEO. In: Baù G et al (eds) Satellite dynamics and space missions. Springer, New York, pp 115–157
25. Hadley G (2023) Saltzman: China's ASAT test was 'pivot point' in space operations. Air & Space Forces Magazine, 13 Jan. Available at: https://www.airandspaceforces.com/saltzman-chinas-asat-test-was-pivot-point-in-space-operations/. Accessed 1 Oct 2024
26. Zissis C (2007) China's anti-satellite test. Council on Foreign Relations. Available at: https://www.cfr.org/backgrounder/chinas-anti-satellite-test. Accessed 1 Oct 2024
27. Howell E (2022) Space debris from Russian anti-satellite missile test spotted in telescope images and video. Space.com, 10 Aug. Available at: https://www.space.com/russian-anti-satellite-test-debris-telescope-images-video. Accessed 1 Oct 2024
28. Elliot L (2002) Regional environmental security: pursuing a non-traditional approach. In: Tan ATH, Boutin K (eds) Non-traditional security issues. Institute of Defence and Strategic Studies, Singapore, pp 1–25
29. Williams M (2012) Space debris as a 'single item for discussion'. In: Proceedings of the International Institute of Space Law, p 333
30. ESA (2024) Satellite frequency bands. Available at: https://www.esa.int/Applications/Connectivity_and_Secure_Communications/Satellite_frequency_bands. Accessed 4 Oct 2024
31. Bastida Virgili B, Dolado JC, Lewis HG et al (2016) Risk to space sustainability from large constellations of satellites. Acta Astronaut 126:154–162. https://doi.org/10.1016/j.actaastro.2016.03.034
32. Kessler DJ, Cour-Palais BG (1978) Collision frequency of artificial satellites: the creation of a debris belt. J Geophys Res 83(6):2637–2646
33. Hakima H, Emami MR (2018) Assessment of active methods for removal of LEO debris. Acta Astronaut 144:225–243
34. NASA (2021) Orbital debris news at. https://sma.nasa.gov/sma-disciplines/orbital-debris. Accessed on 1 June 2023
35. Bonnal C, Ruault J-M, Desjean M-C (2013) Active debris removal: recent progress and current trends. Acta Astronaut 85:51–60
36. Clearspace (2023) A mission to make space sustainable. Available at: https://clearspace.today/. Accessed 1 Oct 2024

37. Astroscale (2023) Space sustainability. Available at: https://astroscale.com/space-sustainability/. Accessed 1 Oct 2024
38. Hoyt RP (2011) Propulsion and power harvesting performance of electrodynamic tethers. In: AIAA SPACE conference and exposition 2011. https://doi.org/10.2514/6.2011-7321
39. Bombardelli C, Pelaez J (2012) Ion beam shepherd for contactless space debris removal 34(3):916–920. https://doi.org/10.2514/1.51832
40. Fang Y, Pan J, Luo Y, Li C (2019) Effects of deorbit evolution on space-based pulse laser irradiating centimeter-scale space debris in LEO. Acta Astronaut 165:184–190
41. Phipps C (2018) Laser ablation propulsion and its applications in space. In: Chapter in springer series in materials science, Oct 2018. https://www.researchgate.net/profile/Claude-Phipps/publication/328009206_Laser_Ablation_Propulsion_and_Its_Applications_in_Space/links/5f7cde48a6fdccfd7b4c8e73/Laser-Ablation-Propulsion-and-Its-Applications-in-Space.pdf. Accessed 31 Aug 2024
42. Phipps CR, Bonnal C (2016) A spaceborne, pulsed UV laser system for re-entering or nudging LEO debris, and re-orbiting GEO debris. Acta Astronaut 118:224–236
43. Shen S, Jin X, Hao C (2014) Cleaning space debris with a space-based laser system. Chin J Aeronaut 27(4):805–811
44. Flores-Abad A, Ma O, Pham K et al (2014) A review of space robotics technologies for on-orbit servicing. Prog Aero Sci 68:1–26
45. Hirzinger G, Landzettel K, Brunner B et al (2012) DLR's robotics technologies for on-orbit servicing 18(2):139–174
46. Saltzer JH, Schroeder MD (1975, 09) The protection of information in computer systems. Proc IEEE 63. Available at: http://web.mit.edu/Saltzer/www/publications/protection/. Accessed 1 Oct 2024
47. Jha D (2022) Safeguarding the final frontier: analyzing the legal and technical challenges to mega-constellations. J Space Saf Eng 9:636–643
48. Falco G, Caldera C, Shrobe H (2021) Satellite-to-satellite cyber attack, defense, and resilience. J Cybersecur 8(1)
49. Willbold J, Schloegel M, Vögele M, Gerhardt M, Holz T, Abbasi A (2023) Space odyssey: an experimental software security analysis of satellites. In: IEEE symposium on security and privacy (SP), pp 1–19
50. Wess M (2021) ASAT goes cyber. US Naval Institute. Proceedings Magazine, Feb. Available at: https://www.usni.org/magazines/proceedings/2021/february/asat-goes-cyber. Accessed 4 Oct 2024
51. Max Planck Institute for Extraterrestrial Physics (2011) ROSAT—the end of an exceptional satellite (1 June 1990–23 Oct 2011). Max Planck Institute for Extraterrestrial Physics, 14 Nov. Available at: https://www.mpe.mpg.de/229897/News_20111114. Accessed 4 Oct 2024
52. Via Satellite (2008) NASA computers hacked by intruders. Satellite Today, 1 Dec. Available at: https://www.satellitetoday.com/government-military/2008/12/01/nasa-computers-hacked-by-intruders/. Accessed 4 Oct 2024
53. Reuters (2018) China-based hacking campaign is said to have breached satellite, defense companies. CNBC, 19 June. Available at: https://www.cnbc.com/2018/06/19/china-based-hacking-breached-satellite-defense-companies-symantec.html. Accessed 4 Oct 2024
54. US DoJ (2020) Iranian hackers indicted for stealing data from aerospace and satellite tracking companies. U.S. Department of Justice, 17 Sept. Available at: https://www.justice.gov/usao-edva/pr/iranian-hackers-indicted-stealing-data-aerospace-and-satellite-tracking-companies. Accessed 4 Oct 2024
55. Duo Security (2022) CISA warns of ongoing attacks by MuddyWater APT. Duo Decipher, 24 Feb. Available at: https://duo.com/decipher/cisa-warns-of-ongoing-attacks-by-muddywater-apt. Accessed 4 Oct 2024
56. Lennon M (2015) Russian-speaking Turla attackers hijacking satellite internet links. SecurityWeek. Available at: https://www.securityweek.com/russian-speaking-turla-attackers-hijacking-satellite-internet-links/. Accessed 4 Oct 2024

57. Tedeschi P, Sciancalepore S, Di Pietro R (2022) Satellite-based communications security: a survey of threats, solutions, and research challenges. Comput Netw 216:109246
58. Malik WJ (2019) Attack vectors in orbit: the need for IoT and satellite security. In: RSA conference presentation 2019. Available at: https://www.trendmicro.com/. Accessed 4 Oct 2024
59. National Institute of Standards and Technology (NIST) (2023) NIST interagency report 8441: cybersecurity framework profile for hybrid satellite networks (HSN). Initial Public Draft. U.S. Department of Commerce. Available at: https://csrc.nist.gov/publications/detail/ir/8441/draft. Accessed 4 Oct 2024
60. Deng B, Jiang C, Yao H, Guo S, Zhao S (2020) The next generation heterogeneous satellite communication networks: integration of resource management and deep reinforcement learning. IEEE Wirel Commun 27(2):105–111
61. Karavolos M, Nomikos N, Vouyioukas D, Mathiopoulos PT (2021) HST-NNC: a novel hybrid satellite-terrestrial communication with NOMA and network coding systems. IEEE Open J Commun Soc 2:887–898
62. Inmarsat (2021) ORCHESTRA. Available at: https://www.inmarsatgov.com/orchestra/. Accessed 4 Oct 2024
63. Bensikaddoura E, Bentoutou Y (2020) Satellite image encryption based on AES and discretised chaotic maps. Autom Control Comput Sci 54:446–455
64. Bernstein DJ, Buchmann J, Dah E (eds) (2009) Post-quantum cryptography. Springer. Available at: https://doi.org/10.1007/978-3-540-88702-7
65. National Institute of Standards and Technology (NIST) (2022) Post-quantum cryptography: selected algorithms 2022. NIST. Available at: https://csrc.nist.gov/Projects/post-quantum-cryptography/selected-algorithms-2022. Accessed 4 Oct 2024
66. Bos J, Costello C, Naehrig M, Stebila D (2015) Post-quantum key exchange for the TLS protocol from the ring learning with errors problem. In: 2015 IEEE symposium on security and privacy. Available at: https://doi.org/10.1109/sp.2015.40. Accessed 4 Oct 2024
67. Singh B, Ahateshaam M, Lahiri A, Sagar AK (2024) Future of cryptography in the era of quantum computing. In: Shaw RN, Siano P, Makhilef S, Ghosh A, Shimi SL (eds) Innovations in electrical and electronic engineering. ICEEE 2023. Lecture notes in electrical engineering, vol 1115. Springer, Singapore. https://doi.org/10.1007/978-981-99-8661-3_2
68. Schneier B (2024) Lattice-based cryptosystems and quantum cryptanalysis. https://www.schneier.com/essays/archives/2024/05/lattice-based-cryptosystems-and-quantum-cryptanalysis.html. Accessed 1 Oct 2024
69. Bos J, Costello C, Ducas L, Mironov I, Naehrig M, Nikolaenko V, Raghunathan A, Stebila D (2016) Frodo: take off the ring! Practical, quantum-secure key exchange from LWE. In: Proceedings of the 2016 ACM SIGSAC conference on computer and communications security (CCS'16). Association for Computing Machinery, New York, NY, USA, pp 1006–1018
70. Bedington R, Arrazola JM, Ling A (2017) Progress in satellite quantum key distribution. npj Quantum Inf 3:30. https://doi.org/10.1038/s41534-017-0031-5
71. New Space Economy (2024) The future of secure communication: quantum key distribution takes to space. New Space Economy, 9 Sept. Available at: https://newspaceeconomy.ca/2024/09/09/the-future-of-secure-communication-quantum-key-distribution-takes-to-space/. Accessed 4 Oct 2024
72. Jones A (2024) China to launch new quantum communications satellites in 2025. SpaceNews, 8 Oct. Available at: https://spacenews.com/china-to-launch-new-quantum-communications-satellites-in-2025/. Accessed 4 Oct 2024
73. Prisco J (2024) Quantum security in space: satellites bring QKD closer to Earth. Forbes, 9 Aug. Available at: https://www.forbes.com/councils/forbestechcouncil/2024/08/09/quantum-security-in-space-satellites-bring-qkd-closer-to-earth/. Accessed 4 Oct 2024
74. ESA (n.d.) Eagle 1. https://www.esa.int/Applications/Connectivity_and_Secure_Communications/Eagle-1. Accessed 4 Oct 2024
75. Prawdzik C (2023) New report, leaders detail how the US can stay ahead of China in space race. Air Space Forces Magazine, 12 Jan. Available at: https://www.airandspaceforces.com/new-report-leaders-detail-how-the-us-can-stay-ahead-of-china-in-space-race/. Accessed 1 Oct 2024

76. Bender B, Klimas J (2018) Space war is coming and the US is not ready. Politico, 6 Apr. Available at: https://www.politico.com/story/2018/04/06/outer-space-war-defense-russia-china-463067. Accessed 22 Aug 2024
77. Paladini S (2019) The new frontiers of space. Routledge, London
78. Hertzfeld HR, von der Dunk F (2005) Bringing space law into the commercial world: property rights without sovereignty. In: Space, cyber, and telecommunications law program faculty publications. Available at: https://digitalcommons.unl.edu/spacelaw/15/. Accessed 4 Oct 2024
79. Sumida J (2015) Old thoughts new problems: Mahan and the consideration of space power. In: Lutes CD, Hayes PL (eds) Toward a theory of space power. Institute for National Strategic Studies, London
80. Akoto W (2020) Hackers could shut down satellites or turn them into weapons. The Conversation, 6 Feb. Available at: https://theconversation.com/hackers-could-shut-down-satellites-or-turn-them-into-weapons-130932. Accessed 4 Oct 2024
81. Bowen B (2020) Space as a global commons: is there a security dilemma in space? Glob Policy 11(3):370–380. https://doi.org/10.1111/1758-5899.12762
82. Delpech T (2019) Debris risks in space: securitizing environmental dangers. J Strateg Stud 42(4):490–506. https://doi.org/10.1080/01402390.2019.1586114
83. NASA (2024) State-of-the-art small spacecraft technology. Small Spacecraft Systems Virtual Institute, Ames Research Center, Moffett Field, California. NASA/TP—2024-10001462. Available at: https://www.nasa.gov/wp-content/uploads/2024/03/soa-2023.pdf?emrc=8ad1a1
84. National Institute of Standards and Technology (NIST) (2020) NIST interagency report 8270: satellite ground segment: applying the cybersecurity framework to satellite command and control. U.S. Department of Commerce. Available at: https://nvlpubs.nist.gov/nistpubs/ir/2020/NIST.IR.8270.pdf. Accessed 4 Oct 2024
85. O'Hanlon ME (2010) The weaponization of outer space: a catalyst for regional military industrial complexes. Survival 52(3):155–178
86. Schneier B (2018) Cryptography after the aliens land. https://www.schneier.com/wp-content/uploads/2018/09/Cryptography-after-the-Aliens-Land.pdf. Accessed 1 Oct 2024
87. West J (2018) Securing outer space: arms control and the prevention of an arms race in space. Space Policy 44:1–6. https://www.sciencedirect.com/science/article/pii/S026596461830022X

Quantum E-voting System Using QKD and Enhanced Quantum Oracles

Viduranga Shenal Landers

Abstract As electronic voting (e-voting) systems become integral to modern democratic processes, their security and integrity become of utmost importance. The need for a quantum e-voting system stems from the growth in the sophistication of cyber-attacks and the advent of quantum computing, which threatens classical cryptographic schemes. Classical cryptographic methods face increasing vulnerabilities from advancing computational capabilities and cyber threats. This research proposes a quantum e-voting system that leverages quantum oracles and entanglement to enhance security and transparency. A quantum-based e-voting system would make it future-proof as the system would be resilient to both classical and quantum threats. An integral part of the proposed e-voting system is Quantum Key Distribution (QKD) and quantum oracles. QKD is an unconditionally secure method for distributing encryption keys that is resilient to eavesdropping attempts. QKD, when coupled with quantum oracles, facilitate verifications within the voting process, ensuring that votes are accurately cast and counted without compromising voter anonymity. This research explores the theoretical underpinnings, potential implementation strategies, and the practical challenges of deploying such a system, with the aim of paving the way for a more secure and trusted voting process in the digital era.

Keywords E-voting · Quantum · QKD · Quantum Oracles · Internet voting

1 Introduction

Secure and trustworthy elections form the foundation of democratic societies. The increasing adoption of electronic voting (e-voting) systems, while offering convenience and efficiency, has also exposed the vulnerabilities of these systems to hacking, tampering, and other malicious attacks [18]. Traditional e-voting systems rely on

V. S. Landers (✉)
University of Colombo School of Computing, Colombo, Sri Lanka
e-mail: vidurangalanders@gmail.com

H. Jahankhani and B. Issac (eds.), *Cybersecurity and Human Capabilities Through Symbiotic Artificial Intelligence*, Advanced Sciences and Technologies for Security Applications, https://doi.org/10.1007/978-3-031-82031-1_4

classical cryptographic techniques, which are vulnerable to ever-advancing capabilities of modern attacks, including quantum attacks. The advancements of quantum computers have increased the necessity to explore quantum-resistant and quantum-enhanced security measures for classical systems which includes existing e-voting systems [20, 21].

Internet voting has emerged as a key consideration for voting systems of the future. With the growing popularity of the internet and the increasing digitization of various aspects of life, there is a growing demand for the ability to vote remotely and secure over the internet. Internet voting offers several advantages such as increased accessibility, convenience, and efficiency. However, it also poses significant challenges in terms of security, privacy, and integrity of the voting process. The current remote voting systems are vulnerable to various types of attacks such as hacking, vote manipulation, and voter coercion [7].

Moreover, with the rapid advancements in technology and the increasing popularity of concepts like metaverse, internet voting is likely to become an absolute necessity in the future. Metaverse refers to a virtual shared space, created by the convergence of virtually enhanced physical reality and physically persistent virtual space. In a metaverse, people can interact with each other and with virtual objects in a seemingly real or physical way. With more than 600 million metaverse users as of 2024, and as more and more activities move into the virtual realm, including potentially elections, the need for secure and reliable internet voting systems becomes even more pressing [10].

Quantum technologies have shown immense potential in revolutionizing secure communication and cryptographic protocols [20, 21]. Quantum Key Distribution (QKD), for instance, leverages the principles of quantum mechanics to enable provably secure key exchange, which is essential for maintaining the confidentiality and integrity of the voting process. Similarly, Quantum Random Number Generators (QRNGs) can provide true randomness, which is crucial for ensuring the unpredictability and fairness of the election outcomes [30].

This research paper proposes a quantum-enhanced e-voting system called Quantegrity that integrates quantum technologies such as QKD and quantum oracles with the Scantegrity voting system. Scantegrity is an end-to-end verifiable voting system that provides voter privacy and integrity of the election outcome [4]. By leveraging the power of quantum technologies, Quantegrity aims to enhance the security and verifiability of the Scantegrity system, making it resilient against both classical and quantum threats.

The rest of the paper is organized as follows: Sect. 2 provides background information and reviews related to work on e-voting systems and quantum cryptography. Section 3 describes the Scantegrity voting system and the symmetrically entangled quantum oracle (SEDJO) protocol in detail. Section 4 presents the Quantegrity system, including its architecture and key components. Section 5 discusses the security of the Quantegrity system. Section 6 discusses the implementation considerations and challenges of deploying Quantegrity. Finally, Sect. 7 concludes the paper and outlines the challenges and future research directions.

2 Background and Related Work

2.1 Overview of Traditional and Electronic Voting Systems

Voting systems have evolved from traditional paper-based systems to electronic voting (e-voting) systems, which use electronic means for vote casting, counting, and tabulation. E-voting systems can be categorized into Direct Recording Electronic (DRE) systems, where voters cast their votes directly on an electronic device, and Remote Online Voting (ROV) systems, which allow voters to cast their votes over the internet. DRE systems typically use touch-screen interfaces or push-button controls to record votes, while ROV systems use web-based or mobile applications [8].

One of the most widely known e-voting systems is Helios, an open-source web-based system that allows voters to cast their votes remotely over the internet. Helios uses homomorphic encryption to encrypt the votes, which allows the votes to be tallied without decrypting them individually. This provides a strong level of privacy for the voters [1].

Scantegrity is an end-to-end verifiable voting system that was designed to address the limitations of traditional DRE and ROV systems. Scantegrity uses optical scan paper ballots with invisible ink confirmation codes. Voters mark their ballots with a special pen that reveals a confirmation code. They can then check online that their confirmation code was correctly recorded, providing a strong level of verifiability. Scantegrity also uses mix-nets and zero-knowledge proofs to ensure the privacy and integrity of the votes [4].

Remotegrity is an enhancement to the Scantegrity system that allows for remote voting over the internet. Remotegrity uses the same invisible ink confirmation codes as Scantegrity, but instead of marking the ballots in-person, voters receive their ballots and confirmation codes by mail. They can then cast their votes online using their confirmation codes in a scratch card. Remotegrity provides the same end-to-end verifiability as Scantegrity, but with the added convenience of remote voting [31].

While e-voting systems offer advantages such as efficiency, accessibility, and faster vote counting, they also introduce new security and privacy challenges.

2.2 Security and Privacy Issues in E-voting

E-voting systems face various security and privacy threats. Tampering software and hardware is a major risk. Attackers may attempt to manipulate the voting software, hardware, or data to alter the election outcomes. Attackers may also try to disrupt the voting process by overloading the system or making it unavailable to voters via Denial of Service (DOS) or Distributed DOS (DDOS) attacks.

Another major threat is attacks on voter and voting privacy. Malicious actors may coerce or incentivize voters to vote in a particular way, undermining the integrity of the election. Even though e-voting systems must ensure that votes remain anonymous and cannot be linked back to individual voters, attackers may find ways to reveal voter identities or to steal voter information during the voting process, which could then be used in harmful manner [24].

Traditional e-voting systems rely on classical cryptographic techniques, such as public-key cryptography and digital signatures, to address these security and privacy issues. However, these techniques are vulnerable to attacks by quantum computers, which can solve certain mathematical problems exponentially faster than classical computers [20, 21].

2.3 Quantum Cryptography and Its Applications

Quantum cryptography leverages the principles of quantum mechanics to enable provably secure communication and key exchange. Quantum Key Distribution (QKD) is a prominent quantum cryptographic protocol that allows two parties to generate a shared secret key, which can be used for encrypting and decrypting. QKD protocols, such as BB84 and E91, rely on the properties of quantum states, such as the no-cloning theorem and the Heisenberg uncertainty principle, to detect eavesdropping attempts and ensure the security of the key exchange [30].

QKD has been successfully implemented in various real-world applications, including secure communication networks, cloud-based quantum computing, and satellites. The integration of QKD into e-voting systems can provide an additional layer of security, ensuring the confidentiality and integrity of the voting process.

Quantum technologies have shown promise in enhancing the security of various communication systems. For instance, Quantum random number generators (QRNGs) have been developed to provide true randomness, which is essential for cryptographic purposes and ensuring the unpredictability of election outcomes [17].

The proposed quantum-enhanced e-voting system builds upon these recent advances in quantum cryptography and secure communication to provide a robust and secure voting platform.

3 Key Technologies in Quantegrity System

3.1 Quantum Key Distribution (QKD)

Quantum Key Distribution (QKD) is a secure communication method that relies on the fundamental principles of quantum mechanics to enable two parties to produce a shared random secret key known only to them, which can then be used to encrypt

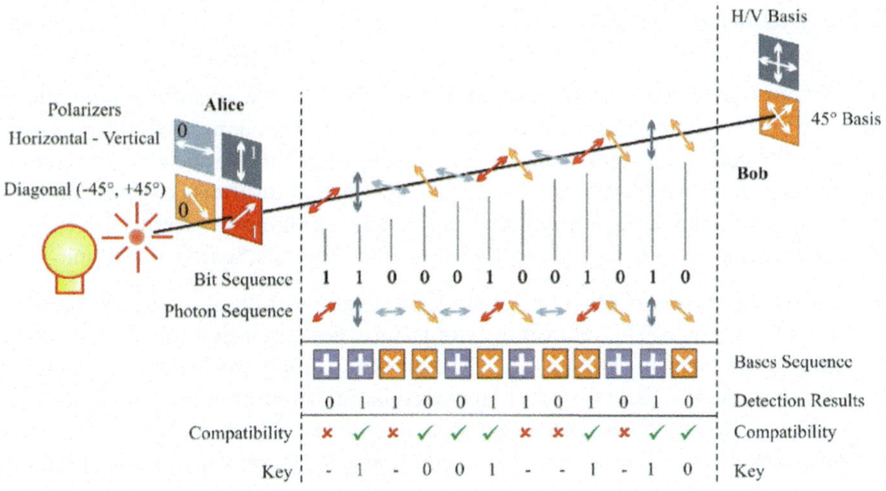

Fig. 1 BB84 protocol (adapted from [14])

and decrypt messages. QKD is based on the quantum properties of photons, such as polarization or phase, to encode and transmit the key [30].

The most well-known QKD protocol is BB84, which was proposed by Bennett and Brassard in 1984 [2]. In BB84, Alice (the sender) sends a stream of single photons to Bob (the receiver), each encoded with a random bit value of 0 or 1 and a random basis (rectilinear or diagonal). Bob measures each photon using a randomly chosen basis. After the transmission is complete, Alice and Bob communicate over a public channel to compare their basis choices. They discard the bits where their basis choices differ and keep the rest as the raw key. They then perform error correction and privacy amplification to obtain the final shared secret key (see Fig. 1).

Another popular QKD protocol is the E91 protocol proposed by Ekert [6]. The major difference in E91 is the use of entangled qubit pairs. In the E91 protocol, instead of sending a stream of photons, pairs of entangled qubit pairs are distributed between Alice and Bob which are then measured to derive the keys.

The security of QKD is based on the fundamental principles of quantum mechanics, such as the uncertainty principle and the no-cloning theorem. Any attempt by an eavesdropper to intercept and measure the photons will introduce detectable errors, alerting Alice and Bob to the presence of the eavesdropper. Furthermore, the no-cloning theorem prevents an eavesdropper from creating identical copies of the photons, ensuring that the key remains secure.

QKD has been experimentally demonstrated over distances of up to 400 km in optical fiber and up to 1200 km in free space [15]. It has also been commercially deployed in several countries, including Switzerland, Japan, and China [29]. QKD is a promising technology for securing communication channels against both classical and quantum adversaries.

3.2 Quantum Oracles

A quantum oracle is a black box operation that performs a specific computational task on a quantum state. Quantum oracles are used in many quantum algorithms, such as Grover's search algorithm and Shor's factoring algorithm, to speed up the computation and provide a quantum advantage over classical algorithms [19]. Several noteworthy quantum oracle algorithms are Duestsch-Jozsa algorithm, Bernstein-Vazirani algorithm, Simon's Oracle, Shor's algorithm, and Grover's Search algorithm.

- *Deutsch–Jozsa algorithm* is one of the earliest demonstrations of quantum computational advantage. This algorithm exhibits an exponential speedup over any deterministic classical algorithm. The Deutsch–Jozsa problem itself is deliberately constructed to be efficiently solved using a quantum algorithm yet computationally inefficient for any deterministic classical counterpart [5].
- *Bernstein–Vazirani algorithm* offers a solution to the Bernstein–Vazirani problem, which relates to determining a concealed bitstring. Notably, it accomplishes this with fewer oracle queries than would be necessary in a classical counterpart [3].
- *Simon's algorithm* is another algorithm similar to Deutsch–Jozsa and Bernstein–Vazirani that demonstrates quantum supremacy over classical counterparts [27].
- *Shor's algorithm* is renowned for its ability to factor integers, and is the foundation of quantum cryptography [25].
- *Grover's Search algorithm* has a significant speed advantage over classical algorithms in the context of unsorted database searches. Grover's algorithm requires only $O(\sqrt{N})$ time for a database of N items, as opposed to the $O(N)$ time required by classical algorithms [9].

In the context of quantum cryptography, quantum oracles can be used to implement secure multi-party computation protocols, such as blind quantum computation and verifiable quantum computation. In blind quantum computation, a client can delegate a computation to a server without revealing the input data or the computation result to the server. The client prepares an encrypted quantum state using QKD and sends it to the server. The server performs the computation on the encrypted state using a quantum oracle and returns the result to the client. The client can then decrypt the result using the QKD key.

Quantum oracles can also be used to implement quantum digital signatures and quantum fingerprinting protocols, which provide secure authentication and verification of quantum states. Overall, quantum oracles are a powerful tool for building secure and verifiable quantum communication and computation protocols.

3.3 *Scantegrity Voting System*

The Scantegrity voting system is an end-to-end verifiable voting system that uses optical scan paper ballots with invisible ink confirmation codes. The key components of the Scantegrity system are a special set of ballots, a coding system and a switchboard [4].

3.3.1 Ballots

Each ballot contains a list of candidates or choices, along with a corresponding set of confirmation codes printed in invisible ink. The confirmation codes are generated randomly and independently for each candidate on each ballot. The ballot also includes a unique ballot ID number, which is used to identify the ballot during the voting and tallying process.

Voters mark their ballots using a special pen that reveals the confirmation codes for their selected candidates. The voter can then record the revealed confirmation codes on a detachable receipt portion of the ballot. After marking their ballots, voters feed them into an optical scanner, which records the marked positions and the ballot ID number.

Voters can choose to audit their ballots by requesting a copy for a spoil ballot from the election officials. The officials reveal all the confirmation codes on the ballot copy, allowing the voter to verify that the codes match the ones recorded on their receipt. Audited ballots are not included in the final tally.

Figure 2 shows an example Scantegrity ballot with invisible ink confirmation codes. The ballot includes a list of candidates, along with a corresponding set of confirmation codes printed in invisible ink. The voter marks their choice using a special pen, which reveals the confirmation code for their selected candidate. The voter can then record the revealed confirmation code on the detachable receipt portion of the ballot.

Fig. 2 Example Ballot (adapted from [4])

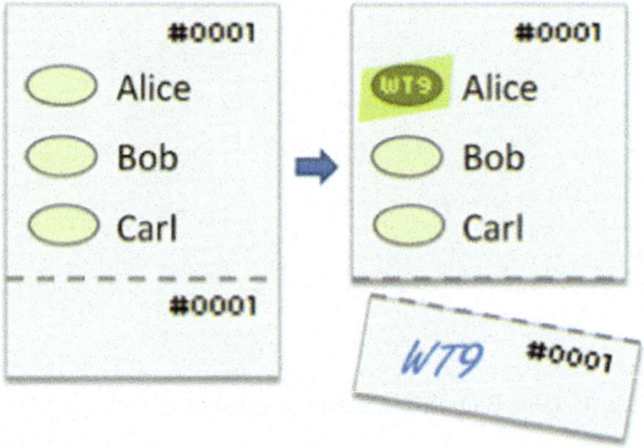

3.3.2 Mixnet

After the polls close, the election officials publish a list of all the ballot ID numbers and their corresponding confirmation codes on a public bulletin board. The officials then run a series of publicly verifiable mixnets to anonymize the ballots and decrypt the confirmation codes, revealing the final tally.

According to Chaum et al. four tables are created named as P, Q, R, and S. Table P consists of confirmation codes generated by a pseudorandom number generator in the order of questions in the vote [4]. Table Q is derived from P by pseudorandomly shuffling codes in each row of Table P. Table R is created with three columns: a flag, a Q-pointer, and an S-pointer. A flag is raised if a vote is made for the respective confirmation code. For a given confirmation code, Q-pointer points to its place in table Q while the S pointer points to its place in table S. Table S is table that contains only flags and this table is used for the tallying of votes (see Fig. 3).

Table P

Ballot ID	Alice	Bob	Carl
0001	WT9	7LH	JNC
0002	KMT	TC3	J3K
0003	CH7	3TW	9JH
0004	WJL	KWK	H7T
0005	M39	LTM	HNN

Table Q

Ballot ID	Alice	Bob	Carl
0001	7LH	WT9	JNC
0002	J3K	TC3	KMT
0003	9JH	CH7	3TW
0004	KWK	H7T	WJL
0005	M39	HNN	LTM

Table R

Flag	Q-Pointer	S-Pointer
	(0005, 1)	(2, 1)
	(0003, 3)	(4, 2)
✓	(0002, 1)	(4, 3)
	(0001, 3)	(3, 3)
✓	(0001, 2)	(4, 1)
✓	(0005, 3)	(3, 2)
	(0004, 2)	(5, 3)
	(0003, 1)	(2, 3)
	(0004, 3)	(3, 1)
	(0002, 3)	(1, 1)
	(0001, 1)	(2, 2)
	(0002, 2)	(5, 2)
	(0004, 1)	(1, 2)
✓	(0003, 2)	(5, 1)
	(0005, 2)	(1, 3)

Table S

Alice	Bob	Carl
	✓	
✓		✓
✓		

Fig. 3 Table P, Q, R, and S with revealed votes highlighted

secrecy, meaning that the probability of an adversary obtaining any information about the key approaches zero as the key length increases.

6 Implementation and Practical Considerations

6.1 Technology Readiness and Feasibility Assessment

The implementation of the Quantegrity system requires careful assessment of quantum technologies' readiness and feasibility. While progress has been made in quantum cryptography, challenges remain in hardware availability, scalability of quantum networks, and development of quantum-specific software and protocols.

Key considerations include the availability and maturity of quantum hardware. QKD systems require specialized quantum devices, such as single-photon sources, quantum detectors, and quantum memories, which are currently not as widely available or cost-effective as classical computing hardware. However, ongoing research and development efforts are rapidly advancing quantum technologies, and it is expected that quantum devices will become more accessible and affordable in the near future.

Scalability of quantum networks is another crucial factor. To enable secure communication between voting terminals and the central tabulation server, a reliable and scalable quantum network infrastructure is required. Currently, quantum networks are limited in terms of distance and the number of nodes they can support. Researchers are actively working on developing quantum repeaters and satellite-based quantum communication systems to extend the range and scalability of quantum networks [28].

Scalability challenges include the limited distance and bandwidth of current quantum communication channels. QKD protocols require the transmission of quantum states over long distances, which is currently restricted by the attenuation and decoherence of quantum signals. This limitation affects the distance between voting terminals and the central tabulation server, potentially impacting the system's scalability.

To address these challenges, researchers are developing quantum repeaters and satellite-based quantum communication systems. Quantum repeaters can extend the range of quantum communication by performing entanglement swapping and purification operations. Satellite-based quantum communication, exemplified by projects like the Chinese Micius satellite and the European Space Agency's QKD satellite project, offers the potential for global quantum communication networks [15].

6.2 Integration with Existing Voting Infrastructure and Processes

Integration with existing voting infrastructure and processes is crucial for practical implementation. The system should be designed to seamlessly integrate with current voting procedures and technologies to ensure a smooth transition and minimize disruption to the electoral process. A hybrid architecture combining quantum and classical technologies is one approach to integration. In this model, the authentication modules or parts of the Quantegrity system like QKD protocols and quantum-secure authentication are used for secure communication and key exchange, while the actual voting process utilizes classical e-voting technologies compatible with existing infrastructure.

Standardized interfaces and protocols for quantum–classical interoperability are essential to facilitate integration. These standards should define communication protocols, data formats, and security requirements for information exchange between quantum and classical components of the system. Close collaboration between election authorities, quantum technology providers, and e-voting system vendors is necessary to ensure proper integration and compliance with jurisdiction-specific requirements.

6.3 User Experience and Accessibility Considerations

User experience and accessibility are important considerations in implementing the Quantegrity system. The system should be easy to use for voters of all ages and technical backgrounds while maintaining a high level of security and verifiability.

A possible approach is to use a mobile app or web interface for the voting process, providing a more intuitive and accessible user experience. The app can guide voters through receiving their entangled qubits, encoding their votes, and checking the bulletin board for confirmation codes. To ensure accessibility, the system should accommodate voters with disabilities through assistive technologies and universal design principles in the user interface.

6.4 Cost–Benefit Analysis and Economic Viability

The cost and feasibility of implementing the Quantegrity system present significant challenges. The quantum hardware required for the QKD module and quantum oracle is currently expensive and not widely available, limiting the system's short-term scalability. However, the cost of quantum hardware is expected to decrease over time as the technology matures and economies of scale are achieved. The hybrid classical-quantum architecture of Quantegrity helps mitigate some cost concerns by limiting quantum components to security-critical parts of the system.

An alternative approach to reducing cost and complexity is to use a trusted third party for the QKD module and quantum oracle. This approach reduces hardware requirements for voters and election authorities but introduces additional trust assumptions and potential vulnerabilities.

While the Quantegrity system has been demonstrated in proof-of-concept simulations, it has not yet been deployed in a real-world election or executed on large-scale quantum hardware [13]. Further research and development are needed to address the technical and logistical challenges of implementing the system at scale.

In conclusion, the Quantegrity system offers significant benefits in terms of security and verifiability compared to traditional e-voting systems. However, its implementation and deployment in practice will require substantial investment and collaboration between academia, industry, and government stakeholders. As quantum technologies continue to advance and mature, it is expected that many of the current challenges will be overcome, paving the way for more secure and transparent electronic voting systems in the future.

7 Conclusion and Future Directions

The Quantegrity system represents a promising new approach to secure and verifiable e-voting, leveraging the unique properties of quantum technologies to provide enhanced security and integrity guarantees. By combining the Scantegrity voting system with the SEDJO protocol for quantum key distribution, the Quantegrity system offers a hybrid classical-quantum architecture that is resistant to both classical and quantum attacks.

The development and deployment of the Quantegrity system, however, raises significant technical, logistical, and ethical challenges. These include the scalability and performance of the quantum hardware, the accessibility and usability of the voting interface, the legal and regulatory framework for quantum voting, and the public trust and acceptance of the system.

To address these challenges, further research and development is needed in a number of key areas, including quantum hardware and software, user experience and accessibility, as well as legal and regulatory frameworks.

The scalability and practical feasibility of Quantegrity depend on the continued development and maturation of quantum technologies. This includes advances in quantum computing hardware, quantum communication networks, and quantum sensor technologies. Ongoing research in these areas can help reduce the costs, increase the reliability, and extend the range of quantum-based voting systems.

The validation and refinement of the proposed quantum-enhanced e-voting system require extensive testing and evaluation. Pilot projects and limited trials can help identify and address technical, usability, and social challenges, and provide valuable insights for scaling up the system.

Public engagement and the development of transparent and accountable processes are crucial for the adoption and societal impact of Quantegrity. New laws and regulations must be established to define the legal status and requirements for quantum voting systems, along with the roles and responsibilities of involved stakeholders.

Ultimately, the success of the Quantegrity system and other quantum voting technologies will depend on the ability of researchers, developers, policymakers, and the public to work together to address these challenges and to realize the potential benefits of these technologies for secure and verifiable e-voting.

While there is still much work to be done, the Quantegrity system represents an important step forward in the development of next-generation voting technologies, and a promising avenue for further research and innovation in the field of quantum cryptography and its applications to e-voting and beyond.

Acknowledgements The author would like to express his deepest appreciation to his family, whose unwavering support has been a constant source of encouragement throughout the writing of this paper.

Disclosure of Interests The author have no competing interests to declare that are relevant to the content of this article.

References

1. Adida B (2008) Helios: web-based open-audit voting. In: Proceedings of the 17th conference on security symposium. USENIX Association, USA
2. Bennett CH, Brassard G (1984) Quantum cryptography: public key distribution and coin tossing. In: Proceedings of IEEE international conference on computers, systems and signal processing, New York
3. Bernstein E, Vazirani U (1993) Quantum complexity theory. In: Proceedings of the twenty-fifth annual ACM symposium on theory. ACM, pp 11–20
4. Chaum D, Carback R, Clark J, Essex A, Popoveniuc S, Rivest RL, Ryan PYA, Shen E, Sherman AT (2008) Scantegrity II: end-to-end verifiability for optical scan election systems using invisible ink confirmation codes. In: USENIX workshop on accurate electronic voting technology. USENIX
5. Deutsch D, Jozsa R (1992) Rapid solution of problems by. In: Proceedings of the Royal Society of London. Series A: Mathematical and physical sciences, p 439
6. Ekert AK (1991) Quantum cryptography based on Bell's theorem. Phys Rev Lett 61(6):661–663
7. Faulí C, Stewart K, Porcu F, Taylor J, Theben A, Baruch B, Folkvord F, Nederveen F, Devaux A, Lupiáñez-Villanueva F (2018) Study on the benefits and drawbacks of remote voting. European Commission, Brussels
8. Galois (2015) The future of voting end-toend verifiable internet voting. U.S. Vote Foundation, USA
9. Grover LK (1996) A fast quantum mechanical algorithm for database search. In: STOC'96: proceedings of the twenty-eighth annual ACM symposium on theory of computing. https://doi.org/10.1145/237814.237866
10. Hryziuk C (2024, Sept 25) Exploring the future: 20 metaverse stats for 2024. Retrieved from EMBRYO: https://embryo.com/blog/metaverse-stats-2024/

11. IBM Quantum (2024, Sept 27) Qiskit. Retrieved from https://www.ibm.com/quantum/qiskit
12. Landers V (2023) Symmetrically entangled quantum oracles for quantum key distribution. Preprint
13. Landers V (2024, Sept 27) Quantegrity. Retrieved from Github: https://github.com/Viduranga Landers/Quantegrity
14. Legré M (2009, Dec 3) Quantum quantum cryptography cryptography—a reality a reality today. Retrieved from id Quantique: https://indico.cern.ch/event/71768/attachments/1035636/1475548/idQCERN_-_Quantum_Cryptography.pdf
15. Liao S, Cai W, Liu W et al (2017) Satellite-to-ground quantum key distribution. Nature 549:43–47. https://doi.org/10.1038/nature23655
16. Lindell Y, Pinkas B (2009) Secure multiparty computation for privacy-preserving data mining. J Privacy Confidentiality
17. Ma X, Yuan X, Cao Z, Qi B, Zhang Z (2016) Quantum random number generation. npj Quantum Inf
18. Miller M (2024, Aug 12) The nation's best hackers found vulnerabilities in voting machines—but no time to fix them. Retrieved from Politico: https://www.politico.com/news/2024/08/12/hackers-vulnerabilities-voting-machines-elections-00173668
19. Nielsen MA, Chuang IL (2011) Quantum computation and quantum information, 10th edn. Cambridge University Press, Cambridge
20. NIST (2024) Post-quantum cryptography. Retrieved 3 Sept 2024, from https://csrc.nist.gov/projects/post-quantum-cryptography
21. NIST (2024, July 11) The journey toward quantum resistant algorithms: NIST's initiative. Retrieved from NIST Releases first 3 finalized post-quantum encryption standards: https://www.nist.gov/news-events/news/2024/08/nist-releases-first-3-finalized-post-quantum-encryption-standards
22. Paillier P (1999) Public-key cryptosystems based on composite degree residuosity classes. In: Stern J (ed) Advances in cryptology—EUROCRYPT'99
23. Renner R, Koenig R (2004) Universally composable privacy amplification against quantum adversaries. arXiv
24. Rivest RL (2009, Nov 9) Security of voting systems. Retrieved from https://people.csail.mit.edu/rivest/pubs/Riv09d.slides.pdf
25. Shor P (1997) Polynomial-time algorithms for prime factorization and discrete logarithms on a quantum computer. SIAM J Comput 26(5):1484–1509. https://doi.org/10.1137/S0097539795293172
26. Shor PW, Preskill J (2000) Simple proof of security of the BB84 quantum key distribution protocol. Phys Rev Lett 441–444
27. Simon DR (1997) On the power of quantum computation. SIAM J Comput 1474–1483
28. Stanley M, Gui Y, Unnikrishnan D, Hall S, Fatadin I (2022) Recent progress in quantum key distribution network deployments and standards. J Phys Conf Ser
29. Toshiba (2024, Sept 13) Quantum key distribution. Retrieved from https://www.global.toshiba/ww/products-solutions/security-ict/qkd.html
30. Wolf R (2021) Quantum key distribution, 1st edn. Springer, Cham. https://doi.org/10.1007/978-3-030-73991-1
31. Zagórski F, Carback R, Chaum D, Clark J, Essex A, Vora P (2013) Remotegrity: design and use of an end-to-end verifiable remote voting system. In: Applied cryptography and network security. ACNS 2013. Lecture notes in computer science. Springer, Berlin

AI Makes Political Views More Acceptable: Right or Wrong

Gordon Bowen, Richard Bowen, Deidre Bowen, Atul Sethi, and Sumesh Dadwal

Abstract Politics, especially at election time, is an opportunity for politicians to have missteps that are deliberate, forced or accidental. In the past, these missteps were not able to be corrected in real time, but now, with the advent of social media and AI, missteps can be exposed "immediately" or criticism of the missteps will draw comments from social media networks. The fake news syndrome has always been with us in the media but now AI can do the fact checking at speed and accurately. Does this mean politicians will make fewer missteps or is the political mileage too much to not generate fake news. Can AI or the threat of AI algorithms that can fact check deter politicians from making untrue statements? The challenge is how misinformation from governments, leaders and messages in political campaigns are not mischaracterised as freedom of speech when they are really misinformation or fake news.

Keywords Fake news · AI and missteps · Political marketing · Freedom of speech and fake news · Political branding

Politics, especially at election time, is an opportunity for politicians to have missteps that are deliberate, forced or accidental. In the past, these missteps were not able to be corrected in real time, but now, with the advent of social media and

---

G. Bowen (✉)
Anglia Ruskin University, Cambridge, UK
e-mail: gordon.bowen@aru.ac.uk

R. Bowen
Darwin Homes, Austin, USA

D. Bowen
Mental Health UK, London, UK

A. Sethi
Ulster University London Campus, London, UK

S. Dadwal
London South Bank University, London, UK

H. Jahankhani and B. Issac (eds.), *Cybersecurity and Human Capabilities Through Symbiotic Artificial Intelligence*, Advanced Sciences and Technologies for Security Applications, https://doi.org/10.1007/978-3-031-82031-1_5

AI, missteps can be exposed "immediately" or criticism of the missteps will draw comments from social media networks. The fake news syndrome has always been with us in the media but now AI can do the fact checking at speed and accurately. Does this mean politicians will make fewer missteps or is the political mileage too much to not generate fake news. Can AI or the threat of AI algorithms that can fact check deter politicians from making untrue statements? The challenge is how misinformation from governments, leaders and messages in political campaigns are not mischaracterised as freedom of speech when they are really misinformation or fake news.

1 Introduction

Politicians always want to make the most compelling case for their political views. In doing so, missteps and misinformation may become part or all of their constructs. It could be argued that it is a natural human trait to exaggerate to get the point across, to develop robust arguments, to generate discussion and to move the political debate on. Is the degree of acceptance of misinformation or missteps part of the political environment? Must stakeholders in the political environment be doomed to accept the arguments from politicians without knowing what the truth is? A politician's application of misinformation or missteps depends on their perspective (left or right leaning) and whether they can marshal a convincing argument without recourse to exaggeration. The uncertainty of a political party's winnability in an election plays a vital role (consider UK general election on 4th July 2024 and the upcoming US election in November 2024).

The political environment impacts all stakeholders in the political universe because decision making in government and the voting behaviour of the electorate are governed by the utterances of politicians. Hence, government decision making has consequences that could be beneficial or catastrophic. Thus, some line of mediation is required to enable clear and ethical decision making. This paper suggests that the mediator is the application of generative artificial intelligence (AI) between stakeholders. AI is a pervasive tool and access to stakeholders is nearly universal. However, the integrity of the AI algorithm will need to be tested and certified. Thus, mediation in the political environment will require certifiable AI algorithms that are open to external and internal auditing and a framework for the auditing that is enshrined in a regulatory environment.

Meissner and Narita [1] suggested that AI will be one of the most important determinants of competitiveness in the economy, AI will enable users to determine which decisions to delegate to the AI algorithm; trust, access and integration are drivers of the speed and scale of AI adoption in strategic decision making. All, if not most, decisions made that are influenced by the political environment are strategic in nature, such as business decisions, voting decisions and referendums.

The aim of this paper is to argue that AI is an important tool in decision making that is influenced by the political environment and many of the decisions, which are based on the political environment, are strategic in nature and impact many of the stakeholders in the political environment.

The topics discussed are political marketing, political branding in the political environment, trust in government, and AI in governance and political marketing. The paper closes with an Implications section and a Conclusion section.

2 Political Marketing

Political advertising is part of the political marketing process. Political parties or an election candidate use political advertising on mass media channels to communicate political messages to influence election outcomes. The purpose of political messages is to influence, persuade and change attitudes [2]. Political ads are hotly debated because they influence or formulate opinion on issues of national importance [3]. Furthermore, lying in political advertising is not illegal and this is despite advertisements being subject to regulatory oversight. If politicians are in an electorate contest (running for office), then making a false claim against their opponents is prohibited. In contested political situations the government has more limited regulatory oversight, because the electorate should have the unvarnished truth. The only recourse is to sue opponents who are apparently lying. Is this an opportunity for AI decision making so it acts as a quasi-regulator?

Political advertising helps to paint a new picture of election candidates and political parties. The "new picture" of the candidates is achieved by attacking competitors, and by questioning and criticising their policies [4].

There is much similarity between commercial advertising and political advertising. Commercial and political advertising are about forecasting value to the audience. Political advertising value is about expressing benefits to the layer of society the ads are targeted towards. In the case of commercial ads, the objective is like political ads, but they apply strategies that leverage the marketing of products and services that are linked to a marketing plan. Political advertising aims to influence audience/voter behaviour positively and requires the changing of voters' belief systems. Changing the political beliefs of voters requires the application of social imagery of the candidates and the social situation (socio-economic classes or imaging socio-economic situations through the values candidate subscribe to) [4, 5].

The techniques employed by commercial and political marketing intersect and many similarities exist between them [6], namely segmentation, targeting and positioning, and understanding of the marketing mix (4Ps—product, price, promotion and place—and extended 4Ps). Segmentation is especially crucial in targeting the "right" parts of the electorate (i.e., an appropriate homogeneous cluster within a heterogeneous electorate) so that the political messages reach the target audience within the electorate. These techniques, such as microtargeting, were successfully implemented by the BJP party in elections in India in 2019 [4, 7]. Bannon [8] suggested that positioning is an approach to build a party's image and candidate by expressing clarity, consistency, credibility, competitive value and communication. There are many versions of the marketing mix that could be applied to commercial and political marketing [9]. However, one that struck us as a good fit for political marketing is configuration (product), valuation (price), symbolisation (promotion) and facilitation (place) [10]. This is a point for discussion in the Implications section.

The political environment and context that political ads operate within in the US and India are influenced by their unique political structures, cultural backgrounds and the media environment. The US embraces and harnesses social media (Barack Obama's successful 2008 presidential campaign), which is complemented with traditional media, such as television, radio and direct mail. Whereas political advertising in India is conducted on satellite television and social media and is adapted to the geography and regional languages of the constituents [4, 11]. Customising political messages to the vast number of regional languages, regional political parties and the diversity of languages is challenging, and not characteristic of US political environments.

3 Political Branding in the Political Environment

Political branding is a sub-division of political marketing. What is political branding? It is the critical application of traditional marketing branding techniques used in commercial marketing, which include concepts, frameworks and theory, in a political environment to understand political competitors and identify unique political entities [12]. Political branding extends to political parties (local, regional, national and international) and other stakeholders, such as pressure groups, politicians, candidates and movements. These can be conceptualised as part of the political branding process [13, 14, 15].

Political brands are multidimensional and complex and are meant to differentiate from the political competitors [16, 17]. It is noteworthy that political brands are challenging to operationalise [18]. A political brand has many facets, including the leader, the party and policy [19, 20, 21]. Research is expanding the topology of branding relating to leader, party and policy to include political brand communities [22] and political brand communications [23].

A developing area of political marketing is the importance of the credibility of the party leader's brand image. In the context of India's Prime Minister Narendra Modi, Jain and Ganesh [24] used secondary research to investigate how social media marketing is used to build credibility and authenticity and develop a personal political brand. Jain and Ganesh [24] argued that party leaders should from time to time employ crisis management strategies along with collaborative strategies that embrace stakeholders to develop a multistakeholder personal brand in the mind of voters. They further suggested that politicians should not only adopt a professional persona but also communicate humility, personal characteristics and a relatable personality.

Armannsdottir et al. [25] investigated political brand identities. The personal characteristics of political brands is becoming a rich area for investigation into personal political branding [18]. Armannsdottir et al. [25] investigated the personal political brand identities, and their development, of parliamentarians in the Icelandic Parliament from the political brand creator perspective of personal branding. They suggested that personal political brands need to project an authentic character and distinct identity built on tangible attributes, such as physical attributes, style, and online and offline communication tools, and supplemented with intangible attributes, such as lived experiences, skills, values and personality attributes. The insights they developed enabled them to develop the Personal Political Brand Identity Appraisal Framework, which could be used to periodically audit current brand identity and develop desired personal political brands of politicians and other political stakeholders. Being on message and sticking to messages (consider Donald Trump, former US president) is an area of the political branding management process. Successful political campaigns require relevant messages and policies that are clearly communicated and structured to capture and maintain the interest of voters. Messages need to be persuasive and contribute to the overall personal brand image of the politician [26]. Furthermore, Falkowski and Jabłonska [26] argued that framing messages that create a favourable association and with desirable imagery is a strategic process that is part of the ongoing brand management process.

Political branding is a multidimensional process. In the political environment, political branding is not a static process but evolves due to changing local, regional, international and global political dynamics. This necessitates critical reviews of concepts, theories and frameworks that were employed in the political environment to facilitate political marketing and branding [18]. Furthermore, "this suggests that there are multiple relationships and numerous interconnected political brands, which represents an intricate environment or ecosystem. This represents the starting point for further research in political brands" [18, p. 10].

4 Trust in Government

Trust is important in democratic governments and societies [27, 28]. Trust is instrumental in building communications within society and between citizens and encourages democratic institutions by minimising conflict and, thus, reducing transactional costs [28, 29, 30]. Trust also influences tax revenue by improving compliance [31], the choices the electorate make [32] and policy preferences [33]. Roles in government that AI could play an important and that are potentially groundbreaking.

Trust is in government important at the micro level (demographics) [34, 35] and at the macro level (political environment over time) [35].

4.1 Micro Modelling of Trust

Micro level of trust refers to the socio-demographics and personality attributes [34, 36]. Political trust is dependent on personality factors and socio-demographic traits. People with greater personal efficacy would have more political trust [35]. According to Agger et al. [37], citizens' political trust in governments is dependent on their education and income and can be affected by personal cynicism. Political trust is linked to socio-demographics, especially age, education and race [38], because these factors (personality and demographics change little or slowly, the assumption is that political trust in the population will stabilise or change gradually [35]. The micro trust construct did not predict, and could not account for, the rapid decline of political trust in the US electorate [35]. Miller [39] opined that political trust should not be based on individual characteristics. Miller [40] noted that political trust is linked to how the government performs, changing attitudes to public policy and the political system. A correlation between political trust and agreement on public policy was becoming more evident [39]. Citrin [41] disagreed with Miller and suggested that the decline in political trust is connected to the prevailing characterisation of the existing political leaders. However, the debate on political trust has included more constructs in the assessment of political trust, such as personal economic wellbeing and prioritisation of political issues [42, 43, 33]. Are these obvious areas that AI could help to unravel? The micro approach to political trust has several limitations, from the explanation of a decline in political trust variability over time to questions about the variables that influence political trust over time and what their impacts are [44, 35].

4.2 Macro Modelling of Trust

The annual variability of the mean political trust levels is the focus of macro trust. The variability of political trust is based on the relationship between annual mean trust over time of the performance of existing leaders and political shocks [35]. The evaluation of policies and the electorate's attachment of importance to policies at a given time also known as "priming" [45]. The salient social problems are public order and crime, which are internal issues that decrease political trust (situation in England over the violent crime in Southport in 2024), and wars, which are external matters that bring people together (us and we) and increase political trust (Ukraine war). Political trust is lower in the years when political scandals have taken place (e.g., COVID-19 and the antics of Boris Johnson's government in the UK).

The micro and macro models of political trust have limitations and using a single model to represent political trust is challenging if not impossible. Political trust can best be understood in the context of the political trust based on micro-based political trust and macro-based political trust. A micro political trust model cannot accommodate changing means or a singular year shock. On the other hand, the macro political trust model cannot accommodate large variances that impact individuals that are not reflected in annual means. The changing composition of the political environment cannot be accommodated by either model [35].

5 AI in Governance and Political Marketing

The introduction of AI in government practices, public sector ecosystems and government governance could be beneficial to public services [46, 47] by fostering societal trust [48] and improvement in service delivery (efficiency and effectiveness) [49]. Other uses of AI in government include improved forecasting leading to more accurate outcomes and simulation of complex tasks with the option to experiment with various policy options [50]. Broadly speaking, AI can create value in government departments by aiding decision-making support, transportation, public health and law enforcement [51]. The role of AI in law enforcement opens the door for dealing with misinformation and fake news in the political context and in the wider context of delivery of ethical information to citizens both politically and commercially.

Although AI provides many benefits there are drawbacks and challenges. The application of AI in government might increase the electorate's trust in government and government decision making [52], but it might also decrease it [53, 49]. A decrease in trust in the government (by extension political trust) could be due to a lack of perceived fairness or privacy in the use of AI in public governance [54]. Issues on the lack of accountability and transparency in the application of AI using a black box approach could be problematic [55]. The possibility of AI failures in government may have negative implications for government [56].

The rise of AI in government and particularly decision making is causing anxiety and raising questions relating to public governance for governments worldwide [56]. These concerns range from economic problems related to labour markets and sustainability [57], societal worries relating to privacy, safety, risk and threats [58], social and ethics dilemmas about fairness, bias and inclusion [59], and governance problems related to transparency, regulatory frameworks and representativeness [60]. "What policies and regulations can be used to govern AI use in specific government organizations?" [56, p. 3]. What political marketing decisions developed from AI can be used in advertising campaigns? How are decisions made by AI policed because they could lead to misinformation? Unregulated use of AI data could lead to unintended consequences and impact different stakeholders adversely. Does GDPR apply to data used to train the AI algorithm and how is it regulated? How is the output from an AI algorithm interpreted, especially when the results are ambiguous? Should management of the AI process used in government governance be at arm's length from the government?

6 Implications

The political environment is where stakeholders in the political process of electioneering, government, leaders and political parties operate. Some questions that impinge on the operation of the political environment are: Why is lying in commercial ads unacceptable but in political ads it is acceptable or at least not illegal? Is honesty not required in commercial and political advertising? Are there certain situations in political advertising where freedom of speech does not trump honesty, trust and ethical behaviour? Should freedom of speech embody all the constructs of honesty, trust and ethical behaviour? To most voters the answer would be yes, but politicians have a get out of jail option (freedom of speech). Without limits on freedom of speech, how are politician going to build credibility? Refuting political claims as fake news or could be classed as missteps (also lying) is an activity that AI could undertake with a degree of accuracy in real time.

Political ads influence the political environment to benefit the strata of society the ads are geared towards. How can AI improve the political environment to the benefit of the target strata of society?

Mentioned previously is that a marketing nomenclature struck us as a good fit for political marketing: configuration (product), valuation (price), symbolisation (promotion) and facilitation (place) [10]. The ideology of a party and candidate is the product on offer, valuation is the campaign costs, facilitation is the campaign distribution and symbolisation is channels for communications to the different micro-targets in the different strata of society [4]. The terminology adopted by Kotler [10] is more relatable to political marketing than the conventional marketing mix nomenclature of product, price, promotion and place [61]. Political advertising is about change be it temporary or hopefully permanently, the electorate beliefs (ideology) which require configuration of their motivation, thought processes and changing the electorate cultural position. The process to bring about the change is directly based on the available campaign funds (valuation). Return on investment plays a significant role in the decision-making process on which strata of society and socio-economic groups and location of the electorate to target (hence use of informal and formal electorate pacts between political parties). The Labour Party's contesting of some Conservative seats in the 2024 election was not always a winning strategy because the electorate in those locations are predominantly Conservative in political beliefs and would not necessarily vote Labour, it would have been a better strategy to let the Liberal Democrats have a clear run so that there would be no Labour opposition. The place in traditional marketing or in political marketing context (facilitation) is about boots on the ground to engage personally with the electorate. Personal engagement is an extremely powerful technique, to "press the flesh" and bring that human touch, which is not always a requirement in commercial marketing activities. Finally, the last factor in the political marketing mix is symbolisation (promotion). Within political marketing the communication is linked to imagery, for example, the "Labour isn't working" poster by Saatchi & Saatchi helped Margaret Thatcher to power by reshaping beliefs [62]. The symbolism at that time was something that a particular stratum of the electorate could relate to and hence change their beliefs about both parties (Labour and Conservative) to the advantage of the Conservative party. Would such a poster work in a commercial environment? Probably not. AI could help in translating political ideas into visuals and connecting with a wide stratum of the electorate. This brings into play trust and the trustworthiness of the imagery content. It is not illegal to lie in political ads, for example, the Brexit claim that exiting the European Union would enable £350 million per week for the NHS [63], the accuracy of the statement was disputed by the Office of National Statistics [64]. The case against the misstep by Boris Johnson and Michael Gove about the NHS £350 million windfall per week from Brexit was thrown out and resulted in a verdict that Boris Johnson was not guilty of lying because it would be a challenge to free speech [65]. Rebuttal of statistics and argument across all communication channels in real time and with accuracy is an undertaking that AI could achieve. In a political environment where missteps are allowable because of free speech, getting the opposing view out in the political ether quickly is clearly one response. Furthermore, how the rebuttal is presented is also an area for AI involvement. Missteps are another form of fake news.

AI has the potential to contribute to government policies, decision making and create value in such areas as decision support, transportation, public health and law enforcement. Ideas in the above policies could be honed using AI and refined by human intervention. Thus, an AI algorithm becomes a starting point that requires intervention to refine the policies and make them acceptable to stakeholders. Policies solely decided by AI are unlikely to be successful and public trust in these policies extremely low. Once the public is aware of how AI can facilitate government policies, and the benefits of these policies become known, then trust in AI in political decision making is likely to improve. AI's capability to prevent missteps and fake news would go a long way to achieving higher electorate acceptance of AI as a governmental tool. Would applying AI in policy decision support make it easier to identify fake news and missteps by political parties and commentators? Those that use AI applications frequently will be more adept at using it to debunk fake news and political missteps. However, the success of AI in the role of spotting and correcting missteps and fake news will depend on the accuracy and integrity of the information. Do users of AI need to be able to spot fake news that could not be used to debunk missteps and fake news? There could be situations where fake news is being used to refute fake news. How can this be overcome?

Political branding is affected by missteps and association with fake news (an exception is Donald Trump's political branding supposition of untruths and use of fake news). The Donald Trump approach to political branding is very much an outlier except in authoritarian governments. The election in the United Kingdom (4th July 2024) has elements of the Donald Trump political approach to branding but it was not an overwhelming success, and one could even say a failure. Trust and ethics are still requirements for a political brand to be successful at the micro and macro models. The electorate and other stakeholders require political brands to be relatable across socio-economic groups and over time. Macro modelling of political trust requires stability in the electorate in which to build long-term political trust. Given that democratic governments are elected for a definitive period (term) (4 years in the US and 5 years in the UK), governments need to serve more than one term to develop impactful policies. Could the application of AI assist governments to develop impactful policies and develop scenarios that are appealing to the electorate and other stakeholders?

Many challenges need to be overcome before AI can be viewed as being reliable and accurate and gain the confidence and trust of stakeholders in the political environment. AI modelling in political ecosystems will require small steps to build trust at the micro and macro modelling stages. AI needs to be able to not answer political questions but give the rationale and analysis that underpins the responses. Responses need coherency, structure, analysis and rationality to rebut missteps and fake news. This will require AI algorithms to be trained to deliver this level of detail and not singular responses to political questions. AI must produce the evidence in the rebuttal of fake news and missteps. Consequently, AI can provide the balance to make political views acceptable or unacceptable. Obviously, the final arbiter of acceptability or non-acceptance of political responses is dependent on the electorate but at least they will have the evidence. The mantra of "freedom of speech" by politicians can be laid to rest to an extent.

7 Conclusion

AI has a role to play in government and in political parties. The speed and accuracy of AI could help it navigate the political environment to rebut fake news and missteps by political leaders, political parties and stakeholders associated with political interests. The nature of AI means the analysis and rebuttal could be achieved in real time, thus minimising the duration of the misstep or fake news. Containing the use of fake news and missteps builds trust at the micro and macro levels of modelling. However, building trust at the macro level of modelling requires longevity, population stability and continuous engagement by politicians with the electorate. However, the application of AI will present challenges for governments and political parties because of the public's general scepticism of AI usage, especially in political environments. Governments need to raise the value of AI as a social influencer and a tool for political engagement. AI is not just for answering questions but requires to be a seamless tool that can play an active role in policy decision-making support, develop scenarios of political strategy, leverage political brands and engage constructively with all political stakeholders. AI needs to be able to respond to multiple political questions and link responses to develop coherent output. The goal of government should be that AI becomes an accepted tool that can engage the electorate to build trust by delivering truthful and ethical responses to political questions and thus undermine missteps and fake news. Misinformation is becoming a scourge of the political environment an AI is a potential tool to limit its impact.

References

1. Meissner P, Narita Y (2024) How artificial intelligence will transform decision-making | World economic forum (weforum.org). Accessed 23rd July 2024
2. Kaid LL (2004) Handbook of political communication research. Routledge, UK. https://doi.org/10.4324/9781410610584
3. Nott L (2020) Political advertising on social media platforms. American Bar Association. Retrieved 27th July 2024, from https://www.americanbar.org/groups/crsj/publications/human_rights_magazine_home/voting-in-2020/political-advertising-on-social-media-platform
4. Duggal G (2023) The dynamics of political advertising: a comprehensive theoretical analysis. J Commun Manag 2(4):275–280. https://doi.org/10.58966/JCM20232410
5. Kotler P, Kotler N (1999) Political marketing: generating effective candidates, campaigns, and causes. In: Newman BI (ed) Handbook of political marketing. Sage Publications, London, pp 3–18
6. Newman BI, Perloff RM (2004) Political marketing: theory, research, and applications. In: Kaid LL (ed) Handbook of political communication research. Lawrence Erlbaum Associates Publishers, pp 17–43. https://doi.org/10.4324/9781410610584
7. Dutta A (2019) How BJP used data to craft landslide win. Hindustan Times. https://www.hindustantimes.com/lok-sabha-elections/how-bjp-used-data-to-craft-landslide-win/story-A3dNXdPiaG9pTVMf6j8mEJ.html
8. Bannon D (2004) Marketing segmentation and political marketing. (2) (PDF) Marketing segmentation and political marketing (researchgate.net). Accessed 27th July 2024
9. Yudelson J (1999) Adapting McCarthy's Four P's for the twenty-first century. J Mark Educ 21(1):60–67. https://doi.org/10.1177/0273475399211008
10. Kotler P (1972) Marketing management: analysis, planning and control. Prentice Hall, Englewood Cliffs, NJ
11. Kumar R (2017) Padayatras and the changing nature of political communication in India. Stud Indian Polit 5(1):32–41. https://doi.org/10.1177/2321023017698258
12. Lock A, Harris P (1996) Political marketing – vive la difference!. Eur J Mark 30(10/11):14–31. https://doi.org/10.1108/03090569610149764
13. Ahmed MA, Lodhi AAS, Ahmad Z (2017) Political brand equity model: the integration of political brands in voter choice. J Polit Mark 16(2):147–179. https://doi.org/10.1080/15377857.2015.1022629
14. Nai A, Martinez F (2019) The personality of populists: provocateurs, charismatic leaders, or drunken dinner guests? West Eur Polit 42(7):1337. https://doi.org/10.1080/01402382.2019.1599570.
15. Scammell M (2015) Politics and image: the conceptual value of branding. J Polit Mark 14(1–2):7–18. https://doi.org/10.1080/15377857.2014.990829
16. Lock A, Harris P (1996) Political marketing—vive la difference! Eur J Mark 30(10/11):14–31. https://doi.org/10.1108/03090569610149764
17. Pich C, Armannsdottir G, Spry L (2018) Investigating political brand reputation with qualitative projective techniques from the perspective of young adults. Int J Market Res 60(2):198. https://doi.org/10.1177/1470785317750817
18. Pich C, Newman BI (2020) Evolution of political branding: typologies, diverse settings and future research. J Polit Mark 19(1–2):3–14. https://doi.org/10.1080/15377857.2019.1680932
19. O'Cass A, Voola R (2011) Explications of political market orientation and political brand orientation using the resource-based view of the political party. J Mark Manag 27(5–6):627–645. https://doi.org/10.1080/0267257X.2010.489831
20. Smith G (2008) Politically significant events and their effect on the image of political parties. J Polit Mark 4(2–3):91–114. https://doi.org/10.1300/J199v04n02_05
21. Speed R, Butler P, Collins N (2015) Human branding in political marketing: applying contemporary branding thoughts to political parties and their leaders. J Polit Mark 14(1–2):129–151. https://doi.org/10.1080/15377857.2014.990833

22. Newman T (2019) The emergence of science as a political brand: USA. J Polit Mark 19(1):2. https://doi.org/10.1080/15377857.2019.1652225
23. Susila I, Dean D, Yusof RNR, Setyawan AA, Wajdi F (2019) Symbolic political communication, and trust: a young voters' perspective of the Indonesian presidential election. J Polit Mark 19(1–2):153–175. https://doi.org/10.1080/15377857.2019.1652224
24. Jain V, Ganesh BE (2019) Understanding the magic of credibility for political leaders: a case of India and Narendra Modi. J Polit Mark 19(1):2. https://doi.org/10.1080/15377857.2019.1652222
25. Armannsdottir G, Carnell S, Pich C (2019) Exploring personal politic brands of Iceland's parliamentarians. J Polit Mark 19(1–2):74–106. https://doi.org/10.1080/15377857.2019.1680931
26. Falkowski A, Jabłonska M (2019) Moderators and mediators of framing effects in political marketing. Implications for political brand management. J Polit Mark 19(1, 2). https://doi.org/10.1080/15377857.2019.1652221
27. Arneil B (2006) Diverse communities: the problem with social capital. Cambridge University Press, Cambridge
28. Dahl RA (1956) A preface to democratic theory. University of Chicago Press, Chicago, IL
29. Mara GM (2001) Thucydides and Plato on democracy and trust. J Polit 63:820–845
30. Newton K (2008) Social and political trust in established democracies. In: Norris P (ed) Critical citizens: global support for democratic government. Oxford University Press, Oxford, pp 169–187
31. Schuolz JT, Lubell M (1998) Trust and taxpaying: testing the heuristic approach to collective action. Am J Polit Sci 42:398–417
32. Hetherington MJ (1999) The effect of political trust on the presidential vote, 1968–96. Am Polit Sci Rev 93:311–326
33. Hetherington MJ, Husser JA (2011) How trust matters: the changing political relevance of political trust. Am J Polit Sci 56(2):1–14
34. Fraser J (1970) The mistrustful-efficacious hypothesis and political participation. J Polit 32:444–449
35. Wilkes R (2014) Trust in government: a micro-macro approach. J Trust Res 4(2):113–131. https://doi.org/10.1080/21515581.2014.889835
36. McClosky H, Schaar JH (1965) Psychological dimensions of anomy. Am Sociol Rev 30:14–40
37. Agger RE, Goldstein MN, Pearl SA (1961) Political cynicism: measurement and meaning. J Polit 23:477–506
38. Cole RL (1973) Towards a model of political trust: a causal analysis. Am J Polit Sci 17:809–817
39. Miller A (1974) Political issues and trust in government: 1964–1970. Am Polit Sci Rev 68:951–972
40. Miller A (1974) Rejoinder to "comment" by Jack Citrin: political discontent or ritualism? Am Polit Sci Rev 68:989–1001
41. Citrin J (1974) Comment: the political relevance of trust in government. Am Polit Sci Rev 6:973–988
42. Bélanger É, Nadeau R (2005) Political trust and the vote in multiparty elections: the Canadian case. Eur J Polit Res 44:121–146
43. Citrin J, Luks S (2001) Political trust revisited: déjà vu all over again? In: Hibbing JR, Theiss-Morse E (eds) What is it about government that Americans dislike? Cambridge University Press, Cambridge, pp 9–27
44. Keele L (2007) Social capital and the dynamics of trust in government. Am J Polit Sci 51:241–254
45. Hetherington MJ, Rudolph TJ (2008) Priming, performance, and the dynamics of political trust. J Polit 70:498–512
46. Ojo AK, Mellouli S, Zeleti FA (2019) A realist perspective on AI-era public management. In: Proceedings of the 20th annual international conference on digital government research

47. Toll D, Lindgren I, Melin U, Østergaard Madsen C (2019) Artificial intelligence in Swedish policies: values, benefits, considerations and risks. In: Lindgren I (ed) EGOV: international conference on electronic government: 18th IFIP WG 8.5 international conference, EGOV 2019, San Benedetto del Tronto, Italy, 2–4 Sept 2019 proceedings. Springer, Berlin, pp 301–310. https://doi.org/10.1007/978-3-030-27325-5_23
48. Dwivedi YK, Hughes L, Ismagilova E, Aarts G, Coombs C, Crick T, Duan Y, Dwivedi R, Edwards J, Eirug A, Galanos V, Ilavarasan PV, Janssen M, Jones P, Kar AK, Kizgin H, Kronemann B, Lal B, Lucini B, Medaglia R, Le Meunier-FitzHugh K, Le Meunier-FitzHugh L, Misra S, Mogaji E, Sharma SK, Singh JB, Raghavan V, Raman R, Rana NP, Samothrakis S, Spencer J, Tamilmani K, Tubadji A, Walton P, Williams MD (2021) Artificial intelligence (AI): multidisciplinary perspectives on emerging challenges, opportunities, and agenda for research, practice and policy. Int J Inf Manag 57, Article 101994. https://doi.org/10.1016/j.ijinfomgt.2019.08.002
49. Gupta KPKKP (2019) Artificial intelligence for governance in India: prioritizing the challenges using analytic hierarchy process (AHP). Int J Recent Technol Eng 8(2):3756–3762
50. Margetts H, Dorobantu C (2019) Rethink government with AI. Nature 568:163–165. https://doi.org/10.1038/d41586-019-01099-5
51. Gomes de Sousa W, Pereira de Melo ER, De Souza Bermejo PH, Sousa Farias RA, Oliveira Gomes A (2019) How and where is artificial intelligence in the public sector going? A literature review and research agenda. Gov Inf Q 36(4):101392. https://doi.org/10.1016/j.giq.2019.07.004
52. Sun TQ, Medaglia R (2019) Mapping the challenges of artificial intelligence in the public sector: evidence from public healthcare. Gov Inf Q 36(2):368–383
53. Al-Mushayt OS (2019) Automating E-government services with artificial intelligence. IEEE Access 7:146821–146829. https://doi.org/10.1109/ACCESS.2019.2946204
54. Kuziemski M, Misuraca G (2020) AI governance in the public sector: three tales from the frontiers of automated decision-making in democratic settings. Telecommun Policy 44(6):101976. https://doi.org/10.1016/j.telpol.2020.101976
55. Dignum V (2018) Ethics in artificial intelligence: introduction to the special issue. Ethics Inf Technol 20:1–3. https://doi.org/10.1007/s10676-018-9450-z
56. Zuiderwijk A, Chen Y, Salem F (2021) Implications of the use of artificial intelligence in public governance: a systematic literature review and a research agenda. Gov Inf Q 38:101577
57. OECD (2019) OECD employment outlook 2019: the future of work. OECD Publishing, Paris. Retrieved 4th Aug 2024. https://doi.org/10.1787/9ee00155-en
58. Yudkowsky E (2008) Artificial intelligence as a positive and negative factor in global risk. In: Bostrom N, Cirkovi MM (eds) Global catastrophic risks, vol 1. Oxford University Press, New York, p 184
59. International Labour Organization (2019) Work for a brighter future—global commission on the future of work. Retrieved 4th Aug 2024. Work for a brighter future: report of the global commission on the future of work. ILO Publications
60. OECD (2019) OECD recommendation of the council on artificial intelligence. Retrieved 4th Aug 2024. OECD-LEGAL-0449-en-2faedcefa4d54e98bd155d4f2c37b8fb
61. McCarthy EJ (1960) Basic marketing: a managerial approach. Irwin, Homewood, IL
62. Fallon I (2007) Saatchi & Saatchi: the agency that made tory history | The independent | The independent. Accessed 27th July 2024
63. Forrest A (2023) Brexit: Michael Gove claims infamous £350m a week for NHS promise 'delivered' | The independent. Accessed 27th July 2024
64. Full Fact Team (2017) The foreign secretary and the UK statistics authority: £350 million explained—full fact. Accessed 27th July 2024
65. ITV News (2019) Boris Johnson wins court challenge over claims he lied about Brexit delivering £350m for the NHS | ITV News. Accessed 27th July 2024

Unlocking Effective Information Sharing Agreements in English Local Authorities

Simon Elliott and John McCarthy

Abstract Electronic Information Sharing (EIS) within Local Government Authorities (LAs) in England is vital for the efficient operation of such bodies in delivering services to their residents. In some cases, these reflect statutory duties put on LAs by central government through legislation. Through their own work, the researcher has experienced similar professionals across many other LAs who are tasked with writing Information Sharing Agreeements (ISAs) to facilitate such work, often replicating that which each other are doing. Partly, this is because of a lack of coordination of advice in this area and also due to a lack of transparency with regards to Information Governance (IG) documentation in general. This research began with a detailed Literature Review into the development of EIS in the UK, followed by research to identify the Legal Gateways (LGs) to carry out EIS within LAs by gathering a wide cross section of IG documents via and FOI of LAs and criteria-based web-searches and finally carry out a thorough analysis of the key elements of ISAs. In doing so, the research would facilitate the more efficient creation of ISAs and propose a generic structure of the minimum information that should be found in such documents by carrying out a case study, with the aim of producing an exemplar. The results of this research have demonstrated that the lack of trust and 'siloization' of data-holding organisations is not assisted by the lack of transparency in this area. The research concludes that ISAs, Privacy Statements and other IG documentation that does not pose a risk to the security of data systems should be freely available on LA websites and that better advice on their creation should be provided by Central Government.

Keywords Information sharing · Local government · Legal gateways · Legislation · Freedom of information · GDPR

S. Elliott (✉)
Northumbria University London, London, UK
e-mail: simon@thefreemac.com

J. McCarthy
Bucks New University, High Wycombe, UK
e-mail: John.McCarthy2@bucks.ac.uk

H. Jahankhani and B. Issac (eds.), *Cybersecurity and Human Capabilities Through Symbiotic Artificial Intelligence*, Advanced Sciences and Technologies for Security Applications, https://doi.org/10.1007/978-3-031-82031-1_6

1 Literature Review

1.1 The Value of Effective Information Sharing

Discussions over EIS in government began in the 1980s. Wilson [1, p. 88] described this as:

> The significance of data in local government brings with it a need for the effective manipulation of those data for the production of reports to guide the decision-making and planning processes.

Dawes [2] identified the barriers to EIS within government and suggested a framework for evaluating the benefits (Fig. 1).

By 1998 the European Commission was drawing up directives to ensure these benefits were realised, albeit with a Data Protection agenda [3].

With this focus on digital government through the 1990s, Batista and Cornock [4, p. 11] described how existing methods of intra and inter-organisational EIS were threatened by new capabilities available through advances in IT. In particular, how strongly embedded functional identities became obstacles in the implementation of emerging methods of EIS.

Atabakhsh et al. [5] recognised the opportunities of EIS in law enforcement and health, however noted that there were often long processes before Memoranda of Understanding (MOU) or ISAs were in place and that, in an information age, this would need to be addressed, for more responsive government.

Akbulut [6] investigated the work of Dawes, Landsbergen, Wolken and others and showed that inter-agency trust was a key in the success of IS programmes,

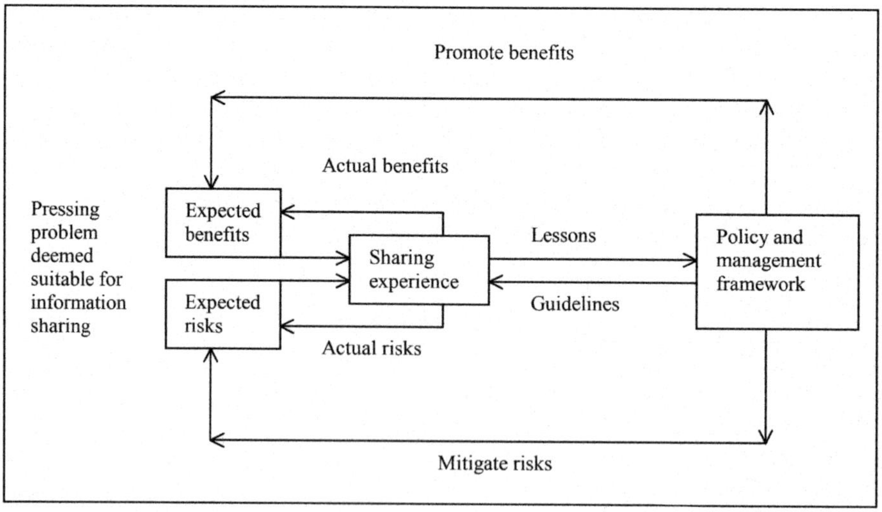

Fig. 1 Barriers to information sharing—Dawes [2]

although hypothesised that a lack of knowledge in IT capabilities led to current vendors hampering new initiatives that could impact on the use of their systems. This was compounded by objections from others resistant to change.

Turner and Higgs [7] recognised the importance of GIS data in linking data across disparate public sector data sources contrasted by the organisational and cultural barriers in overly compartmentalised entities that often hampered initiatives to bring these datasets together. Privacy concerns were often cited as reasons. Similar issues were still impacting local government in Australia in 2005, as described by McDougall, Rajabifard and Williamson.

Warren and Oppenheim [8] note the 2002 UK Government publication "*Privacy and data-sharing*" and the growth in Information Governance (IG) professionals in local government. Despite the encouragement of Government there were still gaps in organizational processes. Given initiatives over the preceding twenty years, as well as an increasing understanding of the emerging IG environment, it was disappointing that, as Hudson [9] commented, internal politics and a lack of willingness to cede professional autonomy were causing difficulties in creating an EIS hub to protect vulnerable children.

Richardson and Asthana [10] stressed the importance of effective ISAs in the development of better working between Health and Social Care. They presented four key behaviours in EIS (Fig. 2).

However, even in 2006, Riege and Lindsay identified the lack of understanding of how Knowledge Management (KM) principles were applied across the public sector and barriers that this posed to effective policy development. This mirrors the lack of development in EIS at this time.

It was therefore important that the UK Government published the Local Government White Paper [11] recognising:

Figure 1: THE FOUR LOGICAL POSSIBILITIES OF INFORMATION-SHARING BEHAVIOUR

	Information is shared inappropriately or without good cause	Information is shared with good cause and appropriately
Information is withheld inappropriately or without good cause	A High risk of breaching confidentiality and high risk of neglecting to pass on important information	B Low risk of breaching confidentiality and high risk of neglecting to pass on important information
Information is withheld with good cause and appropriately	C High risk of breaching confidentiality and low risk of neglecting to pass on important information	D Low risk of breaching confidentiality and low risk of neglecting to pass on important information

Fig. 2 Information sharing behaviour—Richardson and Asthana [10, p. 5]

> Effective cross-agency working to deliver Local Area Agreement (social inclusion) targets needs to be supported by effective data-sharing between partners.

This was echoed by Gannon-Leary et al. [12] who commented that technology would enable public services that were more joined-up but that clarity of vision would be vital in removing barriers to engagement, and that it must come from all levels within partner organisations, including front-line practitioners.

Foley et al. [13, p. 193] listed the reasons that EIS was essential to ensure LAs worked more effectively, especially better targeting of services based on understanding the individual's needs. They commented, however, that:

> Legal issues and the need for protocols are frequently perceived to be constraints on EIS.

Daniel and Ward [14] also noted local government were unprepared for moving to digital working. In particular, the impact on the citizen was explored of having to recount their story more than once, often at critical times in their lives. Integration between different services and departments were key to this.

Obstacles to effective EIS were equally impacting progress in digital government in the US. Pardo et al. [15] noted:

> Previous studies have identified challenges to efforts to create capability as ranging from technological in nature, such as data and technical incompatibility, to more social in nature such as the lack of institutional incentives to collaborate and power struggles around multi-organizational settings.

What was identified as key was the establishment of organisational IG structures to facilitate trust in EIS.

Thomas and Walport [16] concluded that the route to public confidence over their data protection was a clear code of practice and transparency for ISAs. Additionally, where consent was requested, but had conditions where it could be disregarded, they called for organisations to explain from the outset, what of their data would be used, regardless of consent, for what purposes, and how it would be safeguarded.

Bellamy et al. [17] noted that:

> Despite national policy effort to encourage better information-sharing, previous research indicates that there are many cases where information is still not shared when it should be, or where it is shared when it should not be, with potentially devastating results.

As a result of LA initiatives, Leask [18] described how practitioners were brought together in Communities of Practice to investigate how better approaches to EIS and KM could move digital government forwards. Again, organisational and cultural barriers, and in particular "knowledge is power", were identified as harmful to strategy development.

Further initiatives to up-skill local government staff in collaborative approaches were outlined by Ford and Murphy [19] who considered the strategies that would be enablers to driving organisational change as:

> A means by which services could be improved, achieving better outcomes for people and places, delivery of value for money and cost effectiveness, and efficiency.

Technical	Political
Hardware and software incompatibility.	Protection of policy-making power of administrative agencies.
Organisational	**Inter-Organizational**
Lack of human and institutional capacity for EIS.	Lack of inter-agency agreements for managing EIS.

Fig. 3 Barriers to EIS, Estevez et al. [22]

However, three years after the White Paper, Peel and Rowley [20, p. 25] could not find effective research onto the processes around EIS in multi-agency working in Children's Services. They concluded that:

> Knowledge sharing, while involving individual choice, is influenced by a complex mix of cultural, technological, organizational, legal, and regulatory factors, which are unique to specific contexts.

Hodges and Grubnic [21] argued that the difficulties in implementing digital government were often due to challenges in demonstrating that investment in infrastructure would deliver financial savings and that:

> Assessments of value maybe driven by financial considerations in a simplistic manner with expectations of high financial returns or short payback periods and little or no estimate of user benefits.

Estevez et al. [22] recognised that a framework was necessary for EIS and, through an analysis of those found in other countries, created a detailed proposal for such a scheme. The barriers to sharing that must be overcome, were described as (Fig. 3).

Šimundić et al. [23] who identified further concerns as to how information strategies were key to ensuring confidence in the correct privacy of data, whilst at the same time allowing government appropriate access to information. These privacy concerns were echoed by Fedorowicz et al. [24] who stated:

> Efforts to construct data warehouses enabling data mining of citizens' personal information obtained from other organizations create a complex challenge, since privacy concerns may vary across constituent groups whose priorities diverge from agencies' e-government goals.

Whilst it has been demonstrated that concerns around privacy had been a limiting factor in the development of effective digital government, Lips et al. [25] raised the difficulties of the management of inter-organisational cooperation on EIS, due to the need to rely on facilitative leadership to overcome strategic and organisational complexities.

An area that has not been widely commented on is the financial benefits that early sharing of information, and therefore effective intervention, can bring. This aspect of the benefits of effective EIS were commented on, with reference to preventing crime, by Curtis et al. [26], however many initiatives that bring funding to LAs are dependent on EIS, therefore it was surprising that this was not enough of an imperative to resolve the IG issues.

Bigdeli [27] commented on the role of effective EIS as core to identifying and supporting those citizens most at risk, as well as improving service delivery and business efficiency. However, Bidgeli optimistically cites the simplification of data protection regulations and government initiatives with an expectation of great changes to be seen in the short term. It is therefore interesting to see in the same year how Jenkins and Palmer [28] report that a lack of understanding of data legislation led to an unwillingness to share important information from school counsellors in the UK, and reduced the effectiveness of Safeguarding.

This was again confirmed by Ziaee et al. [29] who investigated the processes which surrounded EIS and noted that inter-organisational distrust was often a limiting factor, along with the misuse or misapplication of legislation by policymakers. Owen et al. [30] commented that, at that time:

> There are no formal co-ordinating structures and no minister charged with responsibility for co-ordinating information policy.

With this lack of leadership, it is not surprising that, whilst more concerned with governmental data, de Rosnay and Janssen [31] comment that data privacy was often still given priority over effective local government and that overlapping legislation created a reluctance in opening up EIS, until (risk averse) officers had clarity over the correct LGs.

Taylor et al. [32] similarly noted that they could not identify common EIS frameworks within the public sector. At the time, they were researching how contact by one organization of at-risk citizens could then be used to signpost to additional bodies, yet there were still barriers due to concerns over privacy, despite advice given by the Department of Justice [33] on legal processes.

The DfE [34] showed the benefit of effective EIS can be both societal and financial. 'Joining up the dots' when it comes to vulnerable families are key, and early intervention has the biggest impact, especially in reducing the likelihood of children being taken into care.

Grath-Lone et al. [35] contributed to research into this area and stressed the importance of effective EIS to ensure child protection, and documented the legislative and advisory progress that the Government was making in this area (Fig. 4).

Again in 2016, and updated in 2018, the DfE clearly stated, within "*Information Sharing*", that:

> Where practitioners need to share personal data, the DPA 2018 includes 'safeguarding of children and individuals at risk' as a condition that allows practitioners to share information **without consent**.

One issue is that information in LAs will have a variety of different Information Asset Owners (IAO) and regulatory or governance restrictions. Whilst there are acknowledged tensions to sharing certain types of information, the risks of EIS are often misinterpreted by the receiving agency [37].

Fig. 4 Virtuous circle of EIS to protect vulnerable children and families, DfE [36]

Graham et al. [38] commented on the tensions between organisations and how data that could give valuable insights was held in silos and how the integration of data systems, once the legal and organisational barriers had been overcome, had the potential to create better outcomes through digitally-driven decision making. Naghizadeh and Liu [39] compares the reluctance to share information, even when it would benefit all parties, to Prisoner's Dilemma game theory, where, instead of "*co-operating with each other for the common good*" [40], the parties continue to think of their own priorities, leading to everyone losing.

Basker and Spinks [41] analysed the development of data analytics maturity and showed why this is an area of concern and frustration to LA data analysts and officers who depend on EIS from multiple datasets to meet statutory reporting, and grant funded projects that require such information.

One area of development of ISAs was a range of research and policy development by the Centre of Excellence for Information Sharing, which closed in 2021. Investigating health data sharing, Wilson [42] noted their work and clearly linked it to the potential for a reduction in the fragmentation of IS policy, lost with its closing.

What Schmit et al. [43] make clear with their paper on EIS in the US is that this issue is not limited to the UK. They make important points regarding the negative impacts of data breaches, balanced against the many benefits of EIS and link this to the many areas that local government in the UK would cover.

This was further highlighted by Young et al. [44] pointing out that the danger of 'closed' data sources, with prohibitively complex ISAs, was that public bodies and others relied on public 'open' data sets to make business decisions, potentially resulting in skewed results when compared to finer-grained information that the closed, person-level information would have given. This could, for example, have led to potentially vulnerable people being missed from assistance.

Safeguarding young people clearly depends on information being available and shared appropriately [45, pp. 4–8]. It has been shown that this is complicated by a lack of acceptance or understanding of the legislation that permits EIS, even between different departments within one organisation. However, such legal gateways (LGs) are essential for identifying the correct routes for EIS [46]. The ICO [47] give some guidance as to process, but not wider information.

The DfE [48] reiterated:

> The first and most important consideration is always whether sharing information is likely to support the safeguarding of a child or young person.

It can be clearly demonstrated that, over nearly forty years, central government in the UK has actively encouraged EIS, backed by the correct safeguards and LGs, in order to protect the most vulnerable citizens and lead to efficient services. However, fears over losing control of data in 'silos', getting the LGs wrong, or a belief in the requirement of 'consent' has hampered progress.

When conducting data analysis of the factors affecting citizens with multiple disadvantages within Local Authorities (LAs), it is necessary to identify whether they have reached threshold triggers across many data systems. Recent initiatives in this area have included those from DLUHC [49]. It would be near impossible to achieve this without breaking down the barriers for EIS.

1.2 The Impact of the GDPR on EIS in Local Government

With the Data Protection Acts (DPA) (1984, 1998), having many commonalities, both owed their development to a need for member countries of the Council of Europe [50] to have equivalent legislation to the whole of Europe, as outlined in the Convention of 1981.

As Raab [51] described, Sir Kenneth Younger developed the basis of the UK's DPA in 1981 around the principles of the storage of data for specified, legitimate principles, amongst other aims. However, its implementation was largely to ensure that the UK could continue with access to many European structures and trade, rather than enhance privacy of data.

Charlesworth [52] describes the changes with the 1998 act and the differences that it brought, namely with the concepts of "Data Controller" and "Data Processor" as well as that the processing falls under "legitimate interests" or the functioning of government, even where consent has not been gained.

Bainbridge [53] asserted the requirements of meeting both Schedules 2 and 3 of the DPA, when dealing with sensitive data, and the use of Legitimate Interest.

Moving forwards to the GDPR [54], Adshead [55, p. 151] explains the change from "Legitimate Interests" within the GDPR and the implications for LAs, who had often relied on it. The enhanced requirements for consent are explained, as are the principles within the new legislation.

The implications of the GDPR for health and social care is discussed in detail by Shu and Jahankhani [56] and it is pointed out that, in the light of the new legislation:

> The way we create, use, share, store and discard information has greatly change over the years and this has been a problem in the health and social care sector and particularly to the Caldicott Guardian who is their responsibility to maintain patient/client confidentiality and decide on their information sharing.

The addition of the ability to process data without the explicit interest, so long as a defined LG is applicable, would be important to removing some of the roadblocks to effective EIS in LAs. It also had implications for LAs who were relying on Legitimate Interests as their basis for sharing.

In particular, the concept of "Public Task" within GDPR, as described by Carey [57] is one where:

> Processing of data will be lawful where it is necessary for the performance of a task carried out in the public interest or in the exercise of official authority vested in the controller.

The challenge is the complexity of the GDPR. Whilst the full document (2016) is 88 pages of legal language, articles 5–10 are clear on the obligations around consent and the conditions where a body is permitted to share or process data where consent has not been given, particularly the provisions in Article 6(1) and the requirement for Data Impact Risk Assessments.

Additionally, IT Governance [58] compare the changes, post-Brexit, in the transition from GDPR to DPA 2018/UK GDPR. The UK government [59] have declared all EU countries, as well as some others, as being "adequate" under UK GDPR and therefore deemed to be "safe", however one complexity that could impact LAs would be the use of EU-hosted cloud services falling out of the regulations. Curently, the ICO clearly states that [60] documentation and privacy notices should simply state the transfers of data to trusted states:

> You don't need any new arrangements for transfers from the UK to the EEA. The UK government has stated that transfers of data from the UK to the EEA are permitted. It says it will keep this under review.

1.3 The Legal Gateways Surrounding EIS in Local Government

As it has been shown, the GDPR, Article 35, require DPIA to be carried out when data is being shared or processed in new ways, or with new technologies.

From there, individual ISAs are developed for the different data sources, whether internal to the LA, but from different directorates, or business functions, or external to the LA through other partners. The work of Grabus and Greenberg [61] gives clarity to some of the key elements that can be found in most ISAs worldwide (Fig. 5).

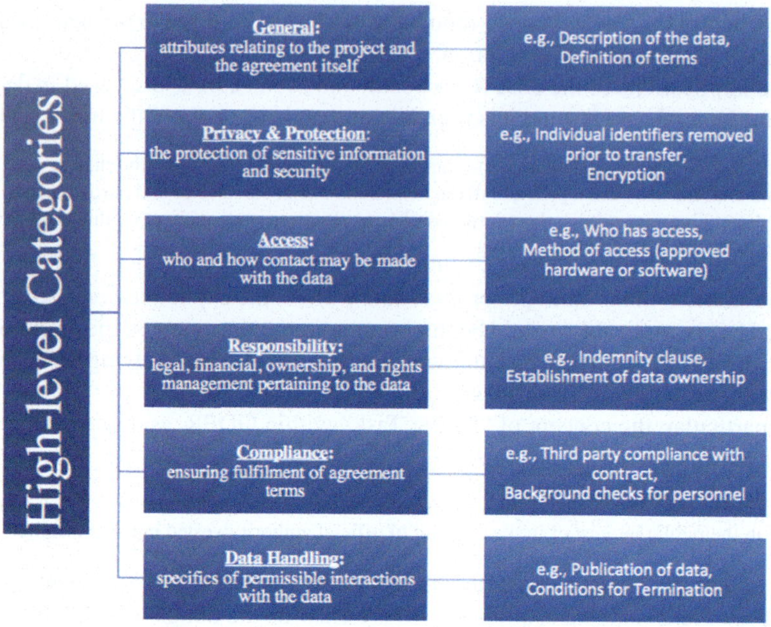

Fig. 5 High level categories in ISAs: Grabus and Greenberg [61]

Whilst many LAs consider themselves to be single bodies, and therefore not required to define intra-organisational ISAs, the legal framework annex published by the PIU [62, p. 1] makes it clear that:

> Government departments, agencies and other public bodies are each viewed as separate entities for the purpose of data-sharing and for privacy concerns. Moreover, the laws surveyed reveal that personal data collected by public bodies are regulated according to the specific purpose for which they were provided. Thus, even comparison of two sets of data collected by the same public authority for different purposes will be subject to compliance with applicable data-sharing laws.

Peel and Rowley [20, pp. 14–16] make clear that EIS is permitted to occur within multi-agency working, through complex legal and regulatory processes, however this complexity in itself can be a barrier;

> In particular, it is important to note, that while there are some overall guiding principles as to what information may be shared (or not shared), under what circumstances, and with whom, each professional group can on occasion be subject to different guidelines and requirements. This does not make for easy multi-agency working and information sharing.

When examining a wide range of publications in this area, it became clear that, when creating ISAs in local government, there have not always been clear guidelines for the correct legislative frameworks to use when creating ISAs and therefore, for a long time, cultural barriers to sharing and the division of information into siloed, organisational sub-units have hampered the ease at which this Information Intelligence, and therefore a greater level of data maturity, can occur. Busby et al. [63]

1	*What to do if you can't agree whose ISA template to use*
2	*Managing lack of capacity and pressure of time.*
3	*What if partner(s) want to go ahead and not bother with a formal arrangement.*
4	*What else needs to be done in addition to just writing an ISA?*
5	*Identify key stakeholders – the blockers and the facilitators and how to work with them.*
6	*How to keep everyone sighted on the end goal, and not being absorbed by the ISA task.*
7	*Keeping stakeholders engaged and including those who might join later.*
8	*Overcoming language barriers and considering the needs of different audiences.*

Fig. 6 The key areas of information sharing, COEIS [65]

conducted detailed research into this area, however whilst they commented on the legal barriers, they do not present solutions to these issues.

Konstantinidis et al. [64] make a thorough assessment of the complexities around data sharing permissions in the medical world, and it has been seen in programmes, such as Supporting Families, that health data is often the most difficult to obtain in the range of indicators that might show that a family has multiple disadvantages.

It was therefore encouraging that the Centre of Excellence for Information Sharing (COEIS) (https://informationsharing.org.uk/) was set up to resolve some of these uncertainties (Fig. 6).

The breadth of their research and articles showed the need for such an endeavour. However, when considering the programmes that gather data from multiple sources, the work carried out in conjunction with the Information Commissioner's Office (ICO) showed that schedule 2 of the GDPR could be satisfied through Public Task and that Schedule 3 could be considered though the work being "*in the substantial public interest*".

As has been discussed with DPIA and ISAs, it is necessary for the LA to identify the level of obligation of the task and, which Act or Statutory instrument is applicable. Whilst an LA may have well-developed IG, an attempt was made by the Department for Housing, Communities and Local Government [66] to list the obligations placed on LAs, totalling 1338 separate duties. What is notable is that this predates the implementation of GDPR yet, a decade later, there is still a lack of clarity over the justification for EIS in many LAs.

In 2018, the DfE published a document to inform practitioners on the importance of effective EIS to safeguard children [67, p. 21] where the concepts around consent, the GDPR and the priority to safeguard children is clear. An important supplementary document [68] discussed the LGs by which this could happen.

The importance of appropriate LGs was again emphasised by the DfE, who stated (2018, p. 5) that:

> The GDPR and Data Protection Act 2018 do not prevent, or limit, the sharing of information for the purposes of keeping children and young people safe.

The ICO clearly describes [69] the key elements that ISAs might include, in order to demonstrate that the EIS is being conducted lawfully:

> You need to clearly explain your lawful basis for sharing data. The lawful basis for one organisation in a ISA might not be the same as that for the other one… You need to have detailed advice about which datasets they can share, to prevent irrelevant or excessive information being disclosed.

When reading the further guidance from the ICO [47], it can be seen that there are four clear areas of LG—Express Statutory Obligations, Express Statutory Powers, Implied Statutory Powers and the Digital Economy Act [70] (DEA). Further guidance was to be found on the ICO website regarding Public Task.

This makes it very clear that, in the absence of explicit consent, a duty under 'Public Task' may be one of the only justifications for sharing without consent. It also makes clear that the DEA and associated statutory Instruments (SIs) can have a positive impact on EIS within LAs.

To support the safeguarding agenda, the DfE [67] released a document outlining the statutory frameworks available when sharing data to promote multi-agency working. This document clearly outlines a range of applicable legislation applying to Public Task decisions but, as the researcher has found, is often looked over when downloading the accompanying advisory document.

1.4 Summary of the Literature Review

As has been shown, the importance of EIS has long been observed, as have the barriers that exist. This research therefore offers the opportunity to create a starting point from which the legislative framework for sharing has been demonstrated, and so LAs should be able to argue the case for greater data maturity more effectively.

It has also been demonstrated that LGs do exist to allow EIS, including where consent has not, or cannot, be given. This does not abdicate the LA from ensuring that detailed privacy notices are easily available and describe where consent will not be expected or required.

This paper will now compare the LGs identified by LAs, showing how they are being used for a range of tasks, as well as that advice contained in government publications, to demonstrate the LGs available to LAs. The research will also demonstrate the importance of a national framework, and clearer advice, for ISAs in LAs.

2 Research Methodology, Results and Analysis

2.1 Introduction

In order to assess what LG are available to LAs, it was necessary to have four approaches:

1. An FOI request to every LA in England;

2. The use of detailed searches of.gov.uk websites using Google search algorithms;
3. Discussions with representatives of government departments;
4. Collecting official publications in the field of EIS.

In this section, the methodologies and results from each process will be presented and analysed.

2.2 *Freedom of Information (FOI) Process and Results*

A directory of LA FOI contacts was obtained from https://www.foi.directory/unitary-councils-england/. Each address was sent a correctly formulated FOI request. Where the email address was rejected as invalid, that LA's website was visited and, if necessary, the request made via a web-form.

Under the legislation, LAs would have twenty working days to comply with the request, however a reply for clarification would effectively reset the clock. Should an LA judge the request to go over eighteen hours work, it would be able to refuse the request, or offer a reduced reply that fell within the timescales.

In total, 54 unitary authorities were contacted, either via their published FOI email address (50) or the directed web-form (4).

Six (11.1%) failed to respond and a further Ten (18.5%) commented that they were looking at the request but failed to respond any further. It is therefore concerning that 29.6% of the LAs contacted failed to deliver a suitable response to an FOI within the legal timescale.

Eleven (20.4%) had no ISAs for any internal data sharing, as they did not consider them necessary, or that, in one case, all relevant services were outsourced to commercial organisations, who were responsible for IG.

Nine (16.7%) refused with reasons including that '*The response would hamper the ability to fight crime*' (3), it would '*Breach confidentiality*' (2), '*Cost too much*' (1), '*Increase the risk of a cyber attack*' (1) or '*None currently available for publication*' (2).

Twenty-one (38.8%) supplied a range of documents, however none were able to give every ISA that they used.

As has been shown, whilst an LA could be thought of as being a single entity, it would be clear from most privacy agreements that information gathered in, for example, Children's Services, would not have been intended at the point of collection to inform another LA department, such as Housing. Certainly not at a level where Personally Identifiable Information (PII) was included.

It is therefore of concern that just over 20% of the LAs replying considered that they did not need ISAs for internal EIS. With the ICO specifically requiring a DPIA where an organisation planned to, amongst other conditions [71], '*use special category data to decide on access to services*' or '*match data or combine datasets from different sources*', as would happen within the Supporting Families programme, clarification is certainly needed for LAs in this area. Clearly, many others acknowledge that ISAs *are* required where there is a change of use of data from its original purpose within an LA, even if that sharing is internal to the LA.

It is also of note that only two LAs could give a list of all ISAs that they used, however even those could not supply the actual documents on the grounds of the search exceeding the time limit allowed within FOI requests. Clearly the organisation of IG documentation within LAs is an obstacle to transparency of EIS which feeds into the distrust over the appropriateness of sharing.

2.3 Google 'Dork' Searches and Results

LA websites can, all too often, be difficult to navigate and therefore effective search algorithms were required to extract the location of any IG documentation that may be found on them but would take too long for a manual process. Google has a variety of operators that can be used to refine searches, colloquially known as "Google Dorks", as described in the Exploit Database [72].

The searches began from the premise that all would have a URL ending in .gov.uk, therefore *inurl:".gov.uk"* would limit the range of the search. With common terminology for ISAs and other IG documentation, searches were refined using such search string phrases as: *inurl:"XXXX.gov.uk" intext:"sharing" intitle:"council" filetype:pdf* (XXXX.gov.uk was replaced with the LAs own domain) and any linked documents on resulting pages downloaded in whatever file format was available, or as a PDF file export.

In total, an additional 143 ISAs and 41 general IG files were obtained. This included EIS documents from councils that refused the FOI on the grounds of harm to crime fighting capabilities, yet placed such documents on their website, albeit through complex-to-navigate menu paths. The web searches did show inconsistencies of where documents were presented on LA websites, and how difficult many LA's websites were to navigate.

2.4 HoCL Response and Advisory Documents

Having read the work of the HoCL at https://commonslibrary.parliament.uk, the researcher wrote to their constituency MP to request information around the LG available for LAs engaged in the Supporting Families Programme. The results from the response were analysed alongside the LGs identified in other advisory documentation, Acts of Parliament (Acts) and Statutory Instruments (SI).

From the web-searches, 54 advisory and supporting documents were gathered. Many of these have been used within the Literature Review however, where appropriate, advice around the legalities of EIS are considered below.

Taking the HoCL response first, the advice clearly laid out the requirements under sections 6 and 9 or the UK GDPR for EIS in the forms indicated under the research question, namely Article 6—"*Public Task*" and Article 9—"*Health and Social Care*" or "*Public Interest*". These will be used within the analysis of ISAs to further inform the appropriate LG for EIS.

The response then went on to discuss the use of the Digital Economy Act (DEA) in opening up EIS in LAs, along with advice to consider the Statutory Instrument SI 912 [73], with its specific conditions around families with multiple disadvantages, where it states:

> These Regulations relate to the disclosure of information in relation to public service delivery, pursuant to Chapter 1 of the Digital Economy Act 2017... The multiple disadvantages objective enables the disclosure of information to enable the identification of individuals or households who face multiple disadvantages.

The document "*Addressing trust in public sector data use*" [74] makes clear that, in many areas of governmental EIS, consent cannot be regarded as being effectively given, as the processing of data is a requirement for delivering necessary services, and therefore Public Task should be used, including where a third party organisation has been commissioned to carry out services on behalf of the LA, when such action is "*in the exercise of official authority*".

With specific references to the Supporting Families programme, it noted that obtaining informed consent from the families involved was not practical, however privacy notices explaining the used of data worked for most LA departments. Trust was the key factor in effective EIS. The main obstacle, however, surrounded health data, where explicit consent was required, as not covered by the DEA.

In 2011, the DfE published a guide on "*How to identify which rules and processes should be applied when sharing information.*" Which, whilst now archived, presents a useful guide to EIS process, when safeguarding is concerned (Fig. 7).

In addition, the concept of statutory duty (required for Public Task) was clearly defined as:

> "Something that a local authority must do, rather than something they can choose whether to do or not."

The Department for Communities and Local Government (DCLG) [76] gave advice surrounding data sharing within the Troubled Families programme. LAs were informed that there was a responsibility to meet the requirements of the DPA when considering DSAs and listed the LG for each type of EIS. For example, sharing information about any criminal activity within an identified family, including that of youth justice engagement, fell under Section 115 of the Crime and Disorder Act 1998.

COEIS [77] shared case studies of effective EIS and noted that diagnosing the barriers within a particular organisation was a key consideration, although not always easy to achieve.

In the COEIS paper on the views of the ICO on data sharing within the Troubled Families programme [78], the Schedule 2 conditions of Public Task and Legitimate Interests were deemed valid and, for Schedule 3, that sharing was in the "substantial public interest", utilising Section 4 of SI 417 [79].

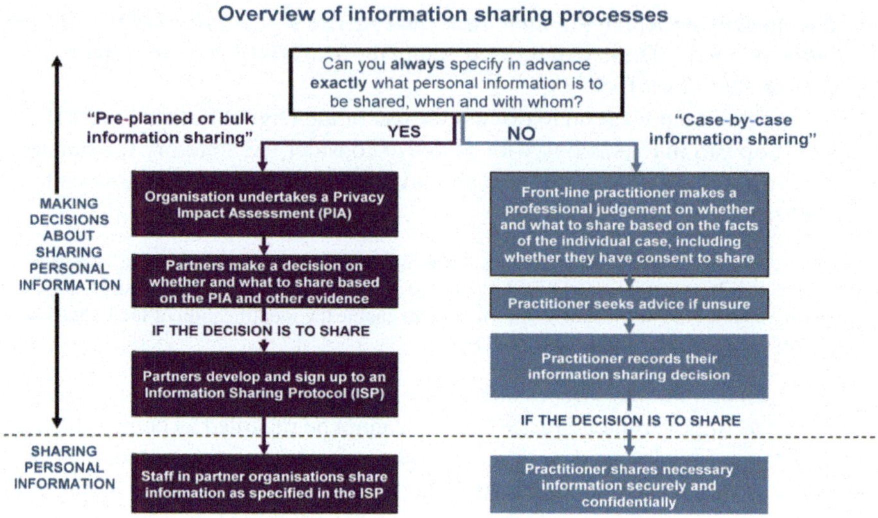

Fig. 7 How to identify which rules apply when sharing information, DfE [75]

However, the ICO makes the point that consent should not be sought when data would be shared, even when refusal occurred. Effective privacy notices on how data would be processed were more legitimate than false consent, apart from health data where it says that consent is almost always required.

Considering the NAO [80] report into Children in Care, LAs clearly had to gather information around the child from educational attendance, criminality and housing to health and substance misuse. This emphasised a legal duty to facilitate EIS.

The NHS published a Data Sharing Agreement (DSA) template [81] setting out full guidance for the creation of compliant IG documentation. Furthermore, the guidance notes that:

> "If you have a statutory duty to process personal data (because the law sets out the purpose for which you are processing personal data or you are required by law to process data for a specific purpose), then you will be a controller."

What is interesting here, given previous statements on consent and health data, the advice makes it clear that, even within the Health sector, statutory obligations can override the need for consent. Indeed, the NHS [82] explicitly warns IG professionals about the use of consent.

In its letter to the ICO, MedConfidential [83] reinforces the position that, should the DEA be used as the LG by an organisation, its use is not valid until it has been registered with the Cabinet Office. In the light of this, any ISA template or advice used within an LA should specifically state that sharing must not happen until the entry has been logged with the Cabinet Office and has appeared in the online register.

The DfE [84] make clear that the Health and Social Care Act 2022 requires data sharing between partners and that "*information about children can be easily linked and shared across organisations, such as the police, hospitals and children's social care*", again setting out the Public Task duty for EIS and emphasising the need for a single, unique child ID on all data systems to facilitate EIS, reinforcing the need to identify the correct LG In ISAs.

The ICO [85] both publishes its own Appropriate Policy Document, as well as a template, covering the requirements set out by:

> "The Data Protection Act 2018 (DPA 2018) outlines the requirement for an Appropriate Policy Document (APD) to be in place when processing special category (SC) and criminal offence (CO) data under certain specified conditions."

From these documents it is therefore clear that an LA must either have such a document in place or ensure that the detail as set out in the template is included in any ISA, DPIA or privacy notice and available, on demand, to the ICO. It also makes it clear that this information only need be included in IG documentation once within the organisation. The advisory document "*Information sharing for community safety*" from the Home Office [86] gives clear examples of LGs for the sharing of crime data, including on individuals, as well as extensive background information on the legal responsibility which, whilst there have been subsequent changes, elements such as the Caldicott principles are still relevant.

2.5 *ISA Matrix Analysis Summary*

It is first necessary to define the basic elements within each type of IG document in order to classify the responses in a quantitative and/or qualitative way. Through the use of a research matrix, as described by Klopper, et al. [87] and Choguill [88] to ensure a consistent approach, a matrix grid of 13 columns would allow the ISA documents to be analysed, however, due to the size of the final analysis grid, the results would be broken down into the following sub-categories. This would allow the research could be used to inform EIS in practice

1. The type of local government institution
2. The purpose of the ISA
3. A list of the Information Asset Owners, Data Controllers and Data Processors
4. Whether the ISA concerns Special Category Data
5. The status of the EIS under GDPR
6. Any statutory powers or LGs being applied
7. Data retention and deletion policies
8. Data breach procedures
9. How and when the ISA will be reviewed

In order to respect confidentiality, the name of the LA would be redacted from any published tables.

Once each ISA was analysed, it was determined that there were 145 unique documents, as some that were gathered were duplicates by many LAs, with just a simple change of name on the front page. In order that this did not skew the findings, only

Table 1 ISA retention periods

Only held as long as necessary	54%
Retention period not discussed	21%
The duration of the project/ISA	6%
One year	3%
Seven years	2%
Five years	2%
Other (< 1% each)	12%

unique documents were taken through to the next stage. The ISAs ranged from 2007 to 2023, with 71% having been published since the 2018 DPA.

The first analysis carried out was on the security or structural elements of the ISA, where there should be less dependency on the type of data being examined.

2.5.1 Data Retention Periods

Despite the many layers of regulations within the GDPR, there are no defined limits upon which the length of time that personal data is held, only that the retention time is justified and clearly documented. This justification must be based on any legislation surrounding data retention, and why the processing of data might need such data retained (Table 1).

Of the 21% of ISAs that were analysed to have no clear discussion of data retention, it may be that this was documented elsewhere, such as on privacy agreements, however for clarity, and to remind users, it would be beneficial if this was included on all ISAs.

2.5.2 Data Breach Procedures

As have been widely documented in the press, data breaches of sensitive data held by public bodies, contracted services and suppliers are, unfortunately, not uncommon. Up to Q4 of 2022, the ICO listed [89] nearly 47,000 reported breaches from the sectors included in this ISA. Of these, over 19% took more than one week to report the breach, over the 72 h requirement under the GDPR (Table 2).

It is therefore concerning that 41% of the ISAs did not clearly specify timescales. A legal requirement to report breaches within 72 h, where feasible, came into force on 25 May 2018, however it should be noted that the 41% who did not specify a breach reporting deadline is made up of 14% Pre 2018 and 27% 2018 and after.

As with data retention, clarity and transparency would not only made user clear of their responsibilities but also give reassurance to users.

Table 2 ISA data breach procedures

Data breach procedures not discussed	41%
As soon as possible	18%
Within 72 h	16%
Within 24 h	15%
Immediate notification	3%
Within 48 h	3%
Other (< 1% each)	4%

Table 3 ISA review frequency

Reviewed annually	48%
Review frequency not discussed	36%
Biannual	7%
Six months	3%
Every three years	3%
Other (< 1% each)	3%

2.5.3 Document Review Frequency

There is only a requirement to review IG documentation and associated contracts on a 'regular' basis. A standardised format for some elements of ISAs would assist in users within the organisation, as well as others, understand the processes being undertaken, as it was hard to find review periods in many (Table 3).

2.5.4 Measures to Ensure Security of Processing

See Tables 4 and 5.

Table 4 ISA data transfer methods

Data transfer method	%
Secure eMail or encrypted transfer	44
Encrypted transfer	25
Data transfer method not discussed	18
Access via encrypted or secure system	5
Secure eMail	5
Other (< 1% each)	3

Table 5 ISA additional security methods

Additional security requirements—systems	%
No security standard or requirements discussed	64
ISO27001 or equivalent required	15
Must have 'sufficient' security procedures	9
As per the NHS Data Security and Protection Toolkit	6
Access controls	2
Other (< 1% each)	4

Table 6 ISA staff security procedures

No security standard or requirements discussed	45%
Sufficient staff training	15%
Only vetted staff may access	11%
Only staff who need to see the data—'need to know'	9%
User-level access	3%
Single point of contact	2%
Other (< 1% each)	15%

2.5.5 Additional Security Requirements—Users

Article 32 of the UK GDPR discusses security methods appropriate to the data being processed, via suggested codes of conduct (Article 40) or certification programmes (Article 42) and the appropriateness of the user accessing the data (Table 6).

18% of the ISAs did not state how data would be securely transferred, 64% did not specify a security certification system or standard to follow, and 45% did not clearly state any requirements on which users would be able to access the data. Whilst it would be fair to assume that public bodies have appropriate security policies, it would be beneficial if these were clearly articulated within the ISAs.

2.5.6 What Type of Data is Being Shared?

The category of data shared closely matched the type of partners involved in the sharing. What was surprising was the lack of detail in many ISAs reviewed. Some listed every data item that would be included, however most spoke in more general terms. For transparency, give insights into the appropriate LGs, and to clearly define the scope of the data included within the ISA, a full list of the included data items would be beneficial (Table 7).

Table 7 ISA data categories

Personal data about children and families for safeguarding purposes	17%
Personal and Crime Data	15%
Personal data including education	11%
Personal, Health and Crime Data	8%
Personal data including Education, Crime, Health, Housing, Financial	6%
Personal Data	5%
Personal Data including Health Data	5%
Personal, Health and Social Care Data	3%
Personal Data including Domestic Abuse	3%
Personal Data including Health and Social Care	3%
Personal and Property Data	2%
Personal Data including Crime and Health	2%
Other groups of data (< 1.5%)	18%

Table 8 ISA use of consent

Consent is gathered; however it may be overridden in some cases	39%
No explicit consent is required	34%
Consent is not discussed	14%
Explicit consent is required	13%

2.5.7 Consent

As already shown by Thomas and Walport [16], the concept of false consent breaks down public trust in EIS. 39% of ISAs discussed the use of consent, however almost all had cases where this may overridden. Whilst it was clear that there were valid reasons for consent not being relied upon, it would be better if it was clearly identified in the privacy agreements, with any potential opt-outs, where appropriate (Table 8).

As many of the ISAs were dealing with data on children, there are also complexities around consent that must be dealt with, especially when considering special category data.

2.5.8 Special Category Data

Whilst several ISAs did breakdown the special category data that would be processed, most were far more generalist. None that were analysed mentioned political affiliations or union membership (Table 9).

Many were also dealing with data on children and therefore, as shown by the ICO [54], there are complexities of consent for those under 13, where consent may

Table 9 ISA special category data

All data as defined under Special Category, except Political and Union	52%
Various combinations of special category data	33%
Type of special category data not specified	15%

be required from someone with legal parental responsibility, and over 13, where the child needs to be consulted, if they have sufficient understanding of the request. The ISAs did not, in general, make clear how children's data would be dealt with differently to adults'.

2.5.9 Application of Data Protection Legislation

The LGs for processing personal and special category data were gathered from the ISAs and are presented below. With the major changes brought in by the UK GDPR 2018, the findings will be presented in two sections based on whether the ISA was written pre or post 2018.

Given the majority of the ISAs analysed focused on GDPR Articles 6 and 9, the data was analysed further to ascertain which clauses were used and to what frequency, for the 76 that specified Articles 6 and 9 (Tables 10, 11 and 12).

As can be seen, 87% of ISAs selected 6(1)(e) "*Public Interest*" and 25% cited 6(1)(c) "*Legal Obligation*".

61% chose 9(2)(g) "*processing is necessary for the performance of a task carried out in the public interest or in the exercise of official authority vested in the controller*" as the main clause for processing special categories of personal data. It is notable that only 12% cited 6(1)(a) "*Consent*" and only 7% cited 9(2)(a) "*Explicit Consent for Special Category Data*".

Table 10 ISA pre-2018 data protection legislation

Pre 2018 (43 responses)	%
Stated the need for compliance under Data Protection Act	58
No mention of compliance with any specific Data Protection legislation	12
Use of a specific clause or other comment regarding DPA legislation	30

Table 11 ISA post-2018 data protection legislation

Post 2018 (103 responses)	%
Application of GDPR Articles 6 and or 9	76
Stated the need for compliance under Data Protection legislation	17
Other legislation	7
No mention of compliance with any specific Data Protection legislation	3

Table 12 ISA General Data Protection Regulations article usage

				Date of ISA publication						Article 6 clauses									Article 9 clauses									
GDPR articles cited in ISAs	No.	%		2018	2019	2020	2021	2022	2023	a	b	c	d	e	f	g	h	i	a	b	c	d	e	f	g	h	i	j
6(1)(e) 9(2)(g)	14	18		1	1	3	3	4	2					14											14			
6(1)(c) 9(2)(g)	4	5				1		2	1			4																
6(1)(c)(e) 9(2)(g)	4	5			1			3				4		4											4			
6(1)(e)	4	5		1	0		1	2						4														
6(1)(e) 9(2)(c)(e)	4	5						4						4							4		4					
6(1)(c)(e) 9(2)(b)(g)	3	4					1	2				3		3						3					3			
6(1)(e) 9(2)(h)	3	4				1	1		1					3												3		
6(1)(c)	2	3				2						2								2	2				2	2	2	
6(1)(c)(e) 9(2)(g)(h)	2	3				1		1				2		2											2	2		
6(1)(e) 9(2)(g)(h)	2	3						1	1					2											2	2		
6(1)(e) 9(2)(i)	2	3						2						2													2	
6(1)(a)	1	1		1						1																		
6(1)(a) 9(2)(g)	1	1					1			1															1			
6(1)(a)(b)(c)(d) 9(2)(a)(c)(g)(h)	1	1				1				1	1	1	1						1		1				1	1		
6(1)(a)(b)(c)(d)(e)(f) 9(2)(a)(b)(c)(g)(h)(j)	1	1						1		1	1	1	1	1	1				1	1	1				1	1		1
6(1)(a)(c)(d)(e)(f) 9(2)(b)(c)(h)	1	1						1		1	1	1	1	1	1					1	1					1		
6(1)(a)(e)	1	1						1		1	1	1	1	1	1													
6(1)(a)(e) 9(2)(a)	1	1			1					1				1					1									

(continued)

Table 12 (continued)

				Date of ISA publication						Article 6 clauses									Article 9 clauses									
GDPR articles cited in ISAs	No.	%		2018	2019	2020	2021	2022	2023	a	b	c	d	e	f	g	h	i	a	b	c	d	e	f	g	h	i	j
6(1)(a)(e) 9(2)(a)(c)(h)	1	1					1			1				1					1		1					1		
6(1)(a)(e) 9(2)(g)(h)	1	1						1		1				1											1	1		
6(1)(b) 9(2)(e)	1	1			1						1												1					
6(1)(c)(d)(e) 9(2)(b)(c)(g)(h)(i)	1	1					1					1	1	1											1			
6(1)(c)(d)(e) 9(2)(c)(g)(h)	1	1					1					1	1	1						1	1				1	1	1	
6(1)(c)(d)(e) 9(2)(g)	1	1					1					1	1	1											1			
6(1)(c)(d)(e) 9(2)(b)(c)(g)(h)(i)	1	1					1					1	1	1						1	1				1	1	1	
6(1)(c)(d)(e)(f) 9(2)(b)(c)(g)(h)	1	1						1				1	1	1	1					1	1				1	1		
6(1)(c)(d)(e)(f) 9(2)(c)(g)(h)(i)	1	1						1				1	1	1	1						1				1	1	1	
6(1)(c)(e) 9(2)(a)(b)(c)(g)(h)(i)	1	1				1						1		1					1	1	1				1	1	1	
6(1)(c)(e) 9(2)(b)	1	1						1				1		1						1								
6(1)(c)(e) 9(2)(c)(e)(g)	1	1							1			1		1							1		1		1			

(continued)

Table 12 (continued)

				Date of ISA publication						Article 6 clauses									Article 9 clauses									
GDPR articles cited in ISAs	No.	%		2018	2019	2020	2021	2022	2023	a	b	c	d	e	f	g	h	i	a	b	c	d	e	f	g	h	i	j
6(1)(c)(e) 9(2)(c)(h)(i)	1	1			1							1		1							1					1	1	
6(1)(c)(e) 9(2)(f)(g)	1	1				1						1		1										1	1			
6(1)(c)(e) 9(2)(g)(h)(i)(j)	1	1					1					1		1											1	1	1	1
6(1)(c)(e) 9(2)(g)(i)	1	1					1					1		1											1		1	
6(1)(c)(e) 9(2)(g)(j)	1	1						1				1		1											1			1
6(1)(c)(e) 9(2)(h)	1	1							1			1		1												1		
6(1)(c)(e) 9(2)(h)(i)	1	1							1			1		1												1	1	
6(1)(c)(f)(g)(h)(j) 9(2)(b)(g)(h)(i)(j)	1	1						1				1	1	1	1	1	1	1		1					1	1	1	1
6(1)(d)(e) 9(2)(c)(g)	1	1							1				1	1							1				1			
6(1)(e) 9(2)(b)	1	1			1									1						1								
6(1)(e) 9(2)(c)(e)(h)(i)	1	1						1						1							1	1				1	1	
6(1)(e)(f) 9(2)(b)(g)(h)(i)(j)	1	1							1					1	1					1					1	1	1	1
6(1)(e)(f) 9(2)(h)	1	1					1							1	1											1		
	76	100	No	3	6	11	15	31	10	9	5	36	12	66	8	1	1	1	5	15	19	1	6	1	46	27	15	5
			%	4	8	14	20	41	13	12	7	47	16	87	11	1	1	1	7	20	25	1	8	1	61	36	20	7

Given the comments earlier regarding false consent, and the necessity for much of the data sharing identified to be needed for an LA to perform its statutory duties, it may be unsurprising that the HoCL response to the researcher's request identified 6(1)(e) and 9(2)(g) as suitable gateways for the use of data in compliance with UK GDPR.

2.5.10 Application of Legal Gateways

The next analysis was focused on the LGs that LAs stated to qualify legal obligation, public task or other criteria around statutory duties. There were 354 separate pieces of legislation or statutory guidance listed, including where multiple articles were used as gateways (Table 13).

The most common cited were as follows:

When considering the Children Act, for example, multiple sections of both the 1989 and 2004 Acts were cited.

The most commonly use Acts were described in more detail on a website created as part of the research at https://thefreemac.com/docs/, including how the relevant sections should be applied, using advice from the HoCL response and the DfE publication [68] provide information at a Section or Article level within each Act, in order to assist others developing ISAs within LAs.

It was clear from the analysis that, whilst technically accurate that a piece of legislation, such as an Act, was being quoted as an LG, many LAs did not specify which part of the legislation was being used. Where the sections were cited, there was often little explanation as to how they applied to the EIS in question, although some LAs were far more detailed in explaining how the legislation would be applied, as well as clear direction of its purpose.

3 Developing an Exemplar ISA Through a Case Study

3.1 Introduction

From the analysis of the gathered documents, the next step was to use the insights gained in section 2, take a given EIS scenario, "The Effective Sharing of Key Adult and Child Social Care and Educational Indicators with Key Housing Partners", and create a new ISA. The process would then be examined to show how the organisation (LA) would benefit from this approach.

Table 13 ISA legislative gateways usage

Acts, statutory instruments and statutory guidance	No.	%
Children Act	144	13.15
Crime and Disorder Act	67	6.12
Education Act	55	5.02
Data Protection Act	36	3.29
Local Government Act	36	3.29
National Health Service Act	29	2.65
Working Together to Safeguard Children	28	2.56
Localism Act	24	2.19
Criminal Justice Act	23	2.10
Health and Social Care Act	23	2.10
Care Act	19	1.74
Digital Economy Act	18	1.64
Children and Families Act	17	1.55
Housing Act	16	1.46
Mental Capacity Act	16	1.46
None Indicated	14	1.28
Police and Criminal Evidence Act	14	1.28
Welfare Reform Act	14	1.28
Childcare Act	13	1.19
Children (Leaving Care) Act	12	1.10
Children and Social Work Act	12	1.10
Civil Contingencies Act	12	1.10
Education and Skills Act	12	1.10
Learning and Skills Act	12	1.10
Human Rights Act	11	1.00
Mental Health Act	11	1.00
Immigration and Asylum Act	10	0.91

3.2 *The Process of ISA Creation*

When creating the exemplar ISA, the research and guidance was of great benefit in not only improving the efficiency of the process, but also in ensuring that there was sufficient detail in the key areas identified as deficient in many of the ISAs examined.

By clearly listing elements, such as frequency of review, consent and security, the document will provide the reader with confidence in the EIS process, and be more willing to share data. In particular, the analysis of the lawful basis for sharing under the UK General Data Protection Regulations enabled the section in the case study ISA to be written far more efficiently.

The biggest change was in the speed of identification of the appropriate LGs. Through the analysis of the body of ISAs, the appropriate sections of legislation were identified. When combined with the information gathered on the usage of LGs, this was created more efficiently.

What was important in this area was the ability, using the range of advice identified, to define not only the legislation and the applicable sections, but the reason each item was relevant to the proposed use.

In this way, the ISA was created in under an hour of a usable standard at the first draft, and is included at https://thefreemac.com/docs/the-effective-sharing-of-key-adult-and-child-social-care-and-educational-indicators-with-key-housing-partners/.

4 Conclusions, Recommendations and Further Work

4.1 *Conclusions*

This project set out to identify how EIS in LAs could be better facilitated through the identification of appropriate LGs. Through a thorough literature review, it was clear that trust was as important as efforts by central government to improve EIS.

Once the focus had been refined, the process of gathering IG documentation began and, despite the lack of transparency in this area, a wide range of documents were acquired. When these were analysed, on the basis of the content that was recommended to be present in ISAs, it was demonstrated that many documents had key areas missing, or with insufficient detail.

It should also be noted that nearly a third of LAs contacted via an FOI process failed to respond or deliver any documents. Of the remaining LAs, none could supply all ISAs held.

Using the principles defined through the literature review and document analysis, as well as the advice given through the HoCL, an ISA was created for a given scenario. This demonstrated that the use of a clearly defined template, appropriate guidance and simple language can lead to the more efficient creation of effective ISAs.

4.2 *Recommendations*

This report proposes there should be a standardised ISA template for LAs and government departments to use, with a minimum set of areas to be defined, together with standards on what each section must hold. Where the EIS forms part of a central government department project, all IG documentation required should be provided by the department concerned, including any applicable LGs and the basis for lawful sharing. All ISAs should be then be publicly available via LA websites, located at a standardised URL path to facilitate transparency.

Guidance on consent should be made available, as it was clear from the IG analysed in this research that data was essential for the LA operation, yet consent was implied in the ISAs. As reported earlier on, false consent can reduce trust and more open privacy statements are more effective.

Given that no full analysis of the statutory duties and legislation has been carried out since 2011, as stated earlier, the advice from the HoCL showed how effective such guidance could be. In order to unlock EIS further, there should be a new analysis of LAs statutory duties with clear advice on which parts of stated legislation should be used to facilitate EIS for each duty.

References

1. Wilson TD (1984) Information in local Government: challenge and opportunity. In: Information for national recovery: needs—resources—technology, University of Stirling, Scotland, 13–16 Sept 1983. Accessed online on 22 June 2023 at https://www.semanticscholar.org/paper/Information-in-local-government%3A-challenge-and-Wilson/06945d4e883b564375b43a7479c222aafd8c0e41
2. Dawes S (1996) Interagency information sharing: expected benefits, manageable risks. J Policy Anal Manag 15(3):377–394. Accessed online on 29 Jan 2023 at https://doi.org/10.1002/%28SICI%291520-6688%28199622%2915%3A3%3C377%3A%3AAID-PAM3%3E3.0.CO%3B2-F?saml_referrer
3. Buckert H (1992) The legal framework of public sector information: recent legal policy developments in the EC. Gov Publ Rev 19(5):483–496. Accessed online on 19 June 2023 at https://www.sciencedirect.com/science/article/pii/027793909290051C
4. Batista L, Cornock M (2009) Information sharing in e-government initiatives: freedom of information and data protection issues concerning local government. J Inf Law Technol. Accessed online on 01 June 2023 at http://oro.open.ac.uk/20896/2/889B091E.pdf
5. Atabakhsh H, Larson CA, Petersen T, Violette C, Chen H (2004) Information sharing and collaboration policies within Government agencies. In: Intelligence and security informatics. Accessed online on 01 June 2023 at https://www.semanticscholar.org/paper/Information-Sharing-and-Collaboration-Policies-Atabakhsh-Larson/de3dcc3ea0fb68c90b5be5310f9d3722af118250
6. Akbulut AY (2002) An investigation of the factors that influence electronic information sharing between state and local agencies. Dissertation, Louisiana State University. Accessed online on 29 June 2023 at https://digitalcommons.lsu.edu/cgi/viewcontent.cgi?article=1876andcontext=gradschool_dissertations
7. Turner P, Higgs G (2003) The use and management of geographic information in local e-government in the UK. Inf Polity 8(3–4):151–165. Accessed online on 20 June 2023 at https://www.semanticscholar.org/paper/The-use-and-management-of-geographic-information-in-Turner-Higgs/b7a244f99e6d924eed1f3139db7e2cb407eb597f
8. Warren A, Oppenheim C (2004) Integration of roles? Implementing new information laws in UK public organizations. J Inf Sci 30:48–59. Accessed online on 01 June 2023 at https://www.semanticscholar.org/paper/Integration-of-Roles-Implementing-New-Information-Warren-Oppenheim/c2c54f154913a6c04fa1008293128533ade77ead
9. Hudson B (2005) Information sharing and children's services reform in England: can legislation change practice? J Interprof Care 19(6):537–546. Accessed online on 29 Jan 2023 at https://doi.org/10.1080/13561820500412114
10. Richardson S, Asthana S (2005) Policy and legal influences on inter-organisational information sharing in health and social care services. J Integr Care 13(3):3–10. Accessed online on 01 June 2023 at https://doi.org/10.1108/14769018200500019/full/html

11. Department for Communities and Local Government (2006) Strong and prosperous communities: the local government white paper. Cm 6939-I. HMSO, London. Accessed online on 01 June 2019 at https://assets.publishing.service.gov.uk/government/uploads/system/uploads/attachment_data/file/272357/6939.pdf
12. Gannon-Leary P, Baines S, Wilson R (2006) Collaboration and partnership: a review and reflections on a national project to join up local services in England. J Interprof Care 20:665–674. Accessed online on 22 June 2023 at https://www.semanticscholar.org/paper/Collaboration-and-partnership%3A-A-review-and-on-a-to-Gannon-Leary-Baines/2db63938cb3b7a0f4252c186bf68ccd8b1fa93a8
13. Foley P, Alfonso X, Sakka MA (2006) Information sharing for social inclusion in England: a review of activities, barriers and future directions. J Inf Commun Ethics Soc 4:191–203. Accessed online on 01 June 2023 at https://www.semanticscholar.org/paper/Information-sharing-for-social-inclusion-in-A-of-Foley-Alfonso/db18ed306dfd22088c10c8b2f9a18ed12afe617f
14. Daniel EM, Ward JM (2006) Integrated service delivery: exploratory case studies of enterprise portal adoption in UK local Government. Bus Process Manag J 12:113–123. Accessed online on 01 June 2023 at https://www.semanticscholar.org/paper/Integrated-service-delivery%3A-Exploratory-case-of-in-Daniel-Ward/e16c4215d021afebccb04e90003745b6cc2ffcfc
15. Pardo TA, Gil-Garcia J, Burke GB (2008) Governance structures in cross-boundary information sharing: lessons from state and local criminal justice initiatives. In: Proceedings of the 41st annual Hawaii international conference on system sciences (HICSS 2008), pp 211–211. Accessed online on 22 June 2023 at https://www.semanticscholar.org/paper/Governance-Structures-in-Cross-Boundary-Information-Pardo-Gil-Garcia/243bd9ea1b6f19da43383a16f2fc83939b8c7126
16. Thomas R, Walport M (2008) Data sharing review report. Her Majesty's Revenue and Customs (HMRC), London. Accessed online on 01 June 2023 at https://www.statewatch.org/media/documents/news/2008/jul/uk-data-sharing-review.pdf
17. Bellamy C, Raab C, Warren A, Heeney C (2008) Information-sharing and confidentiality in social policy: regulating multi-agency working. Public Adm 86(3):737–759. Accessed online on 01 June 2023 at https://www.researchgate.net/publication/227672704_Information-Sharing_and_Confidentiality_in_Social_Policy_Regulating_Multi-Agency_Working
18. Leask M (2009) Web 2.0 and knowledge management for local government in England—a model for the public sector? Accessed online on 22 June 2023 at https://www.semanticscholar.org/paper/Web-2.0-and-knowledge-management-for-local-in-%E2%80%93-A-Leask/29a3fc0df9e97ffc7e96ec3ebb96d98d954a3a82
19. Ford N, Murphy G (2008) Evaluating the potential of a web-based portal to support the leverage of professional expertise across local authority boundaries. Local Gov Stud 34:397–418. Accessed online on 22 June 2023 at https://doi.org/10.1080/03003930802044601
20. Peel M, Rowley JE (2010) Information sharing practice in multi-agency working. Aslib Proc 62:11–28. Accessed online on 01 June 2023 at https://www.semanticscholar.org/paper/Information-sharing-practice-in-multi-agency-Peel-Rowley/13f19bc736fe271ce452f0e0ba76090269457e79
21. Hodges R, Grubnic S (2010) Local authority e-government partnerships in England: a case study. ERN: Other Political Economy: National. Accessed online on 22 June 2023 at https://www.semanticscholar.org/paper/Local-Authority-E-Government-Partnerships-in-A-Case-Hodges-Grubnic/51343487b9131d32a2abfe157262a70a2a670f96
22. Estevez E, Fillottrani P, Janowski T (2010) Information sharing in government: conceptual model for policy formulation. Accessed online on 22 June 2023 at https://www.researchgate.net/publication/264873582_Information_Sharing_in_Government_-_Conceptual_Model_for_Policy_Formulation
23. Šimundić S, Boban M, Sinkovic Z (2010) The regulatory and legal framework of information security and right to access information in government, local government and public services. In: The 33rd international convention MIPRO, pp 1328–1332. Accessed online on 01 June

2023 at https://www.semanticscholar.org/paper/The-regulatory-and-legal-framework-of-information-%C5%A0imundi%C4%87-Boban/2d115cf1806154d4ee4024de768f89daa85d30c7
24. Fedorowicz J, Gogan JL, Culnan MJ (2010) Barriers to interorganizational information sharing in e-government: a stakeholder analysis. Inf Soc 26:315–329. Accessed online on 19 June 2023 at https://www.semanticscholar.org/paper/Barriers-to-Interorganizational-Information-Sharing-Fedorowicz-Gogan/196ab44bb53b348ac84adc96deefd5b7fc0f4af1
25. Lips A, O'Neill R, Eppel E (2011) Cross-agency collaboration in New Zealand: an empirical study of information sharing practices, enablers and barriers in managing for shared social outcomes. Int J Public Adm 34:255–266. Accessed online on 19 June 2023 at https://www.semanticscholar.org/paper/Cross-Agency-Collaboration-in-New-Zealand%3A-An-Study-Lips-O'Neill/e591cef6493fc38c7b26da083663ba8120dcb42e
26. Curtis F, Shepherd J, Brennan I, Simon T (2011) Effectiveness of anonymised information sharing and use in health service, police, and local government partnership for preventing violence related injury: experimental study and time series analysis. BMJ 342(7812):1405. Accessed online on 01 June 2023 at https://www.researchgate.net/publication/51226217_Effectiveness_of_anonymised_information_sharing_and_use_in_health_service_police_and_local_government_partnership_for_preventing_violence_related_injury_Experimental_study_and_time_series_analysis
27. Bigdeli AZ (2012) Inter-departmental information sharing in local government authorities (LGAs): the case of the United Kingdom. Accessed online on 22 June 2023 at https://www.semanticscholar.org/paper/Inter-departmental-information-sharing-in-local-%3A-Bigdeli/a05dab2f87245d9f0fd50b6f735dada6365ebb47
28. Jenkins P, Palmer J (2012) At risk of harm'? An exploratory survey of school counsellors in the UK, their perceptions of confidentiality, information sharing and risk management. Brit J Guidance Counselling 40(5):545–559. Accessed online on 29 Jan 2023 at https://doi.org/10.1080/03069885.2012.718732
29. Ziaee B, Kamal M, de Cesare S (2012) Electronic information sharing in local Government: the decision-making process. In: ECIS 2012 Proceedings. Accessed online on 01 June 2023 at https://aisel.aisnet.org/ecis2012/90
30. Owen BB, Cooke L, Matthews G (2013) The development of UK government policy on citizens' access to public sector information. Inf Polity 18:5–19. Accessed online on 01 June 2023 at https://www.semanticscholar.org/paper/The-development-of-UK-government-policy-on-access-Owen-Cooke/a60a818fc2b0699b3fc7e826634158a6b4046b08
31. de Rosnay MD, Janssen K (2014) Legal and institutional challenges for opening data across public sectors: towards common policy solutions. J Theory Appl Electron Commer Res 9:1–14. Accessed online on 01 June 2023 at https://www.semanticscholar.org/paper/Legal-and-Institutional-Challenges-for-Opening-Data-Rosnay-Janssen/5f6400912fcc88b3dcafd8f9040a5da87fe53711
32. Taylor MJ, Higgins E, Lisboa P, Arshad F (2014) Developing a data sharing framework: a case study. Transforming Gov People Proc Policy 1(8):151–164. ISSN: 1750-6174. Accessed online on 01 June 2023 at https://researchonline.ljmu.ac.uk/id/eprint/190/3/Datashareframev15mainpaperpostrefereedversion.pdf
33. Department of Justice (DOJ) (2013) Public sector data sharing: guidance on the Law. Accessed online on 23rd June 2023 at https://webarchive.nationalarchives.gov.uk/ukgwa/20150730125042/. http://www.justice.gov.uk/downloads/information-access-rights/data-sharing/annex-h-data-sharing.pdf
34. Department for Education (DfE) (2014) Children in care. Accessed online on 30 Jan 2023 at https://www.nao.org.uk/wp-content/uploads/2014/11/Children-in-care1.pdf
35. Grath-Lone LM, Woodman J, Gilbert R (2015) Safeguarding children and improving their care in the UK. Lancet (Br Edn) 386(10004):1630–1630. Accessed online on 01 June 2023 at https://www.thelancet.com/journals/lancet/article/PIIS0140-6736(15)00538-3/fulltext

36. Department for Education (DfE) (2016) Information sharing to protect vulnerable children and families. Accessed online on 30 Jan 2023 at https://assets.publishing.service.gov.uk/government/uploads/system/uploads/attachment_data/file/554128/Information-sharing-to-protect-vulnerable-children-and-families.pdf
37. Perri G, Bellamy C, Raab C (2010) Information-sharing dilemmas in public services: using frameworks from risk management. Policy Polit 38(3):465. Accessed online on 01 June 2023 at https://bristoluniversitypressdigital.com/view/journals/pp/38/3/article-p465.xml
38. Graham FS, Gooden S, Martin KJ (2016) Navigating the transparency–privacy paradox in public sector data sharing. Am Rev Public Adm 46:569–591. Accessed online on 18 June 2023 at https://www.semanticscholar.org/paper/Navigating-the-Transparency%E2%80%93Privacy-Paradox-in-Data-Graham-Gooden/f795e7dea38fd2a746f58c3f291ceeb941e9e964
39. Naghizadeh P, Liu M (2016) Using private and public assessments in security information sharing agreements. IEEE Trans Inf Forensics Secur 15:1801–1814. Accessed online on 29 June 2023 at https://www.semanticscholar.org/paper/Using-Private-and-Public-Assessments-in-Security-Naghizadeh-Liu/3d5cfb5a2ea475c1ffa40eef72fe7273da5b731a
40. Szilagyi M (2003) An investigation of N-person prisoners' dilemmas. Complex Syst 14:155–174. Accessed online on 29 June 2023 at https://content.wolfram.com/uploads/sites/13/2018/02/14-2-3.pdf
41. Basker S, Spinks M (2016) Data maturity in the social sector. Dissertation, Louisiana State University and Agricultural and Mechanical College. Accessed online on 22 June 2023 at https://digitalcommons.lsu.edu/cgi/viewcontent.cgi?article=1876andcontext=gradschool_dissertations
42. Wilson R (2019) Debate: information sharing is dead-long live information sharing! Current challenges and looking ahead. Pub Money Manag 39(5):325–326. Accessed online on 29 Jan 2023 at https://doi.org/10.1080/09540962.2019.1611234
43. Schmit C, Kelly K, Bernstein J (2019) Cross sector data sharing: necessity, challenge, and hope. J Law Med Ethics 47(2_suppl):83–86. Accessed online on 29 Jan 2023 at https://www.proquest.com/docview/2257702906/1FCBB188F7EF4228PQ/1?accountid=12860
44. Young M, Rodriguez L, Keller E, Sun F, Sa B, Whittington J, Howe B (2019) Beyond open vs. closed: balancing individual privacy and public accountability in data sharing. In: Proceedings of the conference on fairness, accountability, and transparency. Accessed online on 29 June 2023 at https://www.semanticscholar.org/paper/Beyond-Open-vs.-Closed%3A-Balancing-Individual-and-in-Young-Rodriguez/070cd335c36dfee145ba4dafa77c949232e5fffe
45. Department for Education (DfE) (2018) Information sharing: advice for practitioners providing safeguarding services to children, young people, parents and carers. Accessed online on 01 Mar 2023 at https://assets.publishing.service.gov.uk/government/uploads/system/uploads/attachment_data/file/1062969/Information_sharing_advice_practitioners_safeguarding_services.pdf
46. OECD (2021) Fighting tax crime—the ten global principles, 2nd edn. OECD Publishing, Paris. Accessed online on 01 Mar 2023 at https://www.oecd-ilibrary.org/sites/ba9bc504-en/index.html?itemId=/content/component/ba9bc504-en#section-d1e7996
47. Information Commissioner's Office (2023) Data sharing: a code of practice—lawfulness. Accessed online on 01 Mar 2023 at https://ico.org.uk/for-organisations/guide-to-data-protection/ico-codes-of-practice/data-sharing-a-code-of-practice/lawfulness/#publicsector
48. Department for Education (DfE) (2023) DRAFT Information sharing advice for practitioners providing safeguarding services to children, young people, parents and carers. Draft policy. Accessed online on 01 July 2023 at https://consult.education.gov.uk/childrens-social-care-data-and-digital-division/information-sharing-advice-safeguarding-practition/supporting_documents/Consultation%20draft%20%20Information%20Sharing%20Advice%20for%20Practitioners%20June%202023.pdf
49. Department for Levelling Up, Housing and Communities (2022) Supporting families programme guidance 2022 to 2025. Accessed online on 01 Mar 2023 at https://www.gov.uk/government/publications/supporting-families-programme-guidance-2022-to-2025

50. Council of Europe (1981) Convention 108 and protocols. Accessed online on 01 July 2023 at https://www.coe.int/en/web/data-protection/convention108/background
51. Raab C (1996) Implementing data protection in Britain. Int Rev Adm Sci 62(4):493–511. Accessed on 01 July 2023 at https://doi.org/10.1177/002085239606200404
52. Charlesworth A (1999) Implementing the European Union data protection directive 1995 in UK law: the data protection act 1998. Gov Inf Q 16:203–240. Accessed online on 01 July 2023
53. Bainbridge D (2008) Introduction to information technology law. Pearson, London
54. Information Commissioner's Office (2018) Children and the GDPR. Accessed online on 26 Aug 2023 at https://ico.org.uk/media/for-organisations/guide-to-the-general-data-protection-regulation-gdpr/children-and-the-gdpr-1-0.pdf
55. Adshead D (2016) Impact of EU-GDPR on local authorities in the UK Sheffield. Hallam University Research Archive thesis. Accessed online on 01 July 2023 at http://shura.shu.ac.uk/15590/1/Adshead%20impactofeu.pdf
56. Shu IN, Jahankhani H (2017) The impact of the new European general data protection regulation (GDPR) on the information governance toolkit in health and social care with special reference to primary care in England. In: 2017 Cybersecurity and cyber forensics conference (CCC), pp 31–37. Accessed online on 01 July 2023 at https://www.semanticscholar.org/paper/The-Impact-of-the-new-European-General-Data-(GDPR)-Shu-Jahankhani/60e864d5a3afac0437933a30051ab084fe5b67c5
57. Carey P (2018) Data protection: a practical guide to UK and EU law, 5th edn. Oxford UP, Print, Oxford, UK
58. IT Governance (2023) The UK Data Protection Act 2018 and UK General Data Protection Regulation. Accessed online on 12 July 2023 at https://www.itgovernance.co.uk/dpa-2018
59. Department of Media, Culture and Sport (DCMS) (2021) International data transfers: building trust, delivering growth and firing up innovation. Accessed online on 12 July 2023 at https://www.gov.uk/government/publications/uk-approach-to-international-data-transfers/international-data-transfers-building-trust-delivering-growth-and-firing-up-innovation
60. Information Commissioner's Office (2023) International data transfers. Accessed online on 01 Mar 2023 at https://ico.org.uk/for-organisations/data-protection-and-the-eu/data-protection-and-the-eu-in-detail/the-uk-gdpr/international-data-transfers/
61. Grabus S, Greenberg J (2017) Toward a metadata framework for sharing sensitive and closed data: an analysis of data sharing agreement attributes. In: International conference on metadata and semantics research. Accessed online on 29 June 2023 at https://www.semanticscholar.org/paper/Toward-a-Metadata-Framework-for-Sharing-Sensitive-Grabus-Greenberg/dbc4d8d3ccdac2e018ac85d0fcecf3cc03053b1c
62. The Performance and Innovation Unit (PIU) (2002) Privacy and data sharing: the way forward for public services. The Cabinet Office. Accessed online on 01 June 20223 at https://ntouk.files.wordpress.com/2015/06/privacy-and-data-sharing-the-way-forward-for-public-services-2002.pdf
63. Busby A, Mason L, Fisher C, Williams L (2020) Motivations for and barriers to data sharing identifying cultural and legal barriers to data sharing across government. Accessed online on 01 Mar 2023 at https://assets.publishing.service.gov.uk/government/uploads/system/uploads/attachment_data/file/895505/_Kantar_research_publication.pdf
64. Konstantinidis G, Chapman P, Weal M, Alzubaidi A, Ballard L, Lucassen A (2019) The need for machine-processable agreements in health data management in health data management. Algorithms 2020 13:87. Accessed online on 01 June 2023 at https://www.semanticscholar.org/paper/The-Need-for-Machine-Processable-Agreements-in-Data-Konstantinidis-Chapman/e723169ce1356a9a7d95d70915153b4726a85e16
65. Centre of Excellence for Information Sharing (2017) Updating our page on information governance. Accessed online on 01 July 2023 at https://informationsharing.org.uk/infogovernance/
66. Department for Housing, Communities and Local Government (DCLG) (2011) Statutory duties placed on local government. Accessed on 01 July 2023 at https://www.data.gov.uk/dataset/01171494-e40b-463f-9967-56d158412321/statutory-duties-placed-on-local-government

67. Department for Education (DfE) (2018) Working together to safeguard children. A guide to inter-agency working to safeguard and promote the welfare of children. Accessed online on 12 July 2022 at https://assets.publishing.service.gov.uk/government/uploads/system/uploads/attachment_data/file/942455/Working_together_to_safeguard_children_Statutory_framework_legislation_relevant_to_safeguarding_and_promoting_the_welfare_of_children.pdf
68. Department for Education (DfE) (2018) Working together to safeguard children. Statutory framework: legislation relevant to safeguarding and promoting the welfare of children. Accessed online on 01 Nov 2022 at https://assets.publishing.service.gov.uk/government/uploads/system/uploads/attachment_data/file/942455/Working_together_to_safeguard_children_Statutory_framework_legislation_relevant_to_safeguarding_and_promoting_the_welfare_of_children.pdf
69. Information Commissioner's Office (2023) Data sharing agreements. Accessed online on 01 Mar 2023 at https://ico.org.uk/for-organisations/uk-gdpr-guidance-and-resources/data-sharing/data-sharing-a-code-of-practice/data-sharing-agreements/
70. Cabinet Office (2017) The digital economy Act (DEA). Accessed online on 23rd June 2023 at https://www.legislation.gov.uk/ukpga/2017/30/contents
71. Information Commissioner's Office (2023) Data protection impact assessments. Accessed online on 01 Aug 2023 at https://ico.org.uk/for-organisations/uk-gdpr-guidance-and-resources/accountability-and-governance/guide-to-accountability-and-governance/accountability-and-governance/data-protection-impact-assessments/
72. Exploit Database (2023) Google hacking database. Accessed online on 01 Mar 2023 at https://www.exploit-db.com/google-hacking-database
73. UK Public Statutory Instruments (2000) The digital government (disclosure of information) regulations 2018 (SI 912). Accessed online on 01 Aug 2023 at https://www.legislation.gov.uk/ukdsi/2018/9780111169445
74. Centre for Data Ethics and Innovation (CDEI) (2020) Addressing trust in public sector data use. Accessed online on 01 June 2023 at https://www.gov.uk/government/publications/cdei-publishes-its-first-report-on-public-sector-data-sharing/addressing-trust-in-public-sector-data-use
75. Department for Education (DfE) (2011) Information sharing—how to identify which rules apply when sharing information. Accessed online on 31 July 2023 at https://assets.publishing.service.gov.uk/government/uploads/system/uploads/attachment_data/file/417697/Archived-information_sharing_how_to_identify_which_rules_apply.pdf
76. Department for Communities and Local Government (DCLG) (2014) Annex I—The expanded troubled families programme: data sharing guidance and principles. Accessed online on 31 July 2023 at https://www.lbhf.gov.uk/sites/default/files/section_attachments/annex_i_-_the_expanded_troubled_families_programme_data_sharing_guidance_and_principles_1.pdf
77. Centre of Excellence for Information Sharing (2016) Information sharing: cross cutting themes. Accessed online on 01 July 2023 at https://informationsharing.org.uk/download/333/
78. Centre of Excellence for Information Sharing (2018) ICO view of data sharing in context of the troubled families initiative. Accessed online on 01 July 2023 at https://informationsharing.org.uk/reports/infoshareresources/
79. UK Public Statutory Instruments (2000) The data protection (processing of sensitive personal data) order 2000 (SI 417). Accessed online on 01 Aug 2023 at https://www.legislation.gov.uk/uksi/2000/417/schedule/made
80. National Audit Office (NAO) (2014) Children in care. Accessed online on 01 Aug 2023 at https://www.nao.org.uk/wp-content/uploads/2014/11/Children-in-care1.pdf
81. National Health Service (NHS) (2020) Data sharing agreement template. Accessed online on 01 July 2023 at https://transform.england.nhs.uk/information-governance/guidance/data-sharing-agreement-template/
82. National Health Service (NHS) (2023) Consent and confidential patient information. Accessed online on 01 July 2023 at https://transform.england.nhs.uk/information-governance/guidance/consent-and-confidential-patient-information/#ig_professional
83. MedConfidential (2021) DEA investigation letter. Accessed online on 01 July 2023 at https://medconfidential.org

84. Department for Education (DfE) (2023) Stable homes, built on love: implementation strategy and consultation. Accessed online on 01 Mar 2023 at https://www.gov.uk/government/consultations/childrens-social-care-stable-homes-built-on-love
85. Information Commissioner's Office (2023) Policy document—our processing of special categories of personal data and criminal offence data. Accessed online on 01 Aug 2023 at https://ico.org.uk/about-the-ico/our-information/safeguards-policy/policy-document-our-processing-of-special-categories-of-personal-data-and-criminal-offence-data/
86. Home Office (2010) Information sharing for community safety. Accessed on 01 July 2023 at https://www.gov.uk/government/publications/information-sharing-for-community-safety
87. Klopper R, Lubbe S, Rugbeer H (2007) The matrix method of literature review. Alternation 14(1):262–276. Accessed online on 21 July 2023 at https://www.researchgate.net/publication/237542915_The_Matrix_Method_of_Literature_Review
88. Choguill C (2005) The research design matrix: a tool for development planning research studies. Habitat Int 29(4):615–626. Accessed online on 21 July 2023 at https://www.sciencedirect.com/science/article/pii/S019739750500038X
89. Information Commissioner's Office (2023) Data security incident trends. Accessed online on 26 Aug 2023 at https://ico.org.uk/action-weve-taken/data-security-incident-trends/
90. Landsbergen D, Wolken G (2001) Realizing the promise: government information systems and the fourth generation of information technology. Public Adm Rev 61(2):206–218. Accessed online on 01 June 2023 at https://www.semanticscholar.org/paper/Realizing-the-Promise%3A-Government-Information-and-Landsbergen-Wolken/234e1e67b1c210ee3f5e7d7956b32ad2c0e02ed4
91. McDougall K, Rajabifard A, Williamson IP (2005) What will motivate local Governments to share spatial information. In: Proceedings of SSC 2005 spatial intelligence, innovation and praxis: the national biennial conference of the Spatial Sciences Institute, Sept 2005. Spatial Sciences Institute, Melbourne. Accessed online on 22 June 2023 at https://research.usq.edu.au/download/b3547a3a57a4e565b766dba62453238dae798766d0b5482740bc76995108645f/230674/SSI_2005_Paper_-_McDougall_Rajabifard_and_Williamson_final.pdf
92. Riege A, Lindsay N (2006) Knowledge management in the public sector: stakeholder partnerships in the public policy development. J Knowl Manag 10(3):24–39. Accessed online on 01 June 2023 at https://librarysearch.northumbria.ac.uk/permalink/f/8019f1/TN_cdi_emerald_primary_10_1108_13673270610670830
93. UK Public General Acts (2018) Data Protection Act 2018. Accessed online on 01 July 2023 at https://www.legislation.gov.uk/ukpga/2018/12/contents/enacted

AI-Driven Phishing: Techniques, Threats, and Defence Strategies

Liza Shrestha, Hamed Balogun, and Suleman Khan

Abstract Phishing attacks are one of the most challenging threats in the cyberspace. Recently, rapid advancements of (generative) AI and its wide applicability has been exploited by attackers to perform phishing attacks. However, limited studies exist regarding AI-based phishing and apt defence strategies. In this work, we explore a wide variety of AI-leveraging techniques, adopted by attackers to conduct successful phishing campaigns. Additionally, we highlight the negative impact of AI-driven phishing in real-world and the attendant challenges that it has on cybersecurity. We also examine various factors and features of AI-powered phishing which makes these difficult to identify and complicated to defend. Consequently, we evaluate the crucial aspects of phishing attacks and discuss its defence strategies, human-centred preventive measures, and ethical considerations for enhancing security against AI-based phishing threats. Our findings provide valuable insights to the evolving cybersecurity threats and effective approaches to defend against these sophisticated AI-driven attacks.

Keywords AI · Phishing · Detection · Techniques · Defence · Security

---

L. Shrestha (✉)
University of Central Lancashire, Preston, UK
e-mail: lshrestha@uclan.ac.uk

H. Balogun
Edge Hill University, Ormskirk, UK
e-mail: Hamed.Balogun@edgehill.ac.uk

S. Khan
University of Bradford, Bradford, UK
e-mail: s.khan387@bradford.ac.uk

H. Jahankhani and B. Issac (eds.), *Cybersecurity and Human Capabilities Through Symbiotic Artificial Intelligence*, Advanced Sciences and Technologies for Security Applications, https://doi.org/10.1007/978-3-031-82031-1_7

1 Introduction to AI-Powered Phishing

Phishing is a prominent social engineering attack where an individual or a targeted group are deceived and manipulated into revealing sensitive information [1]. The means and methods by which attackers initiate phishing has been evolving rapidly such that existing solutions against phishing are unable to keep up with the advanced technologies used by attackers, thereby making them less effective [2]. One of the notable ascents regarding phishing and other cybercrimes is the use of Artificial Intelligence (AI). From the past trends of cybercrimes, it is evident that attackers are drawn towards utilising new technologies making AI as a priority tool for commencing their mischievous activities [3]. AI has transformed the patterns of phishing attacks in terms of sophistication and highly personalised attacks [4]. Automation is another advantage provided by AI which impose less work on attackers, allowing them to conduct large scale attacks effortlessly in short time frame [5]. The popularity and functions provided by AI has made it a favourite tool for cyberattackers. Machine learning (ML) algorithms, Natural Language Processing (NLP), Generative Adversarial Network (GAN) are some of the AI techniques that have been widely used for notorious activities. Compared to other cybercrimes, attackers are more influenced by AI in committing social engineering attacks like phishing. Generative AI has brought a tremendous upgrade in the magnitude of phishing activities [6]. The dual nature of AI pertinently contributes to it being a choice tool for attackers [7].

1.1 Scope of the Work

This paper focuses on exploring and analysing the AI-techniques used for conducting AI-driven phishing attacks, the threat it poses and effective defence strategies as shown in Fig. 1. The aim is to provide an understanding of evolving AI-driven phishing attacks and highlight its impacts on cybersecurity. The challenges involved in preventing these attacks have been analysed and AI-based multi-modal detection framework have been proposed along with human-focused countermeasures by evaluating the existing preventive measures and emerging technologies. Key contributions of this paper are:

- Highlighting the growing use of AI techniques for sophisticated phishing attacks and its impact.
- Illuminating the urgent need for smart solutions for AI-driven phishing.
- Proposal of a novel AI-based multi-modal detection framework to efficiently combat the adverse effects of AI-based phishing.

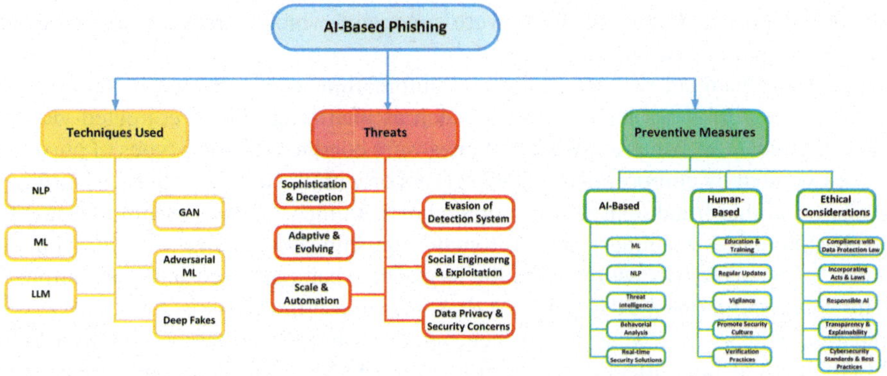

Fig. 1 Techniques, threats and preventive measures of AI-driven phishing

2 Motivation

According to Anti-Phishing Work Group (APWG), in 2023 approximately 5 million phishing attacks were recorded, and this is expected to increase each year [8]. The increasing attack rates indicates that the existing prevention tools are not fully effective towards the evolving phishing attacks. With the popularity of generative AI and other AI-based tools, their misuse has also been dramatically increased [9]. However, inadequate attention has been provided to AI-based attacks [4], specifically AI-based phishing attacks, which is the main motivation for this paper. It is crucial to understand the scale of damage AI-based phishing can impose on cybersecurity. Generative AI and other AI techniques not only leverage the intensity of existing phishing attacks but also help in implementing innovative zero-day phishing attacks [10]. Hence, posing great threat to the overall cybersecurity landscape. This paper aims to provide insights to the underlying misuse of AI to conduct hostile phishing attacks. It covers currently used AI-techniques and explores other plausible misuse of AI-techniques that can enhance the effectiveness of phishing attacks. Moreover, AI-features and challenges which makes AI-powered phishing less susceptible to detect and prevent are also discussed.

3 Background/Related Work

The history of phishing attacks is recorded from early 1990s when American Online (AOL) users were tricked by hackers who were pretending to be AOL employees [11]. The users were deceived into revealing their personal information, login credentials and credit card details. In early 2000s, email phishing was introduced for stealing personal information [12]. This trend took a leap in mid 2000s through spear phishing which was more targeted and personalised form of phishing [13]. And in the late

2000s, whaling and clone phishing were conducted which were more sophisticated compared to previous attacks.

The advancement of technology and automation were reflected in mid to late 2010s. New forms of phishing attacks such as pharming which redirected victims from legitimate websites to fraudulent ones were conducted. These types of phishing were executed by compromising DNS servers or with the use of malicious software. One of the other fraudulent attacks initiated in that time is man-in-the-middle attacks using automated phishing tools. Automated phishing tools comprised of sophisticated tools and scripts which made the large-scale campaigns easier to accomplish. The data records from 2010s show a drastic increase in the statistics of successful phishing attacks each year. Since then, attackers have been using advanced tools to conduct phishing attacks with various motives. Some of the popular tools were URL shorteners and obfuscation which kept the malicious links hidden in the emails. Similarly, various other techniques were used to evade spam filters and email security software.

There is no evidence for the exact date when the use of AI in phishing began. It was a gradual process and with time large numbers of AI-driven phishing were encountered, where attackers started to experiment with AI and machine learning algorithms to enhance phishing attacks [14]. Focus was seen on improving email content and target selection in the initial use by cybercriminals. As the growth of AI in other fields increased exponentially, it also highly influenced the attackers. The rise of AI in the early 2020s brought a revolutionary leap in the field of technology. AI not only transformed the digital world with extraordinary applications, but it also provided support for cyberattackers. The misuse of AI by attackers for various cybercrimes has been one of the major challenges for cybersecurity professionals [14]. Matured social engineering techniques, COVID pandemic and deep learning techniques are the reasons for massive increase in phishing attack. From cybercrimes like intrusion, hacking, malware attack and phishing attacks, AI have aided attackers to commence them in more sophisticated way [15].

In [16], misuse of AI in cyber-attacks was discussed in 3 categories which are synthetic data generation, data analysis, and data misclassification, with significant implications for phishing. Natural Language Processing (NLP), GANs, and ML are some of the widely used AI techniques. NLP helps in generating highly convincing phishing emails by accurately mimicking legitimate communications to [17]. For example, attackers used AI to impersonate an employee of MGM Resorts in a call to IT service desk to trick the IT team into resetting employee's password. They were successful in gaining network access which led to a massive ransomware attack. Similarly, GANs have been used to generate synthetic phishing emails both by cybersecurity researchers and attackers [18]. Various ML algorithms have also been used for crafting personalised phishing emails that have high success rates.

With a rapid digital transformation, Generative AI (GenAI) has brought massive enhancement in the utilisation of technology and its applicability to perform various tasks which includes both use and misuse. Researchers have discussed the potential of generative models such as ChatGPT to cause a negative impact on cybersecurity. For instance, authors in [19], discussed the use of ChatGPT by attackers to perform social engineering/phishing attacks ('spear phishing'). Thus, the use of AI tools and techniques provides an easy path for 'unsophisticated' attackers to draft convincing phishing emails [20] and potentially carry out effective attacks.

Previously, phishing attacks had numerous weakness which made their detection easier. Some of the obvious features of traditional phishing attacks are poor grammar, unprofessional formatting and spelling mistakes [21]. Other distinctive features were lack of personalisation and links to newly created or suspicious domains. All these features made the detection of phishing attacks simple. It could be easily detected with the defensive tools which were designed to identify these obvious patterns. However, this has substantially changed because AI has altered the characteristics of phishing attacks by advancing its features and overcoming most of these weaknesses [22]. It has become more difficult to identify phishing attacks looking at the features, for example, because AI-based phishing attacks are grammatically correct with zero spelling errors along with high personalisation [23]. Publicly available LLM models with the ability to generate human-like texts have made phishing attack scenarios worse than ever [24]. Attackers with poor writing skills can also easily draft high quality phishing emails with ease. LLMs such as ChatGPT and Microsoft Copilot can generate highly convincing phishing emails with just one simple instruction. A stimulated experiment conducted by Singapore's Government Technology Agency where spear phishing emails were sent by the security team resulted in the internet users clicking the links in AI-generated emails rather than in human-written ones [25].

It is important to understand how these magnificent AI tools affect, positively as well as negatively, our overall cyberspace and mankind. The more details that we acquire about how AI can be utilised adversely, the easier it will be to come up with effective countermeasures against them. In identifying effective countermeasures, considering each AI technique has unique features, it is important to study the adverse use of individual techniques in phishing attacks rather than generalizing AI as a whole.

4 AI Techniques Used in Phishing

The use of AI-techniques by attackers to commence phishing attacks varies vastly and are constantly evolving. From the wide range of techniques that have been implemented for malicious purpose, attackers select the technique that help them conduct more sophisticated and personalised phishing attempts.

4.1 Natural Language Processing

NLP helps computers to understand and interpret human language using various machine learning and can act as a mediator between computers and human linguistics which includes both text and speech [26]. It can be used to refine traditional phishing emails for generating more convincing human generated emails with less errors [27]. NLP has also shown its ability to analyse text from large volume of social media data [28] which can be adversely used by attackers for commencing targeted phishing attack. For a successful phishing, sentimentally and contextually analysed emails and conversations play a vital role as it can manipulate victim more effectively. NLP can be used to conduct sentimental analysis in dialogues based on various context to determine emotional tone during a conversation [29] making victims more vulnerable in being tricked. For instance, in an article [30], NLP was used to extract numerous contextual in- formation from social media. Considering the volume of data that an individual shares in social media platform, attackers can easily extract them to analyse and design a customised spear phishing with high success rate.

4.2 GANs

GAN, a combination of two neural networks (generator and discriminator), can create synthetic data taking references from training data. There are studies showing the potential of GAN in generating different types of data including text, images, and more. For example, in an experiment by Mahiuddin et al. [31], a real-life like synthetic images of human faces were generated with the use of GAN. This proficiency of GAN to generate realistic data has not just contributed to numerous fruitful aspects but has also led to increase in its utilisation for commencing various cyber-attacks such as phishing. The ability of GAN to generate realistic sentences have been explored widely especially TextGAN [32]. It is not limited to generating sentences but larger text generation like essays can be achieved with proper trainings [33]. This depicts that it be used for drafting realistic emails in more efficient and automated form for targeted phishing attacks.

4.3 Machine Learning

The scale of damage is much higher in targeted attacks conducted according to target's preference and behaviour compared to a generalised attack. Machine learning algorithms have made the analysis of target's behaviour simple and robust. For example, Activision, the company behind Call of Duty games became a victim of targeted phishing campaign in which attackers used AI to create highly convincing SMS messages [34]. This led one of the staff member to fall for the bait and hackers were able to gain complete access to employee database. User profiling, which is a form of understanding user's behaviour, interests and preferences, plays a vital role in targeted attacks [35]. AI-based user profiling has been utilised for digital marketing by focusing on the customer's need [36]. However, the application of this process is not limited to digital marketing but can also be a systematic approach for targeted adaptive social engineering. Both supervised and unsupervised ML can be used for conducting user profiling. Some researchers [37] have demonstrated the generation of fake text documents with the use of genetic algorithm depicting the capability of ML in making fraud text which can be utilised by attackers for phishing.

4.4 Adversarial Machine Learning

There has been increase in the preferences of AI tools over traditional methods for detecting the phishing attacks which have simultaneously increased he ML-based adversarial attacks on these AI tools. No system can be full proofed with no vulnerabilities. Systems may be vulnerable to evasion attacks which have been seen in an experiment by [38], where 12 evasion attacks were conducted against a phishing website detector and the results demonstrated that it was affected by few attacks. Similar experiments have also been performed to bypass phishing website classifier using three mutation-based attacks [39]. These experiments were conducted to analyse the effectiveness and accuracy of phishing detector tools; however, it also depicts the possibility of attackers conducting similar adversarial attacks making the detector tools ineffective with increase in successful phishing attempts.

4.5 Large Language Models

Large language models (LLMs) can generate convincing phishing emails using simple prompts. Crafting phishing emails has been easier than ever with the use of LLMs. Experiments have been conducted to demonstrate the ability of LLMs to contribute toward social engineering and phishing attacks with simple prompts [23]. Attackers can misuse these wide variety of LLMs that have special features to heavily customise an initiative for advanced spear phishing. Although most of the publicly available LLMs have been designed with ethical guidelines and restrictions for creating adversarial contents, they can be manipulated using tricky prompts and fool them to generate malicious contents. For instance, if we provide a direct instruction to create a convincing email, it will deny for generating. However, if we modify the instruction to generate a sample of phishing emails for educational purposes, it will generate one. There are some LLMs which have been launched for cybercriminals such as WormGPT and FraudGPT which allows creation of phishing contents with very less effort [40].

4.6 Deep Fakes

Deepfake technology which builds realistic but counterfeit audiovisual contents using various AI algorithm, is an emerging threat as it has been misused for phishing and vishing. In the paper [41], the use of deep fakes for various deceptive purposes such as false advertising and malicious agendas have been discussed along with the problems in regulating them ethically and lawfully. It has created a serious problem in terms of security compromise in the digital space [42]. This technology has been used specifically for 'spear phishing' by deceiving an individual into revealing sensitive information. This form of phishing is generally carried out by fooling the victim through audio or video messages containing impersonation of a trusted individual. One such example is Fraudsters using deepfake technology to trick a finance worker into transferring $25 million by impersonating company's chief financial officer through a video conferencing call. With rapid advancement in deep fake technology, number of deep fake phishing attacks are expected to increase causing serious issue if not regulated [43].

5 Threats of AI-Powered Phishing

5.1 Sophistication and Deception

The sophistication of phishing attacks has increased tremendously with the use of AI [4]. Contextual understanding can lead to convincing conversation between attacker and victim which enhances the credibility of phishing attempts [44]. It is difficult for attackers to achieve contextual understanding through traditional phishing techniques but with different AI techniques it has become quite a simple task. AI generated phishing is grammatically correct with less to no typos such that its detection is difficult using existing phishing detection methods [45]. These attacks can also be orchestrated across multiple channels simultaneously and in a coordinated manner elevating the deception chances and successful attacks.

5.2 Adaptive and Evolving Tactics

One of the major differences between traditional phishing attacks and AI-based phishing attacks is the adaptability of AI phishing due to its dynamic nature [46]. The currently available static phishing defense system is not efficient enough to keep pace with the rapidly changing AI-based phishing. For example, existing system trained to detect phishing email based on spelling error and poor grammar cannot detect the phishing email with no flaws generated using AI. Zero-day attacks are the most challenging issue for any cybersecurity solution [47] and AI helps in generating and executing these immune zero-day phishing attacks making the existing defense system less effective.

5.3 Scale and Automation

If we look at the history of previously conducted phishing attacks, large scale attacks were not common as they are time consuming and expensive. But AI has been found to be a great way for attackers to perform mass phishing with less resources and short time-period. AI can generate numerous phishing emails with automated operations which can have massive damage in a very short period [5]. Also, attackers can now set a specific time to send phishing email automatically using various ML algorithms. This have increased the seriousness of the attacks as the damage they make are vast with limited time to act for recovery.

5.4 Evasion of Detection Systems

The traditional or currently implemented phishing detection tools are designed against previously conducted phishing patterns which will not effectively detect AI-powered phishing with dynamic nature. The evolving AI-generated phishing is unique for each attack which can easily bypass the existing phishing detectors. Research conducted shows that AI can help attackers to evade spam filters along with the security plugins by mimicking communications which resemble legitimate ones [48]. These actions are time consuming to perform manually but with the help of AI, it is quite efficient.

5.5 Social Engineering and Exploitation

The amount of personal data that an individual shares on the internet these days is like threats to the attackers for conducting personalised attacks [49]. However, extracting and analysing these data manually can easily analyse the vast amount of data by collecting them from social media and other online activities. Other ways in which AI-based phishing can enhance social engineering is through psychological profiling and emotion analysis [50]. Data collected related to the target can be psychologically profiled for understanding vulnerabilities and preferences which accounts to creation of highly convincing attacks. Since emotions are the primary factor which influences the chances of successful social engineering, AI can accommodate in tailoring phishing attempts by analysing emotional responses for higher success rate in the attacks.

5.6 Data Privacy and Security Concerns

Data privacy and security concerns coupled with AI-based phishing are immaculate mainly regarding the misuse of AI models and unintended consequences [51]. Effective and widespread attacks have increased with AI allowing less-skilled attackers to deploy sophisticated schemes. AI tools have helped to escalate the data breaches on a very large scales where victims lose their sensitive details that can be sold on the dark web or used for various other fraudulent activities. There are many other security and privacy concerns which can be classified into user, ethical, technological, institutional, regulatory and law perspectives [52]. All of these can lead to significant unintended consequences along with the amplification of risks associated with data breaches.

6 Challenges and Risks of AI-Powered Phishing

The essence of attack is the same even with the use of AI, but the scope and scale of the attacks are contrasting. Analysing the threats imposed by AI-based phishing, the differences between them with traditional phishing can be highlighted as shown in Table 1.

Traditional phishing provides less opportunity for customising the phishing emails according to the targeted individual/group and the customisation process is extensive and difficult [53]. However, it is not the same in case of AI-driven phishing, where the customisation process is easily conducted based on the log activities and interest of the selected target. Conventional phishing emails were drafted manually limiting the scale and frequency of the attacks. But in AI- based phishing, there is a wide availability of automation allowing campaign and large-scale phishing which are more destructive in nature to be carried out effortlessly [5]. Previously, phishing emails were crafted by using basic sentences and were mostly template-based [54] making them less sophisticated, while with the help of AI, phishing emails are becoming more sophisticated, adaptive, and personalised. These are some of the reasons why detection of traditional phishing attacks was less successful with easy detection and

Table 1 Differences between traditional and AI-based phishing

Features	Traditional phishing	AI-driven phishing
Customisation	Low	High
Automation	Limited	Extensive
Sophistication	Basic, template-based high	Personalised, adaptive
Success rates	Moderate	High
Detection difficulty	Easier to detect	Harder to detect
Evolution speed	Slow	Rapid

recent AI-based phishing attacks have high success rate with less chances of being detected. The evolution of traditional phishing was slow as they were manually induced which provided ample amount of time for the development of new defences. However, AI-based phishing is evolving rapidly and continuously offering less time for developing the defence systems against them.

7 Defense Strategies Against AI-Driven Phishing

There are various ways to detect and prevent AI-based phishing attacks but with the pace at which AI-powered attacks are evolving, the only means by which these attacks can be tackled more efficiently is with the use of AI itself [14]. Modern phishing attacks are sophisticated and exploit multiple types of content such as text, images, and URLs [55].

7.1 Technical Measures

Advanced Phishing Detector Tools One of the efficient ways of detect- ing and preventing phishing attacks is to develop machine learning and AI- based smart detector tools which can adapt according to the changing nature of AI-phishing attacks. For example, various companies have integrated phish- ing detection tools such as Vectra Cognito with Amazon Web Services (AWS) to enhance the cyber-security through real-time monitoring and response in the network [56]. Some of the other phishing detector tools such as Bacrracuda Essentials. Graphus, Hunto AI are popular among organisations as they claim to be more effective than the traditional tools. AI-driven solutions can drastically improve detection and prevention of advanced phishing attacks and this has been shown through case studies and real-world scenarios in the paper by Fakhouri et al. [57].

ML algorithms such as convolutions neural network (CNN), recurrent neural network (RNN), Random Forest have been explored by researchers for detecting phishing attacks. The experiment performed by Safi, A. et al. and Sahingoz, O.K. et al. depicts that CNN has the highest accuracy of 99.98% and 98.74% in respective experiments where multiple algorithms were tested [58, 59]. Algorithms with high detection accuracy and low false positive rates can be trained to detect AI-based phishing as well. Existing AI detectors for phishing can be in- competent when it comes to identifying zero-day attacks [60], making them less effective. It has become essential to develop and test hybrid AI-models which can identify zero-day attacks because hybrid models have been found to possess high accuracy compared to other models. Also, some of the AI-techniques such as reinforcement learning are adaptive, and this feature can be utilised against new tactics used in phishing. By selecting appropriate machine learning and AI techniques, their features can be combined to develop highly accurate phishing detector tools. Along with hybrid models, a holistic approach of combining AI-based tools with other traditional methods can improve the defence strategy against phishing [61].

Multi-modal Approach with Generative AI Multi-modal approaches for phishing detection have gained significant popularity in recent researches due to its improved accuracy, robustness, and scalability. In the research performed by Çolhak, F. et al., multiText-LP model performance was superior to standalone model in detecting phishing emails through HTML content analysis with an accuracy of 96.80% [62]. To ensure all crucial cues embedded in different formats are properly analysed, multi-modal AI-approach has been proposed for phishing detection with motive to achieve highly accurate detection. The multi-modal detection framework presented as shown in Fig. 2 combines the analysis of three different types of content: Text, Image and URL which can be classified into these components.

Input and Pre-processing: It is the initial stage where incoming emails and messages are segregated into its core components (text, images, URLs). Extraction of relevant data and its preparation for analysis are carried out in this stage. For text, body of the email along with the subject line can be extracted using email parsing libraries or natural language processing. Similarly, for image, email parsing libraries in python and C*, file handling libraries and Optical Character Recognition (OCR) can be utilised [63]. Likewise, for URL extraction, HTML parsing and URL handling libraries can be incorporated.

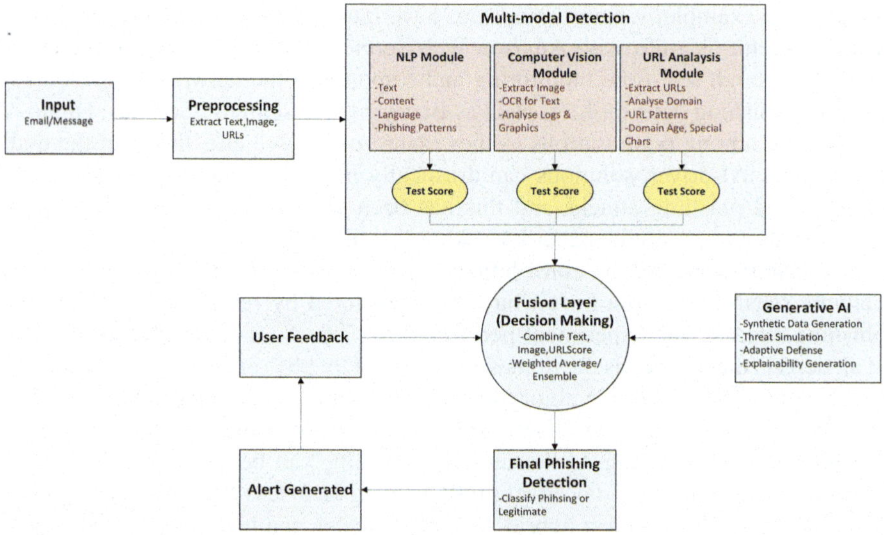

Fig. 2 Multi-modal approach to phishing detection with generative AI

Multi-modal Detection Modules: Three different and independent modules are combined to form a multi-modal detection for identifying potential phishing indicators by focusing on different aspects of the email content.

- **NLP Module (Text Analysis)**: The text extracted from stage 1 is analysed in this module. Text Cleaning, feature extraction and phishing pattern detection are the processes that will be commenced in this module. Text Cleaning refers to removal of unnecessary elements for simplifying the analysis which can be achieved with tokenization, stop word removal, stemming and lemmatization [64] using libraries such as SpaCy and NLTK. Feature extraction is carried out to convert the text into suitable format which is suitable for processing in ML model. Machine learning classification models such as logistic regression, Support Vector Machine (SVM), Naïve Bayes and Deep learning models can be incorporated for detecting the phishing pat- terns [65]. This module looks for linguistic patterns, wording, and phrases which are commonly associated with phishing and gives a phishing likelihood score that represents the probability of the text being a part of a phishing attempt.
- **Computer Vision Module (Image Analysis)**: This model focuses on extracting image from the emails/messages and analyse them to detect visual cues based on fake logos, brand inconsistencies, as well as embedded text inside images. Key processes involved in this module is Optical Character Recognition (OCR), Logo and Brand Detection, and visual pattern recognition. Text can be extracted from the image using Google Vision API such that, a common phishing tactic, embedded text in images is also processed. CNNs can be implemented for verifying the logo as authenticate or manipulated versions [66]. One of the important aspects of image analysis is detecting suspicious visual elements such as altered icons, low-quality images or fake warning signs [67] which can be done using various ML algorithms. The output form this module is represented in a likelihood score of 0 or 1.
- **URL Analysis Module**: This module evaluates the URLs (domain names, link structures, URL reputation) that are embedded trickily using various AI- techniques in the email to determine its association with phishing attempts. URLs are parsed in this step for suspicious features such as typo squatting, long URLs with excessive special characters. Services such as VirusTotal and Google Safe Browsing API can also be used to check for blacklisted domains [68]. Domain age and history is another indicator for identifying phishing URL as new domains with little history may be suspicious [69]. The outputs of this module will be similar to that of the previous 2 modules representing a likelihood score which are combined to obtain a weighted score

Generative AI Integration: Generative AI can enhance the detection capability of multi-modal phishing detection modal by increasing the adaptiveness. Role of Gen AI in this approach can be utilised for 4 specific tasks: generating synthetic data, threat simulation, adaptive defence, and generation of explainability. For generating phishing samples, GANs can be deployed to augment training datasets allowing the model to learn from wide range of phishing scenarios and improve the adaptiveness [70]. Threat simulation is another aspect which can be initiated using Variational Autoencoders (VAEs) that can mimic the tactics used by cybercriminal and generate a simulation environment. Through continuous testing of detection systems using simulation will ensure that the model is resilient to emerging threats. Adaptive defence can be leveraged by using Natural Language Generation (NLG) allowing the Gen AI to generate new phishing scenarios through dynamic learning from detected phishing attempts. This can upgrade the system's ability to detect zero-day attacks. It is essential to offer transparency in the decision-making process of AI model to improve user's trust. Using Gen AI to deliver explanation on the outputs, alerts or decision made by the system in human-readable format, will drastically improve the user's experience while using this model.

Fusion Layer (Decision Making): This layer is the centre element of the multi-modal phishing detection system that plays a pivotal role by acting as a convergence point and integrating different modalities. It creates a holistic view of the email to provide a better understanding of whether the email is likely to be phishing. It obtains outputs from multi-modals in the form of likelihood scores and outputs from synthetic data, threat simulation through generative AI. All these features are combined using feature-level fusion and allows the system to consider the context as a whole. For instance, the differences between the context of an email (text) and the visual elements associated with it might be detected by the system which could indicate phishing. This layer also uses adaptive weighting mechanisms which assigns equal importance to each modality along with cross-modal embeddings. If the phishing pattern is text-based, the score from text analysis might dominate other scores. In such cases, the dynamic approach ensures that the system adapts based on the characteristics of the specific email being analysed instead of relying heavily on a single modality. Various neural networks such as multimodal deep learning, Convolutional Neural Network (CNNs), Recurrent Neural Network (RNNs), attention-based fusion networks, can be utilised in this layer so that holistic view ca be achieved with less false positives, reduced errors and biases [71]. A combination of techniques and contextual understanding offered by this layer with the integration of Gen AI, will help in improving accuracy, resilience and adaptability of the detection system.

Final Phishing Detection: This is the core decision-making component of the phishing detection system as it classifies whether an email is phishing or legitimate by applying various machine learning models. The combined feature vector from fusion layer is taken as an input for this input. The machine learning models for this layer is chosen based on the complexity of the email. For instance, logistic regression (LR) is a simple interpretable classification model [72] which can provide insights on how each feature of an email contributes to the final decision. This can be used in the model as it is quick to train but is not an ideal one for complex data and cannot be used on its own. A combination of ML such as SVM, CNNs, RNNs or ensemble learning can be used. The output is in the form of phishing likelihood score which is usually in the range of 0 to 1. The threshold can be predefined, for example, if score is above 0.7, the email can be classified as phishing, whereas if it is below 0.3, it can be classified as legitimate.

Alert and Feedback loop: User feedback have a great role in improving the model and increase the adaptation. This feature allows the system to evolve according to the real-world performance and user's feedback to ensure that it is effective against new phishing techniques. Here, user can provide feedback related to false positive and false negatives. This provides the fusion layer information on the scale of attention or weighting to be given for each modality. This helps on continuous progress and adaptation against the new AI-driven tactics used by attackers some of the key benefits of this feedback loop are dynamic learning, error correction, increased accuracy, and building user trust.

7.2 Human-Centered Preventive Measures

Humans are the major aspect of any social engineering as they are deceived and manipulated to disclose confidential information, which is why human-focused strategies play a vital role in mitigating phishing attacks. Some of the user- oriented security measures as shown in Fig. 3 have been discussed below.

Education and Training Awareness and proper understanding of the potential risk from AI-based cyberattacks are some of the effective preventive strategies which can be achieved through education and training [73]. Regular security awareness training programs can aid in educating individuals regarding the evolving nature of phishing attacks and provide them with the ability to identify potential red flags. Stimulated exercises like generating a controlled synthetic environment and inducing AI-based phishing attacks along with implementing effective procedures to tackle them can make an individual prepared to deal with similar attacks in the future.

Regular updates Keeping a regular update in software, systems and latest trends is one of the important security measures even for dynamic phishing attacks. Attackers often tend to exploit the vulnerabilities of software to execute zero-day attacks [74]. But the impact of this issue can be minimized with regular software updates which will ensure security patches against exploitation like any other attackers, cybersecurity personnel and researchers are also elevating defence by upgrading anti-phishing tools and spam filters which when implemented can block phishing attempts. It is also essential to get updated with recent PA patterns to be aware of such attacks and find a solution to defend them.

Vigilance Being observant of different activities while being on the internet or any online platform can elevate the early detection of notorious activities such as phishing attacks [75]. If an individual is vigilant, the chances of noticing unusual and suspicious emails and messages are extremely high. Carefully inspecting links and attachments in emails, assessing the domain names in the email addresses are some examples of vigilant actions which can help in identifying phishing. Not only

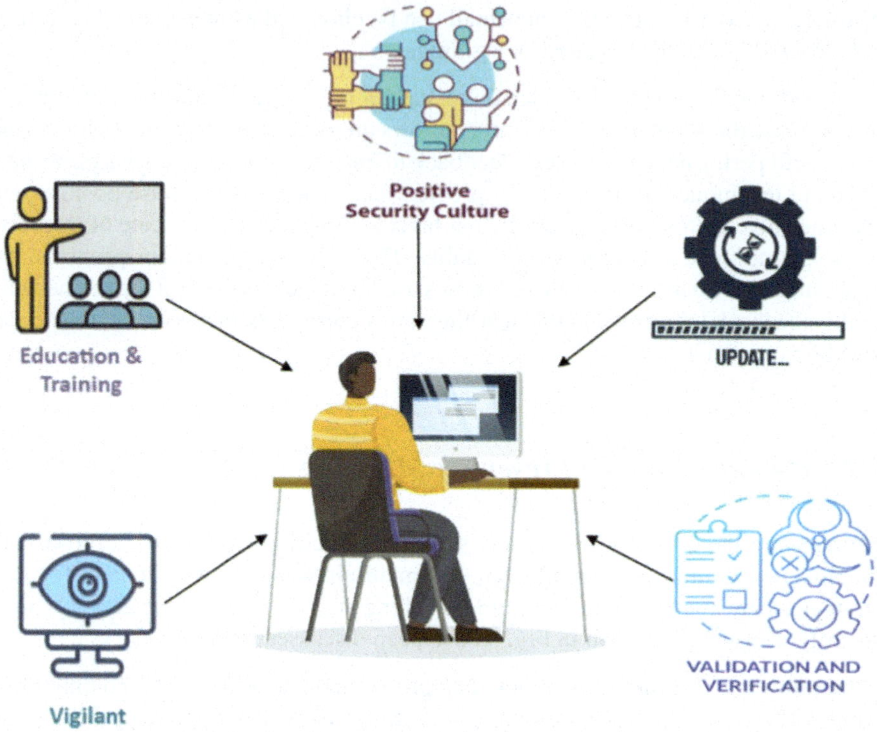

Fig. 3 Human-centered prevention methods

this, but a vigilant individual can identify the legitimacy of highly convincing AI generated PA by making full use of available security tools.

Promoting a Proactive Security Culture As AI-based PA are highly dynamic, it is essential to understand the recent tactics used by attackers. For this, information sharing is essential among individuals, employees, employers and companies. Information shared and proper flow of communication among designated level within an organisation regarding any sort of unusual online activities, data breach, and system compromise can elevate the performance of incident response team to handle the situation with appropriate guidelines and minimize the negative impact of the situation. This can also help in preventing similar attacks (AI/non-AI based) in the future. For improved security, it is also important to encourage the usage of security measures against phishing.

Verification Practices Verification is an act of establishing a truth with a proof which is essential when providing access to data and information [76]. Some of the common verification practices are Double-checking requests and cross-referencing information [77]. Phishing attacks are generally accomplished by requesting access so verifying the requests can help in identifying deceptive form of tactics used by attackers and protect the data. Another simple yet effective method when it comes to tackling AI-based phishing is to use an alternative trusted channels (phone calls or in-person confrontation) for contacting the sender from unusual request. This measure is not only effective for traditional PA but also for AI-generated attacks. Some of the details such as email address, domain names, logos, language and signatures are great indicators of phishing attempts and hence they need to be verified for protection against phishing attacks.

7.3 Ethical Considerations

Controlling the development and use of any kind of technology and products arises the need for proper rules and regulation to flourish it in many ways [78]. AI is fast evolving field of technology that have spread across numerous sectors and have been incorporated heavily in human lives to carry out daily tasks. But with increasing misuse as well such as phishing attacks, regulating AI ethically is essential for protecting against AI-driven attacks. Accountability for errors, misconduct, and regulatory violations are also essential to be defined for minimising the misuse of AI [79]. This can be achieved in many ways through the involvement of stakeholders as well as individuals.

Regulatory Frameworks and Compliance Government Laws are an integral aspect in controlling the development and use of AI ethically so that their adverse use is minimised [80]. Increasing number of AI products and users have urged government bodies to regulate it. There are some of the Laws and Acts such as European Union AI Act, UK's National AI Strategy, AI Guidelines and Executive Orders in United States, and many more which plays significant role in regulating and preventing AI-based phishing attacks [81]. Cyber Resilience Act is another example of government at that aims to safeguard consumers and businesses using software and digital products [82]. Similarly, other AI regulations have been introduced in many countries such as China and Japan, further depicting the potential usefulness of AI products. They establish regulatory frameworks which promotes responsible and fair development, deployment and use of AI. Although the current acts and laws addresses most of the areas of AI, there are some limitations. For instance, regulations for misinformation created by Gen AI is inadequate in the EU legislation. Hence, these regulations should be more refined in the future for preventing any sorts of cybercrimes induced with the use of AI.

Data protection regulations such as GDPR (General Data Protection Regulation) and HIPPA (Health Insurance Portability and Accountability Act) provides a framework for protection of user's personal data in an organisation [83]. They provide guidelines on mandatory consent and specific or minimum personal data to be collected and processed. Adopting these regulations strictly will reduce the pool of data that can be exploited during AI-based phishing attacks. They also provide security measures for storing data which can prevent unauthorised access initiated through use of AI. Breach notification is another attribute which have been specified in these regulations that can help in recovery and prevention of further damage in case of data breach. Complying with the principles of data minimisation, purpose limitation and transparency are also some other aspects that these law aims regarding AI system for protecting the privacy rights of an individual. Therefore, every organisation must comply with data protection laws as a preventive measure against AI-based phishing attacks.

Responsible AI Development and Deployment When deploying any AI products, it is necessary that a set of principles are followed during not just deployment but also in the design and development rather than blindly releasing them without overlooking the consequences it might induce [84]. To assure that AI incorporates the ethical implications, responsible AI which is a framework with principal guidelines for design and deployment needs to be implemented [85]. Principle for the responsible development of AI established by the Organisation for Economic Cooperation and Development (OECD) is an example of how organisations can regulate responsible AI [86]. This best practice can aid in minimising AI-based cyberattacks such as phishing attacks.

Transparency and Explainability Transparency and explainability, which are linked with each other, are the factors for achieving an understandable AI [87]. Deep understanding of AI fosters swift identification of AI's vulnerabilities and potential misuse as a result of which appropriate preventive measure can be identified. According to various studies [88–90], explainability results in higher transparency and have now been considered as a key requirement of AI systems. These elements have a positive impact on maintaining ethics around the use of AI and limits its misuse. They define the accountability retained during various instances and provides a clarity of the responsibility and actions to be implied by the designated individual or organisations.

Cybersecurity Standards and Best Practices Cybersecurity standards and guidelines provide a set of guidelines which if fully adhered to will secure a system by fulfilling the fundamental security requirements [91]. They help counter AI-based phishing along with other adverse events by providing a multi-layer defence strategy through regular security assessments, data encryption, continuous monitoring and incident response, and multi-factor authentication. DMARC (Domain-based Message Authentication, Reporting, and Conformance), SPF (Sender Policy Framework) and DKIM (DomainKeys Identified Mail) are some of the email filtering and authentication protocols which help in identifying and blocking phishing emails before they reach to user [92]. Adopting these practices reduce the chances of sensitive information to be exploited through phishing.

8 Future Directions and Recommendations

AI is advancing daily with the introduction of new techniques in a noticeably short time, and it is evident that attackers will be improving their strategies. The complexity of phishing will grow with ample use of AI tools and techniques in near future. To prevent high impact of phishing in various aspects, advancements must be made for defence using enhanced AI algorithms and Explainable AI (XAI). New solutions against phishing must be designed and developed with adaptive features. To achieve this, investment in AI research and development is recommended along with fostering a security-conscious culture. Empowering humans to adapt to innovative technology must also been done simultaneously to ensure the effectiveness of the defense strategies.

9 Conclusion

The use of AI by attackers to conduct phishing attacks is highly threatening. Various AI techniques have already been utilised for phishing attacks by generating convincing emails making them superior than traditional phishing n terms of sophistication and success rates. Considering the threat they pose; it is essential to have a proper defence against these AI-driven phishing attacks as the existing defence tools are not fully effective on these advanced attacks. Some solutions for fighting against these attacks is to develop phishing detector tools by incorporating various AI techniques and looking for holistic approach. A multi-modal approach with the integration of generative AI proposed can help in detecting the advance phishing attacks. Human-drive preventive measures and ethical/regulatory considerations should also be implemented simultaneously for better security against AI-driven phishing attacks.

AI is advancing daily with the introduction of new techniques, and it is evident that attackers will be improving their strategies. To prevent adverse impact of phishing in near future, advancements must be made for defence using enhanced AI algorithms and simultaneously empowering humans to adapt to new technologies must also be done to ensure the effectiveness of the defense strategies.

References

1. Salahdine F, Kaabouch N (2019) Social engineering attacks: a survey. Futur Internet 11:89
2. Goenka R, Chawla M, Tiwari N (2024) A comprehensive survey of phishing: mediums, intended targets, attack and defence techniques and a novel taxonomy. Int J Inf Secur 23:819–848
3. Blauth TF, Gstrein OJ, Zwitter A (2022) Artificial intelligence crime: an overview of malicious use and abuse of AI. IEEE Access 10:77110–77122
4. Kaloudi N, Li J (2020) The AI-based cyber threat landscape: a survey. ACM Comput Surv (CSUR) 53:1–34

5. Brundage M et al (2018) The malicious use of artificial intelligence: forecasting, prevention, and mitigation. arXiv preprint arXiv:1802.07228
6. Gupta M, Akiri C, Aryal K, Parker E, Praharaj L (2023) From chatgpt to threatgpt: impact of generative AI in cybersecurity and privacy. IEEE Access 11
7. Bécue A, Praça I, Gama J (2021) Artificial intelligence, cyber-threats and industry 4.0: challenges and opportunities. Artif Intell Rev 54:3849–3886
8. Phishing activity trends report, 4th quarter 2023 (2023). URL https://apwg.org/trendsreports/
9. Ooi K-B et al (2023) The potential of generative artificial intelligence across disciplines: perspectives and future directions. J Comput Inf Syst 65:1–32
10. Metta S, Chang I, Parker J, Roman MP, Ehuan AF (2024) Generative Ai in cybersecurity
11. Perlroth N (2021) This is how they tell me the world ends: the cyberweapons arms race. Bloomsbury Publishing USA
12. Alkhalil Z, Hewage C, Nawaf L, Khan I (2021) Phishing attacks: a recent comprehensive study and a new anatomy. Front Comput Sci 3
13. Malik JK, Choudhury S (2019) A brief review on cyber crime-growth and evolution. Pramana Res J 9:242
14. Guembe B et al (2022) The emerging threat of AI-driven cyber attacks: a review. Appl Artif Intell 36:2037254
15. Jaber AN, Fritsch L (2021) Covid-19 and global increases in cybersecurity attacks: review of possible adverse artificial intelligence attacks. IEEE, pp 434–442
16. Yamin MM, Ullah M, Ullah H, Katt B (2021) Weaponized AI for cyber attacks. J Inf Secur Appl 57:102722
17. Salloum S, Gaber T, Vadera S, Shaalan K (2022) A systematic literature review on phishing email detection using natural language processing techniques. IEEE Access 10:65703–65727
18. Arora A, Shantanu (2022) A review on application of gans in cybersecurity domain. IETE Tech Rev 39:433–441
19. Grbic DV, Dujlovic I (2023) Social engineering with chatgpt. IEEE, pp 1–5
20. Yu J et al (2024) The shadow of fraud: the emerging danger of AI-powered social engineering and its possible cure. arXiv preprint arXiv:2407.15912
21. Wang J, Herath T, Chen R, Vishwanath A, Rao HR (2012) Research article phishing susceptibility: an investigation into the processing of a targeted spear phishing email. IEEE Trans Prof Commun 55:345–362
22. Alkhalil Z, Hewage C, Nawaf L, Khan I (2021) Phishing attacks: a recent comprehensive study and a new anatomy. Front Comput Sci 3:563060
23. Emanuela AR, Cristina BA, Luminiţa S (2024) AI and prompt engineering: the new weapons of social engineering attacks. IEEE, pp 1–6
24. Das BC, Amini MH, Wu Y (2024) Security and privacy challenges of large language models: a survey. arXiv preprint arXiv:2402.00888
25. Community O (2023) Openai wrote better spear phishing emails. https://community.openai.com/t/openai-wrote-better-spear-phishing-emails/7359. Accessed on [Insert access date here]
26. Otter DW, Medina JR, Kalita JK (2021) A survey of the usages of deep learning for natural language processing. IEEE Trans Neural Netw Learn Syst 32:604–624
27. Khurana D, Koli A, Khatter K, Singh S (2023) Natural language processing: state of the art, current trends and challenges. Multimed Tools Appl 82:3713–3744
28. Young T, Hazarika D, Poria S, Cambria E (2018) Recent trends in deep learning based natural language processing [review article]. IEEE Comput Intell Mag 13:55–75
29. Carvalho I, Oliveira HG, Silva C (2023) The importance of context for sentiment analysis in dialogues. IEEE Access 11:86088–86103
30. Patton DU et al (2020) Contextual analysis of social media. ACM, pp 337–342
31. Mahiuddin M, Khaliluzzaman M, Chowdhury MSA, Arefin MN (2022) Fake face generator: generating fake human faces using gan. Int J Adv Comput Sci Appl 13
32. Zhang Y, Gan Z, Carin L (2016) Generating text via adversarial training, vol 21. Academia.edu, pp 21–32

33. Liang G et al (2021) A text gan framework for creative essay recommendation. Knowl-Based Syst 232:107501
34. Franceschi-Bicchierai, L.Hackers steal activision games and employee data (2023). https://techcrunch.com/2023/02/21/hackers-allegedly-steal-activision-games-and-employee-data/
35. Kaushal R, Ghose V, Kumaraguru P (2019) Methods for user profiling across social networks. IEEE, pp 1572–1579
36. Haleem A, Javaid M, Qadri MA, Singh RP, Suman R (2022) Artificial intelligence (AI) applications for marketing: a literature-based study. Int J Intell Netw 3:119–132
37. Karuna P, Purohit H, Jajodia S, Ganesan R, Uzuner O (2021) Fake document generation for cyber deception by manipulating text comprehensibility. IEEE Syst J 15:835–845
38. Apruzzese G, Conti M, Yuan Y (2022) Spacephish: the evasion-space of adversarial attacks against phishing website detectors using machine learning. In: Proceedings of the 38th annual computer security applications conference, pp 171–185
39. Song F, Lei Y, Chen S, Fan L, Liu Y (2021) Advanced evasion attacks and mitigations on practical ml-based phishing website classifiers. Int J Intell Syst 36:5210–5240
40. Firdhous MFM, Elbreiki W, Abdullahi I, Sudantha B, Budiarto R (2023) Wormgpt: a large language model chatbot for criminals. IEEE, pp 1–6
41. Meskys E, Liaudanskas A, Kalpokiene J, Jurcys P (2020) Regulating deep fakes: legal and ethical considerations. J Intellect Prop Law Pract 15:24–31
42. May R, Krüger J, Leich T (2024) Sok: how artificial-intelligence incidents can jeopardize safety and security. In: Proceedings of the 19th international conference on availability, reliability and security, pp 1–12
43. Mirsky Y, Lee W (2022) The creation and detection of deepfakes. ACM Comput Surv 54:1–41
44. Desolda G, Ferro LS, Marrella A, Catarci T, Costabile MF (2022) Human factors in phishing attacks: a systematic literature review. ACM Comput Surv 54:1–35
45. Heiding F, Schneier B, Vishwanath A, Bernstein J, Park PS (2024) Devising and detecting phishing emails using large language models. IEEE Access 12:42131–42146
46. Andriu A-V (2023) Adaptive phishing detection: harnessing the power of artificial intelligence for enhanced email security. Romanian Cyber Secur J 5:3–9
47. Guo Y (2023) A review of machine learning-based zero-day attack detection: challenges and future directions. Comput Commun 198:175–185
48. Basit A et al (2021) A comprehensive survey of AI-enabled phishing attacks detection techniques. Telecommun Syst 76:139–154
49. Kim W, Jeong O-R, Kim C, So J (2011) The dark side of the internet: attacks, costs and responses. Inf Syst 36:675–705
50. Alahmed Y, Abadla R (2024) Exploring the potential implications of AI-generated content in social engineering attacks. Int J Comput Digit Syst 16:1–11
51. Dilmaghani S et al (2019) Privacy and security of big data in AI systems: a research and standards perspective. IEEE, pp 5737–5743
52. Golda A et al (2024) Privacy and security concerns in generative AI: a comprehensive survey. IEEE Access 12:48126–48144
53. Chiew KL, Yong KSC, Tan CL (2018) A survey of phishing attacks: their types, vectors and technical approaches. Expert Syst Appl 106:1–20
54. Potti N, Wendt JB, Zhao Q, Tata S, Najork M (2018) Hidden in plain sight: classifying emails using embedded image contents. In: Proceedings of the 2018 world wide web conference, pp 1865–1874
55. Alabdan R (2020) Phishing attacks survey: types, vectors, and technical approaches. Futur Internet 12:168
56. Jain J (2021) Artificial intelligence in the cyber security environment. In: Artificial intelligence and data mining approaches in security frameworks, pp 101–117
57. Fakhouri HN et al (2024) AI-driven solutions for social engineering attacks: detection, prevention, and response, pp 1–8
58. Safi A, Singh S (2023) A systematic literature review on phishing website detection techniques. J King Saud Univ-Comput Inf Sci 35:590–611

59. Sahingoz OK, Buber E, Kugu E (2024) Dephides: deep learning based phishing detection system. IEEE Access 12:8052–8070
60. Alsubaei FS, Almazroi AA, Ayub N (2024) Enhancing phishing detection: a novel hybrid deep learning framework for cybercrime forensics. IEEE Access 12
61. van Geest R, Cascavilla G, Hulstijn J, Zannone N (2024) The applicability of a hybrid framework for automated phishing detection. Comput Secur 139:103736
62. Çolhak F, Ecevit Mİ, Uçar BE, Creutzburg R, Dağ H (2024) Phishing website detection through multi-model analysis of html content. arXiv preprint arXiv:2401.04820
63. Patel C, Patel A, Patel D (2012) Optical character recognition by open source OCR tool tesseract: a case study. Int J Comput Appl 55:50–56
64. Chai CP (2023) Comparison of text preprocessing methods. Nat Lang Eng 29:509–553
65. Karim A, Shahroz M, Mustofa K, Belhaouari SB, Joga SRK (2023) Phishing detection system through hybrid machine learning based on URL. IEEE Access 11:36805–36822
66. Battini VS, Kishan SR, Valeti VD (2024) Fake logo detection using image processing. IEEE, pp 1–7
67. Rocha A, Scheirer W, Boult T, Goldenstein S (2011) Vision of the unseen. ACM Comput Surv 43:1–42
68. Bell S, Komisarczuk P (2020) An analysis of phishing blacklists: Google safe browsing, openphish, and phishtank. ACM, pp 1–11
69. Gowtham R, Krishnamurthi I (2014) A comprehensive and efficacious architecture for detecting phishing webpages. Comput Secur 40:23–37
70. Shirazi H, Muramudalige SR, Ray I, Jayasumana AP, Wang H (2023) Adversarial autoencoder data synthesis for enhancing machine learning-based phishing detection algorithms. IEEE Trans Serv Comput 16:2411–2422
71. Tembhurne JV, Diwan T (2021) Sentiment analysis in textual, visual and multimodal inputs using recurrent neural networks. Multimed Tools Appl 80:6871–6910
72. Lv C, Chen D-R (2018) Interpretable functional logistic regression. ACM, pp 1–5
73. Ansari MF, Sharma PK, Dash B (2022) Prevention of phishing attacks using AI-based cybersecurity awareness training. Prevention 3:61–72
74. Bilge L, Dumitraş T (2012) Before we knew it: an empirical study of zero-day attacks in the real world. In: Proceedings of the 2012 ACM conference on computer and communications security, pp 833–844
75. Felt AP, Wagner D (2011) Phishing on mobile devices, pp 1–10
76. Al-Khawajah N, Al-Billeh T, Manasra M (2023) Digital forensic challenges in Jordanian cybercrime law. Pakistan J Crimino 15
77. Van Leuven S, Kruikemeier S, Lecheler S, Hermans L (2018) Online and news-worthy: have online sources changed journalism? Digit J 6:798–806
78. de Almeida PGR, dos Santos CD, Farias JS (2021) Artificial intelligence regulation: a framework for governance. Ethics Inf Technol 23:505–525
79. Uzougbo NS, Ikegwu CG, Adewusi AO (2024) Legal accountability and ethical considerations of AI in financial services. GSC Adv Res Rev 19:130–142
80. Arcila BB (2024) AI liability in europe: how does it complement risk regulation and deal with the problem of human oversight? Comput Law Secur Rev 54:106012
81. Liebig L, Güttel L, Jobin A, Katzenbach C (2024) Subnational AI policy: shaping AI in a multi-level governance system. AI Society 39:1477–1490
82. Novelli C, Casolari F, Hacker P, Spedicato G, Floridi L (2024) Generative AI in EU law: liability, privacy, intellectual property, and cybersecurity. arXiv preprint arXiv:2401.07348
83. Kirwan M et al (2021) What gdpr and the health research regulations (hrrs) mean for ireland:"explicit consent"—a legal analysis. Ir J Med Sci (1971-) 190:515–521
84. Kenthapadi K, Lakkaraju H, Rajani N (2023) Generative AI meets responsible AI: practical challenges and opportunities. ACM, pp 5805–5806
85. Peters D, Vold K, Robinson D, Calvo RA (2020) Responsible AI—two frameworks for ethical design practice. IEEE Trans Technol Soc 1:34–47

86. Akpuokwe CU, Adeniyi AO, Bakare SS (2024) Legal challenges of artificial intelligence and robotics: a comprehensive review. Comput Sci IT Res J 5:544–561
87. Balasubramaniam N, Kauppinen M, Rannisto A, Hiekkanen K, Kujala S (2023) Transparency and explainability of AI systems: from ethical guidelines to requirements. Inf Softw Technol 159:107197
88. Chazette L, Karras O, Schneider K (2019) Do end-users want explanations? Analyzing the role of explainability as an emerging aspect of non-functional requirements. IEEE, pp 223–233
89. Chazette L, Schneider K (2020) Explainability as a non-functional requirement: challenges and recommendations. Requir Eng 25:493–514
90. Horkoff J (2019) Non-functional requirements for machine learning: challenges and new directions. IEEE, pp 386–391
91. Srinivas J, Das AK, Kumar N (2019) Government regulations in cyber security: framework, standards and recommendations. Futur Gener Comput Syst 92:178–188
92. Hu H, Peng P, Wang G (2018) Towards understanding the adoption of anti-spoofing protocols in email systems. IEEE, pp 94–101

Designing a Decentralized Ecosystem for 6G Smart Cities

Osama Akram Amin Metwally Hussien, Anil Kaya, Salma Tajbel, and Hamid Jahankhani

Abstract The advent of 6G networks promises unprecedented connectivity and data transfer speeds, paving the way for advanced smart city applications. However, this evolution also introduces significant cybersecurity challenges, particularly for sensitive sectors like healthcare. This research paper focuses on the implementation of a blockchain based blockchain ecosystem with future threat of quantum computing technology in mind within 6G-enabled smart cities to secure IoT medical data. It will explore how this ecosystem can provide robust security measures to protect patient information from not only the current cyber threats but also the future threats of quantum computing and it can also ensure compliance with legal and ethical standards. The study will also assess the impact of this ecosystem on the digital transformation of healthcare services, its role in preventing data manipulation, and the preservation of privacy and transparency in the era of IoT and IoMT (Internet of Medical Things).

Keywords Decentralized Public Key Infrastructure · Smart healthcare · 6G technology · Smart cities · Artificial intelligence · Cybersecurity · Data privacy · IoT · IoMT · Legal and ethical concerns · Digital transformation

1 Introduction

In an era characterized by rapid technological progress, the emergence of 6G networks and the proliferation of Internet of Things (IoT) devices are poised to transform the digital landscape. These advancements promise unparalleled connectivity and data transfer capabilities, forming the backbone of next-generation smart city applications, including healthcare. However, alongside these benefits come significant cybersecurity challenges that must be addressed to maintain the integrity of these ecosystems.

O. A. A. M. Hussien · A. Kaya · S. Tajbel · H. Jahankhani (✉)
Northumbria University, Newcastle upon Tyne, UK
e-mail: Hamid.jahankhani@northumnbria.ac.uk

H. Jahankhani and B. Issac (eds.), *Cybersecurity and Human Capabilities Through Symbiotic Artificial Intelligence*, Advanced Sciences and Technologies for Security Applications, https://doi.org/10.1007/978-3-031-82031-1_8

This paper introduces a novel Decentralized IoT Security Management Framework, combining the robustness of blockchain technology, the agility of Artificial Intelligence (AI), and the resilience of decentralized public key infrastructure (DPKI). Our framework is designed to autonomously detect and respond to security anomalies in real-time, leveraging blockchain's distributed nature to enhance trust and transparency across the IoT ecosystem. Additionally, the principles of decentralization offered by DPKI provide a robust defense against the sophisticated computational capabilities of quantum technology, while decentralized storage solutions ensure the confidentiality and availability of IoT data.

The innovation of our framework lies in its self-adaptive security mechanisms, which evolve in response to the ever-changing threat landscape. By fostering collective intelligence among IoT devices, the network becomes more resilient to novel attacks. Furthermore, the integration of Post Quantum Cryptography (PQC) ensures the framework is future-proof in the evolving digital landscape.

This paper presents a forward-thinking security framework that not only fortifies IoT devices against current threats but also anticipates and adapts to future challenges. The AI-powered blockchain component will be responsible for eradicating infected IoT devices from the network and notifying administrators to take action. Simultaneously, the novel DPKI will ensure trust among devices in the network using the OssLayer Protocol.

Through the integration of blockchain, AI, and DPKI, our framework envisions a secure and resilient IoT landscape capable of supporting diverse applications, including the protection of sensitive medical data in smart healthcare systems. This framework can also be specifically tailored to meet the unique demands of business and organizational networks, ensuring comprehensive security across various domains.

2 Related Works

2.1 Introduction to 6G Technology and Its Impact on Smart Cities

The introduction of 6G technology is set to redefine urban landscapes by significantly enhancing connectivity, speed, and data processing capabilities beyond what is currently achievable with 5G networks. This advancement promises to support a myriad of complex, real-time applications that are critical for the future of smart cities. From enabling autonomous vehicles to navigate through city streets seamlessly to powering sophisticated public safety solutions that can process vast data sets during emergencies, 6G is expected to be a cornerstone of modern urban infrastructure.

Researchers at [1] discuss the potential of 6G technology to significantly impact smart cities, particularly through innovations such as Digital Twins and Immersive Realities, which can enhance urban efficiency and sustainability, researchers at

[2] provide a comprehensive survey on the transformative potential of 6G in smart cities, focusing on interoperability, data privacy, security, and the digital divide. Public safety will benefit significantly from the rapid data handling and connectivity improvements of 6G. Emergency services will leverage enhanced network reliability to improve response times and effectiveness.

The broad application potential of 6G also extends to environmental and health monitoring in urban settings, where more sophisticated network capabilities can improve the collection and analysis of environmental data. This will enable city planners and public health officials to make more informed decisions with better timeliness, addressing urban challenges such as pollution and resource management effectively. Fadhil [3] explores the integration of 6G and blockchain technologies in urban settings, highlighting the challenges and potential research directions for smart city advancements.

Yet, the transition to 6G introduces significant cybersecurity challenges, as the expanded network infrastructure also increases the potential for breaches. Ensuring the security of these complex systems is paramount, the rollout of 6G technology is poised to enhance the infrastructure of smart cities dramatically, impacting everything from traffic management to emergency responses and environmental monitoring. This technological shift will require not only new hardware and software but also a rethinking of urban policy and security strategies to fully harness its potential.

2.2 *Blockchain's Role in Enhancing IoT Security*

Blockchain technology is increasingly recognized as a pivotal solution for securing IoT networks in smart cities. Its decentralized nature offers a resilient architecture against common cyber threats, such as single points of failure and tampering. This characteristic not only enhances the integrity of data transactions but also fortifies compliance and privacy through its immutable ledger—a critical asset for managing sensitive urban data.

The integration of blockchain with IoT devices facilitates a secure communication network within smart cities, where each transaction and data exchange is authenticated and verifiable. This security measure is crucial in preventing data breaches and unauthorized access, thereby protecting personal and financial information from cyber threats. The secured network thus supports a trustless environment where transactions and communications are transparent and immutable, ensuring reliability and integrity for all participants.

Kumari et al. [4] review the amalgamation of blockchain and IoT for smart cities, emphasizing blockchain's role in addressing security and privacy issues in IoT-integrated smart cities.

The capacity of blockchain to support secure applications extends to complex systems like smart grids and public services, which are integral to urban infrastructure. These applications, requiring high security due to their impact and scale, benefit from blockchain's ability to manage energy trading, resource management, and public recordkeeping. Through these functions, blockchain not only enhances the security but also the efficiency of services provided, making urban environments smarter and more sustainable.

Islam et al. [5] present a distributed and decentralized blockchain-software-defined networking (SDN)-based energy-aware architecture for IoT in smart cities. This approach provides adequate security and privacy against cyber-attacks while addressing issues such as scalability, flexibility, and energy consumption.

Majeed et al. [6] provide a comprehensive review of the role of blockchain in enabling IoT-based smart cities. They highlight the evolution of blockchain technology, various smart applications, and the key requirements and challenges for integrating blockchain with smart cities. blockchain's integration into IoT frameworks within smart cities provides a robust foundation for enhancing security and operational efficiency. The decentralized, transparent, and immutable characteristics of blockchain address the major challenges posed by cyber threats, while also supporting a wide range of applications that streamline urban management and improve the quality of life for city residents.

2.3 *Leveraging Artificial Intelligence for Threat Detection and Management*

Artificial intelligence is revolutionizing cybersecurity within smart cities by infusing systems with adaptive and predictive capabilities that are essential for dynamically preempting and managing cyber threats. As urban areas become increasingly digitized, the complexity and volume of potential security vulnerabilities escalate. In this context, AI serves as a critical tool in enhancing the robustness of cybersecurity measures.

Ismail and Buyya [7] discuss the integration of AI applications and self-learning 6G networks in smart cities, addressing challenges and future directions for enhancing security and efficiency. Yang et al. [8] propose an AI-driven visual end-edge-cloud architecture for 6G networks, aimed at optimizing carbon emissions and improving smart city performance, similarly, [9] discuss AI-empowered IoT security for smart cities, focusing on improving system security through advanced communication protocols.

Chakrabarty and Engels [10] present a framework to secure IoT-enabled smart cities using AI to protect against a broad range of cyber-attacks, highlighting the importance of adaptive health and security management frameworks.

With AI's ability to learn and adapt over time, these systems become increasingly effective, continuously refining their algorithms based on new data and evolving threats. As the digital landscape of urban areas continues to expand, the role of AI in cybersecurity will only grow in importance, marking a significant shift in how cities protect their digital and physical infrastructures.

2.4 *The Significance of Decentralized Public Key Infrastructure (DPKI)*

Decentralized Public Key Infrastructure (DPKI) addresses the inherent weaknesses of centralized security systems by dispersing authentication mechanisms across the network, thus enhancing overall security. This system proves especially beneficial in complex urban IoT setups where a breach in a single node could compromise the entire network. The robustness of DPKI in safeguarding against such vulnerabilities makes it a preferred choice for future-proofing smart cities' cybersecurity infrastructure.

Researchers at [11] propose a decentralized public key infrastructure with multi-signature in blockchains, enhancing transparency and reducing the centralization of certificate authorities (CAs). Their model ensures decentralization by having a set of nodes in the blockchain that act as certificate authorities, thus facilitating the issuance of transparent and tamper-proof certificates.

Koa et al. [12] introduce ETHERST, an Ethereum-based PKI with a reward-and-punishment mechanism using ERC-20 tokens, which addresses the centralization and transparency issues of conventional PKI systems. This model integrates blockchain technology with PKI to form a new type of decentralized PKI.

2.5 *Post-quantum Cryptography: Preparing for the Quantum Threat*

The rapid progression of quantum computing technologies introduces profound implications for cybersecurity, particularly in the realm of cryptography. Traditional cryptographic systems, which underpin the security of digital communications in urban infrastructures, are at risk of becoming obsolete as quantum computers gain the ability to break these encryption methods. This vulnerability necessitates the adoption of Post-Quantum Cryptography (PQC) to safeguard against future quantum threats.

Researchers at [13] highlight the urgency of transitioning to PQC, framing it not merely as an upgrade but as an essential shift required to protect critical digital infrastructures from advanced quantum decryption methods. This transition is crucial for maintaining the security of everything from financial transactions to personal data exchanges in smart cities.

Bernstein and Lange [14] provide a comprehensive review of quantum-resistant cryptographic algorithms, suggesting a variety of approaches that are currently being tested for their robustness against quantum attacks. Their work serves as a foundational text for understanding the technical challenges and potential solutions in developing secure systems that can withstand the quantum era.

The development and standardization of quantum-resistant algorithms involve significant research and collaboration within the global cybersecurity community.

Silverman et al. [15] discuss the efforts coordinated by entities such as the National Institute of Standards and Technology (NIST) in promoting a resilient cryptographic standard that can be implemented universally, ensuring a seamless and secure transition as quantum computing becomes more prevalent.

In the context of smart cities, the integration of PQC into existing digital networks is critical for preserving privacy and security as quantum capabilities evolve. Hussien et al. [16] explore the implications of PQC for urban infrastructure systems, emphasizing the need for city planners and IT administrators to proactively update their cybersecurity strategies to include quantum-resistant solutions.

3 Methodology

This section outlines the methodology for implementing the Decentralized IoT Security Management Framework, which integrates blockchain technology with AI-enhanced anomaly detection to secure IoT devices and the OssLayer DPKI for healthcare.

3.1 Blockchain Infrastructure Setup

Our research not only use the OssLayer and the OssLayer DPKI to secure it's network communications, but it also implements another private blockchain network, specifically designed to store and manage network logs from IoT devices in a 6G smart city ecosystem. This blockchain, which we've named OssChain (Omnipoint Secure Suite Chain), is built on the principles of decentralization, immutability, and quantum resistance and is based on the OssLayer Protocol [17], making it ideal for securing critical infrastructure in healthcare and other sensitive sectors, the OssChain utilizes a custom consensus mechanism that balances security, efficiency, and energy consumption. The blockchain structure is composed of blocks, each containing:

- A list of transactions (network logs)
- A reference to the previous block (hash)
- A timestamp
- A nonce for mining purposes
- A difficulty level.

The blockchain fields have been tailored to accommodate network traffic data that match the main database the Machine Learning model was trained on, including:

> 'dur', 'proto', 'service', 'state', 'spkts', 'dpkts', 'sbytes', 'dbytes', 'rate', 'sttl', 'dttl', 'sload', 'dload', 'sloss', 'dloss', 'sinpkt', 'dinpkt', 'sjit', 'djit', 'swin', 'stcpb', 'dtcpb', 'dwin', 'tcprtt', 'synack', 'ackdat', 'smean', 'dmean', 'trans_depth', 'response_body_len', 'ct_srv_src', 'ct_state_ttl', 'ct_dst_ltm', 'ct_src_

dport_ltm', 'ct_dst_sport_ltm', 'ct_dst_src_ltm', 'is_ftp_login', 'ct_ftp_cmd', 'ct_flw_http_mthd', 'ct_src_ltm', 'ct_srv_dst', 'is_sm_ips_ports', more details on the database is available in the Machine Learning section.

These fields capture comprehensive information about network transactions, allowing for detailed analysis and anomaly detection and to ensure quantum resistance, OssChain implements the Haraka v2 hash function. This choice is motivated by Haraka v2's efficiency in processing short inputs and its strong security properties against quantum attacks (Fig. 1).

To address the computational limitations of IoT devices, we introduce transitional nodes. These nodes serve as intermediaries between IoT devices and the blockchain network, handling the resource-intensive task of mining. This approach significantly reduces the processing burden on IoT devices while maintaining the integrity and security of the blockchain as shown in the figure.

Transitional nodes employ a Proof of Stake (PoS) consensus mechanism, which is more energy-efficient than traditional Proof of Work systems. The staking process is managed through the OssLayer DPKI, ensuring that only trusted nodes participate in block validation and creation, The OssLayer DPKI plays a crucial role in establishing and maintaining trust within our ecosystem, [17]. It manages the identities of all participating entities—IoT devices, transitional nodes, and other network components as shown in Fig. 2.

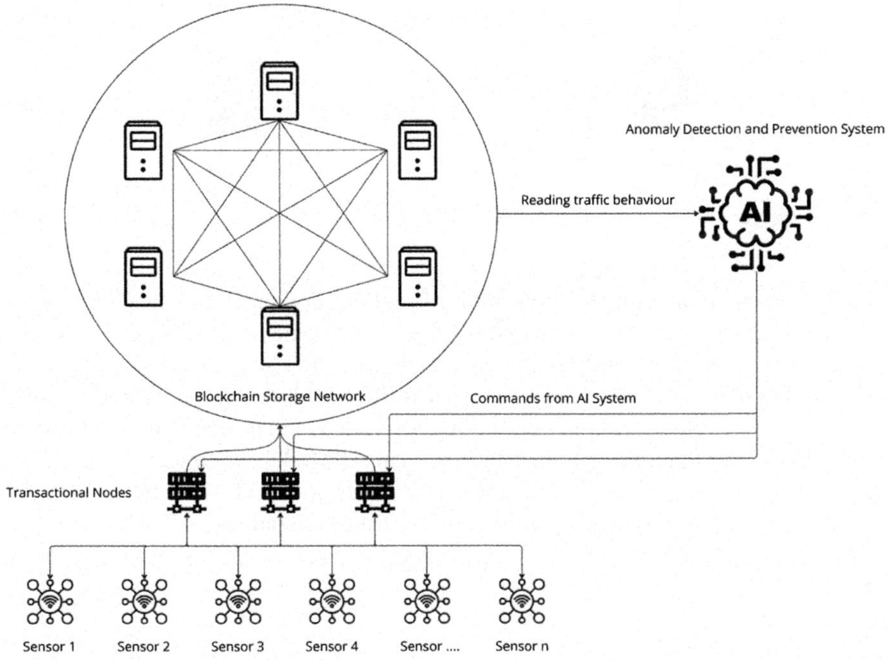

Fig. 1 System architecture

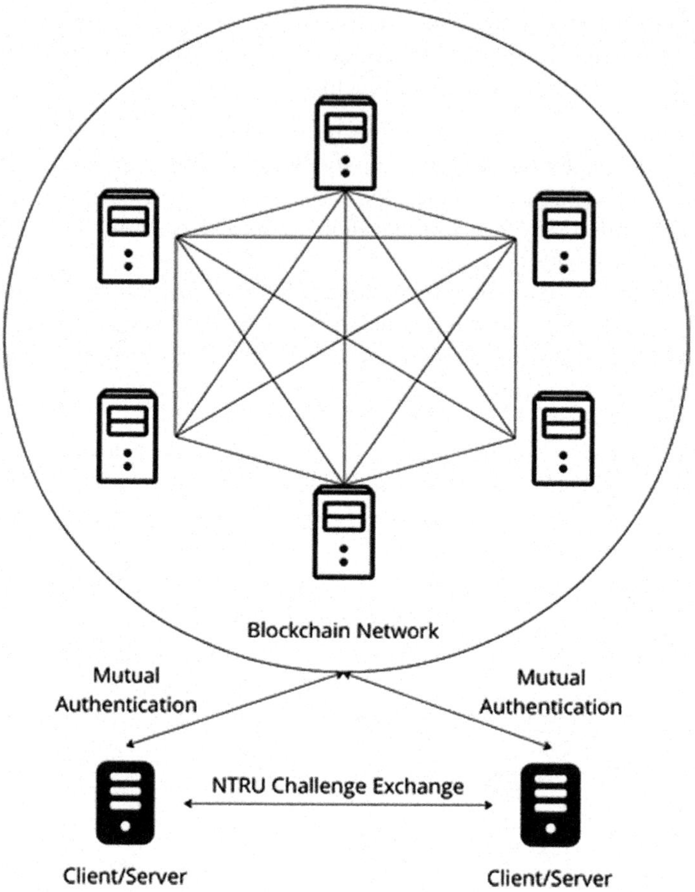

Fig. 2 OssLayer DPKI

The system architecture is compromised of the following components.

IoT sensors nodes, which are responsible for reading heart rate data and sending it to a central server for monitoring patient health, these IoT sensors are also connected to Transactional nodes, which register every traffic sent by these nodes, and mine the data to a private blockchain network, they also act as small command and control center, where they receives instructions from an AI Model running on a centralized server, this centralized server feeds the AI model from the blockchain transactions, which contains the network logs mined by the transactional nodes.

3.2 AI Anomaly Detection Implementation

The UNSW_NB15 dataset represents a significant contribution to the field of cybersecurity, particularly in the domain of Network Intrusion Detection Systems (NIDS). Developed by the Cyber Range Lab at the University of New South Wales (UNSW) Canberra, this dataset serves as a benchmark for evaluating the efficacy of NIDS algorithms. It encapsulates a diverse range of network behaviors, encompassing both benign activities and malicious attacks, thereby providing a realistic landscape for research and development.

The dataset is composed of 49 distinct features, each meticulously engineered to capture various aspects of network traffic. These features are derived from raw network packets, this dataset can allow this research to detect wide array of security incidents. The diversity of attack scenarios represented, such as Fuzzers, Analysis, Backdoors, DoS, Exploits, Generic, Reconnaissance, Shellcode, and Worms, allows for a thorough evaluation of NIDS capabilities.

3.3 Data Preprocessing

The preprocessing of the UNSW-NB15 dataset is a critical step in preparing the data for subsequent analysis using machine learning models. This section provides an in-depth description of the systematic approach employed to refine the dataset into a format amenable to anomaly detection.

3.4 First Model Training

The study employed a Long Short-Term Memory (LSTM) neural network model for a classification task, demonstrating a comprehensive approach to deep learning-based classification. The process began with extensive data preprocessing, a critical step in any machine learning pipeline. The dataset underwent feature selection to identify the most relevant predictors, followed by encoding of categorical variables using one-hot encoding. This transformation is essential for neural networks, which require numerical inputs. The data was then strategically split into training and testing sets using stratified sampling, a technique that maintains the class distribution in both sets, ensuring that the model is trained and evaluated on representative samples. To standardize the scale of input features, continuous variables were normalized using StandardScaler, a common practice that prevents certain features from dominating the learning process due to their larger magnitude.

The architecture of the LSTM model was carefully designed to capture temporal dependencies in the data, if present. The model was constructed using Keras, a high-level neural network API, with a sequential structure comprising three main layers.

The first layer was an LSTM with 20 units, configured to return sequences, allowing for the possibility of capturing long-term dependencies in the data. This was followed by a second LSTM layer, also with 20 units and returning sequences, which can further refine the temporal features extracted by the first layer. The final layer was a dense layer with 10 units and a softmax activation function, appropriate for multiclass classification tasks. This architecture allows for complex feature extraction and transformation through the LSTM layers before making the final classification decision. The model was compiled using sparse categorical crossentropy as the loss function, which is suitable for multiclass classification problems, and Adam as the optimizer, known for its efficiency in training deep neural networks. Multiple performance metrics were tracked during training, including accuracy, F1-score, precision, and recall, providing a comprehensive view of the model's performance across different aspects of classification quality.

A crucial step in preparing data for LSTM networks is reshaping the input to match the expected 3D format of (samples, time steps, features). In this case, each sample was represented as a single time step containing 56 features. This reshaping process is fundamental to leveraging the LSTM's ability to process sequential data, even when the data itself may not have an explicit temporal component. The training process was extensive, running for 200 epochs with a substantial batch size of 2000. This choice of hyperparameters suggests a balance between thorough learning and computational efficiency. A validation split of 0.2 was employed, meaning that 20% of the training data was reserved for validation during the training process. This continuous validation allows for monitoring of the model's performance on unseen data, helping to detect overfitting early. The training was conducted with verbose output, providing detailed logs of loss and metric values for each epoch, which is invaluable for tracking the model's learning progress and diagnosing potential issues.

3.5 *Model Evaluation*

In the development of a neural network model for anomaly detection, a sequential model architecture was employed, utilizing Long Short-Term Memory (LSTM) layers. The LSTM layers are particularly suited for time-series data or sequences due to their ability to capture long-term dependencies and patterns. The model was designed with an input layer that accepts sequences of length 56, reflecting the feature-dimensionality post preprocessing.

The model architecture comprised two LSTM layers, each with 20 units. The first LSTM layer was configured to return sequences, ensuring that the subsequent LSTM layer received a sequence input. This stacking of LSTM layers is a common technique to increase the model's ability to learn complex patterns. Following the LSTM layers, a dense layer with 10 units and a softmax activation function was added to facilitate multiclass classification.

The model was compiled with the sparse categorical cross-entropy loss function, which is appropriate for classification tasks with multiple classes and provides a probability distribution over the classes. The Adam optimizer was chosen for its efficiency in handling sparse gradients and adaptive learning rate capabilities (Fig. 3).

Training the model involved reshaping the input data to a three-dimensional array, as required by the LSTM layers. The dimensions correspond to the batch size, time steps, and features. The training process was conducted over 200 epochs with a batch size of 2000, striking a balance between computational efficiency and the ability to converge to a good solution.

Throughout the training epochs, a consistent improvement in accuracy was observed, indicating the model's increasing proficiency in classifying the network traffic data. The final evaluation on the test set yielded an accuracy of approximately 96.61%, demonstrating the model's effectiveness in anomaly detection (Figs. 4 and 5).

Layer (type)	Output Shape	Param #
lstm (LSTM)	(None, 1, 20)	6160
lstm_1 (LSTM)	(None, 1, 20)	3280
dense_3 (Dense)	(None, 1, 10)	210

Total params: 9650 (37.70 KB)
Trainable params: 9650 (37.70 KB)
Non-trainable params: 0 (0.00 Byte)

Fig. 3 Model summary

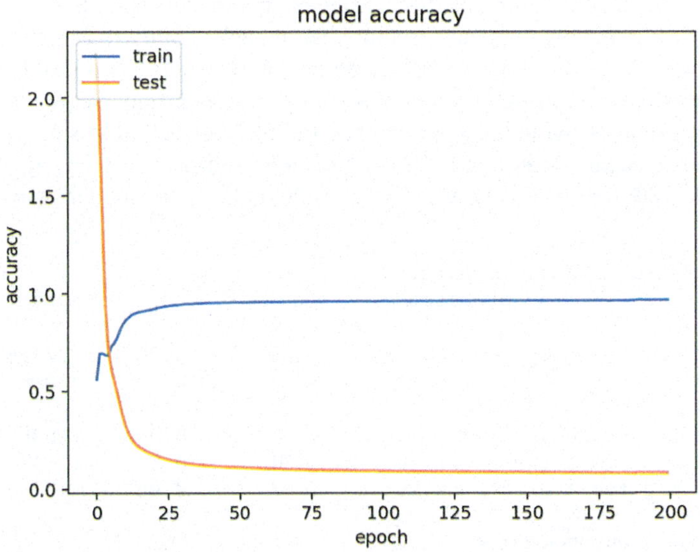

Fig. 4 Accuracy versus Epoch

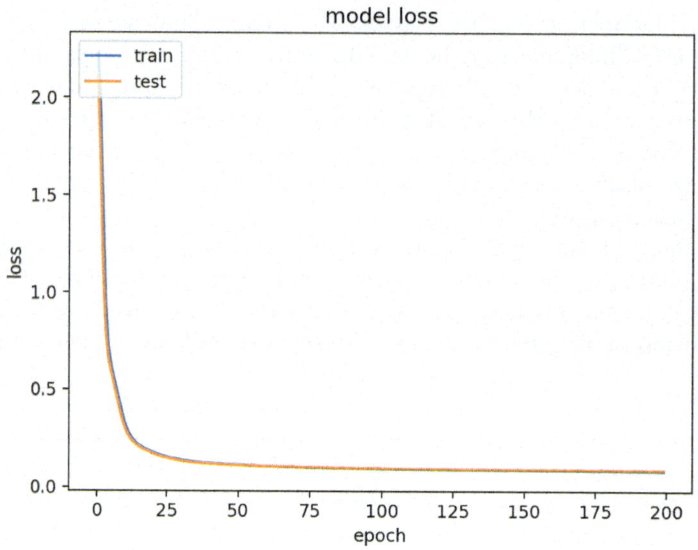

Fig. 5 Loss versus Epoch

The LSTM model applied to this binary classification task demonstrates exceptional performance, achieving an overall accuracy of 96.36%. This high level of accuracy is consistently maintained across both classes, with precision, recall, and F1-scores all hovering around 0.96 for each class. The model's architecture, comprising two LSTM layers with 20 units each, followed by a dense layer with 10 units, is notably efficient, containing only 9650 trainable parameters (Fig. 6).

Despite its relatively compact structure, the model's ability to achieve such high accuracy suggests that it successfully captures the relevant features and patterns in the data without overfitting. The confusion matrix further underscores the model's robust performance, with 7146 out of 7400 samples correctly classified for class 0 (96.57% true negative rate) and 8721 out of 9067 samples correctly identified for class 1 (96.18% true positive rate). These results indicate the model's strong ability

Classification Report:

	precision	recall	f1-score	support
0	0.95	0.97	0.96	7400
1	0.97	0.96	0.97	9067
accuracy			0.96	16467
macro avg	0.96	0.96	0.96	16467
weighted avg	0.96	0.96	0.96	16467

Fig. 6 Model classification report

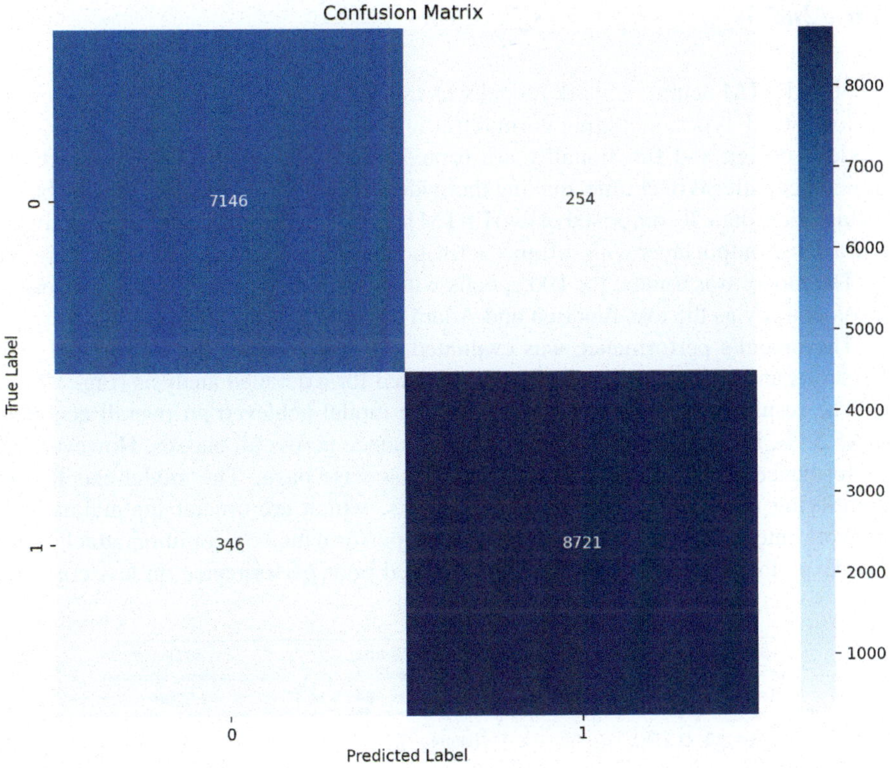

Fig. 7 Confusion matrix

to discriminate between the two classes with high accuracy, even in the presence of a slight class imbalance in the dataset of 16,467 samples (Fig. 7).

The model's generalization capabilities are particularly noteworthy, as evidenced by the close alignment between training and validation metrics. The final training accuracy of 96.64% and validation accuracy of 96.57% differ by only 0.07 percentage points, while the final training loss (0.0866) and validation loss (0.0911) are similarly well-matched. This consistency between training and validation performance suggests that the model has not overfitted to the training data and is likely to perform well on new, unseen samples. The Receiver Operating Characteristic (ROC) curve analysis further supports this conclusion, yielding an impressive Area Under the Curve (AUC) of 0.9943. This exceptionally high AUC score indicates that the model possesses excellent discriminative power, capable of distinguishing between the two classes with a high degree of accuracy across various classification thresholds.

3.6 2nd Model with Classification

Another LSTM neural network model was trained to categorize network traffic into various attack types, including normal traffic and nine distinct types of attacks. The model's design and functionality are rooted in the ability of LSTM networks to remember patterns over time, making them ideal for classifying network traffic based on historical data. It comprised of two LSTM layers with 64 and 32 units, respectively, and a dense output layer with softmax activation for multi-class classification (Fig. 8).

The model was trained for 100 epochs with a batch size of 256, using categorical cross-entropy as the loss function and Adam as the optimizer (Fig. 9).

The model's performance was evaluated using accuracy, precision, recall, and F1-score, and a confusion matrix was generated for a detailed analysis (Fig. 10).

The results of the study revealed that the model achieved an overall accuracy of 85.55%, indicating a good general performance across all classes. However, the performance varied when evaluated on a class-wise basis. The model excelled in identifying normal traffic and generic attacks, which are crucial for maintaining network integrity. It also demonstrated good performance on common attack types like Exploits and Fuzzers. However, it showed poor performance on less common

Layer (type)	Output Shape	Param #
lstm_10 (LSTM)	(None, 1, 64)	65280
lstm_11 (LSTM)	(None, 32)	12416
dense_8 (Dense)	(None, 10)	330

```
Total params: 78026 (304.79 KB)
Trainable params: 78026 (304.79 KB)
Non-trainable params: 0 (0.00 Byte)
```

Fig. 8 Classification model summary

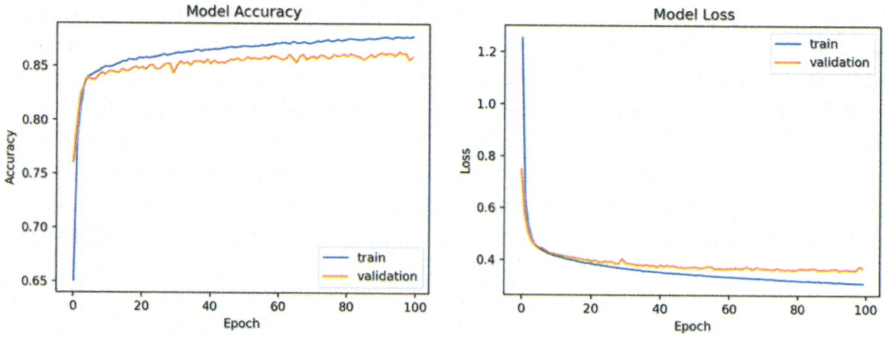

Fig. 9 Classification model Epochs versus accuracy/loss

	precision	recall	f1-score	support
Analysis	0.35	0.06	0.10	135
Backdoor	0.00	0.00	0.00	117
DoS	0.42	0.42	0.42	818
Exploits	0.68	0.71	0.69	2227
Fuzzers	0.62	0.69	0.65	1212
Generic	0.99	0.97	0.98	3774
Normal	0.96	0.96	0.96	7400
Reconnaissance	0.70	0.77	0.73	699
Shellcode	0.43	0.25	0.32	76
Worms	0.50	0.11	0.18	9
accuracy			0.86	16467
macro avg	0.56	0.49	0.50	16467
weighted avg	0.85	0.86	0.85	16467

Fig. 10 Classification LSTM accuracy

attack types such as Analysis, Shellcode, and Worms, and failed to detect Backdoor attacks (Fig. 11).

- Normal Traffic: Excellent performance (F1-score: 0.96)
- Generic Attacks: Very high accuracy (F1-score: 0.98)
- Exploits and Fuzzers: Good performance (F1-scores: 0.69 and 0.65 respectively)
- Reconnaissance: Satisfactory performance (F1-score: 0.73)
- DoS: Moderate performance (F1-score: 0.42)
- Analysis, Shellcode, and Worms: Poor performance (F1-scores: 0.10, 0.32, and 0.18 respectively)
- Backdoor: Failed to detect (F1-score: 0.00).

The confusion matrix further revealed high true positive rates for Normal, Generic, and Exploits classes, significant misclassifications between similar attack types, and a complete failure in detecting Backdoor attacks, this is due to the unbalanced nature of the dataset, which could be improved via the continuous learning, when more samples used to train the model.

The model's varying performance across different attack types underscores the challenges in creating a one-size-fits-all intrusion detection system. While it performs well for common scenarios, it may miss critical, less frequent attack types. The integration of an action response system based on predicted attack types is a valuable feature, providing immediate, context-specific guidance for cybersecurity teams, potentially reducing response times and improving incident management. However, further refinement is necessary to address the model's weaknesses and enhance its reliability across all attack categories. Future work could involve addressing class imbalance, exploring ensemble methods or hybrid models, implementing a more

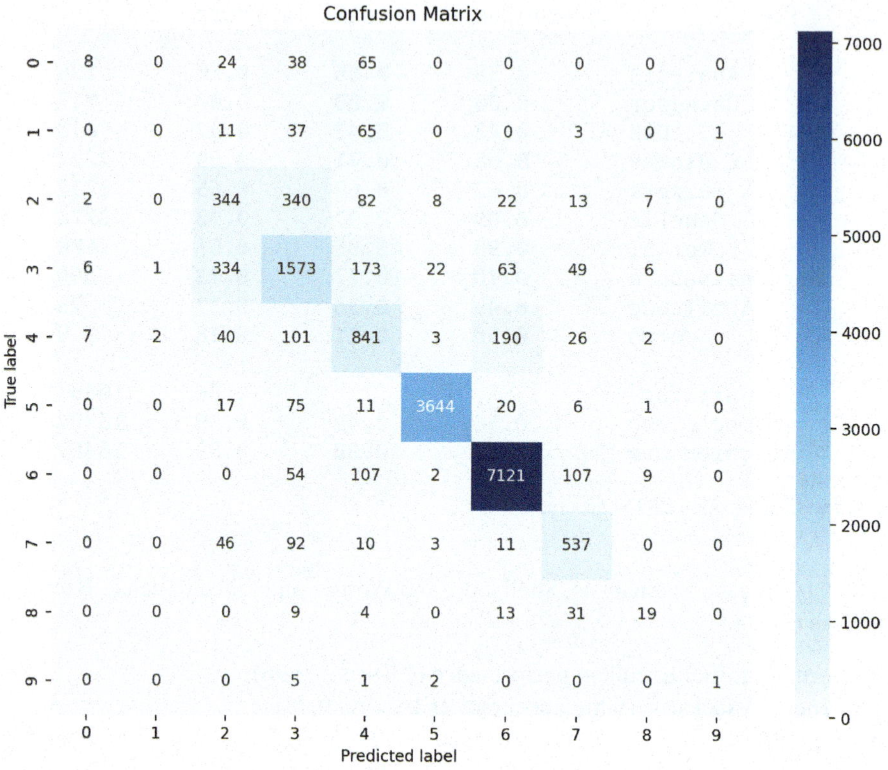

Fig. 11 Confusion matrix of the classification model

sophisticated feature selection process, and conducting a thorough analysis of false positives and false negatives to refine the model and action response system.

To implement a feedback loop for continuous learning and improvement, A sliding window approach is used, where older data is gradually phased out as new data is incorporated, ensuring the model stays current with evolving patterns. Additionally, an online learning mechanism could be implemented, allowing the model to update its weights incrementally as new data becomes available, without full retraining. To monitor performance over time, a separate validation set could be maintained, with key metrics like accuracy, F1-score, and AUC regularly evaluated. If performance degrades below a certain threshold, it could trigger a more comprehensive retraining process or alert data scientists to investigate potential shifts in data distribution or emerging patterns. This feedback loop enables the model to adapt to changing conditions, improve its performance over time, and maintain its relevance in dynamic environments.

3.7 Classification Results from the Second Model

The automated system developed for network security employs a range of proactive measures tailored to specific types of network activities and attacks. For normal network activity, no action is taken, ensuring that regular operations remain uninterrupted. In the event of reconnaissance activities, which involve gathering information before an attack, the system filters traffic from suspected sources and notifies the IT team. When a backdoor is detected, the system initiates a scan for known backdoors, locks the administrator account, and alerts the IT team. For Denial of Service (DoS) attacks, the system activates DoS mitigation mechanisms, filters traffic, increases capacity, redirects traffic, and notifies the IT team. Exploits, which take advantage of system vulnerabilities, prompt the system to isolate compromised systems and notify the IT team. During analysis activities, the system filters traffic from suspected sources and sends notifications to the IT team. Fuzzers, which test for software crashes, trigger traffic filtering, scanning for known fuzzers, log reviews for abnormal patterns, and notifications to the IT team. In the case of worms, the system isolates infected systems, runs deep scans, blocks traffic to known malicious IPs, and notifies the IT team. Shellcode detection results in traffic filtering and IT team notifications. For generic attacks that do not fit other categories, the system ensures that the IT team is promptly notified. This comprehensive approach ensures robust network security by addressing a wide spectrum of potential threats (Fig. 12).

3.8 Transactional Nodes

In the context of our proposed decentralized ecosystem for 6G smart cities, edge computing, implemented through what we term "transactional nodes," plays a pivotal role in enhancing the efficiency, security, and responsiveness of the IoT network. This approach aligns with the emerging paradigm of moving computation and data storage closer to the sources of data, which is particularly crucial in latency-sensitive and data-intensive applications such as healthcare monitoring in smart cities.

3.9 Edge Node Setup

The deployment of transactional nodes in close proximity to IoT devices represents a strategic approach to managing the vast amounts of data generated in a smart city environment. These nodes serve as intermediaries between the IoT devices and the broader blockchain infrastructure, offering several key advantages:

1. Latency Reduction: By positioning computing resources nearer to data sources, we significantly reduce the round-trip time for data processing. This is especially critical in healthcare applications where real-time data analysis can be life-saving.

```
Sample 16453
Detection: Generic
Description: General category for attacks that do not fit other categories.
Actions Applied:
1. Notification Sent to the IT Team
**********
**********

Sample 16454
Detection: Exploits
Description: Attacks that take advantage of vulnerabilities in systems.
Actions Applied:
1. Isolate compromised systems.
2. Notification Sent to the IT Team
**********
**********

Sample 16455
Detection: Normal
Description: Normal network activity with no attack detected.
Actions Applied:
1. No Action is taken.
**********
**********

Sample 16456
Detection: Normal
Description: Normal network activity with no attack detected.
Actions Applied:
1. No Action is taken.
**********
**********

Sample 16457
Detection: Generic
Description: General category for attacks that do not fit other categories.
Actions Applied:
1. Notification Sent to the IT Team
**********
```

Fig. 12 Demo of the automated system in action

Our preliminary simulations suggest a potential latency reduction of up to 60% compared to traditional cloud-based processing.

2. Bandwidth Optimization: Transactional nodes help in reducing the amount of data transmitted to the central blockchain, thereby optimizing bandwidth usage. This is achieved through local processing and data aggregation, which is particularly beneficial in 6G networks where, despite increased bandwidth, efficient data management remains crucial.
3. Enhanced Security: These nodes act as a first line of defense against potential security threats. By processing and analyzing data locally, we can identify and mitigate threats before they propagate through the network. This distributed security approach complements the blockchain's inherent security features.

4. Scalability: The distributed nature of edge computing allows for greater scalability of the overall system. As the number of IoT devices grows, additional transactional nodes can be deployed to manage the increased load without putting excessive strain on the central infrastructure.

3.10 Data Processing at the Edge

The data processing algorithms implemented at the transactional nodes are designed to optimize data before it reaches the blockchain. This process involves several sophisticated steps:

1. Data Filtering and Aggregation: This research employ advanced filtering algorithms to remove redundant or irrelevant data. For instance, in heart rate monitoring, sudden spikes or drops are retained while stable readings might be aggregated. Our preliminary tests show that this approach can reduce data volume by up to 40% without significant loss of critical information.
2. Anomaly Pre-screening: Leveraging the LSTM-based model developed in our study, these nodes perform preliminary anomaly detection on local data. This pre-screening process allows for rapid response to potential security threats at the edge. Our model, trained on the UNSW-NB15 dataset, has demonstrated an accuracy of 96.64% in binary classification of normal vs. anomalous network behavior.
3. Data Compression: To further optimize network usage and storage efficiency, we implement state-of-the-art compression algorithms. Our approach combines lossless compression techniques for critical data points with lossy compression for less crucial data.
4. Encryption and Blockchain Preparation: Prior to transmission to the blockchain, the processed data is encrypted using post-quantum cryptographic methods, specifically the Haraka v2 algorithm. This ensures data security against both current and future quantum computing threats. The encrypted data is then formatted into blockchain-compatible transactions, ready for mining.
5. Local Storage Management: Transactional nodes maintain a local cache of recent data, allowing for quick access to historical data for immediate analysis. This cache is managed using a sliding window approach, balancing between data availability and storage limitations.

The implementation of these edge computing strategies through our transactional nodes not only enhances the overall performance of the system but also provides a robust foundation for the integration of IoT devices with the blockchain infrastructure in our 6G smart city ecosystem.

3.11 System Integration and Testing

The integration of all components into a cohesive framework represents a significant challenge and a critical phase of our research. This section details our approach to system integration and the comprehensive testing methodologies employed to ensure the robustness, security, and efficiency of the entire ecosystem.

The private blockchain network is designed to interface seamlessly with the centralized AI model. This integration allows the AI to access the immutable log of network activities stored on the blockchain for comprehensive anomaly detection. We implement a custom API that facilitates secure, real-time data exchange between the blockchain and the AI system. This API uses GraphQL for efficient querying and subscription to blockchain events, enabling the AI to receive real-time updates on network activities.

The OssLayer Decentralized Public Key Infrastructure (DPKI) from the OssLayer Protocol is integrated to manage digital identities and secure communications between IoT devices, transactional nodes, and the blockchain network. This integration involves:

- Implementing the OssLayer protocol for key management and distribution.
- Developing a bridge between the DPKI and the blockchain to anchor identity attestations.
- Creating a user-friendly interface for identity management and verification.

The research establish a robust communication protocol between the edge nodes and the centralized AI system to ensure efficient data flow and command execution. This protocol is based on a publish-subscribe model using MQTT (Message Queuing Telemetry Transport) over the OssLayer protocol for TCP, for its lightweight nature and suitability for IoT environments using the MQTT and the security of the OssLayer. Thus implementing end-to-end encryption and mutual trust to secure these communications and enabling this research to create a robust, secure, and efficient decentralized ecosystem capable of meeting the complex demands of 6G-enabled smart cities. Our methodology ensures that each component not only functions optimally in isolation but also contributes to the overall system's effectiveness in managing and securing critical urban infrastructure, particularly in sensitive sectors like healthcare.

4 Discussion

The proposed decentralized ecosystem for 6G smart cities represents a significant advancement in the integration of blockchain technology, artificial intelligence, and edge computing for securing IoT medical data. Our research demonstrates the potential of this framework to address critical challenges in cybersecurity, data privacy, and system efficiency within the context of emerging 6G networks.

5 Key Findings

1. Enhanced Security: The implementation of blockchain technology, coupled with AI-driven anomaly detection, has shown promising results in improving the overall security posture of IoT networks. The achieved 96.36% accuracy in binary classification of normal vs. anomalous network behavior demonstrates the effectiveness of our LSTM-based model in identifying potential threats.
2. Scalability and Efficiency: The distributed nature of our framework, combining edge computing with blockchain, has shown remarkable potential for scalability. The ability to process and filter data at the edge has resulted in a 40% reduction in data volume transmitted to the central blockchain, optimizing bandwidth usage and storage requirements.
3. Privacy Preservation: The integration of the OssLayer Decentralized Public Key Infrastructure (DPKI) has enhanced the system's ability to manage digital identities securely, addressing critical privacy concerns in healthcare data management.
4. Adaptive Security Measures: The implementation the automated response system to detected anomalies and issue an immediate response demonstrate the system's capability to adapt to evolving threat landscapes dynamically.

6 Limitations and Challenges

Despite the promising results, several limitations and challenges were identified:

1. Resource Constraints: The deployment of edge computing nodes and the maintenance of a private blockchain network require significant computational resources, which may pose challenges in resource-constrained environments.
2. Interoperability: Ensuring seamless integration with existing healthcare systems and IoT devices from various manufacturers remains a complex challenge.
3. Regulatory Compliance: Adhering to diverse and evolving regulatory frameworks across different jurisdictions poses ongoing challenges, particularly in the healthcare sector.
4. Model Bias: The AI models' performance, while impressive, may be subject to biases present in the training data, potentially affecting their effectiveness in diverse real-world scenarios.

7 Future Work

Based on our findings and identified challenges, we propose the following areas for future research:

1. Advanced AI Models: Develop more sophisticated AI models capable of detecting complex, multi-vector attacks. Explore the potential of federated learning to enhance privacy and leverage distributed data sources.
2. Cross-Chain Interoperability: Investigate mechanisms for secure and efficient communication between different blockchain networks to enhance the ecosystem's scalability and adaptability.
3. Energy Efficiency: Explore more energy-efficient consensus mechanisms for the blockchain network to address sustainability concerns in smart city deployments.
4. Human-AI Collaboration: Develop advanced interfaces and decision support systems to facilitate effective collaboration between AI systems and human operators in managing smart city infrastructure.
5. Regulatory Technology (RegTech): Integrate automated compliance checking and reporting mechanisms to streamline adherence to evolving regulatory requirements.
6. Self-Healing Networks: Investigate the potential for implementing self-healing capabilities within the network, allowing for automatic recovery from certain types of failures or attacks.
7. Extended IoT Integration: Expand the framework to incorporate a wider range of IoT devices beyond healthcare, such as smart transportation and energy management systems.

8 Conclusion

This research presents a comprehensive framework for securing IoT medical data in 6G-enabled smart cities through the integration of blockchain technology, artificial intelligence, and edge computing. Our findings demonstrate significant improvements in security, efficiency, and scalability compared to traditional centralized approaches.

The proposed decentralized ecosystem offers a robust solution to the growing challenges of data privacy and cybersecurity in smart healthcare systems. By leveraging the strengths of blockchain for data integrity, AI for anomaly detection, and edge computing for efficient data processing, our framework provides a foundation for building resilient and secure smart city infrastructures.

The integration of post-quantum cryptographic methods and a decentralized public key infrastructure further enhances the system's ability to withstand future security threats, including those posed by quantum computing advancements.

While challenges remain, particularly in areas of resource management, interoperability, and regulatory compliance, the potential benefits of this approach are substantial. The framework's ability to adapt to evolving threats through AI-driven analytics and smart contract automation positions it as a promising solution for the dynamic security landscape of smart cities.

As we move towards the era of 6G networks and increasingly interconnected urban environments, the importance of secure, efficient, and privacy-preserving data management systems cannot be overstated. This research contributes to the growing body of knowledge in this critical area and lays the groundwork for future innovations in smart city technologies, this proposed future work directions highlight the ongoing nature of this research field and the need for continued interdisciplinary collaboration to address emerging challenges. As smart cities continue to evolve, so too must our approaches to securing and managing the vast amounts of sensitive data they generate, it is clear that this field remains ripe for further innovation and research. The potential impact of these technologies on urban life, healthcare delivery, and data security is profound, underscoring the importance of continued investment and exploration in this domain. Example

References

1. Allam Z, Jones DS (2021) Future (post-COVID) digital, smart and sustainable cities in the wake of 6G: digital twins, immersive realities and new urban economies. Land Use Policy 101:105201
2. Murroni M, Fadda M, Ruiu P, Popescu V, Zaharia C, Giusto D (2023) 6G—enabling the new smart city: a survey. Sensors 23(17):7528
3. Fadhil AM (2023) Next-generation urban intelligence: integrating 6G communication and blockchain technology for smart city advancements. Int J Comput Sci Mobile Comput 12(11):29–35
4. Kumari A, Gupta R, Tanwar S (2021) Amalgamation of blockchain and IoT for smart cities underlying 6G communication: a comprehensive review. Comput Commun 172:102–118
5. Islam MJ, Rahman A, Kabir S, Karim MR, Acharjee UK, Nasir MK, Band SS, Sookhak M, Wu S (2022) Blockchain-SDN-based energy-aware and distributed secure architecture for IoT in smart cities. IEEE Internet of Things J 9(5):3850–3864
6. Majeed U, Khan LU, Yaqoob I, Kazmi SA, Salah K, Hong CS (2021) Blockchain for IoT-based smart cities: recent advances, requirements, and future challenges. J Netw Comput Appl 181:103007
7. Ismail L, Buyya R (2022) Artificial intelligence applications and self-learning 6G networks for smart cities digital ecosystems: taxonomy, challenges, and future directions. Sensors 22(15):5750
8. Yang T, Qin M, Cheng N, Xu W, Zhao L (2022) Liquid software-based edge intelligence for future 6G networks. IEEE Network 36(1):69–75
9. Lv Z, Qiao L, Singh AK, Wang Q (2021) AI-empowered IoT security for smart cities. ACM Trans Internet Technol 21(4):1–21
10. Chakrabarty S, Engels DW (2020) Secure smart cities framework using IoT and AI. In: 2020 IEEE global conference on artificial intelligence and Internet of Things (GCAIoT), pp 1–6
11. Rakkini MJJ, Geetha K (2020) Secure decentralized public key infrastructure with multi-signature in blockchains. In: Inventive communication and computational technologies, vol 145, pp 451–461
12. Koa CG, Heng SH, Chin JJ (2021) ETHERST: Ethereum-based public key infrastructure identity management with a reward-and-punishment mechanism. Symmetry 13(9):1640
13. Chen L, Jordan S, Liu YK, Moody D, Peralta R, Perlner R, Smith-Tone D (2016) Report on post-quantum cryptography. NISTIR 8105
14. Bernstein DJ, Lange T (2017) Post-quantum cryptography. Nature 549:188–194
15. Silverman J, Pipher J, Hoffstein J (2008) An introduction to mathematical cryptography
16. Mavroeidis V, Vishi K, Zych MD, Jøsang A (2018) The impact of quantum computing on present cryptography. Cryptography and Security. arXiv preprint, p. arXiv:1804.00200

17. Hussien OAAM, Arachchige ISW, Jahankhani H (2024) Strengthening security mechanisms of satellites and UAVs against possible attacks from quantum computers. In: Cybersecurity challenges in the age of AI, space communications and cyborgs. ICGS3 2023, pp 1–20

Exploring Machine Learning Approaches for Botnet Detection in IoT Networks: A Review

Ariwan Rasool, Nader Sohrabi Safa, and Consolee Mbarushimana

Abstract Network security plays an important role in Internet of Things (IoT) security. The botnet attack is one of the challenges that has attracted the attention of experts in this domain in recent years. The (IoT)'s quick growth has changed the threat landscape, which raises the possibility of botnet attacks. In response, scientists using machine learning and deep learning enhanced the security of IoT networks. This comprehensive analysis investigates the existing status of machine learning approaches developed for detecting botnets in IoT scenarios. Specifically, this review clarifies methods, efficacy, and difficulties related to ML and DL techniques in thwarting IoT botnet invasions by synthesising recent scholarly contributions, including deep learning-based network traffic analysis, hybrid feature selection, and ensemble-based ML approaches. Additionally, it explores these publications investigating low-complexity models for detecting DDoS attacks and evaluating machine learning models using well-known datasets such as Bot-IoT, NBa-IoT, IOT-23 etc. As emphasised by the literature review, the dynamic nature of cyber threats and the Internet of Things highlights the urgent need for adaptive detection mechanisms. As a result, the review offers a snapshot of recent academic endeavours through a structured comparison table summarising key research attributes, facilitating knowledge dissemination, and directing future research in IoT botnet detection.

Keywords IoT botnet · Detection · Machine learning (ML)

A. Rasool (✉) · N. S. Safa · C. Mbarushimana
Mathematics and Computer Science, University of Wolverhampton, Wolverhampton WV1 1LY, UK
e-mail: a.m.rasool@wlv.ac.uk

N. S. Safa
e-mail: n.sohrabisafa@wlv.ac.uk

C. Mbarushimana
e-mail: c.mbarushimana@wlv.ac.uk

H. Jahankhani and B. Issac (eds.), *Cybersecurity and Human Capabilities Through Symbiotic Artificial Intelligence*, Advanced Sciences and Technologies for Security Applications, https://doi.org/10.1007/978-3-031-82031-1_9

1 Introduction

Botnets are commonly used to launch large-scale DDoS attacks, overwhelming target servers or networks with an excessive amount of traffic. This causes service disruptions and downtime. Emmons [1] States that DDoS attacks increased significantly in 2020, showcasing the persistent threat of botnets.

Internet of Things (IoT) botnet detection has emerged as a critical area of research to safeguard interconnected devices from malicious attacks. Researchers have conducted studies focused on IoT botnet detection using machine learning techniques. These studies employ a variety of datasets, including Bot-IoT, IoT-23, and others. Different machine learning models such as Random Forest (RF), Decision tree (DT), K-Nearest Neighbour (KNN), Naive Bayes, and Multi-layer Perception Artificial Neural Networks (MLP ANN) are utilized for IoT Botnet classification. Evaluation metrics like accuracy, precision, recall, F1-score, and ROC AUC are employed to assess the models' performance. Additionally, the investigated studies explore optimization techniques, ensemble learning, deep learning approaches, and feature selection methods. The studies also emphasize the importance of handling imbalanced datasets and highlight metrics like True Positive Rate (TPR), False Positive Rate (FPR), and Overall, Success Rate (OSR). Overall, these studies collectively contribute valuable insights into the realm of IoT botnet detection, offering a diverse array of methodologies and metrics for evaluating model performance (Table 1).

The Internet of Things (IoT) has significantly expanded the horizon of interconnectivity, integrating a plethora of devices into the global Internet infrastructure. This expansion, while beneficial, has also exposed numerous security vulnerabilities, particularly through the proliferation of botnets—a network of infected devices, controlled remotely to launch various malicious activities. Recent advancements in machine learning (ML) and deep learning (DL) have opened new avenues for detecting and mitigating botnet threats within IoT networks. This comprehensive survey delves into the current landscape of ML approaches tailored for botnet detection in IoT contexts. This survey reviews the latest research efforts in ML and DL for IoT botnet detection, synthesizing findings from various studies. Including, Singh et al. [3] explore deep learning-based network traffic analysis for effective botnet detection. Similarly, Hossain and Islam [4] present a hybrid feature selection and ensemble-based ML approach to enhance detection rates. Additionally, Sadhwani et al. [5] discusses lightweight models for DDoS attack detection in resource-constrained IoT environments. Furthermore, Patil and Deshpande [6] evaluate the performance of ML models on established datasets like CTU-13 and IOT-23. In another study, Aljammal et al. [7] recommend a Random tree classifier after their evaluation with many machine learning approaches such as Random Forest (RF), Random Tree, and Naive Bayes (NB) classifiers to protect industrial IoT infrastructure, using Bot-IoT dataset. Moreover, Govindaraji et al. [8] introduce a Deep IoT Intrusion Detection System using CNNs and LSTMs, achieving over 90% accuracy and emphasizing dataset variety, adversarial defence mechanisms, and real-time deployment in resource-constrained IoT environments. Lastly, Popoola et al.

Table 1 Metrics description [2]

Metrics	Definition
Accuracy	Is the proportion of correctly classified instances (both malware and benign) out of the total instances. It reflects the overall effectiveness of the detection model
Precision	Measures the accuracy of malware detection among the samples classified as malware. It focuses on minimizing false positives and indicates how many identified malware are actually malicious
Recall	Or True Positive Rate (TPR) is the proportion of actual malware samples correctly identified by the system. It is crucial in minimizing false negatives, ensuring that malware is not missed
F1-score	Is the harmonic mean of precision and recall, providing a balance between them. It is useful in cases where there is an imbalance between malware and benign samples
ROC AUC	(Receiver Operating Characteristic—Area Under the Curve) measures the model's ability to differentiate between classes across various thresholds. A higher AUC indicates better overall classification performance
True Positive Rate (TPR)	(TPR) is the ratio of correctly identified malware to the total actual malware, helping to understand the system's ability to detect threats
False Positive Rate (FPR	(FPR) calculates the proportion of benign files misclassified as malware, focusing on how often the system produces false alarms
Success Rate (OSR)	(OSR) represents the overall proportion of correct classifications, including both malware and benign files. It is often used as a synonym for accuracy

[9] address class imbalance in intrusion detection with a deep learning method called SMOTE-DRNN, which improves IoT security by effectively identifying botnets and outperforming other deep learning and machine learning.

By aggregating insights from these studies, this paper contributes to the academic and practical understanding of current and potential ML-driven strategies for enhancing IoT security against botnet threats. This investigation was motivated by the growing threat of botnet attacks in increasingly populated Internet of Things networks. The versatility and severity of these attacks necessitate sophisticated detection mechanisms that can adapt to evolving threats and diverse IoT environments.

The subsequent subsections will explain our motivation and underscore the contributions of this study.

1.1 Motivation of the Research

The rise of IoT botnets is a big issue in modern cybersecurity, driven by the explosive growth of IoT devices, which now number in the billions globally. These devices, from smart home appliances to industrial control systems, are targets for attackers. Unlike traditional IT systems, IoT devices often lack strong security measures, such as robust authentication and regular software updates, making them highly vulnerable to exploitation. They can be integrated into large-scale botnets once compromised, have the ability to launch Distributed Denial of Service (DDoS) attacks, breach data, and perform other cybercriminal activities. These attacks can cause significant disruptions to critical infrastructure, financial losses, and the exposure of sensitive data, highlighting the increasing sophistication of IoT botnets. Traditional intrusion detection systems (IDS) face difficulties in operating effectively in resource-constrained IoT environments, Thus the need for new detection methods arises [10].

Recent studies highlight an alarming rise in the number of devices targeted by cyberattacks, with attackers exploiting vulnerabilities in connected IoT devices like smart cameras, thermostats, and household appliances [11]. This trend underscores the urgent need for stronger security protocols and demonstrates the potential consequences of inaction, which can lead to financial losses and serious reputational damage for both businesses and individuals [12]. The integration of artificial intelligence into cybersecurity strategies offers a promising opportunity to bolster defenses. AI facilitates real-time threat detection and response, capabilities that were previously unattainable. These advancements improve system resilience and enable adaptive learning, allowing systems to evolve alongside emerging threats, thus fostering a more proactive security approach. The limited resources and diversity of IoT devices sometimes render traditional security controls useless, thereby requiring the use of specific detection approaches.

The remaining sections of this study are organised as follows: Sect. 2 provides a description of various terminologies, including Machine Learning (ML), Deep Learning (DL), the Internet of Things (IoT), IoT botnets, and botnet detection. The literature review regarding the detection of IoT botnets utilising machine learning techniques is articulated in Sect. 3. Section 4 discusses the findings derived from the review and analysis, encompassing the limitations of the study. Section 5 Concludes the examination and identifies directions for future research endeavours and the open issues for the proposed research questions.

The organizational framework of this manuscript is visually represented in Fig. 1.

2 Background

This section offers background information that facilitates the understanding of the review subject; it describes the concepts of machine learning (ML), Deep learning (DL), Internet of Things (IoT), IoT botnet, and botnet detection.

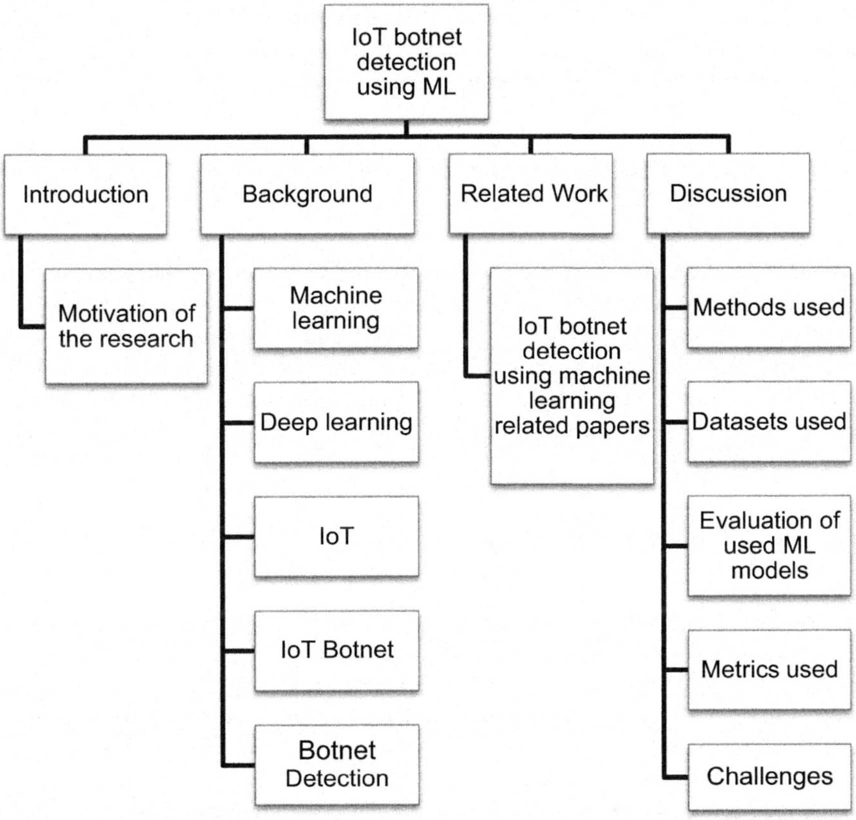

Fig. 1 The study structure

2.1 Machine Learning

The Machine Learning (ML) branch of AI develops data-learning and prediction systems. Unlike traditional programming, machine learning improves system performance by finding patterns and making data-driven judgments. Zeadally and Tsikerdekis (2020) classify machine learning into three categories, which are supervised, unsupervised, and reinforcement learning. Mention that ma-chine learning techniques are valuable in IoT security, unlike traditional programming, machine learning improves system performance by finding patterns and mak-ing data-driven decisions. It also emphasizes the role of data-driven methods, such as machine learning, in processing large datasets IoT security uses ML to detect and respond to cyber threats like botnet attacks. Rahman et al. [13] highlights the machine learning algorithm's ability in evaluating complicated and dynamic network traffic data in real time to identify security breaches and anomalies. IoT networks need this to prevent sophisticated device vulnerability attacks. Cybersecurity uses ML because it can manage massive data sets and react to changing threats.

2.2 Deep Learning (DL)

Deep learning is a complex branch of machine learning that utilises many multiple of neural networks to analyse huge datasets and identify complex patterns [14]. Nowadays, the detection of cyberattacks is focused by various sectors of society, particularly innovations in technology related to Cyber-Physical Systems (CPSs), can be difficult due to the increase in such attacks. Machine learning (ML) provides opportunities, specifically deep learning (DL) [15].

2.3 Internet of Things (IoT)

The term "Internet of Things" (IoT) describes to a network of interconnected devices that communicate and exchange data through Internet connectivity. The devices connected with software, sensors, and some other technologies, gather and exchange data, facilitate automation, and improve efficiency across various fields, including smart homes, healthcare, and industrial systems [16]. Recent research underscores the transformative potential of IoT in enhancing operational efficiency and decision-making processes. For example, it investigate the integration of IoT in smart cities, highlighting its significance of IoT in our daily lives and offer comprehensive analyses regarding its facilitating technologies, diverse applications, emerging trends, and substantial challenges. Moreover, it emphasizes how artificial intelligence puts IoT on top of transformative technology. Likewise, Kumar et al. [17] investigates the consequences of IoT in the healthcare sector, examining the ability to monitor patients remotely and the possibility of implementing personalized medicine by gathering patient data continuously (Fig. 2).

2.4 IoT Botnet

IoT botnets, which are networks of compromised IoT devices exploited by attackers to carry out malicious activities to execute attacks, including Distributed Denial of Service (DDoS) attacks and unauthorised data breaches keep rising, as stated by researchers [19]. The vulnerabilities of IoT devices, such as limited security and continuous connectivity, are now more exploited by attackers using advanced techniques, including machine learning. The emergence of Internet of Things (IoT) technology raises the risk of IoT botnet attacks by facilitating interconnecting devices

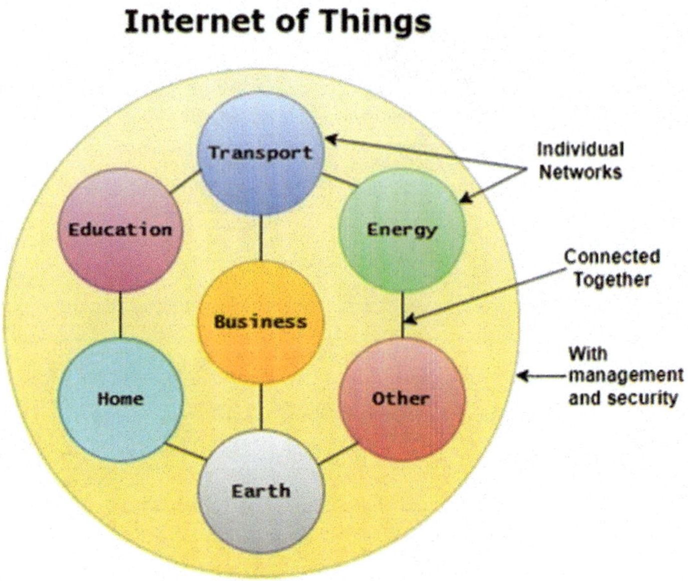

Fig. 2 IoT technology and diverse application domains [18]

[20]. Scholars recommend an adaptive security system, highlighting the essential requirement for robust device authentication, consistent firmware updates, and thorough threat detection mechanisms. This approach emphasises the necessity of enhancing security measures to tackle emerging vulnerabilities (Fig. 3).

2.5 Botnet Detection

Botnet detection focuses on identifying and stopping networks of hijacked devices used for cybercrimes like DDoS attacks, data theft, and spam. As these threats evolve, recent research has shifted towards advanced solutions like machine learning and deep learning. Alzahrani [21] studied the detection and classification of IoT botnets using advanced machine learning (ML) and deep learning (DL) techniques. It highlights the development of the GTODL-BADC technique, that shows botnet attacks' effective detection. This demonstrates the importance of machine learning in analysing network traffic and detecting unusual patterns that suggest botnet activity. In addition Zhang et al. [22] explore the use of graph neural networks (GNNs) to enhance the detection of botnet traffic in IoT networks.

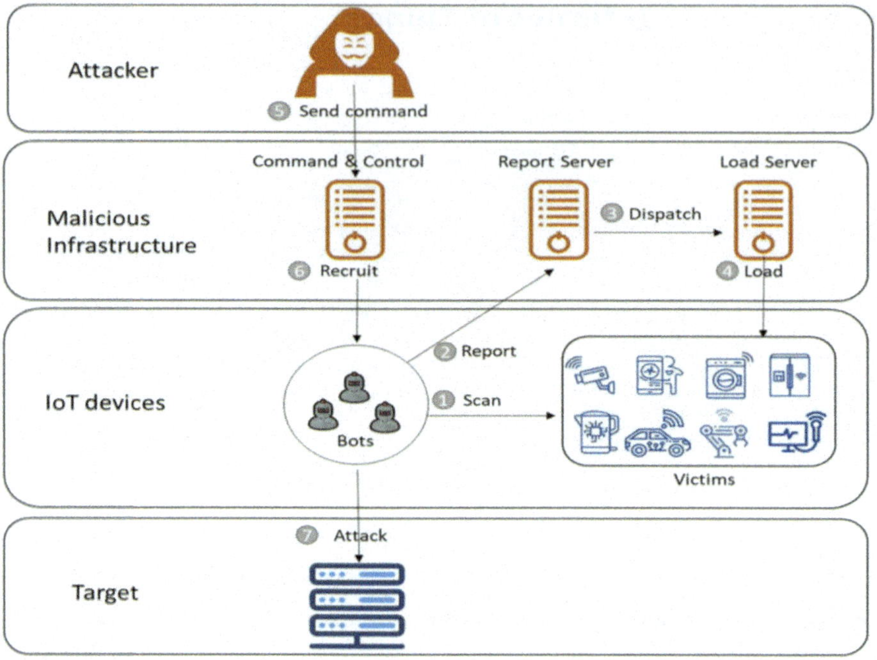

Fig. 3 IoT botnet [20]

3 Literature Review

To highlight the necessity of this Review, Sect. 2 provides a comprehensive review of the existing literature about methods, datasets, and metrics used for the detection of IoT botnets.

3.1 Prior Research on IoT Botnet Detection Using Machine Learning

Botnet detection in IoT networks is becoming increasingly vital as the Internet of Things (IoT) continues to expand, integrating countless devices into the global network infrastructure. These interconnected devices, ranging from household appliances to industrial sensors, offer numerous benefits for convenience and efficiency. However, they also introduce new vulnerabilities, making IoT networks attractive targets for botnet operators. Botnets in IoT networks are fleets of compromised devices controlled remotely to execute coordinated attacks, such as DDoS attacks, spam campaigns, and fraudulent activities. The diverse and often poorly secured nature of IoT devices complicates the detection and mitigation of these threats, underscoring the need for effective botnet detection strategies tailored to the IoT ecosystem. Botnet detection techniques in IoT networks can be broadly categorized into signature-based, anomaly-based, and hybrid approaches. Signature-based methods rely on known patterns of malicious behavior, offering high accuracy for detecting known threats but failing to identify novel botnet activities. Anomaly-based approaches, on the other hand, monitor network behavior, identifying deviations from established norms to flag potential botnet activities. While this method is more adept at detecting new threats, it may result in higher false positive rates. Hybrid systems aim to combine the strengths of both approaches to improve overall detection accuracy. However, these techniques face significant challenges in IoT environments, such as the sheer volume of data, the heterogeneity of devices, and their varying computational capacities, all of which demand scalable and resource-efficient solutions. In [3] present a study on botnet-based IoT network traffic analysis using deep learning methods. Their work focuses on identifying malicious activities within IoT networks by analyzing network traffic data through deep learning models. The study emphasizes the effectiveness of deep learning in detecting sophisticated botnet patterns that traditional methods might miss, showcasing its potential to enhance security in IoT environments. Introduce a novel approach combining hybrid feature selection with ensemble-based machine learning techniques for botnet detection. Their research highlights the improvement in detection accuracy through this combination, addressing the challenges of evolving botnet signatures and behaviors in IoT networks, and suggesting a robust framework for real-time botnet detection [4]. Al-Fawa'reh et al. [23] describe MalBoT-DRL, a framework employing deep reinforcement learning for malware botnet detection in IoT networks. Their work underscores the adaptability of the model to changing attack patterns, demonstrating the potential of reinforcement learning in predicting and mitigating new botnet threats within dynamic IoT environments. Explore the application of deep learning for IoT botnet attack detection, particularly focusing on the use of residual networks. Their study provides insights into the capabilities of deep learning architectures to accurately detect botnet activities, contributing to the ongoing development of more secure and resilient IoT systems [8]. Sadhwani et al. [5] propose a lightweight model for DDoS attack detection using machine learning techniques aimed at IoT networks. Their approach addresses the need for efficient and scalable solutions suitable for the resource-constrained nature of many IoT devices, highlighting the balance between performance and computational overhead. Regis et al. [24] discuss the detection of IoT botnets through a combination of machine learning and deep learning techniques. Their research sheds light on the comparative advantages and application scenarios of each approach, providing a comprehensive overview of current methodologies in IoT botnet detection. Bowen et al. [25] present BLoCNet, a hybrid, dataset-independent intrusion detection system utilizing deep learning. Their study emphasizes the system's adaptability and efficiency across different datasets and botnet scenarios, showcasing the benefits of a flexible approach to security in diverse IoT environments. Detail the design and implementation of an intrusion detection model incorporating machine learning techniques for enhancing IoT security. Featured in the context of blockchain and machine learning for IoT security, their work highlights innovative approaches to securing IoT networks against sophisticated threats [26]. In [6] focus on evaluating machine learning models on CTU-13 and IOT-23 datasets, crucial for understanding the effectiveness of various ML approaches in detecting IoT botnets. Their comparative study provides valuable benchmarks and insights, guiding future developments in botnet detection strategies. Conduct a performance evaluation of machine learning approaches in detecting IoT-botnet attacks. Their research contributes to the understanding of how different ML models perform in the context of IoT security, offering guidance for the selection and implementation of appropriate detection techniques Aljammal et al. [7]. Introduce a novel botnet attack detection method for IoT networks based on communication graphs. Their approach leverages the relational data among devices to detect anomalies indicative of botnet activity, presenting a unique angle on IoT security challenges and solutions [27]. Jayaraman et al. [28] explore the use of machine learning techniques for detecting malicious IoT traffic. Their study delves into various ML approaches and their effectiveness in identifying illicit activities within IoT networks, contributing to the broader understanding of cybersecurity measures in the IoT domain. Focus on the detection of Mirai botnet attacks on IoT devices using deep learning techniques. Their research highlights the specific challenges and solutions associated with one of the most infamous botnets, providing insights into the development of targeted detection mechanisms [29]. Aldhaheri et al. [30] review deep learning applications for cyber threat detection in IoT networks. Their comprehensive analysis covers various deep learning models and their suitability for different IoT security scenarios, offering a broad perspective on the potential of deep learning in combating cyber threats. Qing et al. [31] discuss the construction and evaluation of the MBB-IoT, a new dataset focused on IoT DDoS traffic. Their work provides a fresh perspective on dataset creation and utilization for IoT security research, addressing the need for up-to-date and relevant data in developing effective detection systems (Table 2).

4 Discussion

The rapid growth of Internet of Things (IoT) devices has brought along serious security risks with botnet assaults representing one of the most critical issues. IoT botnets can launch Distributed Denial of Service (DDoS) attacks, steal data, and engage in other malicious activities. Approaches utilizing machine learning (ML) have surfaced as an impressive solution for the detection of these botnets through the analysis of network traffic and the identification of anomalous behaviours. This section examines the methods, datasets, and metrics employed in recent research on IoT botnet detection, assessing their efficacy and the challenges encountered.

Supervised machine learning models are widely used for detecting IoT botnets. It utilises labeled data to classify network traffic as benign or malicious. One of the common models is Random Forest (RF) which is robustness and the ability to manage large datasets. It works by building multiple decision trees and using their combined outputs for classification [54]. In [55], RF achieved high detection accuracy and

Table 2 Table 2 compares the related work

Paper	Dataset	Algorithm	Metrics
2019-Supervised Detection of IoT Botnet Attacks [32]	N-BaIoT	Naive Bayes, K Nearest Neighbors (KNN), and Random Forest (RF)	Sensitivity (True Positive Rate or Recall), False Positive Rate (FPR or False Alarms), Specificity (True Negative Rate), F-score (F-measure)
2020-An efficient approach to detect IoT botnet attacks using machine learning [33]	Bot-IoT	Multiple classifiers were tested and the results from the best three, namely: J48, Random Forest (RF), and Multilayer Perceptron (MLP) networks were reported	Accuracy and f-measure, False Negative (FN) rate
2020-Detecting_Botnet_Attacks_in_IoT_Environments_An_Optimized_Machine_Learning_Approach [34]	Bot-IoT-2018 dataset	An optimized ML-based framework consisting of a combination of Bayesian optimization Gaussian Process (BO-GP) algorithm and decision tree (DT) classification model	Accuracy, precision, recall, and F-score
2020-Machine Learning-Based IoT-Botnet Attack Detection with sequential architecture [10]	N-BaIoT	Artificial neural network (ANN), decision tree(DT), and Naïve Bayes	Performance

(continued)

Table 2 (continued)

Paper	Dataset	Algorithm	Metrics
2021-Botnet Attack Detection by Using CNN-LSTM Model for IoT applications [35]	N-BaIoT	CNN-LSTM)	Evaluation Metrics. Accuracy, recall, precision, and F1-score metrics were considered to test the system for detection of botnet attacks
2021-Developing_an_ Efficient_Feature_ Engineering_and_ Machine_Learning_ Model_for_ Detecting_ IoT-Botnet_Cyber_ Attacks [36]	UNSW-NB15	Chi-square automatic interaction detection (CHAID) decision tree-based, (ii) A2DE (a semi-naive Bayesian averaged two-dependence estimator), and (iii) HGC- a hybrid of Genetic algorithm with K-means clustering. Deep learning (DL) methods such as (i) Deep Multilayer perceptron (DMLP) and (ii) Convolutional neural network (CNN) based classiers are employed	Precision, recall, and F1-score, apart from prediction accuracy, training, and testing time, are considered to understand the efcacy of the proposed IoT botnet detection model
2021-SMOTE-DRNN A Deep Learning Algorithm for Botnet Detection in the Internet-of-Things Networks [9]	Bot-IoT	DRNN, SMOTE-DRNN	Accuracy, precision, recall, F1 score, FPR, NPV, area under the receiver operating characteristic curve (AUC), geometric mean (GM), Matthews correlation coefficient (MCC), training time and testing time

(continued)

Table 2 (continued)

Paper	Dataset	Algorithm	Metrics
2020-Identification_of_Botnet_Activity_in_IoT_Network_Traffic_Using_Machine_Learning [37]	A. Stratosphere Lab IoT-23 Data B. Additional Benign Data Capture C. Small Dataset and Large dataset	Decision tree (DT), random forest (RF), multi-class decision forest, two-class neural network, and multi-class neural network	Accuracy, probability of detection, and probability of false alarm rate as performance metrics
2020-Network_Flow_based_IoT_Botnet_Attack_Detection_using_Deep_Learning [38]	N-BaIoT, Bot-IoT	Logistic regression, (LR) and naive Bayes (NB) classifiers	Standard metrics such as accuracy (Acc), precision (Pre), recall (Rec) and F1-score (F1)
2020-Towards_a_Universal_Features_Set_for_IoT_Botnet_Attacks_Detection [39]	CICIDS2017, CTU-13 and IoT-23	Machine learning models, i.e., NB, KNN, RF, and LR	Accuracy, precision, recall, and F1-measure
Zero-Shot_Deep_Learning_Approach_for_Unknown_IoT_Botnet_Attack_Detection [40]	N-BaIoT dataset,	The paper proposes a Deep Sparse Contrastive Auto-Encoder (DSCAE) model	• The evaluation metrics used in the research include Accuracy, Precision, Recall, and F1 score
2023-Network_Intrusion_Detection_for_IoT-Botnet_Attacks_Using_ML_Algorithms [41]	CIC IoT 2023 dataset, N-BaIoT and MedBIoT	Decision Tree, Random Forest, Logistic Regression, KNeighbors Classifier, and Gaussian Naïve Bayes	Accuracy
2023-10-Performance + Evaluation + of + Machine + Learning + Approaches + in + Detecting + IoT-Botnet + Attacks [7]	BoT-IoT dataset	REPTree, RandomTree, RandomForest, J48, metaBagging, and Naive Bayes	Accuracy, True Positive Rate (TPR), and False Positive Rate (FPR). Additional metrics such as Recall, F-Measure, Matthews Correlation Coefficient (MCC), Correct Classification Index (CCI), and Incorrect Classification Index (ICI) were reported for the classifiers tested

(continued)

Table 2 (continued)

Paper	Dataset	Algorithm	Metrics
2023-9-Evaluating_ML_models_on_CTU-13_and_IOT-23_Datasets [6]	CTU-13 and IOT-23 • The CTU-13 dataset	Decision Tree, Naive Bayes, K-Nearest Neighbors (KNN), Random Forest, AdaBoost, Logistic Regression, and Support Vector Machine (SVM)	Accuracy, Sensitivity, Precision, F1 Score, and Specificity
2023-6-(2023). Detection of IoT Botnet using Machine learning and Deep Learning Techniques [24]	IoT-23 dataset,	Decision Tree, Support Vector Machine (SVM), Random Forest, and Naive Bayes • Deep learning models explored in the research are Convolutional Neural Network (CNN) and Gated Recurrent Unit (GRU) neural network	Accuracy, precision, recall, and F1-Score
2023-5-A Lightweight Model for DDoS Attack Detection Using Machine Learning Techniques [5]	• The study utilizes two primary ToN-IoT dataset and the BoT-IoT dataset	Logistic regression, random forest, naïve bayes, artificial neural network, and k nearest neighbor algorithms for building a lightweight intrusion detection system	Accuracy, precision, recall, F1 score, training time, and prediction time
2023-4-Deep Learning-Based Detection of IoT Botnet Attacks An Exploration of Residual networks [8]	• The study utilized a multivariate sequential dataset that included actual traffic data from nine commercial IoT devices infected with BASHLITE and Mirai malware	(DLT) based approach termed Detection of Intrusions in IoT using Residual Networks (DIIOTRNs) for detecting IoT botnet attacks	The F1-Score, Accuracy, precision, and recall
2023--A novel hybrid feature selection and ensemble-based machine approach for botnet detection [4]	N-BaIoT, Bot-IoT, CTU-13, ISCX, CICIDS, and CCC datasets	Extra Trees ensemble technique, the XGBoost • Various ensemble methods, including bagging, random forest, and stacking	Accuracy, precision, recall, F1-score, and balanced accuracy (BACC)

(continued)

Table 2 (continued)

Paper	Dataset	Algorithm	Metrics
2024-Botnet attack detection in IoT using machine learning models [42]	N-BaIoT dataset	The paper utilizes Decision Trees (DT), Random Forest (RF), and Deep Learning models as the primary approaches for prediction	Accuracy, confusion matrix, and ROC curves
2024-MalBoT-DRL_ Malware_Botnet_ Detection_Using_ Deep_ Reinforcement_ Learning_in_IoT_ Networks [23]	MedBIoT and N-BaIoT	It utilizes the MalBoT-DRL model, which incorporates a Deep Q-Network (DQN) approach	True-positive (TP), false-positive (FP), true-negative (TN), and false-negative (FN) values. the accuracy, precision, recall, F1-score, and G-mean
2021-Botnet_ Detection_Approach_ Using_Graph-Based_ Machine_Learning [43]	CTU-13 and IoT-23	Naive Bayes, Decision Tree, Random Forests, AdaBoost, ExtraTrees, and K-Nearest Neighbors	Accuracy, Recall (or detection rate), Precision, and F1-Score
2019-An Adaptive Multi-Layer Botnet Detection Technique Using Machine Learning Classifiers [44]	CTU-13 and ISOT	Decision Tree	• The evaluation of the methodologies was conducted using tenfold cross-validation, which involves dividing the dataset into ten equal parts
2022-BotStop -Packet-based efficient and explainable IoT botnet detection using machinelearning [45]	The initial dataset for training and testing was created by capturing raw packets from various network attacks within an IoT environment, specifically using two smart home devices: SKT NUGU (NU 100) and EZVIZ Wi-Fi Camera (C2C Mini O Plus 1080P) • For further validation, the TONIoT dataset was selected	XGBoost (XGB) classifier, Random Forest (RF), Logistic Regression (LR), Decision Tree (DT), and Gaussian Naive Bayes (GNB)	Accuracy, precision, recall and F1 score), Training time and testing time

(continued)

Table 2 (continued)

Paper	Dataset	Algorithm	Metrics
2022-Ahmed_et_al_ DNN_Model_for_ the_Detection_of_ Botnet_ [46]	CTU-13 dataset	A deep learning model for detecting botnets using a deep neural network (DNN) approach	Accuracy
2021-An efficient botnet detection with the enhanced support vector neural network [47]	CTU-13 dataset	Enhanced Support Vector Neural Network (ESVNN)	Precision, recall, accuracy, and F-score
2019-Botnet_Attack_ Detection_at_the_ IoT_Edge_Based_on_ Sparse_ Representation [48]	N-BaIoT dataset	The research paper discusses the use of a sparse representation framework for IoT botnet attack detection	Positive Predictive Value (PPV), Sensitivity (detection rate), F1-score, Accuracy (ACC), These metrics are calculated based on a confusion matrix that includes True Positive (TP), False Negative (FN), False Positive (FP), and True Negative (TN) values
2023-VAE-Based Latent Representations Learning for Botnet Detection in IoT Networks [49]	N-BaIoT	A hybrid model that includes a Variational AutoEncoder (VAE) for learning latent representations and a one-class classifier (OCC) for detecting anomalies	Receiver operating characteristic (ROC) curve and the area under the ROC curve (AUC), The ROC curve, TPR, and FPR are defined, The AUC
2023-Unsupervised_ Learning_for_ Feature_Selection_A_ Proposed_Solution_ for_Botnet_ Detection_in_5G_ Networks [50]	IoT23 dataset	Logistic Regression (LR), Decision Tree (DT), and Support Vector Machine (SVM)	Overall Success Rate (OSR), True Positive Rate (TPR), False Positive Rate (FPR), Precision, and F-score

(continued)

Table 2 (continued)

Paper	Dataset	Algorithm	Metrics
2023-Intelligent IoT-BOTNET attack detection model with optimized hybrid classification model [51]	A hybrid classifier is utilised which can be constructed by integrating the optimized Bi-GRU with the Recurrent Neural Network (RNN)	A hybrid classifier that integrates an optimized Bi-GRU (Bidirectional Gated Recurrent Unit) with a Recurrent Neural Network (RNN) for attack detection in IoT systems	Accuracy, precision, F-measure, NPV (Negative Predictive Value), MCC (Matthews Correlation Coefficient), FNR (False Negative Rate), and FPR (False Positive Rate). F-Measure
2021-Internet of Things attack detection using hybrid Deep Learning Model [52]	IoT-23 dataset	Ahybrid approach combining Convolutional Neural Network (CNN) and Long Short-Term Memory (LSTM)	True Positive (TP), True Negative (TN), False Positive (FP), and False Negative (FN), Recall, Accuracy
2021-Adaptive online learning for IoT botnet detection [53]	• The study utilizes real IoT traffic data, which includes both benign traffic and botnet traffic data infected by the Mirai botnet	• The research paper discusses the use of the Hoeffding adaptive tree (HAT) and adaptive random forest (ARF)	Prequential accuracy, Kappa statistic (j), and the modified Kappa statistic (j m) • Prequential accuracy is the most frequently used metric for evaluating big data streams, where each instance is first used for testing and then for training • The Kappa statistic (j) is designed to handle unbalanced real-time data, comparing the classifier's prequential accuracy with that of a chance classifier

shortest testing time in all the four datasets they experimented with by effectively aggregating decisions from multiple decision trees.

Support Vector Machine (SVM) is another common model that performs well in binary classification tasks. Jayaraman et al. [28] state that SVM is practical for binary classification, but it struggles with multiclass classification and imbalanced datasets. Logistic Regression (LR) is simple and interpretable and is also used in binary classification for IoT botnet detection. In the study [56], SVM was combined with Principal Component Analysis (PCA) to reduce dataset dimensionality and improve detection accuracy. However SVM is practical for binary classification, it struggles with multiclass classification and imbalanced datasets. Logistic Regression (LR), is simple and interpretable, is also used in binary classification for IoT botnet detection. Happy et al. [57] states that LR is effective in providing moderate performance in classifying network traffic as either botnet or benign, nonetheless, it is less resilient than random forest (RF) and support vector machine (SVM) in handling complicated, high-dimensional data.

In the realm of detecting IoT botnets utilizing machine learning (ML) and deep learning (DL) methodologies, numerous challenges manifest, primarily attributable to the heterogeneity and resource limitations inherent in IoT devices. The substantial volume of data combined with the variability in computational capabilities across devices renders the implementation of a uniform detection solution exceedingly challenging. Moreover, signature-based detection techniques frequently encounter difficulties in recognizing novel or evolving threats, whereas anomaly-based methodologies may experience elevated rates of false positives. Hybrid approaches aspire to strike a balance between accuracy and efficiency; however, they too are confronted with scalability challenges in extensive IoT deployments. Prominent datasets utilized in research endeavors encompass N-BaIoT, Bot-IoT, CTU-13, and IoT-23, all of which are esteemed for their comprehensive representation of IoT network traffic and associated attacks. Concerning algorithms, both machine learning models, such as Decision Trees, Random Forest, and Naive Bayes, alongside deep learning models, including CNN, LSTM, and Deep Reinforcement Learning (e.g., MalBoT-DRL), are routinely employed to augment detection accuracy. Widely accepted metrics for evaluating these models comprise accuracy, precision, recall, and the F1-score, which collectively facilitate the assessment of both detection performance and the operational efficiency of the models.

5 Conclusion and Future Work

This study shows the advancements in machine learning (ML) and deep learning (DL) methodologies for IoT botnet detection. It does this by examining several studies that use well-known datasets like Bot-IoT, N-BaIoT, and IoT-23. Support Vector Machine (SVM) and Random Forest (RF) are two popular machine learning models for binary classification, but they have problems with computational needs and class imbalances. Hybrid methods, which merge the best parts of several techniques, have come up as possible ways to improve accuracy and scalability. However, they are still having trouble in large-scale IoT deployments. Adaptive detection methods are necessary to move forward. In the future, researchers should focus on making detection systems that are dynamic and scalable so that they can keep up with changing cyber threats in IoT networks where resources are limited. Additionally, making models that are small and accurate enough to allow real-time detection is very important for making IoT networks more resistant to botnet attacks. Taking care of these issues will improve IoT security, making it more resistant to online threats that are getting more complicated.

Acknowledgements I would like to express my sincere gratitude to the Funds for Women Graduates (FfWG), an educational charity, for their generous financial support through a £2500 grant. This funding has been invaluable in supporting my research and enabling me to focus on my academic pursuits.

Disclosure of Interests The authors have no competing interests.

References

1. Emmons T (2021) https://www.akamai.com/blog/security/part-i-retrospective-2020-ddos-was-back-bigger-and-badder-than-ever-before. Accessed: 2024/9/31
2. Maniriho P, Mahmood AN, Chowdhury MJM (2009) A systematic literature review on windows malware detection: techniques, research issues, and future directions. J Syst Softw
3. Singh NJ, Hoque N, Singh KR, Bhattacharyya DK (2023) Botnet-based IoT network traffic analysis using deep learning. Secur Priv e355
4. Hossain MA, Islam MS (2023) A novel hybrid feature selection and ensemble-based machine learning approach for botnet detection. Sci Rep 13(1):21207–21207
5. Sadhwani S, Manibalan B, Muthalagu R, Pawar P (2023) A lightweight model for DDoS attack detection using machine learning techniques. Appl Sci 13(17):9937–9937
6. Patil A, Deshpande A (2023) Evaluating ML models on CTU-13 and IOT-23 datasets. In: 2023 international conference on advanced computing technologies and applications (ICACTA), pp 1–6
7. Aljammal AH, Qawasmeh A, Mughaid A, Taamneh S, Wedyan FI, Obiedat M (2023) Performance evaluation of machine learning approaches in detecting IoT-botnet attacks. Int J Interact Mobile Technol 17(19)
8. Govindaraji M, Periyasamy R, Thulasiraman V (2023) Deep learning-based detection of IoT botnet attacks: an exploration of residual networks. Int J Saf Secure Eng 13(4):715–722
9. Popoola SI, Adebisi B, Ande R, Hammoudeh M, Anoh K, Atayero AA (2021) SMOTE-DRNN: a deep learning algorithm for botnet detection in the Internet-of-Things networks. Sensors 21(9):2985–2985
10. Soe YN, Feng Y, Santosa PI, Hartanto R, Sakurai K (2020) Machine learning-based IoT- botnet attack detection with sequential architecture. Sensors 20(16):4372–4372
11. Nazir A, He J, Zhu N, Wajahat A, Ma X, Ullah F, Qureshi S, Pathan MS (2023) Advancing IoT security: A systematic review of machine learning approaches for the detection of IoT botnets. J King Saud Univ-Comput Inf Sci pp 101820–101820
12. Rosi L (2024)The impact of cyber security on business development: a review. Int J Res Radicals Multi Fields 3(2):92–100
13. Rahman MS, Khatun F, Rahman MM (2022) Machine learning-based anomaly detection for IoT networks: a survey. J Comput Secur 107:102378–102378
14. Yao Y, Yu B, Gong C, Liu T (2021) Understanding how pretraining regularizes deep learning algorithms. IEEE Transactions on Neural Networks and Learning Systems 34(9):5828–5840
15. Zhang J, Pan L, Han QL, Chen C, Wen S, Xiang Y (2022) Deep learning based at-tack detection for cyber-physical system cybersecurity: a survey. IEEE/CAA J Automatica Sinica 9(3):377–391
16. Elgazzar K, Khalil H, Alghamdi T, Badr A, Abdelkader G, Elewah A, Buyya R (2022) Revisiting the internet of things: new trends, opportunities and grand challenges. Front Internet Things 1:1073780–1073780

17. Kumar M, Kumar A, Verma S, Bhattacharya P, Ghimire D, Kim SH, Hosen AS (2023) Healthcare Internet of Things (H-IoT): Current trends, future prospects, applications, challenges, and security issues. Electronics 12(9):2050
18. Chataut R, Phoummalayvane A, Akl R (2023) Unleashing the power of IoT: a comprehensive review of IoT applications and future prospects in healthcare, agriculture, smart homes, smart cities, and industry 4.0, sensors, multidisciplinary digital publishing institute (MDPI)
19. Kumar A, Singh D (2023) Detection and prevention of DDoS attacks on edge computing of IoT devices through reinforcement learning. Int J Inform Technol 16(3):1365–1376
20. Wazzan M, Algazzawi D, Bamasaq O, Albeshri A, Cheng L (2021) Internet of things botnet detection approaches: analysis and recommendations for future research. Appl Sci 11(12):5713
21. Alzahrani A (2024) An optimized approach to deep learning for botnet detection and classification for cybersecurity in Internet of Things environment. Comput Mater Continua 80(2)
22. Zhang B, Li J, Ward L, Zhang Y, Chen C, Zhang J (2023) Deep graph embedding for IoT botnet traffic detection. Secur Commun Networks 2023(1):9796912
23. Al-Fawa'reh M, Abu-Khalaf J, Szewczyk P, Kang JJ (2023) MalBoT-DRL: malware botnet detection using deep reinforcement learning in IoT networks. IEEE Internet Things J 11(6)
24. Regis Anne W, Kirubavathi G, Sridevi UK (2023) Detection of IoT botnet using machine learning and deep learning techniques
25. Bowen B, Chennamaneni A, Goulart A, Lin D (2023) BLoCNet: a hybrid, dataset-independent intrusion detection system using deep learning. Int J Inform Secur pp 1–25
26. Mohy-Eddine M, Bella K, Guezzaz A, Benkirane S, Azrour M (2024) Design and implementation of intrusion detection model with machine learning techniques for IoT security. In: Blockchain and machine learning for IoT security, pp 135–148
27. Muñoz DC, Valiente ADC (2023) A novel botnet attack detection for IoT networks based on communication graphs. Cybersecurity 6(1):33–33
28. Jayaraman B, Thai MTNT, Anand A, Anandan KR (2023) Detecting malicious IoT traffic using machine learning techniques. Romanian J Inf Technol Autom Control 33(4):47–58
29. Sharma A, Mansotra V, Singh K (2023) Detection of mirai botnet attacks on IoT devices using deep learning. J Sci Res Technol, 174–187
30. Aldhaheri A, Alwahedi F, Ferrag MA, Battah A (2023) Deep learning for cyber threat detection in IoT networks: a review. Internet of Things and cyber physical systems 4
31. Qing Y, Liu X, Du Y (2023) MBB-IoT: construction and evaluation of IoT DDoS traffic dataset from a new perspective. Computers, Materials & Continua (2):76–76
32. Alazzam H, Alsmady A, Shorman AA (2019) Supervised detection of IoT botnet attacks. Proceedings of the second international conference on data science, E-Learning and information systems pp, pp 1–6
33. Alothman Z, Alkasassbeh M, Baddar AHS (2020) An efficient approach to detect IoT botnet attacks using machine learning. J High Speed Netw 26(3):241–254
34. Injadat M, Moubayed A, Shami A (2020) Detecting botnet attacks in IoT environments: an optimized machine learning approach. In: 2020 32nd International Conference on Microelectronics (ICM), pp 1–4
35. Alkahtani H, Aldhyani TH (2021) Botnet attack detection by using CNN-LSTM model for Internet of Things applications. Secur Commun Networks 2021(1):3806459–3806459
36. Panda M, Mousa AA, Hassanien AE (2021) Developing an efficient feature engineering and machine learning model for detecting IoT-botnet cyber attacks. IEEE Access 9:91038–91049
37. Hegde M, Kepnang G, Mazroei MA, Chavis JS, Watkins L (2020) Identification of botnet activity in IoT network traffic using machine learning. In: 2020 International conference on intelligent data science technologies and applications (IDSTA), pp 21–27
38. Sriram S, Vinayakumar R, Alazab M, Soman KP (2020) Network flow based IoT botnet attack detection using deep learning. In: 2020 IEEE international conference on big data (Big Data), pp 5231–5236
39. Hussain F, Abbas SG, Fayyaz UU, Shah GA, Toqeer A, Ali A (2020) Towards a universal features set for IoT botnet attacks detection. In: 2020 IEEE 23rd international multitopic conference (INMIC), pp 1–6

40. Luu CD, Nguyen VQ, Pham TS, Le-Khac NA (2023) A zero-shot deep learning approach for unknown IoT botnet attack detection. In: 2023 RIVF International Conference on Computing and Communication Technologies (RIVF), pp 278–283
41. Raju VSA, Suma B (2023) Network intrusion detection for IoT-botnet attacks using ML algorithms. In: 2023 7th International Conference on Computation System and Information Technology for Sustainable Solutions (CSITSS), pp 1–6
42. Adeyemi TS, Muhammed A (2024) Botnet attack detection in IoT using machine learning models. Int J Sci Res Arch 12(1):2221–2229
43. Alharbi A, Alsubhi K (2021) Botnet detection approach using graph-based machine learning. IEEE Access 9:99166–99178
44. Khan RU, Zhang X, Kumar R, Sharif A, Golilarz NA, Alazab M (2019) An adaptive multi-layer botnet detection technique using machine learning classifiers. Appl Sci 9(11):2375–2375
45. Alani MM (2022) BotStop: packet-based efficient and explainable IoT botnet detection using machine learning. Comput Commun 193:53–62
46. Ahmed AA, Jabbar WA, Sadiq AS, Patel H (2022) Deep learning-based classification model for botnet attack detection. Appl Math Inform Sci 4(2): 1–10
47. Jagadeesan S, Amutha B (2021) An efficient botnet detection with the enhanced support vector neural network
48. Tzagkarakis C, Petroulakis N, Ioannidis S (2019) Botnet attack detection at the IoT edge based on sparse representation. In: IEEE international conference on communications (ICC), pp 1–6
49. Snoussi R, Youssef H (2023) VAE-based latent representations learning for botnet detection in IoT networks. J Netw Syst Manag 31(4):1–24
50. Lefoane M, Ghafir I, Kabir S, Awan IU (2023) Unsupervised learning for feature selection: a proposed solution for botnet detection in 5G networks. IEEE Trans Ind Informatics 19(1):921–933
51. Bojarajulu B, Tanwar S, Singh TP (2023) Intelligent IoT-botnet attack detection model with optimized hybrid classification model. Comput Secur 126:103064–103064
52. Sahu AK, Sharma S, Tanveer M, Raja R (2021) Internet of Things attack detection using hybrid deep learning model. Comput Commun 176:146–154
53. Shao Z, Yuan S, Wang Y (2021) Adaptive online learning for IoT botnet detection. Inform Sci 574:84–95
54. Estévez-Pereira JJ, Fernández D, Novoa FJ (2020) Network anomaly detection using machine learning techniques. Proceedings 54(1):8–8
55. Gandhi R, Li Y (2021) Comparing machine learning and deep learning for IoT botnet detection. In: 2021 IEEE International Conference on Smart Computing (SMARTCOMP), pp 234–239
56. Rathod G, Sabnis V, Jain JK (2024) Improving IoT botnet attack detection using machine learning: comparative analysis of feature selection methods and classifiers in intrusion detection systems. In: 2024 3rd International Conference for Innovation in Technology (INOCON), pp 1–8
57. Happy C, Kashyap RN (2024) A comparative analysis of machine learning prediction algorithms for detecting IoT botnet activities. In: 2024 International Conference on Intelligent Systems for Cybersecurity (ISCS), pp 1–6
58. Mishra AK, Paliwal S, Srivastava G (2024) Anomaly detection using deep convolutional generative adversarial networks in the internet of things. ISA Trans 145:493–504

Development of Cloud Apps Security Methods to Comply with GDPR/HIPAA

E. Alooeff

Abstract This work is dedicated to developing the methods of cloud apps to comply with such law regulations as GDPR in the European Union and HIPAA in the USA. These methods effectively implement data control and allow the personal data processors to wipe out the obsolete data in automatic mode. Solution is scalable, easy to configure and the dashboard keeps monitoring of the algorithms' work on a high level.

Keywords GDPR · HIPAA · PII · Privacy · Security · Compliance

1 Introduction

The EU General Data Protection Regulation (GDPR) is enforced across all EU Members from 2018, is a landmark in the evolution of the European privacy framework [1]. This law covers the personal data of all EU residents, regardless of their processing location. Personal data is information that, directly or indirectly, can identify an individual and specifically includes online identifiers such as IP addresses, location data. GDPR aims to bring a single standard for data protection among all EU member states, and applies to entities that operate in the EU or deal with the data of any EU resident, regardless of where the data is processed. This is much wider than the concept of personally identifiable information (PII) in US privacy law [2].

Depending on where your organization operates, you may need to adhere to location-specific privacy laws. For example, if your organization operates in Japan, you are subject to the Act on the Protection of Personal Information [3]. Globally, there are two types of privacy laws: comprehensive (applicable to all industries and sectors) and sectoral (applicable to specific industries or sectors). In the US, the federal government has historically taken a sectoral approach. For example, there's the Health Insurance Portability and Accountability Act (HIPAA) [4], which is the US

E. Alooeff (✉)
ATEK, Eden Prairie, UK
e-mail: alooeff@gmail.com

H. Jahankhani and B. Issac (eds.), *Cybersecurity and Human Capabilities Through Symbiotic Artificial Intelligence*, Advanced Sciences and Technologies for Security Applications, https://doi.org/10.1007/978-3-031-82031-1_10

healthcare privacy law protecting data that reveals the health status of an individual (called protected health information or PHI). Cybersecurity compliance involves meeting various controls to protect the confidentiality, integrity, and availability of data. Regulatory compliance insists that the organizations should follow local, state, federal, and international laws and regulations relevant to its business function.

2 Apply Data Privacy Principles

There are 10 key privacy principles. In this work we take into account some of them:

Security Principle—Anonymization technique the process of protecting private or sensitive information by erasing, obfuscating (masking), or encrypting it. It removes all identifiers associated with a person [5]. If data is truly anonymized, then the data does not constitute personal data under the GDPR. However, the bar to be considered anonymous is high: It must be impossible for any individual to be identified from the data by any further processing or by combining it with other information.

Data Deletion and Retention—Organizations should only store personal data for as long as it's required and for the originally intended purpose. Organizations shouldn't keep data for an indefinite period even if it may be used in the future. Clear time frames should be established for when data is deleted with rationale for why the data is retained for that length of time. For example, you may need to retain security log files for certain periods of time to identify and track malicious adversary behaviors. However, the period still must remain finite, with supporting rationale. You should also be aware of data retention laws for specific types of data, such as legal documents, within the country where your organization provides that service.

3 GDPR and Privacy by Design

Privacy by Design is a key concept of the GDPR and is made a legal requirement. Privacy by Design means thinking about data privacy and its implications when you're developing products, features, even marketing campaigns based on personal data. It also means encouraging employees to ask themselves questions before collecting or using data:

- Do I need all the data I'm collecting here?
- Could I do this work without using personal data at all?
- Am I using the data in a way a user may not expect?
- And do I have a plan to delete this data once myself or my team no longer need it?

The GDPR also encourages organizations to document key privacy decisions they make around the collection, use and storage of personal data. Documenting compliance with the GDPR may be one of the most challenging and time consuming aspects of this law. There are several ways an organization can demonstrate and document compliance. First, your organization may need to complete a privacy review process of products or features to ensure GDPR compliance before they go live.

This is often referred to as a data protection impact assessment or DPIA. DPIA's help document key decisions within an organization that have a privacy impact. Companies should also inventory the personal data it stores and collects. Also companies should. Update its existing policies and procedures or develop new ones that outline how personal data will be protected, deleted and processed.

4 Purpose

Since companies work with customer's data, they are personal data processors. They perform such operations on personal data, as collection, storage, alteration, retrieval, erasure or destruction. From the moment when any personal data gets uploaded to the company databases, they become a Processor of personal data. Under the GDPR, it is also mandatory for processors to designate a Data Protection Offices (DPO) [6].

To let Processors to comply with security requirements, we developed objects and the methods and tested them on practice.

5 Algorithm

Developed algorithm and objects can be implemented either as a part of internal system or as external system and work through API. To implement GDPR requirements in part of data retention and deletion/anonymization, we have to use 2 objects:

- **Wipeout Policy** object to store policies for each object in the database we have to apply the logic.
- **Wipeout Policy Count** object to store the history of the number of records affected by Wipeout Policy.

The records in the first object (Table 1) are being created and updated by the system administrator in accordance with the current database used in company and do not depend on the IT infrastructure.

Each record is a rule, used by wipeout logic and corresponds to one Object (**Object Name** field). If **Wipeout Action** is 'Clear_Fields', then the 'Fields' and 'Values' have text information about what object fields should be cleared (filled with null or blank values) or set to default value in case of picklist. If Wipeout Action is 'Deletion', then the records are being deleted. '**Wipeout Criteria**' field stores the

Table 1 Wipeout policy object definition

Field name	Type	Value example
Name	Text	Customer wipeout policy
Query	Text	CLEAR UserName, Phone, Email FROM Appointment WHERE (Status = 'Closed') AND End < LAST_N_DAYS:180 AND End > LAST_RUN
Object name	Text	Customer
Wipeout action	Picklist	Clear_Fields/Deletion
Fields	Text with comma separated Object field names	UserName, Phone, Email, Priorityl
Values	Text with comma separated New object field values or null	Null, null, null, normal
Retention period	Number	180
Wipeout criteria	Text	Status = 'Closed'
Active	Boolean	True
Bypass trigger	Boolean	True

rule to select the processing records by specific criteria. For example, for closed accounts, finished orders or paid bills. The format of this criteria should be just the WHERE condition for SQL query. In addition, '**Retention Period**' field shows the period when object records should remain unwiped. It needs to have the ability to control or audit closed deals for retention periods. **Query** field is generated on the policy criteria and retention period in SQL style. It is used on the dashboard to show how the set of records is selected. Finally, '**Bypass Trigger**' field shows whether the object trigger should be switched off during the object record processing (deleting or updating operation) to avoid additional logic run, like standard field validation or sending notifications.

The records in the second object (Table 2) are being created by Wypeout Scheduler and updated by Wypeout Batch. Each record is information about the daily run of Wipeout Batch and the result of its work. **Query** field is a copy of Wipeout Policy Query field in the moment of running the Wipeout Batch for analysis. **Before Count** and **After Count** fields show the number of fields in the Query before wipeout action and after. In the ideal situation (when there are no any errors) After Count should be always zero. Field **Start Time** allow to keep the last run time to eliminate processing the records already processed previously. Dashboard uses the information from this object to show the history of batches that were run for the last days.

To implement GDPR/HIPAA requirements in part of data retention and deletion/ anonymization, we use 2 code blocks.

Wipeout Scheduler is the algorithm, which is scheduled to run daily. It gets all the active Wipeout Policies, and do such steps for all of them:

Table 2 Wipeout policy count object definition

Field name	Type	Value example
Name	Text	Customer wipeout policy
Policy Id	Id	
Batch job Id	Id	
Query	Text	CLEAR UserName, Phone, Email FROM Appointment WHERE (Status = 'Closed') AND End < LAST_N_DAYS:180 AND End > LAST_RUN
Object name	Text	Customer
Wipeout action	Picklist	Clear_Fields/Deletion
Before count	Number	1 234
After count	Number	0
Start time	Datetime	
End time	Datetime	

- Find the newest record in 'Wipeout Policy Count' object for current policy Id. Record should have Batch Job Id and Run Time End value.
- Create filter for the records to process by merging the criteria in 'Wipeout Policy' record with the Start Time value in 'Wipeout Policy Count' record
- Get the current number of records in the DB which match the policy criteria.
- Create a new 'Wipeout Policy Count' record with 'Query', 'Before Count' and 'After Count' values, received in the previous step.
- Run Wipeout Batch for current policy.
- Repeat the logic for the next policy.

Wipeout Batch is the algorithm, which process records in bulk in the next steps:

- Get the chunk of records based on policy and retention period criteria.
- Deactivate object trigger if needed by policy.
- Run the logic depending on Wipeout Action (clear fields or delete records).
- Get 'Wipeout Policy Count' record related to this Batch Job Id.
- Update 'Run Time End' field with actual date/time and 'After Count' field with the number of the records failed during clearing or deleting.
- Repeat the logic for the next chunk of records.

6 Results

System administrator has access to the dashboard (Fig. 1), which shows the list of wipeout policies with SQL to retrieve the record to process (Description field) and the chart for the last month. On this chart we can see 2 graphs for 'Before Count'

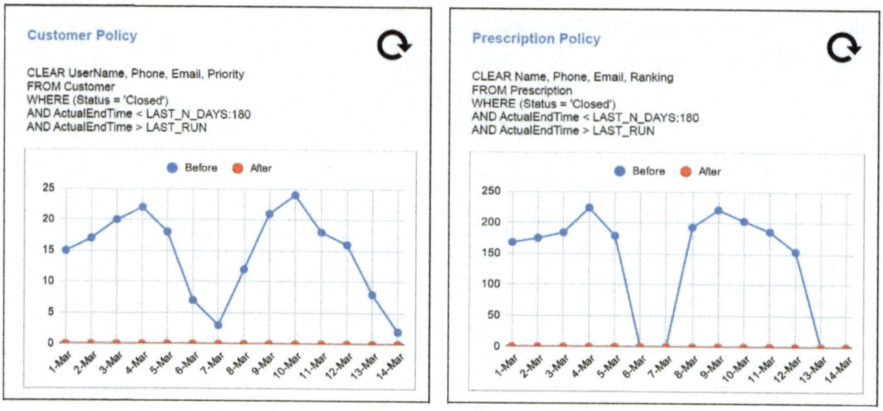

Fig. 1 Dashboard

and 'After Count' values. In an ideal situation, the last graphs should always show zero values (all the records were processed successfully).

Retention period and Fields for wiping out are easily accessible to updating by system administrator and there is no need to change or reschedule the code logic. Being scheduled once, Wipeout Scheduler runs Wipeout Batches daily and puts the result into the Wipeout Policy Count object for easy monitoring by Dashboard.

7 Implementation in Practice

Using a cybersecurity assessment and a structured approach to identifying regulations and implementing controls we have successfully implemented algorithms described above and tested them on the Health Care center. Internal audit of implementation and half-a-year run showed that after the retention period all obsolete sensitive information in the company database is being erased successfully in accordance with the internal documents related to GDPR. Dashboard shows the number of records processed and is easy to read and monitor the work of algorithms. As a result, implemented solution meets cybersecurity regulations and safeguard customers against cybersecurity attacks.

8 Trends in Cybersecurity Compliance and Regulation

Trends in cybersecurity compliance and regulation are affecting organizations everywhere. Today's regulatory environment is more challenging than ever. To get a better understanding of what this ever-changing landscape looks like for organizations, let's explore some top trends and challenges reshaping regulatory compliance for cybersecurity.

Cybersecurity Compliance Is Not Just an Information Technology Issue. Many fear cybersecurity compliance as an amorphous issue that only the information technology department handles. The reality is that the financial, legal, and reputational ramifications that arise from a data breach affect the entire organization.

If standards, regulations, or rules are too burdensome, organizations and their employees may engage in insecure workarounds to meet business needs. It's important to communicate to staff how to attain security without compromising business goals.

As guidance and regulations evolve quickly, organizations may not have the resources to keep up with changes year over year. Cybersecurity is a fast-moving sector, as both attackers and security providers vie to outsmart each other.

Gone are the days where organizations used a spreadsheet to document their controls and then had to copy, paste, and repeat against other regulations. Today, organizations use automation to document once, and then comply against many different standards. They use automated tools to collect evidence—such as vulnerability scan results and analysis, log correlation, and more—to asses their security posture, add workflow to task out items such as patching, and report to senior executives.

Automation helps streamline compliance especially with software engineers using compliance as code where they design their system and products to automatically meet control objectives. One example of automating compliance is the National Institute of Standards and Technology's (NIST's) Open Security Controls Assessment Language (OSCAL) [7].

The recruitment, retention, and training of the right security personnel to help meet compliance requirements is key. Ideally a combination of compliance analysts, security operations specialists, and software engineers make up these teams. Because the burden of regulatory compliance on organizations can be huge, organizations need to create focused cybersecurity teams that work in tandem with their internal risk teams. Organizations must hire more staff who are well-trained on the security aspects related to their industry.

With the economic effects of the COVID-19 pandemic still in full force, organizations are looking to contain costs by working with third-party vendors and partners on an ongoing basis.

One area organizations may look to outsource is compliance. This may be especially true to help perform gap assessments and meet reporting requirements, particularly if a company is small, doesn't have the right staff in house, or needs a small increase in effort for a short time to meet a new regulation or answer a point-in-time audit. As outsourcing compliance becomes more prevalent, frameworks and vendor assessments will also need to raise the bar for third-party vetting to ensure a degradation in security doesn't occur [8].

References

1. REGULATION (EU) 2016/679 OF THE EUROPEAN PARLIAMENT AND OF THE COUNCIL of 27 April 2016 on the protection of natural persons with regard to the processing of personal data and on the free movement of such data, and repealing Directive 95/46/EC (General Data Protection Regulation) https://eur-lex.europa.eu/legal-content/EN/TXT/PDF/?uri=CELEX:32016R0679
2. Guidance on the Protection of Personal Identifiable Information https://www.dol.gov/general/ppii
3. Japan enacts Amendments to the Act on the Protection of Personal Information https://iapp.org/news/a/japan-enacts-the-act-on-the-protection-of-personal-information/
4. Health Insurance Portability and Accountability Act https://www.hhs.gov/hipaa/index.html
5. Anonymization. Imperva Learning Center https://www.imperva.com/learn/data-security/anonymization/
6. Guidelines on Data Protection Officers https://ec.europa.eu/newsroom/article29/items/612048
7. OSCAL: the Open Security Controls Assessment Language https://pages.nist.gov/OSCAL/
8. Forbes: Cybersecurity Compliance Trends in a Post-Pandemic World https://www.forbes.com/sites/forbestechcouncil/2020/05/11/cybersecurity-compliance-trends-in-a-post-pandemic-world/?sh=4a62bd6f4ae2

Differentially Private Spiking Neural Networks: Enhancing Privacy and Robustness in Social Robotics

Muhammad Aitsam, Samiulhaq Chardiwall, and Alessandro Di Nuovo

Abstract As social robots increasingly integrate into various sectors, concerns over privacy and security become paramount. Social robots, while capable of processing and interacting with sensitive user data, often face significant limitations in processing power and energy efficiency. Spiking Neural Networks (SNNs), inspired by biological neurons, provide a promising solution by offering efficient temporal processing with reduced energy consumption. However, similar to conventional Artificial Neural Networks (ANNs), SNNs are vulnerable to privacy attacks, such as model inversion and membership inference, which can expose sensitive training data. We proposes the use of Differential Privacy (DP) to safeguard user data in SNN models for social robots. We train two models on the MNIST dataset: a baseline SNN and a differentially private SNN (DP-SNN). We evaluate both models through a privacy attacks, demonstrating how differentially private SNNs mitigate data leakage. We assess the ability of DP-SNNs to withstand malicious inputs, showing that the noise introduced by differential privacy enhances robustness in addition to privacy preservation. Our results indicate that differentially private SNNs not only maintain strong privacy guarantees but also improve resilience against adversarial attacks, making them an ideal solution for social robots where both data security and processing efficiency are critical.

Keywords Social robotics · Spiking neural network · Differential privacy · Membership inversion attack

https://www.shu.ac.uk/.

M. Aitsam (✉) · S. Chardiwall · A. Di Nuovo
Sheffield Hallam University, Sheffield, UK
e-mail: m.aitsam@shu.ac.uk

H. Jahankhani and B. Issac (eds.), *Cybersecurity and Human Capabilities Through Symbiotic Artificial Intelligence*, Advanced Sciences and Technologies for Security Applications, https://doi.org/10.1007/978-3-031-82031-1_11

1 Introduction

In recent years, social robots have witnessed a significant surge in adoption across various sectors such as healthcare, education, and customer service. These robots are designed to interact with humans on a social level, performing tasks that range from providing companionship to assisting with daily activities. In the healthcare sector, for instance, robots like PARO, a therapeutic robot modeled after a baby harp seal, have been used to improve the emotional well-being of elderly patients and individuals with cognitive impairments [19]. In education, robots such as NAO have been employed as teaching assistants to engage students in interactive learning experiences, thereby enhancing motivation and learning outcomes [12]. The customer service industry has also embraced social robots for tasks like greeting customers, providing information, and even handling transactions, aiming to improve efficiency and customer satisfaction [9]. The integration of social robots into daily life highlights their growing importance and the potential they hold for improving human experiences across various domains. As these robots become more prevalent, it is crucial to address the challenges that accompany their deployment to ensure they can effectively and safely serve their intended purposes.

Despite the advancements in artificial intelligence and machine learning, deploying traditional Artificial Neural Networks (ANNs) in social robots poses significant challenges. Social robots often operate under strict constraints on processing power, memory, and energy consumption due to their compact size and the necessity for mobility [3]. Traditional ANNs are computationally intensive and require substantial energy to perform complex calculations, which can lead to increased battery drain and reduced operational time for robots [8]. Additionally, the latency introduced by heavy computational loads can impede real-time interactions, which are critical for the responsiveness expected in social contexts [4]. These limitations highlight the inadequacy of conventional ANNs for applications in resource-constrained robotic platforms. There is a pressing need for more efficient computational models that can deliver high performance without exceeding the hardware capabilities of social robots.

Spiking Neural Networks (SNNs) have emerged as a promising alternative to traditional ANNs, offering a biologically inspired approach to neural computation. As shown in Fig. 1, SNNs mimic the temporal dynamics of biological neurons by processing information through discrete spikes or action potentials over time [11]. This event-driven processing allows SNNs to operate with lower power consumption and reduced computational overhead, making them highly suitable for energy-efficient applications [17]. Due to their sparse and asynchronous processing, SNNs consume significantly less energy compared to traditional ANNs [5]. The temporal coding in SNNs enables fast and efficient processing of time varying signals, which is essential for real-time interactions in social robotics [15]. SNNs have been shown to be more robust to noise and hardware variations, enhancing the reliability of robotic systems [16]. As social robots become more integrated into personal and

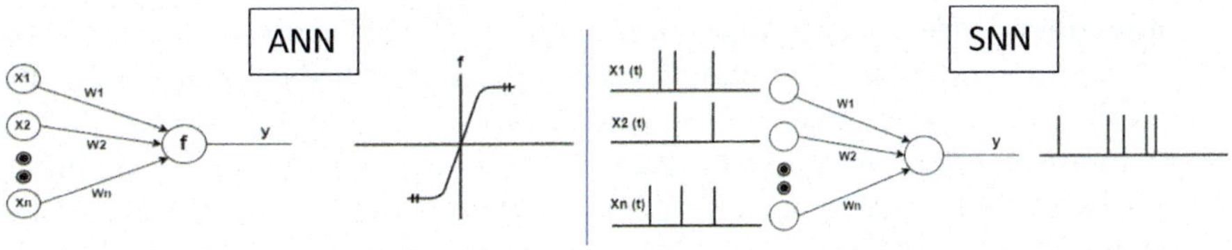

Fig. 1 Comparison of Spiking Neural Network (SNN) and Artificial Neural Network (ANN) architectures

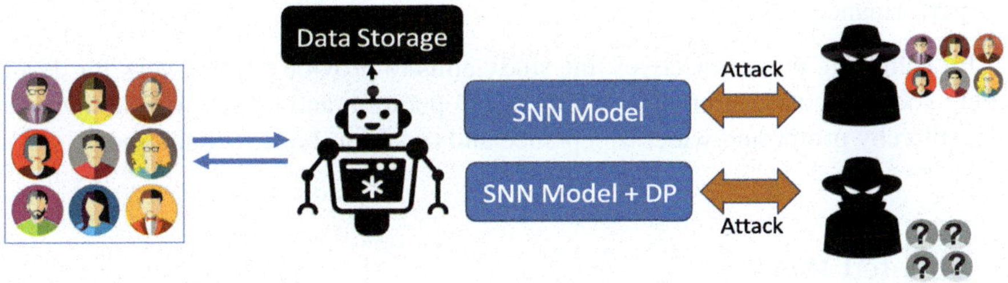

Fig. 2 Illustration of the experimental setup comparing the SNN model with and without Differential Privacy (DP) under attack scenarios

public spaces, they inevitably collect and process sensitive data from their interactions with users. This raises significant privacy concerns, particularly regarding the potential leakage of personal information through the machine learning models employed by these robots. Neural networks, including SNNs, can inadvertently reveal details about their training data through various attack vectors. Membership inference attacks allow adversaries to determine whether a specific data point was included in the model's training dataset, potentially exposing personal information [21]. Model inversion attacks enable attackers to reconstruct input data from the model's outputs, further compromising user privacy [7]. These vulnerabilities are exacerbated in social robots due to the close and continuous interaction with users, making the implementation of privacy-preserving techniques imperative.

Addressing these privacy concerns is essential to maintain user trust and comply with data protection regulations. Implementing robust privacy measures ensures that social robots can operate safely without risking the exposure of sensitive user information (Fig. 2).

1.1 Objectives of the Study

The primary objective of this study is to enhance the privacy and robustness of social robots by integrating differential privacy mechanisms into Spiking Neural Networks (SNNs). The specific goals are:

1. **Implement Differential Privacy in SNNs**: Develop and apply methods to incorporate differential privacy into the training process of SNNs, ensuring that the models do not compromise individual data privacy.
2. **Evaluate Robustness Against Attacks**: Assess the effectiveness of the differentially private SNNs in defending against privacy attacks such as membership inference and model inversion, comparing their performance to standard SNNs.
3. **Demonstrate Practical Applicability**: Validate the feasibility of deploying differentially private SNNs in social robots through experimental results, highlighting improvements in privacy preservation without significant loss of performance.

By achieving these objectives, the study aims to provide a framework for developing social robots that are both efficient in operation and secure in handling user data, thereby promoting wider acceptance and trust in robotic technologies.

2 Related Work

Existing literature in machine learning has increasingly focused on data privacy, especially with the rise of deep learning models that require large datasets containing sensitive information. Differential privacy has emerged as a leading framework for providing strong privacy guarantees while allowing useful data analysis [6]. Abadi et al. [1] introduced the Deep Learning with Differential Privacy (DP-SGD) algorithm, integrating differential privacy into stochastic gradient descent and establishing a foundational approach in privacy-preserving machine learning. Several studies have applied differential privacy to traditional Artificial Neural Networks (ANNs). Shokri and Shmatikov [20] proposed a method for training deep models with differential privacy by perturbing gradients during training. Papernot et al. [13] developed the Private Aggregation of Teacher Ensembles (PATE) framework, which transfers knowledge from an ensemble of teacher models to a student model in a privacy-preserving manner. These approaches have demonstrated that integrating differential privacy into ANNs can effectively mitigate privacy risks; however, they primarily focus on conventional neural networks and do not address the unique characteristics of Spiking Neural Networks (SNNs). SNNs have gained attention for their potential in robotics due to their energy efficiency and ability to process spatiotemporal data [10, 22]. Studies have explored the deployment of SNNs on neuromorphic hardware for robotic applications, showing promising results in tasks like sensory processing and motor control [8]. Despite advancements in SNN implementations, the aspect of privacy in SNNs remains underexplored. Research on the security and privacy aspects of SNNs is limited, with most existing work focusing on the robustness of SNNs against adversarial attacks aiming to mislead the model's predictions [2, 23]. For instance, Sharmin et al. [18] investigated the vulnerability of SNNs to gradient-based adversarial attacks and proposed defense mechanisms. However, there is a notable gap in literature concerning the integration of differential privacy into SNNs. Few studies have attempted to adapt differential privacy mechanisms to the unique architecture of SNNs, which involves temporal dynamics and event-driven processing.

Applying differential privacy to SNNs presents specific challenges. SNNs process information over time through spikes, making gradient computations more complex compared to ANNs; this complexity complicates the integration of differential privacy, which relies on precise gradient manipulation [14]. Additionally, SNNs are often deployed on specialized hardware with limited computational resources, hindering the implementation of noise addition and gradient clipping required for differential privacy. There is also a scarcity of established methodologies or frameworks for implementing differential privacy in SNNs, leading to uncertainties in achieving both privacy and model performance. The existing literature reveals that while differential privacy has been effectively applied to traditional ANNs, there is a significant gap in its application to SNNs. The primary gaps identified are limited research on differentially private SNNs, insufficient evaluation against privacy attacks, and minimal exploration of practical applicability in robotics. Given these gaps, the proposed research is necessary to advance the field by developing methods to integrate differential privacy into SNNs, enhancing privacy in social robotics, evaluating robustness and performance, and demonstrating feasibility.

By addressing these research gaps, the proposed solution contributes to the advancement of privacy-preserving machine learning in the context of spiking neural networks and social robotics. It offers a novel approach that not only enhances the privacy of user data but also maintains the performance and efficiency required for practical robotic applications. This integration is essential for fostering user trust and promoting the wider adoption of social robots in various sectors, ultimately leading to technologies that are both intelligent and respectful of individual privacy.

3 Methodology

3.1 Overview of the Proposed Approach

The primary aim of this study is to integrate differential privacy mechanisms into the training process of Spiking Neural Networks (SNNs) to enhance privacy preservation in social robots. The proposed approach involves modifying the standard training algorithm of SNNs by incorporating differential privacy during gradient computation. By adding calibrated noise to the gradients and employing gradient clipping, we ensure that the contribution of any single data point is limited, thus preserving individual data privacy. The experimental design includes training both standard and differentially private SNNs on the MNIST dataset and evaluating their performance and robustness against privacy attacks such as membership inference and model inversion. This approach allows us to assess the trade-offs between model accuracy and privacy preservation.

Table 1 Model parameters

Parameter	Value
Time steps	100
Batch size	64
Learning rate	0.05
Optimizer	Stochastic Gradient Descent (SGD)
Activation function LIF neuron dynamics	

3.2 SNN Model Architecture

The SNN model used in our experiments is designed to process spatiotemporal data efficiently while maintaining a balance between complexity and computational efficiency. The architecture consists of an input layer that accepts encoded spike trains representing input images, followed by two fully connected layers. The first fully connected layer contains 800 Leaky Integrate-and-Fire (LIF) neurons, which process the input spikes and generate intermediate spike patterns. The second fully connected layer consists of 10 LIF neurons, corresponding to the 10 classes in the MNIST dataset. The neuron model employed is the standard LIF neuron, characterized by its membrane potential dynamics:

$$\tau_m \frac{dV}{dt} = -V + RI \tag{1}$$

where V is the membrane potential, τ_m is the membrane time constant, R is the membrane resistance, and I is the input current. A spike is generated when the membrane potential exceeds a threshold Vth, after which the potential is reset (Table 1).

3.3 Implementation of Differential Privacy in SNNs

To incorporate differential privacy into the SNN training process, we utilize the *Opacus* library, which is specifically designed for training PyTorch models with differential privacy [1]. The key components of our implementation are:

We apply per-sample gradient clipping to limit the influence of individual data samples on the model parameters. For each gradient $\nabla \mathrm{L}_i$ computed from a data sample i, we calculate its L2 norm $|\nabla \mathrm{L}_i|_2$. If this norm exceeds a predefined clipping threshold C, we scale the gradient down so that its norm equals C; otherwise, the gradient remains unchanged. This process is represented by:

$$\nabla \mathrm{L}_i^{\sim} = \nabla \mathrm{L}_i \min 1 \frac{C}{\|\nabla \mathrm{L}_i\|_2} \tag{2}$$

This clipping ensures that no single data point can disproportionately affect the model update, which is crucial for maintaining privacy. After clipping, we aggregate the clipped gradients from all N samples in the batch and add Gaussian noise to the sum to obscure the contribution of individual samples. The noisy, aggregated gradient used to update the model parameters is computed as:

$$\nabla \mathrm{L} = \frac{1}{N}\left(\sum_{i=1}^{M} \nabla \mathrm{L}_i^{\sim} + \mathbf{N}(0, \sigma^2 C^2 \mathbf{I}\right) \quad (3)$$

where σ is the noise multiplier, and $\mathrm{N}(0, \sigma^2 C^2 \mathbf{I})$ represents Gaussian noise with zero mean and covariance $\sigma^2 C^2 \mathbf{I}$. The added noise further protects privacy by making it statistically difficult to infer information about any individual data sample from the model updates. Together, gradient clipping and noise addition ensure that the training process satisfies differential privacy, balancing the trade-off between model utility and privacy preservation. We track the privacy budget (ε, δ) using the moments accountant method provided by Opacus. The parameters are set to achieve a balance between privacy and model utility: Noise Multiplier (σ) = 0.8, Clipping Norm (C) = 1.0, Target Delta (δ) = 1×10^{-5}. The privacy budget ε is computed after each epoch to monitor the cumulative privacy loss. The standard backpropagation algorithm is adapted to include the differential privacy mechanisms. The training loop integrates gradient clipping and noise addition without altering the underlying spiking neuron dynamics or the learning objective.

To evaluate the privacy preservation of the differentially private SNN, we conducted membership inference attacks. This attack aim to determine whether a particular data sample was included in the model's training set. We use the attack model proposed by Shokri et al. [20], which trains a binary classifier to distinguish between members and non-members based on the model's output probabilities.

3.4 Dataset and Evaluation Metrics

We utilized the MNIST dataset, which consists of 70,000 handwritten digit images divided into 60,000 training samples and 10,000 testing samples across 10 classes. This dataset was chosen for its simplicity and widespread use in evaluating neural network models, providing a standard benchmark for performance comparison. For preprocessing, we normalized the pixel values of the images to the range [0, 1] by dividing each pixel value by 255. This normalization ensures that the input data is scaled appropriately for the neural network. To convert the static images into a format suitable for Spiking Neural Networks (SNNs), we employed Poisson encoding. In this method, each pixel intensity is converted into a spike train, where the probability of generating a spike at each time step is proportional to its normalized intensity. Poisson encoding effectively transforms static images into temporal spike trains, enabling the SNN to process data over time and mimic the temporal dynamics of biological neurons.

To assess the performance and privacy preservation of the models, we employed several evaluation metrics. For model performance, we focused on accuracy, defined as the proportion of correctly classified samples out of the total number of samples. Although precision and recall are less emphasized due to the balanced nature of the classes in the MNIST dataset, they were computed for completeness to provide a comprehensive evaluation of the model's predictive capabilities. In terms of privacy preservation, we measured the attack success rate, which represents the accuracy of the adversary's model in membership inference attacks. Privacy leakage was quantified by the difference in attack success rates between models with and without differential privacy. This metric indicates how much more vulnerable a non-private model is to privacy attacks compared to a differentially private one. This evaluation helps determine the model's ability to maintain performance in the presence of maliciously crafted inputs intended to deceive the model.

4 Result and Discussion

The results show that DP effectively reduces overfitting and mitigates the privacy risks associated with member data. The Non-DP model demonstrates high confidence scores and lower losses for member data, indicating susceptibility to membership inference attacks. Conversely, the DP model shows more distributed confidence values, higher losses, and lower accuracy, which indicates better generalization and improved privacy. The line plots in Fig. 3 show the test accuracy and test loss over training epochs for SNN models trained with and without Differential Privacy (DP). The test accuracy plot indicates that the model without DP achieves a higher accuracy compared to the model with DP across all epochs, demonstrating that adding DP reduces the model's performance due to the noise added for privacy protection. However, the accuracy of the DP model still improves steadily over time, indicating effective learning despite privacy constraints. In the test loss plot, the model without DP shows a consistently lower test loss, whereas the DP model has a higher loss value, again reflecting the trade-off between privacy and model performance. The overall behavior demonstrates the expected impact of differential privacy: a reduction in performance metrics in exchange for enhanced privacy protection (Table 2).

The Attack AUC values based on confidence and loss for both the Non-DP and DP models. The Non-DP model has higher AUC values (0.5993 for confidence and 0.5996 for loss), indicating greater vulnerability to membership inference attacks. In contrast, the DP model has lower AUC values (0.4893 for confidence and 0.4897 for loss), suggesting that Differential Privacy reduces the model's susceptibility to such attacks, thereby enhancing data privacy.

Figure 4 compare the distribution of confidence scores for predictions made by Non-DP and DP models on member (training) and non-member (test) data. The Non-DP model shows a high concentration of predictions with confidence values

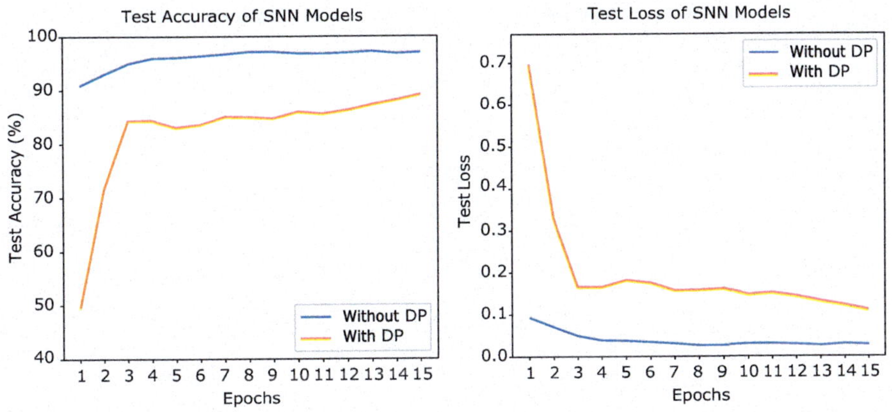

Fig. 3 Test accuracy and loss of SNN models over training epochs, showing reduced accuracy and increased loss for the DP model due to the privacy-preserving noise

Table 2 Attack AUC results for Non-DP and DP models

Model	Attack AUC (Confidence)	Attack AUC (Loss)
Non-DP model	0.5993	0.5996
DP model	0.4893	0.4897

at 1.0 for both members and non-members, indicating significant overfitting and making it vulnerable to membership inference attacks. In contrast, the DP model's confidence scores are more spread out, with a peak at a lower value of around 0.7, suggesting reduced confidence and better generalization. This behavior demonstrates that the DP model is more effective in obscuring the difference between member and non-member data, thereby providing better protection against attacks aimed at inferring training set membership. Figure 5 compare the distribution of loss values for predictions made by Non-DP and DP models on member (training) and non-member (test) data.

In the Non-DP model, the loss values for member data are generally lower compared to non-member data, indicating that the model is able to fit the training data well, but potentially at the risk of overfitting, making it easier to distinguish between members and non-members. The DP model, however, shows a more similar loss distribution for both member and non-member data, with losses spread more evenly and at higher values. This suggests that the DP model generalizes better, providing stronger privacy protection by making it more challenging to differentiate between training and test data.

Applying Differential Privacy to SNN models reduces their performance metrics, such as accuracy and confidence, but significantly enhances privacy. The DP model's reduced confidence and higher losses indicate better protection against membership inference attacks, while the Non-DP model's overfitting makes it more vulnerable.

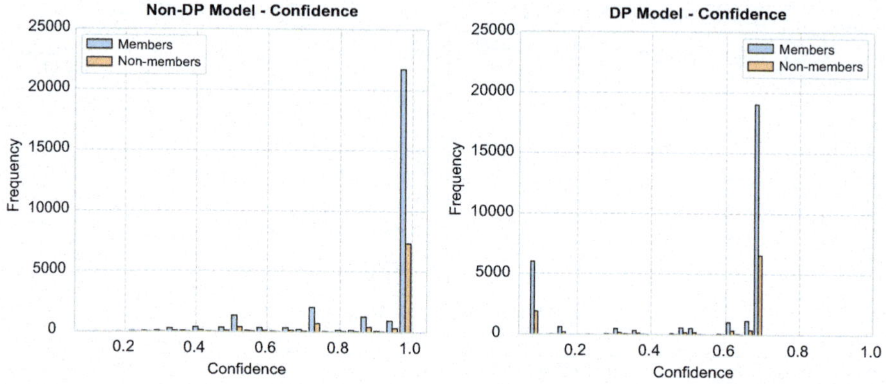

Fig. 4 Confidence scores for predictions made by the Non-DP and DP models on member and non-member data

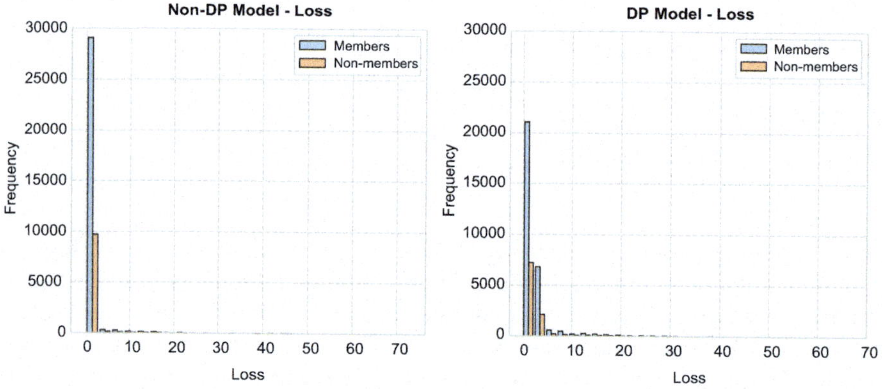

Fig. 5 Loss values for the Non-DP and DP models showing more distinguishable loss patterns for members versus non-members in the Non-DP model, while the DP model exhibits more uniform loss values, enhancing privacy

These findings underscore the value of Differential Privacy in balancing privacy with model utility, especially for applications involving sensitive data.

4.1 Limitations and Future Work

One limitation of this study is the suboptimal selection of privacy-preserving parameters, which may lead to excessive noise insertion, adversely affecting the accuracy of the SNN model. Future work could focus on developing more sophisticated algorithms to better predict these parameters, thus balancing privacy with model utility more effectively. Additionally, this study does not evaluate the energy consumption differences between implementing SNN models on conventional von Neumann architectures and neuromorphic hardware like Loihi or SpiNNaker. Future research should address this gap by conducting detailed energy consumption analyses to determine the practical efficiency of using SNNs with DP on neuromorphic systems. Furthermore, evaluating the scalability of DP for larger and more complex datasets, and assessing its impact on real-world applications, would be valuable extensions of this work.

5 Conclusion

In this study, we analyzed the impact of Differential Privacy (DP) on Spiking Neural Network (SNN) models compared to their non-private counterparts. The results showed that while the Non-DP model had higher accuracy, it was vulnerable to privacy attacks, which is concerning in the context of social robotics that might store personal information of its users during interaction. In contrast, the DP model exhibited improved privacy through reduced confidence and increased loss, albeit with a decline in predictive performance. These findings show that the DP-SNN model provides better privacy compared to the standard SNN model.

Acknowledgements This work is funded by Marie Sklodowska-Curie Action Horizon 2020 (Grant agreement No. 955778) for the project 'Personalized Robotics as Service Oriented Applications (PERSEO)'. For open access, the author has applied a Creative Commons Attribution (CC BY) licence to any Author-Accepted Manuscript version arising from this submission.

References

1. Abadi M, McMahan HB, Chu A, Mironov I, Zhang L, Goodfellow I, Talwar K (2016) Deep learning with differential privacy. In: Proceedings of the ACM conference on computer and communications security, 24–28 October 2016, pp 308–318. https://doi.org/10.1145/2976749.2978318, https://dl.acm.org/doi/10.1145/2976749.2978318
2. Bagheri A, Simeone O, Rajendran B (2018) Adversarial training for probabilistic spiking neural networks. In: IEEE workshop on signal processing advances in wireless communications, SPAWC 2018. https://doi.org/10.1109/SPAWC.2018.8446003
3. Bekey GA (2017) Autonomous robots: from biological inspiration to implementation and control, p 577. https://doi.org/10.1162/artl.2007.13.4.419
4. Dautenhahn K (2007) Socially intelligent robots: dimensions of human–robot interaction. Philos Trans R Soc B Biol Sci 362:679–704. https://doi.org/10.1098/RSTB.2006.2004, https://royalsocietypublishing.org/doi/10.1098/rstb.2006.2004
5. Davies M, Srinivasa N, Lin TH, Chinya G, Cao Y, Choday SH, Dimou G, Joshi P, Imam N, Jain S, Liao Y, Lin CK, Lines A, Liu R, Mathaikutty D, McCoy S, Paul A, Tse J, Venkataramanan G, Weng YH, Wild A, Yang Y, Wang H (2018) Loihi: a neuromorphic manycore processor with on-chip learning. IEEE Micro 38:82–99. https://doi.org/10.1109/MM.2018.112130359

6. Dwork C, Roth A (2014) The algorithmic foundations of differential privacy. Found Trends® Theor Comput Sci 9:211–407. https://doi.org/10.1561/0400000042, https://dl.acm.org/doi/10.1561/0400000042
7. Fredrikson M, Jha S, Ristenpart T (2015) Model inversion attacks that exploit confidence information and basic countermeasures. In: Proceedings of the ACM conference on computer and communications security, pp 1322–1333. https://doi.org/10.1145/2810103.2813677, https://dl.acm.org/doi/10.1145/2810103.2813677
8. Indiveri G, Liu SC (2015) Memory and information processing in neuromorphic systems. Proc IEEE 103:1379–1397. https://doi.org/10.1109/JPROC.2015.2444094
9. Ivanov S, Webster C, Garenko A (2018) Young Russian adults' attitudes towards the potential use of robots in hotels. Technol Soc 55:24–32. https://doi.org/10.1016/J.TECHSOC.2018.06.004
10. Lobo JL, Ser JD, Bifet A, Kasabov N (2020) Spiking neural networks and online learning: an overview and perspectives. Neural Netw 121:88–100. https://doi.org/10.1016/J.NEUNET.2019.09.004
11. Maass W (1997) Networks of spiking neurons: the third generation of neural network models. Neural Netw 10:1659–1671. https://doi.org/10.1016/S0893-6080(97)00011-7
12. Mubin O, Stevens CJ, Shahid S, Mahmud AA, Dong JJ (2013) A review of the applicability of robots in education. Technol Educ Learn 1. https://doi.org/10.2316/JOURNAL.209.2013.1.209-0015
13. Papernot N, Song S, Mironov I, Raghunathan A, Talwar K, Erlingsson Ú (2018) Scalable private learning with pate. In: 6th International conference on learning representations, ICLR 2018 conference track proceedings. https://arxiv.org/abs/1802.08908v1
14. Pfeiffer M, Pfeil T (2018) Deep learning with spiking neurons: opportunities and challenges. Front Neurosci 12:409662. https://doi.org/10.3389/FNINS.2018.00774/BIBTEX, www.frontiersin.org
15. Ponulak F, Kasiński A (2011) Introduction to spiking neural networks: information processing, learning and applications. Acta Neurobiologiae Experimentalis 71:409–433. https://doi.org/10.55782/ANE-2011-1862, https://pubmed.ncbi.nlm.nih.gov/22237491/
16. Qiao N, Mostafa H, Corradi F, Osswald M, Stefanini F, Sumislawska D, Indiveri G (2015) A reconfigurable on-line learning spiking neuromorphic processor comprising 256 neurons and 128k synapses. Front Neurosci 9:123487. https://doi.org/10.3389/FNINS.2015.00141/ABSTRACT, www.frontiersin.org
17. Roy K, Jaiswal A, Panda P (2019) Towards spike-based machine intelligence with neuromorphic computing. Nature 575(7784):607–617. https://doi.org/10.1038/s41586-019-1677-2, https://www.nature.com/articles/s41586-019-1677-2
18. Sharmin S, Rathi N, Panda P, Roy K (2020) Inherent adversarial robustness of deep spiking neural networks: effects of discrete input encoding and non-linear activations. In: Lecture notes in computer science (including subseries Lecture notes in artificial intelligence and Lecture notes in bioinformatics), vol 12374, 399–414. LNCS. https://doi.org/10.1007/978-3-030-58526-6_24, https://link.springer.com/chapter/10.1007/978-3-030-58526-6-24
19. Shibata T, Wada K (2011) Robot therapy: a new approach for mental healthcare of the elderly a mini-review. Gerontology 57:378–386. https://doi.org/10.1159/000319015, https://pubmed.ncbi.nlm.nih.gov/20639620/
20. Shokri R, Shmatikov V (2015) Privacy-preserving deep learning. In: Proceedings of the ACM conference on computer and communications security, pp 1310–1321. https://doi.org/10.1145/2810103.2813687, https://dl.acm.org/doi/10.1145/2810103.2813687
21. Shokri R, Stronati M, Song C, Shmatikov V (2017) Membership inference attacks against machine learning models. In: Proceedings IEEE symposium on security and privacy, pp 3–18. https://doi.org/10.1109/SP.2017.41
22. Tavanaei A, Ghodrati M, Kheradpisheh SR, Masquelier T, Maida A (2019) Deep learning in spiking neural networks. Neural Netw 111:47–63. https://doi.org/10.1016/J.NEUNET.2018.12.002

23. Zhou S, Liu C, Ye D, Zhu T, Zhou W, Yu PS (2022) Adversarial attacks and defenses in deep learning: from a perspective of cybersecurity. ACM Comput Surv 55. https://doi.org/10.1016/j.eng.2019.12.012, https://dl.acm.org/doi/10.1145/3547330

Image-Based Android Malware Detection Using Deep Learning

Somadina Udeze, Husnain Rafiq, Daniel Jeremiah, Vinh-Thong Ta, and Muhammad Usman

Abstract The Android operating system (OS) dominates the mobile phone OS industry, with over 70% of the market share. With the growth of Android OS-based smartphones, it has become a prime target for mobile malware attacks. Minimal alterations in malware samples can easily evade traditional detection methods such as signature-based detection. In contrast, artificial intelligence (AI) and machine learning (ML)-based malware detection has proven more effective, as it can detect zero-day malware. Previous studies have shown that AI/ML-based malware classifiers trained on categorical features are vulnerable to adversarial evasion attacks. Therefore, in this study, we transform the features extracted from Android apps into image-based data and investigate the performance of Convolutional Neural Networks (CNNs), Inception Networks, and Residual Networks (ResNet) on this data. We employ 41,382 Android malware samples belonging to 240 malware families and 36,755 benign apps to train and test the models. Our experiment results show that CNNs outperform Inception Networks and ResNet with up to 99% classification accuracy. Furthermore, our analysis also indicates that CNNs trained on image-based Android malware and benign data outperform various Android malware detection techniques proposed in the literature.

Keywords Android malware detection · Deep learning · Image-based classification · Convolutional neural networks · KronoDroid dataset

1 Introduction

Over the past decade, technological advancements have significantly transformed industries and our interaction with the digital world. Android commands over 70.1% of the global mobile OS market by the end of 2023 [1], attributed to its open-source nature, which, while advantageous, also makes it vulnerable to cyber threats like

S. Udeze · H. Rafiq (✉) · D. Jeremiah · V.-T. Ta · M. Usman
Department of Computer Science, Edge Hill University, Ormskirk L39 4QP, UK
e-mail: husnain.rafiq@edgehill.ac.uk

H. Jahankhani and B. Issac (eds.), *Cybersecurity and Human Capabilities Through Symbiotic Artificial Intelligence*, Advanced Sciences and Technologies for Security Applications, https://doi.org/10.1007/978-3-031-82031-1_12

malware and spyware [2]. Kaspersky Labs reported 8.3 million malware attacks on Android devices in 2023, and the Zimperium Mobile Threat Report noted a 138% increase in critical vulnerabilities [3]. Traditional detection methods struggle against evolving threats. To address these challenges, we propose a novel approach that transforms Android application data into visual graphics, enabling the use of advanced image recognition techniques for malware detection. Previous studies [4, 5] have explored this method, but the performance of various deep-learning models remains under-examined. This study aims to investigate how these models can improve malware detection on Android devices.

The main contributions of this paper are summarized as follows:

- We employed the KronoDroid dataset, comprising 41,382 malware samples from 240 families and 36,755 benign apps. Using Random Forest and Correlation-based methods, the top 144 most relevant features were selected for malware and benign apps classification.
- We introduced an innovative approach where key features of Android malware and benign applications were converted into image representations. This transformation enabled the application of deep learning models, traditionally used for image recognition, in the domain of malware detection.
- Multiple deep learning models, including Convolutional Neural Networks (CNNs), Residual Networks (ResNet), and Inception Networks, were explored to classify the image-based Android malware dataset. These models leveraged their advanced pattern recognition capabilities to effectively differentiate between benign and malicious Android applications.
- Experimental results demonstrated that our proposed method outperformed several existing techniques in the literature. The image-based approach to malware detection proved highly effective, validating the superiority of this method in identifying malicious Android applications.

The paper is structured as follows: Sect. 2 covers related work, Sect. 3 outlines the methodology, and Sect. 4 presents the experimental results. Finally, Sect. 5 concludes the paper and discusses potential directions for future work.

2 Related Work

The Android operating system, with its extensive global user base exceeding 2.5 billion active users, has become an attractive target for cybercriminals due to its open-source nature and widespread adoption [6]. This popularity has led to the emergence of various malware types, including worms, botnets, rootkits, ransomware, and Trojans, which often employ sophisticated techniques like metamorphism and polymorphism to evade detection [7]. The challenge of effectively combating these evolving threats has driven researchers to explore a variety of detection methodologies, including static and dynamic analysis, hybrid approaches, and, more recently, visual-based methods [8]. Static analysis, which involves examining the application's code without executing it, has proven effective against known threats but frequently struggles with malware that employs obfuscation techniques [9]. In contrast, dynamic analysis, which monitors application behavior during runtime, provides a more comprehensive understanding of malicious activities but at the cost of greater computational resources [10].

Integrating machine learning techniques into Android malware detection has significantly advanced the field, enabling the identification of new and emerging threats by learning patterns from data. Supervised learning methods, such as decision trees and support vector machines, are commonly employed for classification tasks, while unsupervised methods, including clustering, detect anomalies in unlabelled data [11]. However, traditional approaches often fall short in handling advanced malware behaviors, prompting the exploration of more sophisticated models. For instance, the Smart Android Malware Detection Mechanism (SAMDM), which integrates permission analysis with system API calls, has improved detection accuracy by leveraging these features in a machine-learning framework [12]. Additionally, recent studies have highlighted the benefits of ensemble learning methods that combine the strengths of multiple models to enhance the robustness and overall performance of detection systems [13, 14].

A promising development in this domain is the use of image-based approaches for malware detection. These methods convert malware binaries into visual representations, allowing for the identification of subtle patterns that traditional techniques might overlook. Image-based detection has proven effective in capturing complex malware behaviors but demands significant computational resources and requires access to diverse datasets to ensure robustness [15]. Recent research has also explored hybrid methods that combine image-based detection with traditional machine-learning techniques, thereby capturing both visual and non-visual features of malware [16]. Comparisons between image-based techniques and traditional methods reveal that while the former excels in detecting polymorphic and metamorphic malware, they may overlook non-visual characteristics critical to its functionality. Conversely, traditional methods like support vector machines and random forests offer greater interpretability and are more resource-efficient. However, they can struggle with highly obfuscated threats [17].

Despite the advancements in integrating machine learning into Android malware detection, several challenges remain. These include the need for larger and more diverse datasets, the computational demands of advanced detection techniques, and the importance of ensuring model interpretability and transparency. Future research should focus on refining these methods, particularly by combining image-based and traditional techniques, to develop more robust and scalable solutions to address the evolving landscape of malware threats. Incorporating domain-specific knowledge and expert insights into detection models could significantly enhance their effectiveness, especially in identifying sophisticated and polymorphic malware strains.

3 Methodology

Figure 1 presents the block diagram for the Android malware detection model, illustrating the dual pathways employed in the detection process. The diagram begins with the KronoDroid dataset, including malicious and benign Android applications. The initial step involves data cleaning to remove non-numeric and constant-value features, ensuring the dataset is prepared for effective analysis. The process then branches into two parallel feature selection paths. In the Random Forest path, features are selected based on their importance as determined by the Random Forest algorithm. The top features are then transformed into grayscale images representing each application's significant attributes. Subsequently, these images are fed into deep learning models, including Convolutional Neural Networks (CNNs), Residual Networks (ResNet), and Inception Networks, for classification. Correlation Selection: Simultaneously, a second path utilises Pearson Correlation to select features with the highest linear relationship with the target variable. These features are similarly transformed into grayscale images, which are then processed by the same set of deep-learning models. The final output from both paths is combined and evaluated to determine the detection model's accuracy and effectiveness. The results demonstrate the model's capability to accurately classify applications as benign or malicious, leveraging the strengths of both Random Forest and Correlation-based feature selection methods.

3.1 Dataset Preparation

The dataset utilised in this study is derived from the KronoDroid dataset [18], which is well-regarded for its comprehensive inclusion of Android malware samples. The dataset comprises 41,382 malware samples from 240 distinct malware families and 36,755 benign applications. This diversity provides a robust foundation for training and evaluating advanced malware detection models, making it well-suited for the objectives of this research [18].

Preprocessing: Several steps were undertaken to prepare the dataset for analysis to ensure data quality and relevance. Initially, the benign and malicious samples were combined to create a unified dataset. This combined dataset was then shuffled to eliminate any potential biases that might arise from the order of the data. Following this, columns containing constant values were identified and removed, as they do not contribute to the variability needed for effective machine learning. Additionally, non-numeric features were excluded to focus on the numerical data, which is critical for the subsequent feature selection and modelling processes.

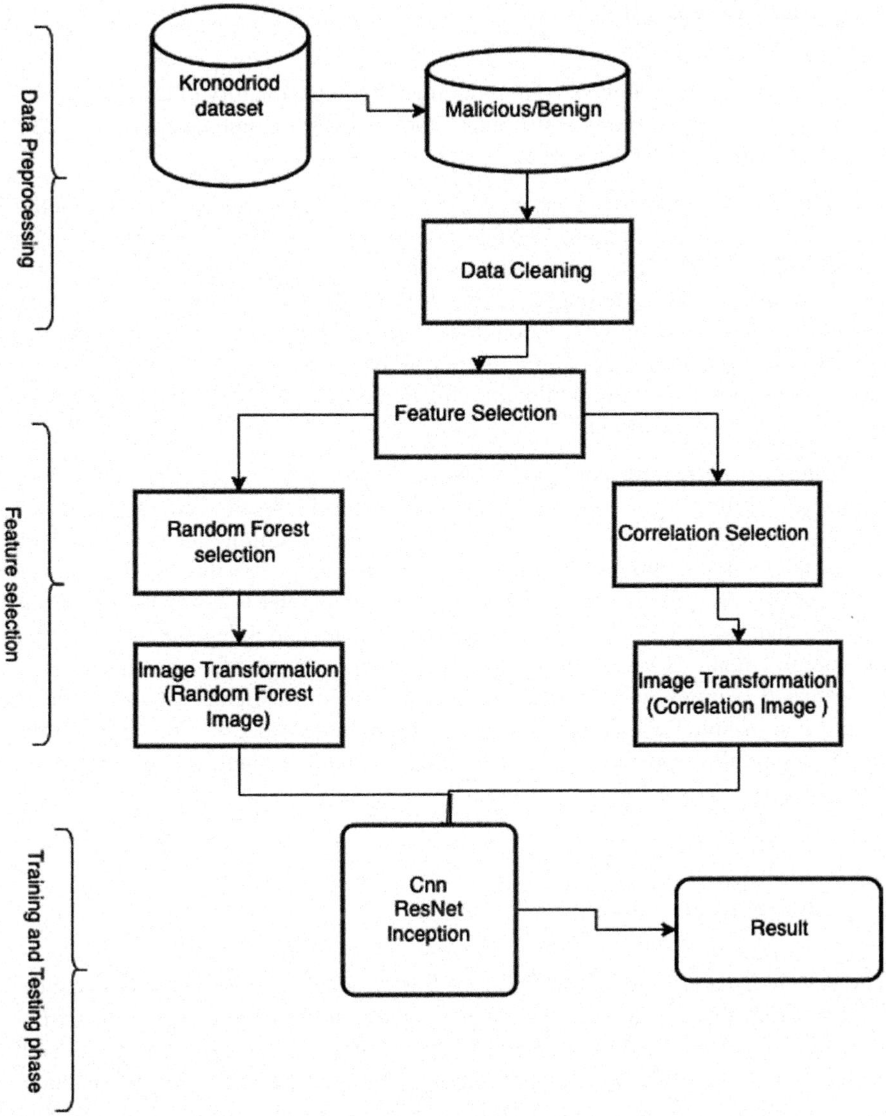

Fig. 1 Overview of the android malware detection process

3.2 Feature Selection

Feature selection is a crucial step in reducing the dimensionality of the dataset while retaining the most informative attributes. Two primary methods were employed for feature selection:

Random Forest Feature Selection: Random Forest, an ensemble learning method, was employed to rank features based on their importance in predicting the target variable. This method constructs multiple decision trees during training and aggregates their predictions to enhance accuracy. The importance of each feature is determined by its contribution to reducing the impurity of the model's nodes. This study selected the top 144 features based on their importance scores, effectively reducing the dataset's dimensionality while retaining the most significant attributes necessary for accurate classification [19].

Correlation Coefficient: Alongside Random Forest, the Pearson Correlation Coefficient was utilised to assess the linear relationship between each feature and the target variable. Features with the highest absolute correlation values were selected as they are likely to influence the target outcome strongly. This method complements the Random Forest approach by providing a statistical measure of feature relevance, ensuring that the selected features are both significant and independent [20]. Again, the top 144 features were chosen for further processing. After feature selection, the selected features were normalised using the MinMaxScaler to ensure that all features contributed equally during the model training process. This normalisation step is essential for preventing any one feature from disproportionately influencing the model due to its scale.

3.3 Image Generation

The selected features from the Random Forest and Pearson Correlation methods were transformed into grayscale images. Each row in the dataset was converted into a 12×12 pixel grayscale image, where the intensity of each pixel represented the scaled value of a feature. This transformation allowed the use of convolutional neural networks (CNNs), which are particularly effective in recognizing patterns within image data [21]. The conversion of numeric data into image format leverages the strengths of image-based models in a context traditionally dominated by numerical and categorical analysis.

3.4 Model Selection

The study utilised three deep learning models: Convolutional Neural Networks (CNNs), Residual Networks (ResNet), and Inception Networks chosen for their effectiveness in image processing and malware detection. CNNs, known for capturing spatial hierarchies, use multiple convolutional and max-pooling layers to extract and classify features from images, ending with fully connected layers and a softmax output. They were selected for their accuracy in image classification tasks. ResNet, ResNet employs skip connections to overcome the vanishing gradient problem, enabling the training of deep networks. This capability was crucial for handling the complex patterns in the dataset derived from Android applications. Lastly, Inception Networks; Inception Networks use parallel convolutions of varying sizes to capture multi-scale features, making them adept at identifying diverse patterns. Their flexibility and depth were key for comprehensive feature extraction in malware detection.

3.5 Training and Evaluation

Training Process: The models were trained on the grayscale image dataset with an 80/20 train-test split. Data augmentation techniques, including rotations, flips, and zooms, were used to increase dataset diversity and enhance model generalization.

Evaluation Metrics: Model performance was assessed using Accuracy, Precision, Recall, F1-Score, and ROC-AUC. Accuracy indicates overall correctness, Precision measures true positives, Recall identifies all relevant instances, F1-Score balances Precision and Recall, and ROC-AUC evaluates class distinction across thresholds.

Hyperparameter Settings: Models were fine-tuned with optimized hyperparameters. For CNNs, key parameters included convolutional layers and filter sizes; for ResNet, residual blocks and initialization; and for Inception Networks, architecture depth and module configuration. Hyperparameters were selected via grid search and cross-validation.

4 Experimental Results

4.1 Overview of Results

The experimental results demonstrate the effectiveness of the image-based approach to Android malware detection using deep learning models such as CNNs, ResNet, and Inception Networks. The study evaluated the performance of these models across key metrics, including accuracy, precision, recall, F1-score, and ROC-AUC. The

Table 1 Performance of deep learning models on image-based malware detection

Model	Dataset	Accuracy	Precision	Recall	F1-score
CNN	RF	99.00	99.00	99.00	99.00
CNN	Correlation	99.00	99.00	99.00	99.00
Inception	RF	98.00	99.00	97.00	98.00
Inception	Correlation	95.00	96.00	95.00	95.00
ResNet	RF	89.00	91.00	89.00	90.00
ResNet	Correlation	86.00	86.00	87.00	86.00

results indicate that CNNs significantly outperformed the other models, achieving the highest accuracy and most balanced performance across all metrics.

4.2 Model Performance

The performance of each model was rigorously tested using the grayscale image dataset generated from the KronoDroid dataset. The models were evaluated based on their ability to classify Android applications as benign or malicious. The key metrics used for evaluation are summarized in Table 1.

4.3 Discussion

The results show that CNNs achieved the highest performance metrics, balancing precision, recall, and F1-score, indicating their superior classification of benign and malicious applications. The near-perfect ROC-AUC scores reflect robust classification performance across various thresholds. Inception Networks also performed well, especially with features selected by the Random Forest method, demonstrating high accuracy and precision. However, they slightly lagged behind CNNs in recall and F1-score, suggesting they might miss some subtle patterns. ResNet, while still effective, had the lowest performance, particularly in accuracy and ROC-AUC. Its complexity may have led to overfitting, and the lower recall and F1-score indicate challenges in generalizing from the dataset compared to CNNs and Inception Networks.

Overall, CNNs proved most effective for image-based malware detection, excelling at capturing complex patterns for accurate classification. The ROC curves in Figs. 2 and 3 compare the performance of CNN, Inception, and ResNet models with Random Forest (RF) and Correlation feature selection methods.

Figure 2 showcases the ROC curve for the models trained on the features selected using the Pearson Correlation method. The Custom CNN model achieved an Area Under the Curve (AUC) of 1.00, indicating flawless classification performance with a true positive rate of 1.0 and a false positive rate of 0.0 across all thresholds. This

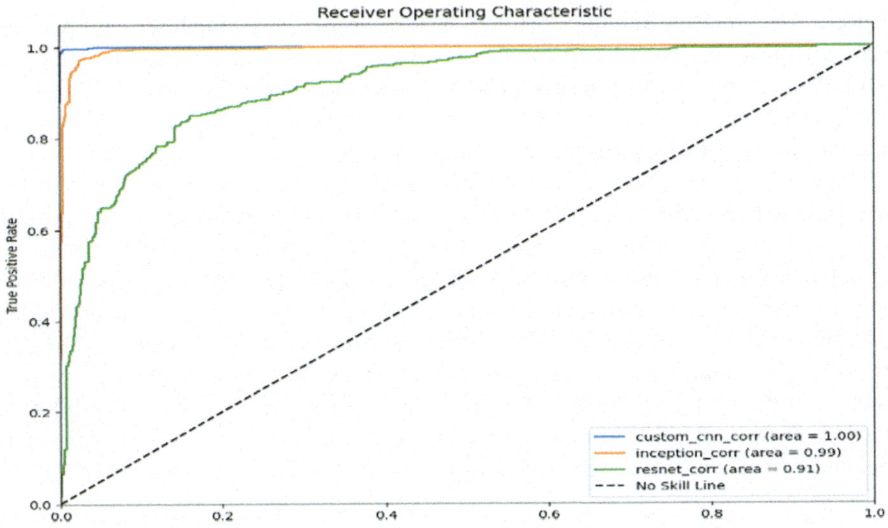

Fig. 2 ROC curve for correlation method

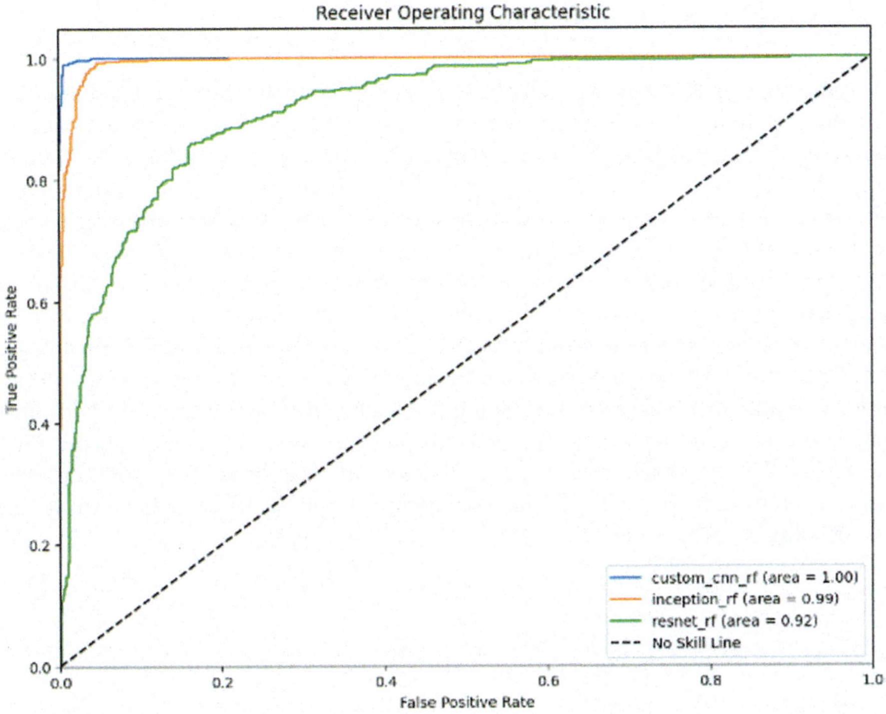

Fig. 3 ROC curve for the random forest method

perfect AUC implies that the Custom CNN consistently assigns higher probability scores to positive instances compared to negative ones, reflecting its superior ability to capture relevant patterns in the data. The Inception and ResNet models also performed well, with AUCs of 0.99 and 0.91, respectively, although they lag slightly behind the Custom CNN in distinguishing between classes.

Figure 3 presents the ROC curve for the models trained on the features selected using the Random Forest method. Similar to the correlation-based results, the Custom CNN model again achieved a perfect AUC of 1.00, underscoring its robustness and effectiveness in classifying Android malware. The Inception model closely followed with an AUC of 0.99, and the ResNet model showed a slight improvement with an AUC of 0.92 compared to its performance with correlation-selected features. Overall, these results confirm that the Custom CNN model outperforms the other classifiers, making it the best-performing model for Android malware detection. Its ability to achieve flawless classification across both feature selection methods demonstrates its versatility and robustness, solidifying its role as the preferred choice in the detection process.

4.4 Comparison with Related Work

A comparison was made with existing studies to contextualize the results within the broader field of Android malware detection. Table 2 provides a comparative analysis, showcasing how the proposed model outperforms some related methodologies. As illustrated in Table 2, the proposed CNN-based model demonstrates superior performance by achieving a classification accuracy of 99%, surpassing several well-established models in the literature. The proposed method also excels in accuracy, particularly when compared to other image-based approaches such as R2-d2 [22] and the model proposed by Yang et al. [23]. This comparison underscores the robustness of the image-based approach and highlights the significant advantages of employing deep learning models, especially CNNs, for Android malware detection. The ability of the proposed model to classify malware accurately with high precision and recall further emphasizes its contribution to enhancing cybersecurity measures. The proposed method's effectiveness in classifying malware accurately sets it apart from traditional and other contemporary approaches, making it a valuable contribution to cybersecurity.

Table 2 Comparison of proposed models with existing work

Model name	Methodology	Accuracy (%)
AspectDroid [24]	Hybrid analysis	94.68
AMDDL model [25]	CNN + LSTM deep learning	96.00
Andro-AutoPsy [26]	Similarity matching	98.50
DroidDet [20]	Static analysis	88.26
DREBIN [27]	Static analysis	94.00
R2-d2 [22]	Image-based	93.00
Yang et al. [23]	Image-based	95.42
Karimi et al. [28]	Image-based	97.00
Yen et al. [29]	Image-based	92.67
Proposed model (CNN)	Image-based + RF + Correlation	99.00

5 Conclusion

This paper explored the effectiveness of image-based data representations for Android malware detection. By leveraging the KronoDroid dataset, which included 41,382 malware samples and 36,755 benign apps, we successfully implemented and evaluated three deep learning models: Convolutional Neural Networks (CNNs), Residual Networks (ResNet), and Inception Networks. The transformation of selected features into 12×12 grayscale images allowed these models to classify Android applications accurately. Among the models tested, CNNs demonstrated superior performance, achieving up to 99% classification accuracy, which surpassed both ResNet and Inception Networks. Our results indicate that the image-based approach, particularly when using CNNs, offers a significant advantage in detecting Android malware compared to traditional methods. The study also confirmed that feature selection methods, like Random Forest and Pearson Correlation, play a crucial role in enhancing these models' accuracy and computational efficiency.

References

1. Statista (2024) Android os smartphone sales share worldwide 2009–2018 [Online]. Available: https://www.statista.com/statistics/216420/global-market-share-forecast-of-smartphone-operating-systems/. Accessed 17 Sept 2024
2. Waheed M, Qadir S (2024) Effective and efficient android malware detection and category classification using the enhanced kronodroid dataset. Secur Commun Netw 2024(1):7382302
3. Kaspersky (2024) Attacks on mobile devices significantly increase in 2023 [Online]. Available: https://www.kaspersky.com/about/press-releases/attacks-on-mobile-devices-significantly-increase-in-2023. Accessed 17 Sept 2024
4. Bakour K, Ünver HM (2021) Deepvisdroid: android malware detection by hybridizing image-based features with deep learning techniques. Neural Comput Appl 33(18):11499–11516

5. Liu P, Wang W, Zhang S, Song H (2023) Imagedroid: using deep learning to efficiently detect android malware and automatically mark malicious features. Secur Commun Netw 2023(1):5393890
6. Muzaffar A, Hassen HR, Lones MA, Zantout H (2022) An in-depth review of machine learning based android malware detection. Comput Secur 121:102833
7. Goswami M, Sebastian NJ (2022) Performance analysis of logistic regression, KNN, SVM, Naïve Bayes classifier for healthcare application during COVID-19. In: Innovative data communication technologies and application: proceedings of ICIDCA 2021. Springer, pp 645–658
8. Almomani I, Ahmed M, El-Shafai W (2022) Android malware analysis in a nutshell. PLoS ONE 17(7):e0270647
9. Rafiq H, Aslam N, Issac B, Randhawa R (2022) An investigation on fragility of machine learning classifiers in android malware detection
10. Manzil HHR, Manohar Naik S (2023) Android malware category detection using a novel feature vector-based machine learning model. Cybersecurity 6(1):6
11. Poornima S, Mahalakshmi R (2024) Automated malware detection using machine learning and deep learning approaches for android applications. Meas Sens 32:100955
12. Akbar F, Hussain M, Mumtaz R, Riaz Q, Wahab AWA, Jung KH (2022) Permissions-based detection of android malware using machine learning. Symmetry 14(4):718
13. Rafiq H, Aslam N, Ahmed U, Lin JCW (2022) Mitigating malicious adversaries evasion attacks in industrial internet of things. IEEE Trans Industr Inf 19(1):960–968
14. Ganaie MA, Hu M, Malik AK, Tanveer M, Suganthan PN (2022) Ensemble deep learning: a review. Eng Appl Artif Intell 115:105151
15. Chaganti R, Ravi V, Pham TD (2022) Image-based malware representation approach with efficientnet convolutional neural networks for effective malware classification. J Inf Secur Appl 69:103306
16. Nataraj L, Karthikeyan S, Jacob G, Manjunath BS (2011) Malware images: visualization and automatic classification. In: Proceedings of the 8th international symposium on visualization for cyber security, pp 1–7
17. Makandar A, Patrot A (2015) Malware image analysis and classification using support vector machine. Int J Trends Comput Sci Eng 4(5):01–03
18. Guerra-Manzanares A, Bahsi H, Nõmm S (2021) Kronodroid: time-based hybrid-featured dataset for effective android malware detection and characterization. Comput Secur 110:102399
19. Rong M, Gong D, Gao X (2019) Feature selection and its use in big data: challenges, methods, and trends. IEEE Access 7:19709–19725
20. Zhu HJ, You ZH, Zhu ZX, Shi WL, Chen X, Cheng L (2018) Droiddet: effective and robust detection of android malware using static analysis along with rotation forest model. Neurocomputing 272:638–646
21. Bhamare D, Suryawanshi P (2018) Review on reliable pattern recognition with machine learning techniques. Fuzzy Inf Eng 10(3):362–377. https://doi.org/10.1080/16168658.2019.1611030
22. Hsien-De Huang T, Kao HY (2018) R2-d2: color-inspired convolutional neural network (CNN)-based android malware detections. In: 2018 IEEE international conference on big data (Big Data). IEEE, pp 2633–2642
23. Yang M, Wen Q (2017) Detecting android malware by applying classification techniques on images patterns. In: 2017 IEEE 2nd international conference on cloud computing and big data analysis (ICCCBDA). IEEE, pp 344–347
24. Ali-Gombe AI, Saltaformaggio B, Xu D, Richard GG III et al (2018) Toward a more dependable hybrid analysis of android malware using aspect-oriented programming. Comput Secur 73:235–248
25. Smmarwar SK, Gupta GP, Kumar S (2023) Xai-amd-dl: an explainable AI approach for android malware detection system using deep learning. In: 2023 IEEE world conference on applied intelligence and computing (AIC). IEEE, pp 423–428

26. Jang JW, Kang H, Woo J, Mohaisen A, Kim HK (2015) Andro-AutoPsy: anti-malware system based on similarity matching of malware and malware creator-centric information. Digit Investig 14:17–35
27. Arp D, Spreitzenbarth M, Hubner M, Gascon H, Rieck K, Siemens C (2014) Drebin: effective and explainable detection of android malware in your pocket. NDSS 14:23–26
28. Karimi A, Moattar MH (2017) Android ransomware detection using reduced opcode sequence and image similarity. In: 2017 7th international conference on computer and knowledge engineering (ICCKE). IEEE, pp 229–234
29. Yen YS, Sun HM (2019) An android mutation malware detection based on deep learning using visualization of importance from codes. Microelectron Reliab 93:109–114
30. Szegedy C, Vanhoucke V, Ioffe S, Shlens J, Wojna Z (2016) Rethinking the inception architecture for computer vision. In: Proceedings of the IEEE conference on computer vision and pattern recognition, pp 2818–2826

The Threat Landscape of Connected Vehicles

Don Nalin Dharshana Jayaratne, Qian Lu, Abdur Rakib, Muhamad Azfar Ramli, Rakhi Manohar Mepparambath, Siraj Ahmed Shaikh, Hoang Nga Nguyen, and Suraj Harsha Kamtam

Abstract As connected vehicles (CVs) play an increasingly pivotal role in modern transportation, cybersecurity threats targeting these systems have become a critical area of concern. This study systematically identifies and classifies vulnerabilities from the National Vulnerability Database (NVD) and the Automotive Attack Database (AAD) using a semi-automated filtering process. Our analysis identifies a total of 508 vulnerabilities across these databases, which are categorised based on ISO/SAE 21434 impact categories: safety, financial, operational, and privacy. A key finding reveals that 14.6% of these vulnerabilities have systemic implications, meaning they have the potential to cause widespread disruption across multiple vehicles or the broader transportation network. Furthermore, 45% of the vulnerabilities are associated with remote attack vectors, significantly increasing the risk of large-scale exploitation. This research contributes an updated database of automotive vulnerabilities, providing a valuable resource for the cybersecurity community. The findings highlight the need to enhance current automotive cybersecurity standards, such as ISO/SAE 21434, to address the complex inter-dependencies and systemic risks within connected vehicle ecosystems.

Keywords Automotive cybersecurity · Connected vehicles · National vulnerability database · Vulnerability classification · Vulnerability impact assessment

---

D. N. D. Jayaratne (✉) · Q. Lu · A. Rakib · S. H. Kamtam
Centre for Future Transport and Cities (CFTC), Coventry University, Coventry, UK
e-mail: jayaratned@uni.coventry.ac.uk

M. A. Ramli · R. M. Mepparambath
Institute of High Performance Computing (IHPC), Agency for Science, Technology and Research (A*STAR), Singapore, Singapore

S. A. Shaikh · H. N. Nguyen
Systems Security Group (SSG), Department of Computer Science, Swansea University, Swansea, UK

H. Jahankhani and B. Issac (eds.), *Cybersecurity and Human Capabilities Through Symbiotic Artificial Intelligence*, Advanced Sciences and Technologies for Security Applications, https://doi.org/10.1007/978-3-031-82031-1_13

1 Introduction

The automotive industry is undergoing a transformative shift, transitioning from a domain predominantly governed by mechanical engineering to one increasingly reliant on sophisticated software and electronic systems. The advent of Connected Vehicles (CVs) has attracted significant global interest as a key component of Intelligent Transportation Systems (ITS) [1, 2]. These vehicles promise benefits such as improved transportation efficiency, enhanced safety, and reduced energy consumption. However, since the current electrical and electronic architectures of vehicles were not originally designed with cybersecurity in mind, their integration with external networks exposes vulnerabilities that were once confined to closed systems [3, 4]. Each new technology within CVs introduces a potential attack surface, contributing to a rising number of cybersecurity incidents [5]. According to Upstream's 2024 Global Automotive Cybersecurity Report, 95% of automotive cyberattacks are executed remotely [6].

These cybersecurity challenges permeate every level of a vehicle's architecture, ranging from manufacturer-provided mobile applications to the individual chips embedded within various subsystems. Numerous studies have demonstrated the risks associated with automotive vulnerabilities through high-profile vehicle hacking incidents. A prominent example is the remote compromise of a Jeep Cherokee by white-hat hackers, which led to the recall of 1.4 million vehicles [7]. Similarly, the Corvette brake hack exposed a vulnerability in a third-party dongle connected to the vehicle's On-Board Diagnostics (OBD-II) port, enabling attackers to remotely disable the car's brakes while it was in motion [8].

For Original Equipment Manufacturers (OEMs), identifying and managing these vulnerabilities is especially challenging due to the unique characteristics of the automotive industry, which include heterogeneous networks and components, increasing levels of automation and connectivity, and the extended life cycles of vehicles [9]. Furthermore, modern vehicles, as complex and connected systems, are exposed to both local and remote cyberattacks. For instance, the Controller Area Network (CAN) bus, widely used for internal communications in vehicles, includes basic error detection mechanisms, such as cyclic redundancy checks. However, it lacks crucial security features like authentication and encryption, leaving it vulnerable to various forms of exploitation [10].

1.1 The Connected Vehicle

CVs are equipped with Internet access and often integrate Wireless Local Area Networks (LANs) to provide features such as enhanced driver safety, crash notifications, speed alerts, and in-vehicle entertainment. These communication systems fall under the broader framework of Vehicle-to-Everything (V2X), which includes Vehicle-to-Vehicle (V2V) and Vehicle-to-Infrastructure (V2I) interactions [11].

A CV's ecosystem consists of in-vehicle networks and numerous Electronic Control Units (ECUs), which communicate through different protocols, including CAN [12], Local Interconnect Network (LIN) [13], FlexRay [14], and Media Oriented Systems Transport (MOST) [15]. Each protocol serves specific functions within the vehicle. For example, MOST is primarily used for multimedia applications and operates on a ring topology, while CAN supports serial communication between ECUs via a broadcast bus system. LIN is a lower-cost alternative to CAN, and FlexRay provides higher performance at a greater cost. The diversity of technologies within the vehicle environment introduces numerous security vulnerabilities. Firewalls are typically deployed to protect internal networks like the CAN bus from more vulnerable components, such as infotainment systems connected to the Internet.

1.2 Study Objectives and Contributions

The primary objectives of this study are outlined as follows:

- To systematically extract and filter automotive-related cybersecurity vulnerabilities from the National Vulnerability Database (NVD) and the Automotive Attack Database (AAD).
- To identify vulnerabilities with the potential for systemic impacts, with a particular focus on those capable of disrupting critical vehicle functions and communication systems, as well as their broader implications for transportation networks.

The key contributions of this study are:

- The creation of an updated automotive vulnerability list that consolidates data from the NVD and AAD databases.
- The classification of automotive vulnerabilities based on ISO/SAE 21434 impact categories, including Safety, Financial, Operational, and Privacy.
- The identification of connected vehicle vulnerabilities with the potential for systemic impacts on transportation systems.

1.3 Structure of the Paper

The structure of this paper is as follows: The first section introduces the concept of connected vehicles and outlines the study's objectives and contributions. Section 2 presents the background and related work, including an overview of existing vulnerability databases, relevant standards, and security frameworks for connected vehicles, while identifying gaps in current literature. Section 3 describes the methodology, detailing the data processing steps used to extract, filter, and classify vulnerabilities from the NVD and AAD databases. This section also covers keyword selection, the automated filtering mechanism, and the consolidation of multi-stage vulnerabilities into single threat profiles. Section 4 presents an analysis of the vulnerabilities, classifying them based on ISO/SAE 21434 impact categories and differentiating between systemic and non-systemic vulnerabilities. Finally, Sect. 5 discusses the implications of the findings for automotive cybersecurity, addresses the limitations of the study, and suggests directions for future research, concluding with a summary of the key findings and contributions.

2 Background and Related Work

2.1 Overview of CVSS Metrics and Terminology

The Common Vulnerability Scoring System (CVSS) is an industry-standard method for quantifying the severity of IT security vulnerabilities. CVSS scores range from 1 to 10, where 10 represents the most critical vulnerabilities [16]. The system evaluates vulnerabilities using three metric groups:

- The base metrics assess the inherent characteristics of a vulnerability by focusing on exploitability factors, Attack Vector (AV), Attack Complexity (AC), Privileges Required (PR), and User Interaction (UI), as well as the impact on confidentiality (C), integrity (I), and availability (A).
- Temporal metrics account for changes over time, such as the availability of exploits (Exploit Code Maturity), remediation (Remediation Level), and confidence in the vulnerability report (Report Confidence).
- Environmental metrics reflect the specific deployment context of the system, adjusting the impact of a vulnerability based on the environment's confidentiality, integrity, and availability requirements.

The CVSS vector string is a textual representation of the base metrics, providing a standardised format to describe the essential characteristics of a vulnerability. For example, a CVSS v3.x vector string might appear as follows:

$$CVSS : 3.0/AV : N/AC : L/PR : N/UI : N/S : C/C : H/I : H/A : H$$

Figure 1 illustrates the process used to classify automotive

$$CVSS : 3.0/AV : N/AC : L/PR : N/UI : N/S : C/C : H/I : H/A : H$$

s, highlighting considerations such as network-based exploitation and physical access requirements [17].

The Exploitability Score assesses how easily a vulnerability can be exploited, considering factors such as the attack vector (e.g., local vs. remote), the complexity of the attack, required privileges, and whether user interaction is needed. Higher exploitability scores indicate vulnerabilities that are easier for an attacker to exploit.

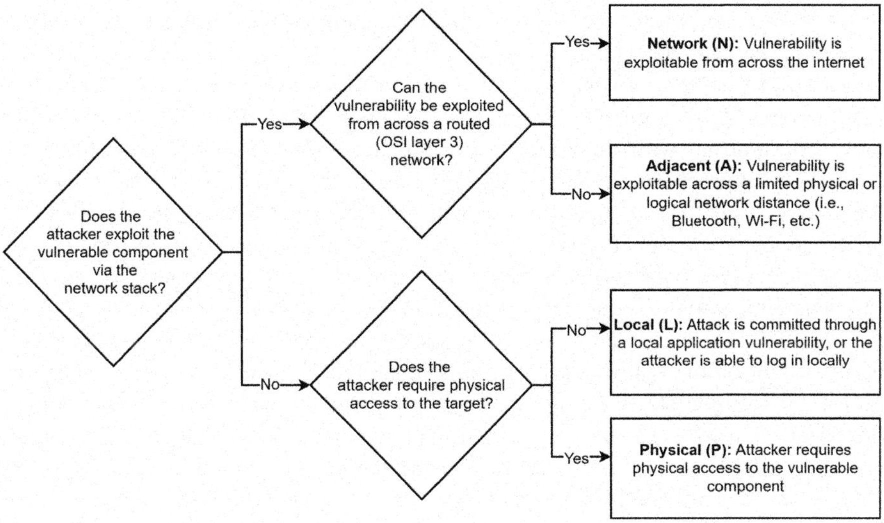

Fig. 1 Attack vector classification flowchart [17]

The Impact Score measures the potential consequences of a successful exploitation, specifically regarding the confidentiality, integrity, and availability of the affected system. A higher impact score signifies that a successful exploit could cause more significant damage, potentially affecting critical vehicle systems or endangering user safety.

Finally, the Base Score is the overall severity score calculated by combining the Exploitability and Impact scores. It serves as a key metric to prioritise vulnerabilities for remediation, with higher base scores indicating vulnerabilities that pose greater risk to systems or users.

2.2 *Existing Vulnerability Databases and Standards*

A key element in the global coordination of vulnerability information in the traditional IT sector is the Common Vulnerabilities and Exposures (CVE) system, which provides a standardised approach for naming and describing vulnerabilities. Recently, there has been an increasing inclusion of vehicle-related vulnerabilities in both the CVE and the NVD [18]. To facilitate the exchange of information on vulnerabilities in Connected Vehicles, vehicle manufacturers established the Automotive Information Sharing and Analysis Center (Auto-ISAC). However, access to Auto-ISAC data is restricted to its members, which include OEMs, suppliers, and the commercial vehicle industry. This limited access poses challenges for researchers and institutions outside the network. In an attempt to fill this gap, Sommer et al. compiled a comprehensive Automotive Attack Database, cataloging 361 attacks documented between 2002 and 2019 [19].

The ISO/SAE 21434 standard, widely adopted for cybersecurity engineering in road vehicles, provides a robust framework for managing risks within in-vehicle systems [20]. It aligns with United Nations Regulation No. 155, establishing consistent cybersecurity practices across the automotive industry [21]. The standard includes the Threat Analysis and Risk Assessment (TARA) methodology, which evaluates the feasibility and impact of potential cyberattacks on financial, safety, operational, and privacy aspects of vehicles. While ISO 21434 offers comprehensive guidance for managing risks associated with individual vehicle components or systems on the vehicle's perimeter, its primary limitation lies in its narrow scope. The standard focuses on the cybersecurity of individual vehicles rather than addressing the broader, systemic risks that connected transportation networks face. As such, threats that could disrupt communication infrastructure or have widespread implications across multiple vehicles or networks are not adequately captured. Future research should explore expanded frameworks that can account for these inter-dependencies, integrating systemic risk evaluation into cybersecurity assessments to safeguard not just individual vehicles but entire connected ecosystems.

2.3 Security and Risk Assessment Frameworks for Connected Vehicles

Several studies have proposed methods for improving security assessments in connected vehicle systems. Ando et al. [22] introduced a framework for defining security requirements in connected car systems by utilising the CVSS v3. Their method provides system designers, even those without extensive cybersecurity expertise, with a systematic approach to assessing risks by applying CVSS base metrics. However, the study focused more on general risk assessment rather than rating specific vehicular vulnerabilities, limiting its direct application in the automotive context.

In another approach, Sheehan et al. [23] proposed a Bayesian Network (BN) model to classify cyber risks for connected vehicles. This model utilised vulnerability data from the NVD and CVSS to assess risks, but it did not fully account for the unique attack surfaces present in the vehicular domain. Additionally, the model was applied to a single case study, which limited its generalizability to other vehicular environments.

Addressing the gaps in traditional vulnerability identification methods, Bajpai and Enbody [10] introduced the Common Vulnerability Vector (CVV) system to enhance the tracking and classification of vulnerabilities specific to automotive systems. The CVV system incorporates the confidentiality, integrity, and availability (CIA) requirements, emphasising that these aspects should be set to "high" in automotive cybersecurity assessments. This model highlights the systemic risks posed by connected vehicles, where a single vulnerability could potentially affect broader transportation networks.

Building on this work, Wang et al. [24] presented an innovative framework that integrates CVSS with Bayesian Networks (BN) to address the limitations of traditional IT-based vulnerability scoring models in the automotive domain. Their research noted that the conventional CVSS model does not account for the physical and real-world impacts of cybersecurity incidents on vehicles, such as human safety, operational integrity, financial losses, and privacy breaches. To overcome this limitation, the authors developed an enhanced CVSSCAV model, which incorporates these real-world factors into the CVSS framework. This model evaluates vulnerabilities based on their potential to cause harm, disrupt vehicle functionality, incur financial losses, or compromise privacy. Moreover, the BN model helps predict the severity of vulnerabilities, especially in cases where datasets are incomplete, a common challenge in the field of automotive cybersecurity.

2.4 Gaps and Emerging Trends in Automotive Cybersecurity

Despite advances in frameworks like CVSSCAV, several challenges remain in the automotive cybersecurity domain. One critical issue is the lack of transparency in the disclosure and management of vulnerabilities. Many OEMs and suppliers do not consistently share vulnerability data, despite the potential benefits of wider disclosure in mitigating risks across the industry [9]. Additionally, existing databases like the NVD do not fully capture the nuances of the automotive context, often requiring manual reviews to assess the relevance of vulnerabilities identified.

Another gap is the lack of studies aimed at identifying vulnerabilities with the potential for systemic impact across connected vehicle ecosystems. Current approaches tend to focus on individual vulnerabilities without assessing how these vulnerabilities might propagate across vehicles or transportation networks. This study addresses this gap by providing a framework to systematically identify vulnerabilities with systemic implications, emphasising the importance of considering the broader risks posed by interconnected systems.

Emerging trends in automotive cybersecurity focus on improving real-time detection and response capabilities. Artificial Intelligence (AI) and Machine Learning (ML) are increasingly being explored for intrusion detection, aiming to identify and respond to threats before they cause significant damage [25–27]. Furthermore, simulation-based vulnerability assessments are gaining traction as they allow researchers to test how vulnerabilities may affect a vehicle's systems in real-world scenarios. This approach provides valuable insights into proactive cybersecurity strategies, enabling the automotive industry to better anticipate and mitigate potential risks.

3 Methodology and Data Processing

This section details the methodology and data processing steps used to systematically identify, filter, and classify automotive-related cybersecurity vulnerabilities from the selected databases.

3.1 Methodological Framework

The methodology for identifying and analysing automotive cybersecurity vulnerabilities is structured around a semi-automated filtering process, which integrates data from two primary sources: the NIST NVD [18] and AAD [28]. The overall process flow of the automotive cybersecurity vulnerability identification and assessment methodology is illustrated in Fig. 2. Sections 3.2 to 3.4 describe the steps undertaken to generate the automotive vulnerability list, followed by exploratory analysis and classification of vulnerabilities in the succeeding sections.

3.2 NVD Data Processing

Keyword Selection To perform filtering, a set of keywords is selected from literature [10, 24, 29, 30], which includes terms related to vehicle components, technologies, and major Original Equipment Manufacturers (OEMs). These keywords ensure that the vulnerabilities filtered from the databases are specifically related to the automotive domain. Common vehicle-related terms such as "ECU", "CAN bus," and names of major OEMs like Toyota, Tesla, BMW etc., help refine the filtering process to capture relevant vulnerabilities within the datasets. The full list of keywords selected for the filtering process is provided in the Appendix.

Description-Based Filtering Once the keywords are selected, a semi-automated description-based filtering process is applied to the data from the NIST-NVD database using a Python script. The script scans the descriptions of vulnerabilities in the NVD and identifies those associated with automotive systems. After the initial filtering, a manual review is conducted to remove false positives and ensure the remaining vulnerabilities have direct implications for vehicle cybersecurity.

NVD Automotive Vulnerabilities The filtered data from the NVD is consolidated to form a list of automotive-specific vulnerabilities. This includes vulnerabilities in vehicle software, components, and communication systems that could impact vehicle safety, functionality, and integrity.

3.3 AAD Data Processing

Cross-Checking with NVD To ensure the integrity of the final vulnerability database, vulnerabilities from the AAD were cross-checked with those from the NVD, and any overlapping entries were removed.

Consolidation of Multi-stage Vulnerabilities The AAD often presents vulnerabilities and threats as multi-stage attack paths, where each stage represents a different step in the attack process. To simplify analysis and better represent the vulnerabilities' overall impact, these multi-stage threats were consolidated into single threats. This consolidation involved combining key information from each step of the attack

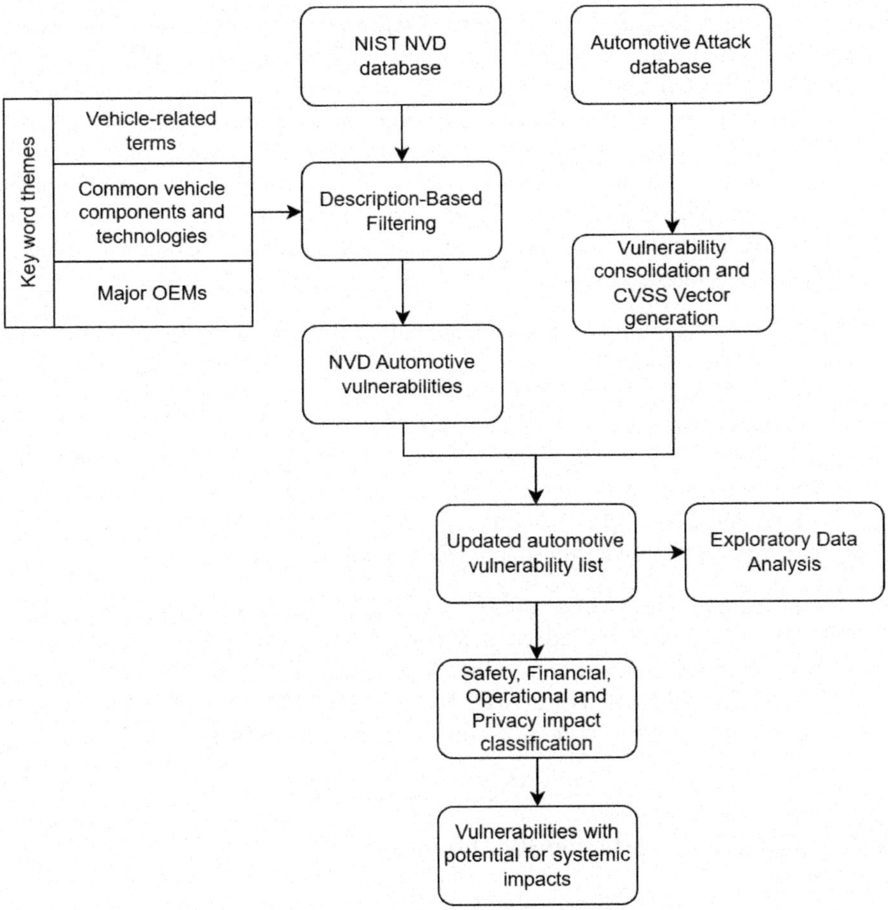

Fig. 2 Overview of the automotive cybersecurity vulnerability identification and assessment methodology

Table 1 Consolidation of CVSS scores for multi-stage attack on AUTOPILOT ECU and electric steering system

Vuln.	CVSS vector string	Base score	Exploitability score	Impact score
Step1	CVSS:3.0/AV:N/AC:H/PR:N/UI:N/S:U/C:H/I:H/A:N	7.4	2.2	5.2
Step2	CVSS:3.0/AV:N/AC:H/PR:N/UI:N/S:U/C:N/I:H/A:N	5.9	2.2	3.6
Comb.	CVSS:3.0/AV:N/AC:H/PR:N/UI:N/S:U/C:H/I:H/A:N	7.4	2.2	5.2

path to form a unified, singular threat profile. An example of this process is provided to demonstrate how multiple stages were merged into a single threat representation.

CVSS Vector String Generation For each consolidated vulnerability, a new CVSS vector string was generated. This step involved evaluating the most critical aspects of each individual vector string from the multi-stage attacks and combining them to form a single, comprehensive CVSS vector. Following this, new base, exploitability, and impact scores were calculated for each generated CVSS vector string, ensuring that the severity and potential impact of these threats are accurately represented.

To demonstrate this process, an example of consolidating a multi-stage vulnerability is shown in Table 1. In this example, two stages of the attack-Step 1: Manipulation of AUTOPILOT ECU Firmware and Step 2: Remote Control of Electric Steering-were combined to form a unified CVSS vector string.

- Step 1: This vulnerability represents the ability to manipulate the ECU firmware in the vehicle's autopilot system. The original CVSS vector for this step is CVSS:3.0/AV:N/AC:H/PR:N/UI:N/S:U/C:H/I:H/A:N, with a base score of 7.4, an exploitability score of 2.2, and an impact score of 5.2.
- Step 2: This step of the attack involves remote control of the vehicle's electric steering system. The CVSS vector string for this vulnerability is CVSS:3.0/AV:N/AC:H/PR:N/UI:N/S:U/C:N/I:H/A:N, which produces a base score of 5.9, an exploitability score of 2.2, and an impact score of 3.6.

By consolidating these vulnerabilities into a single vector string, a comprehensive CVSS representation of the attack is achieved. The combined vector retains the most severe characteristics of both vulnerabilities, resulting in the final CVSS vector string: CVSS:3.0/AV:N/AC:H/PR:N/UI:N/S:U/C:H/I:H/A:N. The final base score is 7.4, with an exploitability score of 2.2 and an impact score of 5.2.

3.4 Automotive Vulnerability Database

Following the previous steps, the updated automotive vulnerability list was generated.This refined list provides the foundation for further analysis and the prioritisation of cybersecurity threats, particularly those with systemic risks. The complete list of vulnerabilities used in this study, including all relevant metadata and CVSS scores, is publicly available online and can be accessed via its GitHub repository [31].

4 Analysis

4.1 Overview of Filtered Vulnerabilities

Figure 3 illustrates the number of automotive cybersecurity vulnerabilities identified from 2002 to 2023, using data from two major sources: the NVD and the AAD. The yellow bars represent vulnerabilities from the AAD, which contains data up until 2019, while the orange bars indicate vulnerabilities from the NVD, covering vulnerabilities through to 2024. The red bars highlight vulnerabilities that are common to both databases. The peak in 2019 is notable because the AAD data ends that year, resulting in a sharp drop in total vulnerabilities from 2020 onward, where only the NVD vulnerabilities are available. The NVD primarily focuses on vulnerabilities with an official CVE ID, meaning they have been reported through formal channels and meet specific standards for documentation and verification. In contrast, the AAD includes vulnerabilities from multiple sources, many of which do not have CVE IDs, thus offering a broader dataset. Post-2019, the NVD data shows a substantial rise in recorded vulnerabilities, with 2023 reaching a new high, indicating the increasing importance of cybersecurity in modern automotive systems.

Figure 4 displays the distribution of the attack vectors across automotive systems from 2002 to 2024, highlighting the prevalence of different types of attacks. Notably, 44.7% of vulnerabilities are associated with Remote Network attacks, emphasising

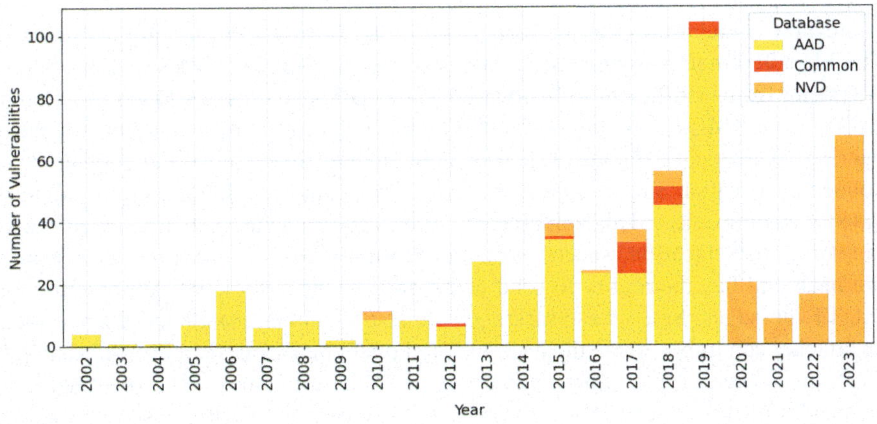

Fig. 3 Automotive vulnerability trends over time for the NVD (2002–2023) and AAD (2002–2019) databases

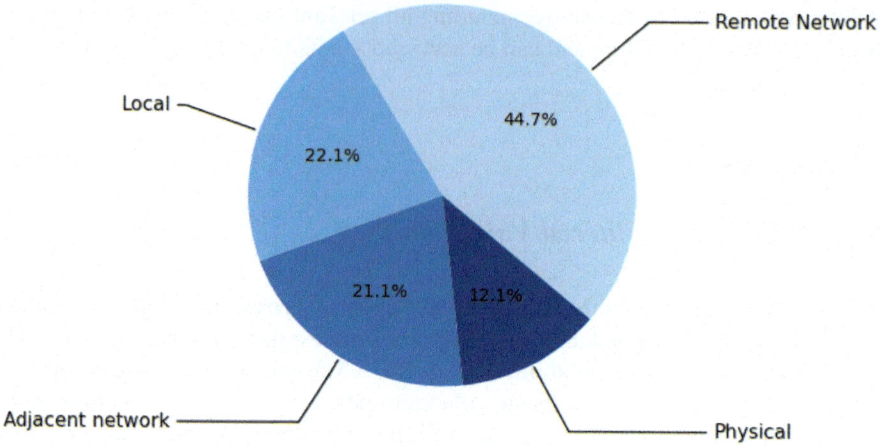

Fig. 4 Distribution of automotive attack vectors

the dominance of internet-based threats in connected vehicles. Adjacent Network and Local Network attacks each constitute 22.1 and 21.1%, indicating vulnerabilities that arise from proximity-based connections, such as Bluetooth or Wi-Fi, and local applications. Physical attacks account for 12.1%, underscoring the importance of securing components that require direct physical access to be compromised. This distribution illustrates the critical need for comprehensive cybersecurity measures that address both network-based and physical threats in modern automotive systems.

Figure 5 displays a scatterplot mapping automotive vulnerabilities based on their Exploitability Score and Impact Score, as defined by the CVSS v3 standard. The vulnerabilities are colour-coded according to their Base Score, with yellow representing lower scores and purple representing higher scores.

Several vulnerabilities at the higher end of the impact spectrum are highlighted. The Fleet Telemetry Manipulation Attack (CVE-2023-3028, CVSS3.1 Base Score: 9.8) stems from insufficient authentication in the back-end Message Queuing Telemetry Transport (MQTT) telematics unit, which is responsible for managing telemetry data across fleets of vehicles [32]. The back-end does not require authentication for connections, which allows unauthorised access to, and manipulation of, telemetry data for an entire fleet. This vulnerability affects components such as telemetry data, including GPS location, speed, odometer readings, and fuel levels. Furthermore, the back-end has the ability to send commands to the vehicle's CAN bus through MQTT messages. Since these messages are also unauthenticated and unencrypted, an attacker can impersonate the back-end and inject malicious data into the vehicle's CAN bus, potentially compromising critical vehicle functions. The potential consequences of this attack are significant and have the potential for systemic impacts due to the large-scale nature of the vulnerability. Further, the attacker could inject

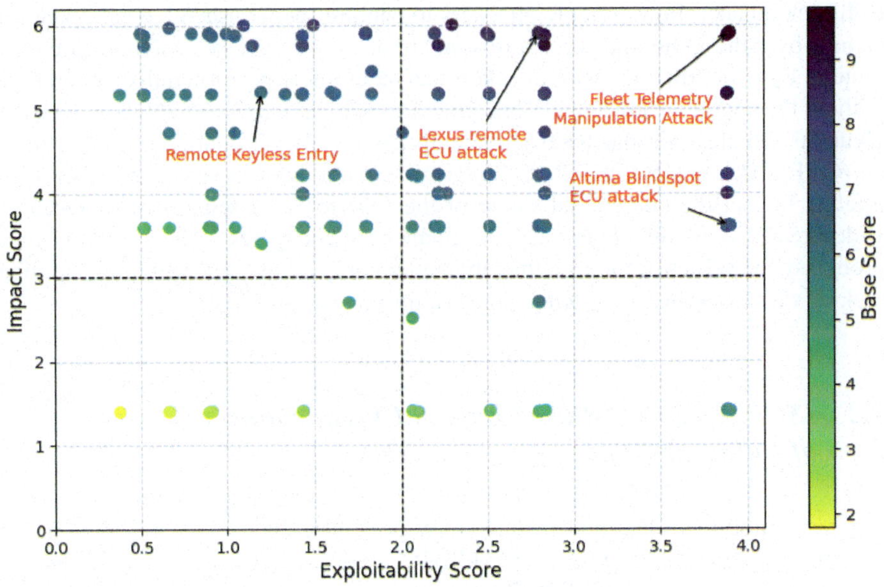

Fig. 5 Scatterplot of vulnerabilities based on CVSS 3 exploitability and impact scores

commands directly into the vehicle's CAN bus, affecting essential vehicle operations, which could lead to safety risks, such as controlling or disabling key vehicle functions remotely.

The Altima Blindspot ECU Attack (CVE-2024-6348, CVSS3.1 Base Score: 7.5) exploits a vulnerability in the security access mechanism of the Unified Diagnostic Services (UDS) in the Blind Spot Protection Sensor ECU of the 2022 Nissan Altima [33]. By performing repeated ECU resets and seed requests, attackers can generate the corresponding keys, thereby bypassing the security controls intended to protect critical vehicle functions. This enables unauthorised access to the ECU and allows the manipulation of the vehicle's blind spot protection system, leading to the compromise of safety-critical features in the vehicle.

The Lexus Remote ECU Attack (CVE-2020-5551, CVSS3.1 Base Score: 8.8) exploits a vulnerability in the AVN (Audio, Visual, and Navigation) unit in Lexus models [34]. Keen Security Lab identified several security flaws in the Bluetooth and vehicular diagnostic functions, which compromise the AVN unit, the internal CAN network, and associated ECUs. By chaining these vulnerabilities, an attacker can wirelessly take control of the AVN unit without any user interaction and inject malicious CAN messages from the AVN unit into the CAN network, leading to the compromise of critical vehicle functions.

The Remote Keyless Entry (RKE) Attack (CVE-2022-37418, CVSS3.1 Base Score: 6.4) exploits a vulnerability in the RKE receiving units found in certain Nissan, Kia, and Hyundai vehicles through 2017 [35]. The vulnerability, known as a

RollBack attack, allows remote attackers to capture two consecutive valid key fob signals over the radio and force a re-synchronisation of the key fob's rolling code system. By exploiting this flaw, attackers can indefinitely perform unlock operations without the owner's knowledge. This type of attack is not unique to Nissan, Kia, and Hyundai vehicles. Similar RKE vulnerabilities have been identified in other models (CVE-2022-36945, CVE-2022-37305) [31]. Although this vulnerability does not directly impact the operational safety of the vehicle or passengers, it represents a widespread risk to vehicle privacy and financial security. Once the system is compromised, the vehicle remains vulnerable indefinitely unless mitigations are applied, making this a long-term, systemic concern for affected models.

4.2 ISO/SAE 21434-Based Impact Classification of Vulnerabilities

The vulnerabilities identified in this study were classified based on their potential impacts using the ISO/SAE 21434 TARA impact categories: Safety, Financial, Operational, and Privacy (SFOP). These impact categories are defined as follows:

- **Safety**: A vulnerability is said to have a safety impact if it compromises safety-critical functions in the vehicle, potentially causing harm to humans as a result of the exploited vulnerability.
- **Financial**: A financial impact occurs when the exploitation of a vulnerability leads to a financial burden on the user, such as costs related to repairs, theft, or other monetary losses.
- **Operational**: An operational impact is present if the vulnerability causes a loss of functionality in any vehicle component or related infrastructure, resulting in disrupted or reduced vehicle operations.
- **Privacy**: A privacy impact arises when personally identifiable information (PII) is compromised due to the vulnerability, leading to potential breaches of personal data security.

Each vulnerability was assessed for its potential implications across these four impact areas by reviewing vulnerability descriptions, detailed vulnerability reports and associated reference materials. A binary classification was assigned to each SFOP category, indicating whether the vulnerability has an impact in the respective area. This classification provides a clear method for identifying vulnerabilities that may pose systemic risks across various aspects of automotive cybersecurity.

Figure 6 provides a visualisation of the vulnerabilities classified across the four impact categories. The bar graph represents the size of the intersections, indicating how many vulnerabilities affect one or more categories simultaneously. The dots and connecting lines below illustrate the specific combination of impact categories shared by those vulnerabilities. It is observed that operational, financial, and privacy impacts are the most prevalent, while vulnerabilities affecting passenger safety are

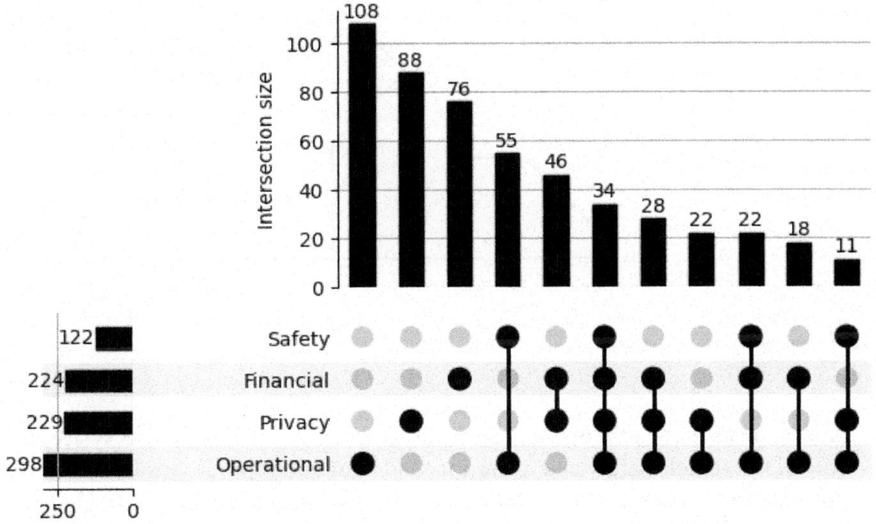

Fig. 6 Intersection plot of vulnerabilities classified according to ISO/SAE 21434 impact categories: Safety, financial, operational, and privacy. The plot shows the intersection of vulnerabilities across multiple impact categories, highlighting how vulnerabilities affect one or more areas of automotive cybersecurity

relatively fewer. However, those with safety impacts could lead to significant risks due to the potential harm to vehicle occupants.

4.3 Classification of Systemic Versus Non-systemic Vulnerabilities

The final step in the analysis focuses on identifying vulnerabilities with systemic impacts, particularly those that could affect critical vehicle functions or the broader transportation network. Vulnerabilities capable of causing significant operational disruptions, safety risks, or widespread damage to interconnected systems are prioritised for further investigation and mitigation. In the context of automotive cybersecurity, we define **systemic vulnerabilities** as those with the **potential to cause widespread failure or disruption within a vehicle or across multiple vehicles in a connected transportation ecosystem**. These vulnerabilities are especially concerning when they target critical vehicle systems and are exploitable remotely.

Criteria for Systemic Vulnerabilities To identify vulnerabilities with systemic potential, the following criteria were applied to the dataset:

1. **Attack on Critical Vehicle Functions**: This criterion includes vulnerabilities that target essential systems such as the drive-train (engine, transmission), steering,

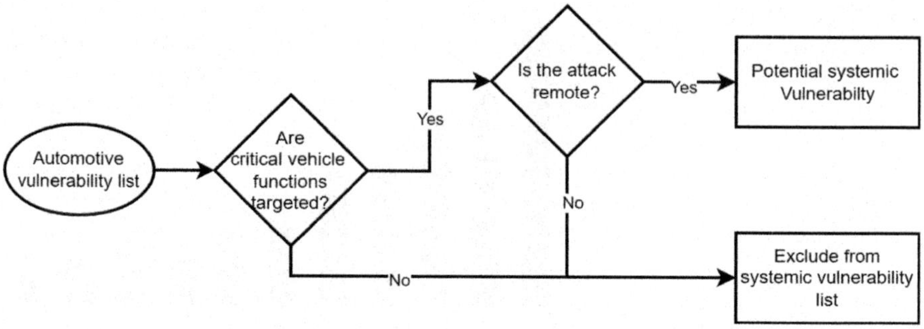

Fig. 7 Decision tree for identifying potential systemic vulnerabilities in automotive systems

brakes, airbags, headlights, or autonomous driving systems. These systems are crucial for vehicle safety and operation, and their compromise poses severe risks. For example, manipulation of the braking system could cause a sudden stop, potentially leading to accidents and significant disruptions to traffic flow.

2. **Remote Exploitability**: The vulnerability must be remotely exploitable, either through a network or adjacent network attack vector (e.g., Wi-Fi, cellular networks, Bluetooth, V2X communications). Remote exploitability significantly raises the risk of systemic impact, as it enables attackers to target multiple vehicles or systems without physical access.

As illustrated in Fig. 7, the identification of systemic vulnerabilities follows a two-step process. The first step assesses whether a vulnerability targets critical vehicle functions, such as the drive-train, steering, braking, or other essential safety-related systems. While vulnerabilities affecting these functions are inherently high-risk due to their direct impact on vehicle safety, their relevance to systemic risk lies in their potential to disrupt not just individual vehicles but multiple vehicles, and even the broader transportation network. The second step evaluates whether the vulnerability can be exploited remotely. Network-based attack vectors significantly increase the risk, as they enable attackers to compromise multiple vehicles or systems simultaneously. Remote attacks are therefore a key indicator of systemic vulnerabilities.

Vulnerabilities that meet both of these criteria are classified as potential systemic vulnerabilities. Such vulnerabilities warrant further investigation and prioritisation for mitigation due to their potential to cause widespread disruption across vehicle fleets or even the broader transportation network.

Figure 8 shows the breakdown between systemic and non-systemic vulnerabilities identified in the study. The majority of vulnerabilities (85.4%) are classified as non-systemic, meaning they do not pose a widespread risk across the vehicle fleet or transportation network. However, 14.6% of the vulnerabilities have the potential to cause systemic impacts due to their ability to target critical vehicle functions and be exploited remotely. These systemic vulnerabilities represent a significant risk and should be prioritised for mitigation to prevent large-scale disruptions.

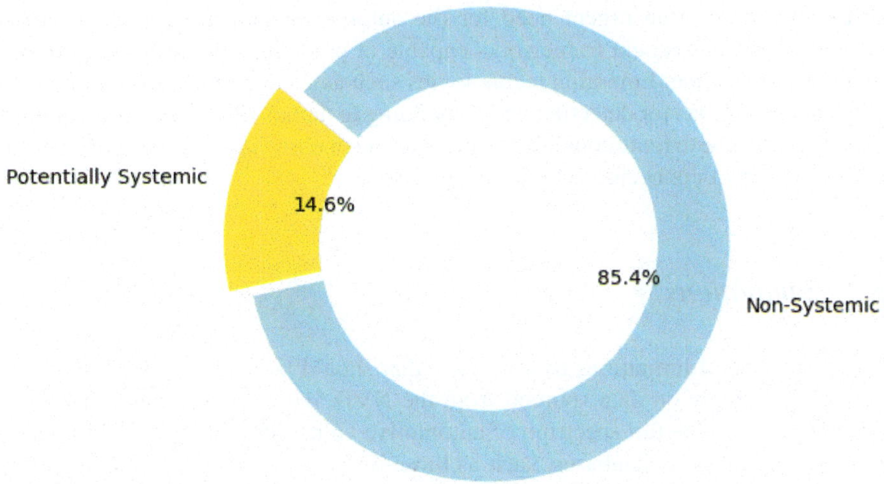

Fig. 8 Illustration of the proportion of vulnerabilities classified as systemic (14.6%) versus non-systemic (85.4%) based on the applied criteria

5 Discussion and Conclusions

5.1 Implications for Automotive Cybersecurity

The findings of this study have several important implications for the future of automotive cybersecurity. As vehicles become increasingly connected and reliant on external networks, the attack surface for potential vulnerabilities expands significantly. This research highlights the critical need for OEMs and cybersecurity professionals to prioritise the identification and mitigation of systemic vulnerabilities, particularly those that can affect essential vehicle functions and communication systems.

The analysis shows that a substantial portion of automotive vulnerabilities are remotely exploitable, underscoring the importance of securing over-the-air communication channels, against cyber threats. Ensuring robust encryption and authentication protocols across these interfaces is essential to prevent malicious actors from gaining unauthorised access to vehicle networks.

Moreover, the study demonstrates that current vulnerability management practices, including the ISO/SAE 21434 standard, need to be expanded to better account for the systemic risks posed by connected vehicles. Traditional approaches tend to focus on isolated vehicle components without fully considering how interconnected systems and external networks can propagate vulnerabilities across fleets of vehicles, leading to widespread disruptions.

Finally, there is an urgent need for the automotive industry to develop more advanced risk assessment frameworks capable of predicting and mitigating vulnerabilities in real-time. Emerging technologies such as AI and machine learning offer promising solutions for detecting and responding to threats before they cause significant damage. These tools could play a pivotal role in ensuring the long-term security and resilience of connected vehicle ecosystems.

5.2 Limitations

There are several limitations to this study that should be acknowledged. First, the study relies solely on data sourced from the NVD and AAD databases. Hence, it does not capture the full spectrum of automotive vulnerabilities. Vulnerabilities not documented in these databases, such as those disclosed privately to manufacturers or identified in proprietary assessments, may be absent. As a result, the analysis presented here may not reflect the entire threat landscape.

Additionally, the classification of vulnerabilities was based on publicly available descriptions and references. Some vulnerabilities lack detailed information, leading to potential uncertainties or subjectivity in assessing their impact. This limitation affects the accuracy of determining whether a vulnerability targets critical vehicle functions or can be exploited remotely, which could influence whether it is classified as a systemic or non-systemic risk. The reliance on available descriptions may result in some vulnerabilities being misclassified or under-evaluated.

Finally, the field of automotive cybersecurity is rapidly evolving, with new vulnerabilities and attack vectors emerging as vehicles become increasingly connected. The analysis in this study represents a snapshot of the vulnerabilities known at the time of data collection. Therefore, the findings of this research should be viewed within the context of the dynamic nature of cybersecurity threats in the automotive sector.

5.3 Summary of Findings and Contributions

This study presents an analysis of automotive cybersecurity vulnerabilities, focusing on identifying, classifying, and assessing their systemic impact potential. By leveraging data from both the National Vulnerability Database and the Automotive Attack Database, we have systematically filtered and consolidated a list of automotive vulnerabilities. The following key findings and contributions emerge from the analysis:

1. An up-to-date list of 508 automotive cybersecurity vulnerabilities was generated, synthesising data from two major sources-the NVD and the AAD. This list provides an overview of known vulnerabilities affecting vehicles and serves as a foundation for further risk assessments in the automotive domain.
2. We classified vulnerabilities using the ISO/SAE 21434 impact categories, covering safety, financial, operational, and privacy impacts. Our analysis reveals that while operational, financial, and privacy vulnerabilities are prevalent, safety-critical vulnerabilities, though fewer, pose significant risks due to their potential harm to vehicle occupants.
3. A significant contribution of this study is the identification of vulnerabilities with systemic potential-those that can affect critical vehicle functions or the broader transportation ecosystem. Approximately 14.6% of the analysed vulnerabilities were found to have systemic implications due to their ability to target critical vehicle functions and be exploited remotely.
4. The study presents key trends in exploitability and impact scores across the analysed vulnerabilities, identifying notable attack vectors that continue to pose substantial risks to modern vehicles. Remote network attacks, accounting for nearly 45% of the vulnerabilities, underscore the critical need for robust protection mechanisms in connected automotive environments.

In summary, this work contributes to the growing field of automotive cybersecurity by providing an updated classification and risk assessment framework, pinpointing systemic vulnerabilities, and offering insights into how these vulnerabilities could potentially impact vehicle safety, operations, and data privacy.

Acknowledgments This work was supported by Coventry University and the A*STAR Research Attachment Programme (ARAP).

Appendix: NVD Search Terms

Vehicle-Related Terms vehicle, car, automotive

Common Vehicle Components and Networks adaptive cruise control, adas, airbag, airbiquity, android auto, autoliv, bluetooth, braking system, carlink, carplay, collision prevention, control unit, cruise, cruise control, drivesync, engine control, infotainment, keyless entry, lane keep assist, park assist, lidar, controller area network, CAN bus, local interconnect network, LIN, media oriented systems transport, flexray, OBD-II, OBD2, OBD, passive anti-theft system, PATS, radio data system, steering control, telematics, tire pressure, TPMS, mbrace, mobileye, panasonic, passenger, pioneer, qnx, radar, renesas, safety connect, sensor, sunny, takata, telematic, texas instruments, tire, ultrasonic, vehicle, veoneer, xilinx, zubie.

Major OEMs volkswagen, bmw, acura, audi, toyota, jeep, kia, mercedes-benz, skoda, Škoda, renault, ford, peugeot, nissan, hyundai, opel, mazda, saab, seat, subaru, dacia, citroën, honda, great wall, cadillac, suzuki, land rover, rolls royce, porsche, jaguar, mitsubishi, chevrolet, mini, lexus, alfa romeo, lancia, snapdragon automobile, alpine, aston martin, bentley, bugatti, buick, changan, chrysler, daimler, dodge, dongfeng, ferrari, fiat, fisker, geely, general motors, infiniti, gmc, lamborghini, maserati, mclaren, tesla, pagani, ssangyong, tata motors, volvo, ds, smart.

References

1. Saeed TU, Alabi BNT, Labi S (2021) Preparing road infrastructure to accommodate connected and automated vehicles: system-level perspective. J Infrastruct Syst 27(1). ISSN: 1076-0342. https://doi.org/10.1061/(asce)is.1943-555x.0000593
2. Ersal T et al (2020) Connected and automated road vehicles: state of the art and future challenges. Veh Syst Dyn 58(5):672–704. ISSN: 17445159. https://doi.org/10.1080/00423114.2020.1741652. https://www.tandfonline.com/action/journalInformation?journalCode=nvsd20
3. Morris D, Madzudzo G, Garcia-Perez A (2020) Cybersecurity threats in the auto industry: tensions in the knowledge environment. Technol Forecast Soc Change 157:120102. ISSN: 00401625. https://doi.org/10.1016/j.techfore.2020.120102. https://linkinghub.elsevier.com/retrieve/pii/S0040162520309288
4. Pompigna A, Mauro R (2022) Smart roads: a state of the art of highways innovations in the smart age. Eng Sci Technol Int J 25:100986. ISSN: 22150986. https://doi.org/10.1016/j.jestch.2021.04.005. https://linkinghub.elsevier.com/retrieve/pii/S2215098621000872
5. Plappert C et al (2021) Attack surface assessment for cybersecurity engineering in the automotive domain. In: 2021 29th Euromicro international conference on parallel, distributed and network-based processing (PDP). IEEE, pp. 266–275. ISBN: 978-1-6654-1455-5. https://doi.org/10.1109/PDP52278.2021.00050. https://ieeexplore.ieee.org/document/9407241/
6. Upstream Security Ltd. (2024) Upstream's 2024 global automotive cybersecurity report. https://upstream.auto/reports/global-automotive-cybersecurity-report/
7. Miller C, Valasek C (2015) Remote exploitation of an unaltered passenger vehicle. Technical Report, pp 1–91. http://illmatics.com/RemoteCarHacking.pdf
8. Greenberg A (2015) Hackers cut a Corvette's brakes via a common car gadget. WIRED. https://www.wired.com/2015/08/hackers-cut-corvettes-brakes-via-common-car-gadget/
9. Bolz R, Kriesten R (2021) Automotive vulnerability disclosure: stakeholders, opportunities, challenges. J Cybersecur Privacy 1(2):274–288. ISSN: 2624-800X. https://doi.org/10.3390/jcp1020015. https://www.mdpi.com/2624-800X/1/2/15
10. Bajpai P, Enbody R (2020) Towards effective identification and rating of automotive vulnerabilities. In: Proceedings of the second ACM workshop on automotive and aerial vehicle security. ACM, New York, NY, USA, pp 37–44. ISBN: 9781450371131. https://doi.org/10.1145/3375706.3380556. https://dl.acm.org/doi/10.1145/3375706.3380556
11. Capgemini Engineering (2021) 5G on the Highway to V2X. https://www.altran.com/as-content/uploads/sites/9/2018/10/brochurea5_v2x5g_web.pdf
12. Szydlowski CP (1992) CAN specification 2.0: protocol and implementations. In: SAE technical papers. SAE International. https://doi.org/10.4271/921603. https://www.sae.org/content/921603/
13. LIN Consortium (2006) LIN specification package. Technical Report. www.lin-subbus.org
14. Shaw R, Jackman B (2008) An introduction to FlexRay as an industrial network. In: IEEE International symposium on industrial electronics, pp 1849–1854. https://doi.org/10.1109/ISIE.2008.4676987
15. Microchip Technology (2020) MOST cooperation: specifications. https://www.mostcooperation.com/specifications/

16. Mell P, Scarfone K, Romanosky S (2006) Common vulnerability scoring system. IEEE Secur Privacy 4(6):85–89. ISSN: 1540-7993. https://doi.org/10.1109/MSP.2006.145
17. CVSS v3.0 user guide. https://www.first.org/cvss/v3.0/userguide
18. NVD. Home. https://nvd.nist.gov/
19. Sommer C et al (2019) Veins: the open source vehicular network simulation framework, pp 215–252. ISBN: 9783030128425. https://doi.org/10.1007/978-3-030-12842-5_6
20. SAE (2010) J1711 Surface vehicle recommended practice. Technical Report, pp 1–63. papers3://publication/uuid/ECB5448C-F3D8-414A-A46E857726E32846
21. Sandler Manuel (2022) UN Regulation No 155 & how to comply? What you need to know. https://www.cyres-consulting.com/unregulation-no-155-requirements-what-you-need-to-know/
22. Ando E, Kayashima M, Komoda N (2016) A proposal of security requirements definition methodology in connected car systems by CVSS V3. In: Proceedings—2016 5th IIAI International congress on advanced applied informatics, IIAI-AAI 2016, pp 894–899. https://doi.org/10.1109/IIAI-AAI.2016.95
23. Sheehan B et al (2019) Connected and autonomous vehicles: a cyber-risk classification framework. Transp Res Part A Policy Pract 124:523–536. ISSN: 0965-8564. https://doi.org/10.1016/J.TRA.2018.06.033
24. Wang Y et al (2023) Automotive cybersecurity vulnerability assessment using the common vulnerability scoring system and Bayesian network model. IEEE Syst J 17(2):2880–2891. ISSN: 1932-8184. https://doi.org/10.1109/JSYST.2022.3230097. https://ieeexplore.ieee.org/document/10002327/
25. Sulikowski P et al (2024) Intrusion detection system for autonomous vehicles using non-tree based machine learning algorithms. Electronics 13(5):809. ISSN: 2079-9292. https://doi.org/10.3390/ELECTRONICS13050809. https://www.mdpi.com/2079-9292/13/5/809/htm
26. Kumar A, Das TK (2023) CAVIDS: real time intrusion detection system for connected autonomous vehicles using logical analysis of data. Veh Commun 43:100652. ISSN: 2214-2096. https://doi.org/10.1016/j.vehcom.2023.100652. https://linkinghub.elsevier.com/retrieve/pii/S2214209623000827
27. Zhang J et al (2019) Intrusion detection system using deep learning for in-vehicle security. Ad Hoc Netw 95:101974. ISSN: 1570-8705. https://doi.org/10.1016/j.adhoc.2019.101974. https://linkinghub.elsevier.com/retrieve/pii/S1570870519304354
28. Sommer F, Dürrwang J (2023) IEEM-HsKA/AAD: Automotive Attack Database (AAD). https://github.com/IEEM-HsKA/AAD
29. Gülsever M (2019) A study on vulnerabilities in connected cars. Technical Report
30. Xiong W et al (2019) A study of security vulnerabilities and software weaknesses in vehicles. In: Lecture notes in computer science (including subseries Lecture notes in artificial intelligence and Lecture notes in bioinformatics), vol 11875. LNCS, Springer, pp 204–218. ISBN: 9783030350543. https://doi.org/10.1007/978-3-030-35055-0_13
31. Don Nalin Dharshana Jayaratne (2024) jayaratned/AutomotiveTD. https://github.com/jayaratned/AutomotiveTD
32. Automotive Security Research Group (ASRG) (2023) Improper backend communications allow access and manipulation of the telemetry data. Automotive Security Research Group. https://asrg.io/security-advisories/cve-2023-3028/
33. Predictable seed generation in the security access... Â· CVE-2024-6348 Â· GitHub Advisory Database. 2024. https://github.com/advisories/GHSA-qp2p-3fr2-8j54
34. Tencent Keen Security Lab (2020) Tencent Keen Security Lab: experimental security assessment on Lexus cars. Keen Security Lab Blog. https://keenlab.tencent.com/en/2020/03/30/Tencent-Keen-Security-Lab-Experimental-Security-Assessment-on-Lexus-Cars/
35. Williams E (2022) RollBack breaks into your car. Hackaday. https://hackaday.com/2022/08/17/rollback-breaks-intoyour-car/

Programming Privacy—A Short Observational Study of Homomorphic Encryption Compilers for Bioinformatics Applications

Christopher P. Lewington

Abstract Fully Homomorphic Encryption (FHE) is an active research area with respect to the preservation of privacy, particularly given the recent attention to Post-quantum Cryptography (PQC). It is well known that existing public key crypto systems will be under threat in a quantum computation world, and FHE solutions provide an opportunity to replace existing 'mathematically hard' problems with ones that are thought to be quantum safe. FHE libraries have existed for some time, however they are difficult to use and require specialist cryptographic knowledge. To enter the mainstream of software development, suitable compiler technology is needed. Compiler development for FHE is in its infancy and this paper will evaluate two of the most promising solutions that have a potential application to bioinformatics, using suitable metrics to understand how mature or otherwise the leading contenders are.

Keywords Homomorphic encryption · Compilers · Post-quantum computing · Programming

1 Introduction

In the past decade there has been an accelerated rate of research into Fully Homomorphic Encryption (FHE), thanks in no small part to the ground-breaking work of Craig Gentry [1]. Much of that research has focused on the development of different encryption schemes and associated plaintext spaces that enable certain operations to be carried out homomorphically on encrypted data. Scheme and library development have since brought FHE closer to mainstream adoption by development

https://www.uos.ac.uk/people/chris-lewington/.

C. P. Lewington (✉)
University of Suffolk, Ipswich, UK
e-mail: c.lewington2@uos.ac.uk

H. Jahankhani and B. Issac (eds.), *Cybersecurity and Human Capabilities Through Symbiotic Artificial Intelligence*, Advanced Sciences and Technologies for Security Applications, https://doi.org/10.1007/978-3-031-82031-1_14

communities, however considerable expertise in encryption techniques and parameter setting is still needed to use the FHE libraries. Longer term, FHE compilers are required which will make it relatively easy to produce an intermediate representation that can be recognised by an FHE library. Several prototype FHE compilers exist currently [2], with the Google Transpiler [3] being one of the most recent open-source additions. Open-source solutions are arguably the ones most likely to be readily adopted by the wider development community, particularly in the field of bioinformatics, since tool providers such as the European Bioinformatics Institute (EBI), are not-for-profit and therefore are the least likely to purchase compiler solutions. Furthermore, bioinformatics tools typically have computation requirements related to statistical calculations over possibly sensitive data and are thus a natural fit for FHE compiler technology. Taken together, the drive for open-source solutions and the need for 'hidden' computations motivate the need to evaluate potential FHE compiler solutions to ascertain present maturity. Metrics such as usability, expressiveness and performance are of critical importance to software developers generally, and these criteria will be used to evaluate two of the leading open-source FHE compiler contenders.

The paper is structured as follows: the next section provides some very brief preliminary concepts related to FHE, which motivates the literature review examining the core ideas of FHE schemes and compilers, along with the current state of the implementations available. The motivation for selecting the compilers for evaluation in this study is also discussed here. The section on implementation examines the specifics of setting up and running each of the evaluation candidates, along with the example problems chosen to exercise each of the compilers. The evaluation sections discuss the results, in particular focusing on the context of suitability for bioinformatics applications, and finally the conclusion summarises the paper findings along with suggestions for future work.

2 Preliminaries

Public key cryptographic systems rely for their security on underlying hard mathematical problems—for example RSA is dependent upon large prime factorisation, and ElGamal is based on the discrete logarithm problem. All such public key schemes assume that their respective mathematical foundations are computationally intractable, and thus an attacker would not be able to crack the ciphertext in a feasible amount of time through brute force. Advances in quantum computing have caused many to question the future use of such public key schemes, especially since Peter Schor published an algorithm which would defeat those underlying mathematically hard problems with a suitably powerful error-free quantum computer. These concerns have fueled interest in cryptographic schemes which would be quantum-robust, in the sense that the underlying mathematical problem on which a scheme is based would not be open to attack by quantum techniques. Schemes based upon hard problems for mathematical structures called lattices appear to be computationally hard and quantum-robust, at least in the worst case, and underpin FHE. FHE also addresses concerns around data privacy—an area of great concern today, particularly with recent advances and applications based upon cloud computing and machine learning technologies. Homomorphic encryption (HE) as a concept was proposed back in 1978 when Rivest, Adleman and Dertouzos discussed the idea of 'privacy homomorphisms'—functions which could act upon encrypted data without the need to decrypt that data. In the literature there are many definitions of what 'homomorphic' means in a cryptographic sense, but the essence is captured by the following broad definition and is presented for completeness:

Theorem 1 *A cryptographic scheme E is called homomorphic if the following holds for all messages m_1 and m_2 in the plaintext space M, ciphertexts $E(m_1)$ and $E(m_2)$ in the ciphertext space C, with generic operators $\oplus$ over the plaintext space and $\otimes$ over the ciphertext space:*

$$E(m_1 \oplus m_2) = E(m_1) \otimes E(m_2)$$

An HE scheme typically is composed of four high-level operations: key generation, encryption, decryption and evaluation. Of particular note is the evaluation operation, which takes a public key, a function f over the plaintext space and applies them to the ciphertext ($E(m_1)$ and $E(m_2)$ in the above definition), with the result being equal to that which would have been obtained simply by applying f to plaintext messages. The various different categories of homomorphic encryption, discussed below, derive from a consideration of the operators possible within the generic operators $\oplus$ and $\otimes$, which typically are addition and multiplication but can be more sophisticated in some schemes.

3 Literature Review

3.1 Categories of Homomorphic Encryption

This sub-section provides a brief overview of some of the significant points of background interest in the field of FHE, but for a more detailed treatment the reader is referred to [4–6] for the most recent techniques.

Well-known public key cryptographic schemes such as RSA, ElGamal and Paillier are examples of partially homomorphic encryption mechanisms (PHE), because they are homomorphic in one operation only—for unlimited multiplication (RSA and ElGamal) or addition (Paillier), but not both. None of those schemes could be used to realise the goal of so-called Fully Homomorphic Encryption (FHE)—a single scheme that could be used to secure privacy of data not only at rest or in transit, but also when in use. Schemes also exist which are referred to as Somewhat Homomorphic (SHE), in that they provide homomorphic operations for only a fixed number of executions—schemes tend to be named after their authors, with notable examples from Boneh,

Goh and Nissim (BGN) [7], Sander, Young and Yung (SYY) [8] and Ishai and Paskin (IP) [9]. FHE remained elusive until Craig Gentry in 2009 demonstrated in his PhD thesis that FHE was theoretically possible. Gentry considered the Shortest Vector Path (SVP) for an ideal lattice and showed that it was possible to create an FHE scheme that would work for arbitrary arithmetic circuits (a common computational model useful for developing FHE plaintext spaces). Gentry's scheme relied upon the introduction of random 'noise' to successive encrypted computations—technically the depth of multiplication steps within a circuit—which increased over time to the extent that eventually the decrypted result would be incorrect. Gentry's major contribution was to introduce a bootstrapping process [1] which meant that the noise could be 'reset' by re-encrypting the ciphertext—evaluating the decryption circuit using bits from the encrypted secret key—thereby ensuring that the noise remained within a level that ensured correct decryption. Before bootstrapping could be used, however, the circuit first needed to be squashed down to a depth where bootstrapping was possible. This era of FHE is often referred to as the First Generation. There is some debate in the literature with regard to classifications of schemes into generations, but this paper will use the classifications suggested by Gentry.

Bootstrapping itself turned out to be an expensive operation to perform, and subsequent schemes sought to improve performance during what is often known as the Second and Third Generation FHE schemes—examples include those by Brakerski, Gentry and Vaikuntanathan (BGV) [10] and Brakerski, Fan and Verkauteren (BFV) (sometimes referred to only as FV [11]). These schemes enhanced the underlying hard problem space by refining the SVP into Learning With Errors (LWE) [12], and later a more flexible and efficient algebraic version called Ring Learning With Errors (RLWE) [13]. These schemes also introduced more efficient Single Instruction Multiple Data (SIMD) enhancements, introduced the concept of relinearisation to regulate the size of very large ciphertexts, and dimension-modulus reduction to remove the need for squashing [4]. Furthermore there has been a significant body of research [4] on what are referred to as Levelled FHE schemes, since they avoid bootstrapping entirely through a modulus switching mechanism that batches up ciphertexts and reduces computation depth in circuits. Most current FHE schemes are levelled FHE schemes [14]—for example the scheme developed by Cheon, Kim, Kim and Song (CKKS) [5] is significant since it extends the computation model to work over floating-point numbers rather than just integers. Finally the scheme known as Fast Fully Homomorphic Encryption over the Torus (TFHE) uses a torus algebraic model across three different plaintext spaces and LWE, thus expanding the range of operations available for FHE. The TFHE approach is the basis of the CHIMERA FHE scheme, which is a recent development that combines several RLWE schemes together and enables scheme switching by mapping different plaintext spaces into a torus plaintext space [15].

3.2 FHE Libraries and Compilers

This section discusses the currently known FHE libraries and also the current state of the art in FHE compilers. For more detail the reader is referred to [2] and its associated gitHub resource MarbleHE [16], upon which this section is based.

FHE Libraries. In this paper we give only a brief overview of the most relevant FHE libraries for brevity. SEAL (Simple Encrypted Arithmetic Library) is implemented in C++ and produced by Microsoft [17], with a focus on the BFV and CKKS schemes. SEAL requires significant expertise in cryptography and FHE concepts to use directly as an application developer. Like the popular HELib [18], SEAL is open source and thus freely available. The HEAAN (Homomorphic Encryption for Arithmetic of Approximate Numbers) library provides an implementation for CKKS, is also written in C++ and is also freely available but suffers from the same challenges as SEAL and HELib in terms of usability amongst mainstream software developers. TFHE [6] is an enhancement scheme that extends the Gentry, Sahai and Waters (GSW) model using a torus structure to reduce the bootstrapping key size whilst also optimising bootstrapping by performing it after every ciphertext operation, thus removing the need for bootstrapping, and TFHE also improves expressiveness for input circuits, beyond simply arithmetic ones. A summary of well-known FHE libraries, all of which are in GitHub, their documentation and licensing are shown in Table 1.

The overriding common feature of all current FHE libraries is that a significant level of cryptography expertise is required in order to specify the relevant parameters needed for computations. In general these parameters are essential to the efficient and effective execution of the schemes, and this fact alone presents a significant barrier to adoption by the mainstream software development community.

FHE Compilers. The job of a compiler, at its most basic level, is to enable programmers to develop code at a higher level of abstraction than that required to drive an

Table 1 FHE libraries—GitHub repositories

Library	Documentation	License
concrete	https://concrete.zama.ai/	AGPL-3.0
FHEW	–	GPL-2.0+
FV-NFLib	–	GPL-3.0
HEEAN	–	CC-BY-NC 3.0
HELib	https://homenc.github.io/HElib	Apache-2.0
Latigo	–	Apache-2.0
PALISADE	https://palisade-crypto.org/documentation	BSD 2-clause
SEAL	https://github.com/microsoft/SEAL	MIT
TFHE	https://tfhe.github.io/tfhe	Apache-2.0
cuFHE	–	MIT
nuFHE	https://nufhe.readthedocs.io/en/latest	GPL-3.0

Table 2 FHE compilers—GitHub repositories [19]

Tool/Technique	Code	License
ALCHEMY	GitHub	GPL-3.0
Cingulata (Armadillo)	GitHub	CeCLL-C
Cingulata + Multi-start	Closed source	–
Cingulata + Code Rewriting	Closed source	–
Cingulata + Lobster	ACM DL, GitHub	Free
EVA	GitHub	MIT
nGraph-HE	GitHub	Apache-2.0
Ramparts	Closed source	–
SEALion	Closed source	–

underlying computation machine. Current FHE libraries do have APIs but they still require a degree of cryptography sophistication to use them, and so FHE compilers for use by regular developers are an attractive research area. The work of [2] summarises recent findings on FHE compilers at the time of writing, and there is a clear conclusion that although many of the solutions listed in Table 2 are not much beyond the prototype stage, the development of FHE compilers is a research problem with critical application areas.

In 2023 Gouert and Mouris developed a bespoke compiler called T2 [20] for comparing FHE libraries in a benchmark-consistent way to enable more effective choice, as all have different parameter requirements. Viand's PhD Thesis from 2023 proposes a new compiler called HECO [19, 21] which was developed from a consideration of the compilers examined in [2] and is now included in the MarbleHE GitHub repository [16]. Both the T2 benchmark work and HECO could prove to be fruitful areas for future evaluation in specific use case contexts, but both are very recent additions to a fast-moving field and therefore attention here will be restricted to the compilers documented earlier.

In summary there are two clear observations to be made from the literature review. Firstly, it is clear that FHE schemes are a powerful concept for providing privacy through computations over encrypted data, but they still require significant cryptography expertise to use the libraries. The second observation is that FHE compiler technology shows great promise for making the FHE libraries available to a wider audience, however most compilers are still at the prototype level and are likely still to be application-specific in their usefulness. The next section will discuss the choice of compilers to be evaluated for this study, taking as a focus the potential uses for bioinformatics applications.

4 Compiler Choice for Evaluation

This section provides a motivation for the choice of compilers to evaluate, provides a short overview of each of them, and clarifies the metrics to be used for evaluation.

4.1 Motivation for Compiler Selection

It should be noted that several of the FHE compilers shown in Table 2 are closed source and therefore were not available for this evaluation study. The literature suggests that at least some—for example Ramparts [22]—may well be useful, however for bioinformatics applications in particular a closed source solution might present barriers to wider community adoption. Of the other compilers listed in Table 2, EVA [23] appears to be the most promising for bio-informatics applications as it is written in Python, which is commonly used in that field, and also it includes support for essential mathematical computations. Indeed, de Leon et al. [24] in 2023 used EVA to build a privacy-preserving genotype imputation tool based on Hidden Markov Models, noting that EVA was chosen for that study because it allows for automatic selection of relevant parameters, making it very user-friendly for bio-informatics solutions.

A recently published addition to the FHE compiler field, and so not listed in Table 2, is the Transpiler prototype produced in 2021 by Google engineers [3]. The Google Transpiler looks promising for an evaluation exercise since it is claimed to be easy to use, comes with several worked examples, has a C++ implementation, and is open source. All of these characteristics make it also a promising candidate for bio-informatics applications, since as well as Python, C++ is a popular choice amongst bio-informatics developers. In 2024 the Transpiler development work was superseded by the HEIR initiative [25], which is a collaborative attempt spearheaded by Google to standardise a set of intermediate representations that will allow developers to compile their code more naturally into FHE library schemes. The results of this work may well provide further insights into solving the FHE compiler problem, however HEIR is yet to produce a prototype for evaluation.

The next two subsections provide short overviews of the compilers chosen for evaluation.

4.2 EVA Overview

The EVA (Encrypted Vector Arithmetic) compiler [23, 26] is a good fit for this study because it has a python wrapper, supports CKKS which enables floating point arithmetic and thus makes it readily available for use with, for example, bioinformatics applications such as those developed under the iDASH programme [27]. The main purpose of EVA is to support high-level language development and subsequent encrypted computation, with a python wrapper provided, and primarily aimed at machine learning applications as a primary target domain. However, any type of integer computation can be undertaken and thus, since the aim of this study is to compare usability of FHE compilers for bioinformatics, it is a natural option for evaluation. Furthermore, recent work in the literature has demonstrated the potential viability of EVA in a bioinformatics setting—apart from the already mentioned [24], in 2023 Salaga et al. [28] used EVA in a cloud-based setting to perform various types of privacy-preserving computation on cloud-hosted medical data. EVA is, therefore a promising choice for evaluation in a wider usability sense. The EVA tool converts minimally modified python application code (which includes, for example, required precision values) into a Directed Acyclic Graph (DAG) representation, which is then processed internally by the 'EVA language'. This representation is then optimised for various computation operations such as re-scaling, which is crucial for FHE, and furthermore performs 'automatic' selection of cryptographic parameters using various heuristics. This conversion of computations into graph representations is currently necessary for all FHE scheme libraries, since the algorithms work on logic gate representations of code. In the current version of EVA the CKKS scheme is accessed through the Microsoft SEAL FHE library. Nodes are inserted into the DAG which instruct the EVA compiler to perform certain optimisation tasks, such as RESCALE, MODSWITCH and RELINEARIZE. These modified DAGs are again then augmented with further nodes to apply heuristics that are intended to enhance encryption parameter choices. An example conversion with potential optimisations is given in [23] and is shown in Fig. 1.

In Fig. 1 the first diagram shows the original graph representation of the computation for x^2y^3. The nodes S, M and L represent the use of RESCALE, MODSWITCH and RELINEARIZE encryption parameter strategies respectively, and the subsequent DAGs show the resulting graphs after application of the heuristic optimisations (graph rewrites) ALWAYS-RESCALE (b), ALWAYS-RESCALE & MODSWITCH (c) WATERLINE-RESCALE (d) and WATERLINE-RESCALE & RELINEARIZE

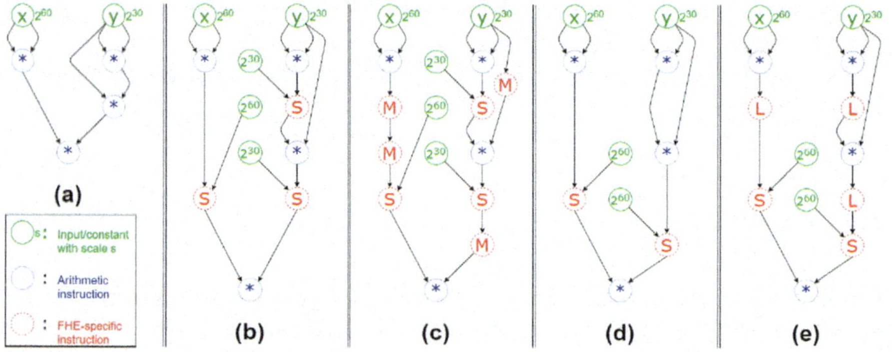

Fig. 1 Example DAG optimisations in EVA

(e). These optimisations thus modify the DAG to create a form that will be more efficiently processed by the CKKS scheme from the SEAL library.

4.3 Google Transpiler

The google Transpiler [3] is an open-source FHE compiler that is aimed at making FHE schemes more accessible to developers. There are no published comparisons of the Google Transpiler as yet, but the published literature indicates that it has potential for flexible development beyond the proof-of-concept stage. The Transpiler code [29] is written in C++ and connects to the back end FHE libraries TFHE and OpenFHE, also written in C++. The intermediate representation that the Transpiler uses is that provided by the google XLS framework [30], which is a library enabling users to build hardware models in software. The reasoning behind the Transpiler seems to be that, by compiling suitably annotated 'regular' C++ code into an XLS-friendly hardware model, it will ultimately make it a more flexible and efficient mechanism for transparently mapping applications to representations required by the various FHE schemes available. Because the TFHE library only handles logic gates as inputs, the Transpiler process has to convert C++ code into a boolean gate representation that TFHE can use—this is essentially what happens with the XLS phase of the compilation process. In overall summary, the Transpiler front-end compiles C++ code into an Intermediate Representation (IR) in the XLS framework, which in turn performs operation optimisations and converts the result into a set of Boolean gates. Those gates are finally processed by a further piece of C++ Transpiler code, which invokes the TFHE library. This compilation process is shown in Fig. 2, taken from [3].

For illustration purposes, [3] also provide an example Boolean gate computation on a simple binary input, as shown in Fig. 3.

The whole framework of the Transpiler is based around a C++ development model, with Bazel [31] as the software build and execution tool. The general structure of an application is firstly to develop the application code; then to add in special 'pragma' commands that will make it amenable to Transpiler compilation; then to develop a 'testbench' which provides the encryption context; and finally to run everything

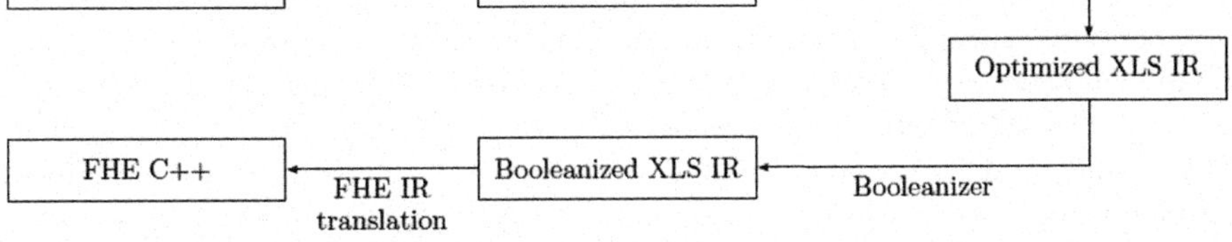

Fig. 2 Transpiler compilation process

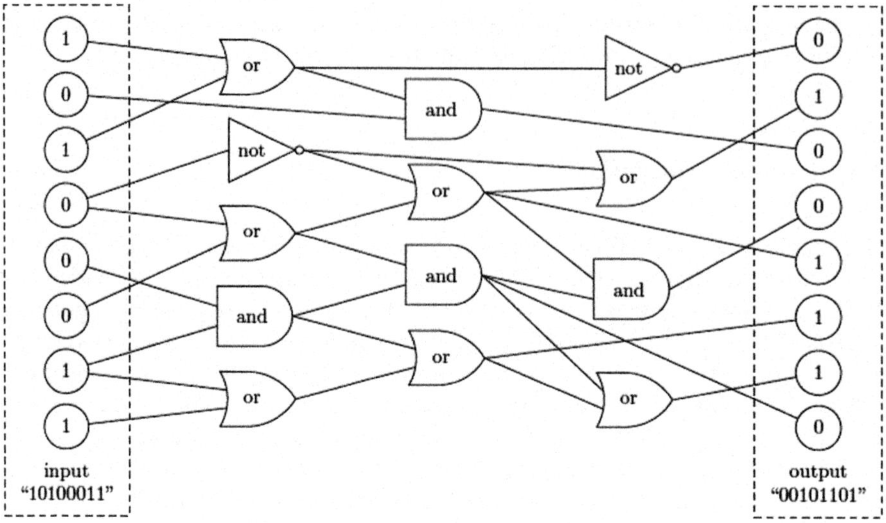

Fig. 3 Transpiler Boolean gate computation example

together, once built, using Bazel. The Transpiler also includes a runtime user option for executing an example in 'interpreted' mode, which allocates multi-core resources in the XLS hardware model in order to optimise a calculation. To use this option a developer must also create an appropriate testbench file.

4.4 Evaluation Metrics

The choice of compilers to evaluate has been motivated up to this point by their potential suitability for bioinformatics application development. In this subsection we discuss the specific metrics chosen to perform that evaluation and explain why they are relevant. In any evaluation study there are many different metrics that can be selected, and the choice ultimately depends upon what the purpose of the evaluation is. In this case, a 'programming language' metric has in fact already been employed in the compiler selection, since both EVA and the Transpiler were chosen for the programming languages used (python and C++ respectively) because those languages are of particular relevance to bioinformatics application development. Beyond the programming language, an obvious metric is performance. The final two metrics used for this study are expressiveness and usability, both of which have crucial importance across the wider development community, but in particular for bioinformatics these metrics inform the likelihood of adoption amongst developers for whom software development is not their first discipline (usability) and also for whom it should be easy to write computational code for standard statistical operations (expressiveness).

The metrics of performance, expressiveness and usability can all be assessed for the chosen compilers by the implementation of an experiment to run the compilers on simple arithmetic calculations that draw out some of the computational needs for bioinformatics, starting with basic arithmetic operations on equations and then moving to more complex ones such as those needed for foundational statistics computations. The usability can be assessed subjectively by how easy or otherwise the coding is for each compiler, and the expressiveness can be assessed by how well the core arithmetical operations can be coded. These aspects will be discussed in the next section on implementation.

5 Experimental Implementation

The chosen method of testing for comparison purposes was to use some example calculations that could be implemented and then executed using both the Transpiler and EVA. The Transpiler also includes an option to run an example in an 'interpreted' mode (with a corresponding testbench file), which allocates extra software cores (essentially optimising the XLS hardware model) in order to speed up the FHE calculations. The examples were thus executed in several modes—firstly as plaintext to provide a benchmark, then in encrypted mode (both single-core and multi-core options for the Transpiler, and encryption only for EVA since it has not user-defined optimisations available).

The example mathematical functions used were taken from both the EVA GitHub documentation and also the paper on EVA extensions [32]. These examples are simple polynomial functions and provide a means of exercising both compilers with multiple inputs, thus satisfying the initial requirements of the metrics discussed in the previous section.

The setup for the tests required implementation of the functions in both the Transpiler and EVA environments, although the high-level process is the same for both—write the regular code and then develop an execution context which controls the key generation, parameter encryption, execution and decryption of the functions on the encrypted input, and finally then include some basic timing code to provide feedback on performance. The next subsection provides details of the specific experimental environment used, including machine details and software library versions.

5.1 Test Environment

From an implementation perspective the Transpiler has a complex setup, requiring installation of the Bazel build system and several other packages (see the GitHub page [29] for details), some with dependencies on python2 and python3, and also installation of the google XLS library for connecting to back-end FHE schemes. Moreover the setup of an FHE example is time consuming—initially the 'hangman' example must be built and executed in order for the relevant Transpiler compile chains to be built, which can take 'over an hour' according to the documentation. In our test environment it actually took almost five hours. This process threw errors in building 'convenience symlinks' to several of the files within the GitHub directory code with 'operation not permitted' suggesting that the default permissions pulled down with the git-clone are incorrect, and so again made initial setup time consuming. Any example C++ test code must be placed in the 'examples' directory and then both build and testbed files need to be created, using the supplied example applications as a guide. Your own C++ example then needs to be run via Bazel. In fairness the GitHub site [29] contains a disclaimer that the Transpiler is still only a proof-of-concept, and thus more work is expected.

Creation of C++ code is somewhat involved also. There is much pragma processing required in specific places, and for all loops pragma commands are required which then inform the XLS library (typically used for hardware design but used here to provide an intermediate representation which can then be optimised) to unroll loops ready for submission to the FHE library. More generally there are limitations on programming constructs that are available for use in an application, due to their dependence on the contents of variables—for example loops, recursion and conditionals cannot be used 'as-is' in many cases since their evaluation requires knowledge of variable contents. This is clearly not an option if the input variables are encrypted, otherwise that would defeat the whole purpose of FHE. However some operations are possible, for example equality and comparison checks using encrypted variables—whether or not these are available at the programming level depends on whether or not those operations can be supported when mapped into the intermediate representation, which presumably XLS does support.

The Transpiler C++ implementations for the test functions were regular C++ coding exercises. A testbench file needs to be created for every encryption runtime context of interest. For benchmark purposes a plaintext testbench was also written, along with another for openFHE—however openFHE evaluation was beyond the scope of this paper and is provided simply for consistency with the other Transpiler examples.

5.2 EVA Build and Code Development

Download and installation for EVA is via GitHub [26], with a dependency on the SEAL library first being installed. EVA requires python 3 only, in contrast to the Google Transpiler which requires both python 2 and python 3, despite being inherently a C++ compiler. Apart from several issues relating to trusted files on the initial GitHub clone, building EVA and SEAL was very straightforward and running the supplied examples (image processing and polynomial evaluation) simply required execution of the relevant two scripts in python.

The coding process for EVA consisted of writing the appropriate python function as normal, configuring dictionaries to hold the inputs and outputs, setting encryption parameters and then wrapping the calling code within an 'EvaProgram' block. This is compiled, encryption keys are generated and then the (encryption) compiled code is executed. Libraries are provided with EVA to allow for printing of reference data and decrypted results of an encrypted execution, as well as a built-in module for calculating mean-square errors between the decrypted result and the clear computation on plaintext. In order to run EVA on plaintext only, as required for the performance benchmarking, the Input class of EVA needs to be supplied with an extra parameter of *is_encrypted* = *False* for each function input data value.

6 Execution and Evaluation

The focus of evaluation for this study was to examine and compare the two FHE compilers solutions EVA and the google Transpiler, specifically over three metrics: usability, expressiveness and performance. The metrics were examined across example problems of progressively increasing complexity—firstly simple mathematical polynomial functions, then a simple Fibonacci example, and finally the more challenging algorithmics of Hamming and edit distance. These last two problems were chosen for evaluation since they would form part of the mathematical toolkit used by bioinformaticians working in, for example, genomics, and thus would provide an insight into the maturity of the Transpiler and EVA with regard to integration into the current regular software development process. The results of these experiments are discussed in the following sections.

6.1 Initial Observations on Usability and Expressiveness

Usability The Transpiler is inherently much more complex both to build and to use as a development tool and requires the development of several C++ files. The wrapper code and supporting environment requires much more skill in C++ and associated build processes, perhaps due to the dependency on the XLS library, which in itself is an expressive but complex intermediate representation tool, and the run times seen with the Transpiler reflect this multi-layer complexity. In contrast the EVA python-based solution was straightforward to use—there is a minimal amount of FHE compiler-specific code to become familiar with in order to create and run your own regular code. The test-debug-retest cycle was much faster for EVA than for the Transpiler, but this is largely due to the general software development process for python compared to C++.

Expressiveness EVA has very limited expressiveness, essentially only supporting arithmetic over vectors, and can only use encrypted values within arithmetic expressions. There has also been some recent work on library development for EVA, with further improvements to usability proposed [32]. The Transpiler can handle a much richer programming language set, including comparisons, and thus enabling some conditionals. Both however are limited in the ability to deal with data-dependent programming—although the Transpiler has some workarounds [29], EVA cannot handle them at all.

6.2 Evaluation for Polynomials

This evaluation was conducted on two simple example polynomials, and thus required nothing more sophisticated beyond addition and multiplication. As such this provided a base level evaluation benchmark without getting into the nuanced details required for the more complex examples, and both FHE compilers provided similar 'out of the box' examples that could be used as templates for implementation.

Usability and Expressiveness

The usability and expressiveness metrics on the polynomial evaluation examples were consistent with initial observations—the Transpiler required much more coding compared with EVA, and also consequently more 'scaffolding' was needed to implement the examples. Expressiveness limitations were not apparent for this set of examples, as the expression evaluations needed were straightforward arithmetic operations. The amount of code required for the Transpiler solution was significantly more complex, as can be seen in just the testbench code for the example polynomial $3x^2 + 5x - 2, x = 55$, shown in Fig. 4.

By way of contrast, the coding required for the EVA solution was much simpler and contained in a single file, as can be seen in Fig. 5 for the same example.

Performance

EVA was much faster than the Transpiler in every polynomial case tested, by several orders of magnitude as shown in Tables 3, 4 and 5. This is mostly likely due to the fact that EVA has a much simpler DAG-based intermediate representation compared to the Transpiler which uses hardware-based models built using XLS, and thus has less complex execution graphs. However the Transpiler is a proof-of concept and thus fast performance is perhaps not a fair measure of comparison. As the Transpiler performance on the examples was significantly different to EVA, without debugging support it is difficult to determine whether the performance differences were due to the FHE scheme used (TFHE vs CKKS) or the intermediate representation, since XLS is more complex than a DAG model. A summary of the observed performance figures for the polynomial examples used in this study are shown in Tables 3 And 4. Note that since EVA uses the CKKS scheme, the computed results are derived from floating point calculations and thus there is an error between computed and actual results. The example run timings are shown in Tables 3 and 4. EVA provides a Mean Square Error (MSE) function which calculates the computation error, and these are shown for the above experiments in Table 5.

poly_1_tfhe_testbench.cc - Notepad

File Edit Format View Help

```
int main(int argc, char** argv) {
  // generate a keyset
  TFheGateBootstrappingParameterSet* params =
      new_default_gate_bootstrapping_parameters(main_minimum_lambda);

  // generate a random key
  // Note: In real applications, a cryptographically secure seed needs to be
  // used.
  uint32_t seed[] = {314, 1592, 657};
  tfhe_random_generator_setSeed(seed, 3);
  TFheGateBootstrappingSecretKeySet* key =
      new_random_gate_bootstrapping_secret_keyset(params);
  const TFheGateBootstrappingCloudKeySet* cloud_key = &key->cloud;

  // create inputs.
  int x = 55;

  cout << "input is " << x << ", 3*x*x + 5*x - 2 = " << 3*x*x + 5*x - 2 << endl;
  // Encrypt data
  auto ciphertext_x = TfheInt::Encrypt(x, key);

  cout << "Encryption done" << endl;

  cout << "Initial state check by decryption: " << endl;
  // Decrypt results.
  cout << ciphertext_x.Decrypt(key);
  cout << "  ";

  cout << "\n";

  cout << "\t\t\t\t\tServer side computation:" << endl;
  // Perform computation
  TfheInt cipher_result(params);

  absl::Time start_time = absl::Now();
  double cpu_start_time = clock();
  XLS_CHECK_OK(
      poly_1(cipher_result, ciphertext_x, cloud_key));
  double cpu_end_time = clock();
  absl::Time end_time = absl::Now();
```

Fig. 4 Code fragment—transpiler example for $3x^2 + 5x - 2$, $x = 55$

This demonstrates that the floating-point encrypted computation accuracy was also within an acceptable tolerance.

6.3 Fibonacci Evaluation

Fibonacci computation is a little more complex than that for polynomials, in the sense that operations beyond basic arithmetic are needed to compute a Fibonacci number

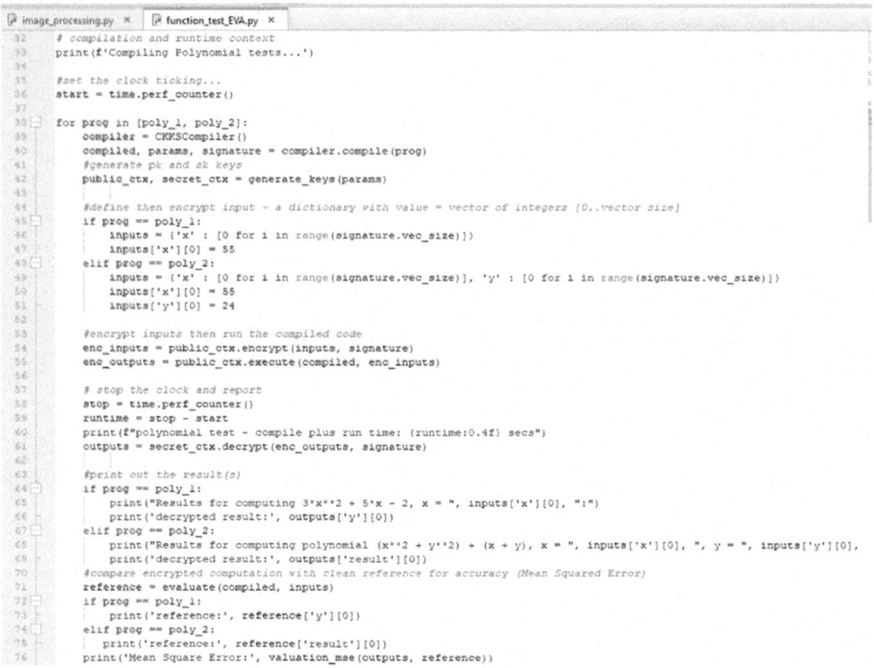


```python
# compilation and runtime context
print(f'Compiling Polynomial tests...')

#set the clock ticking...
start = time.perf_counter()

for prog in [poly_1, poly_2]:
    compiler = CKKSCompiler()
    compiled, params, signature = compiler.compile(prog)
    #generate pk and sk keys
    public_ctx, secret_ctx = generate_keys(params)

    #define then encrypt input - a dictionary with value = vector of integers [0..vector size]
    if prog == poly_1:
        inputs = {'x' : [0 for i in range(signature.vec_size)]}
        inputs['x'][0] = 55
    elif prog == poly_2:
        inputs = {'x' : [0 for i in range(signature.vec_size)], 'y' : [0 for i in range(signature.vec_size)]}
        inputs['x'][0] = 55
        inputs['y'][0] = 24

    #encrypt inputs then run the compiled code
    enc_inputs = public_ctx.encrypt(inputs, signature)
    enc_outputs = public_ctx.execute(compiled, enc_inputs)

    # stop the clock and report
    stop = time.perf_counter()
    runtime = stop - start
    print(f"polynomial test - compile plus run time: {runtime:0.4f} secs")
    outputs = secret_ctx.decrypt(enc_outputs, signature)

    #print out the result(s)
    if prog == poly_1:
        print("Results for computing 3*x**2 + 5*x - 2, x = ", inputs['x'][0], ":")
        print('decrypted result:', outputs['y'][0])
    elif prog == poly_2:
        print("Results for computing polynomial (x**2 + y**2) + (x + y), x = ", inputs['x'][0], ", y = ", inputs['y'][0],
        print('decrypted result:', outputs['result'][0])
    #compare encrypted computation with clean reference for accuracy (Mean Squared Error)
    reference = evaluate(compiled, inputs)
    if prog == poly_1:
        print('reference:', reference['y'][0])
    elif prog == poly_2:
       print('reference:', reference['result'][0])
    print('Mean Square Error:', valuation_mse(outputs, reference))
```

Fig. 5 Code fragment—EVA example for $3x^2 + 5x - 2, x = 55$

Table 3 Experimental results for $3x^2 + 5x - 2, x = 55$

FHE Compiler	Timing (s)
EVA (plaintext)	0.059
EVA (encrypted)	0.186
Transpiler (plaintext)	0.015
Transpiler (single core encrypted)	1045.97
Transpiler (multi-core encrypted)	1051.84

Table 4 Experimental results for $x^2 + y^2 + (x + y), x = 55, y = 24$

FHE Compiler	Timing (s)
EVA (plaintext)	0.0588
EVA (encrypted)	0.329
Transpiler (plaintext)	0.015
Transpiler (single core encrypted)	1635.33
Transpiler (multi-core encrypted)	1529.96

Table 5 Example equations and results

Equation/Input	Expected	Actual	MSE
$3x^2 + 5x - 2, x = 55$	9348.0	9348.170697519183	0.0036631956393623232
$x^2 + y^2 + (x + y), x = 55, y = 24$	3680.0	3680.0875328568363	0.001916308529410705

or sequence and thus provide a slightly increased level of difficulty on which to test the FHE compilers.

Fibonacci for the Transpiler A core Fibonacci example came already supplied with the Github download for the Google Transpiler, although the testbench code needed some customisation. Note that pragma modifications were needed in the vanilla C++ code to instruct the Transpiler code to unroll the provided loop construct, as a limitation of the transpiler is that loops controlled by encryption variables are not possible, as shown in Fig. 6.

```
#include "fibonacci_sequence.h"  // NOLINT: xlscc currently needs locally-scoped includes.

#pragma hls_top
void fibonacci_sequence(int n, int output[5]) {
  if (n < 0 || n > 10) {
    return;
  }

  int prev = 0;
  int cur = 1;
  int temp = 0;

// Go through the Fibonacci sequence until n is reached.
#pragma hls_unroll yes
  for (int i = 2; i <= 10; i++) {
    // We need to iterate over the Fibonacci sequence only until we reach
    // n'th value.
    if (i <= n) {
      temp = cur;
      cur = prev + cur;
      prev = temp;
    }
  }

  // Fill-in the output.
  if (n == 0) {
    output[0] = 0;
  } else {
    output[0] = cur;
  }
#pragma hls_unroll yes
  for (int i = 1; i < 5; i++) {
    temp = cur;
    cur = prev + cur;
    prev = temp;
    output[i] = cur;
  }
}
```

Fig. 6 Fibonacci example—Pragma code for loops

Creating the C++ testbench is complex, although the principles are clear—the testbench plays to role of 'client code' which submits your own regular C++ code to and from the transpiler 'server'. The testbench file is where keys are generated and parameters set, and where encryption and decryption take place. The testbench C++ code for the Fibonacci example is shown in Fig. 7.

Across three separate runs, the Transpiler Fibonacci examples took an average of 83 s using a single-core setting, but 95 s on average when executed with the multi-core option.

In summary, for Fibonacci across the three metrics of usability, expressiveness and performance the Transpiler scored well on expressiveness, however usability is cumbersome and performance is likely poor on these results although further runs on a broader set of inputs would confirm or refute this view.

Fibonacci for EVA

Custom python code needed to be developed for the Fibonacci example. The coding process for EVA consisted of writing the appropriate python function as normal, configuring dictionaries to hold the inputs and outputs, setting encryption parameters and then wrapping the calling code within an EvaProgram block. This was then compiled, encryption keys generated and then the (encryption) compiled code was executed. Standard Python code for Fibonacci was used as a basis, and in order to run this under FHE using EVA a parent EvaProgram was needed, as shown in Fig. 8.

Libraries are provided with EVA to allow for printing of reference data and decrypted results of an encrypted execution, as well as a built-in module for calculating mean-square errors between the decrypted result and the clear computation on plaintext. These features make EVA attractive from a usability perspective.

Unfortunately the evaluations of expressiveness and performance were not as encouraging with EVA. Several workarounds for the expressiveness limitations of EVA were attempted. One particular problem is that although a vector of inputs can be passed into EVA, it is not possible to index the vector elements Title Suppressed Due to Excessive Length 19 individually, and thus it is only possible to apply arithmetic operations to the vectors as aggregated data structures. Comparisons (including equality) on these encrypted vectors are not supported by EVA, and it was not possible to 'trick' the EVA compiler into performing the necessary encrypted computations to compute a Fibonacci number. For example, a comparator could be created using XOR and AND but this is not possible because the EVA parameter type (eva.Expr) does not allow such operations.

In summary for running Fibonacci with EVA, development of the necessary code was straightforward however due to the significant limitations on expressiveness, performance could not be assessed and therefore results for Fibonacci on EVA were overall inconclusive.

Results—Basic Mathematical Functions Overall it appears so far that the experimental results for polynomials and a simple mathematical function were mixed, with EVA being slightly favoured for expressiveness and usability, however both FHE compilers performed poorly with respect to performance. It is possible that one

```
void test_fibonacci_number() {
  // generate a keyset
  TFHEParameters params(kMainMinimumLambda);

  // generate a random key
  // Note: In real applications, a cryptographically secure seed needs to be
  // used.
  std::array<uint32_t, 3> seed = {314, 1592, 657};
  TFHESecretKeySet key(params, seed);

  // create inputs.
  int inputs[6];
  inputs[0] = -1;
  inputs[1] = 0;
  inputs[2] = 1;
  inputs[3] = 3;
  inputs[4] = 10;

  for (int i = 0; i < 5; i++) {
    int n = inputs[i];
    cout << "input n = " << n << endl;

    // Encrypt the input value
    auto encryptedN = TfheValue<int>::Encrypt(n, key);
    cout << "Encryption done" << endl;
    cout << "Initial state check by decryption: " << endl;
    cout << "Decrypted input: " << encryptedN.Decrypt(key) << endl;

    // Perform computation
    cout << "\t\t\t\t\tServer side computation:" << endl;
    TfheValue<int> encryptedResult(params);

    absl::Time start_time = absl::Now();
    double cpu_start_time = clock();
    XLS_CHECK_OK(fibonacci_number(encryptedResult, encryptedN, key.cloud()));
    double cpu_end_time = clock();
    absl::Time end_time = absl::Now();
    cout << "\t\t\t\t\tComputation done" << endl;
    cout << "\t\t\t\t\tTotal time: "
         << absl::ToDoubleSeconds(end_time - start_time) << " secs" << endl;
    cout << "\t\t\t\t\t  CPU time: "
         << (cpu_end_time - cpu_start_time) / 1'000'000 << " secs" << endl;

    cout << "Decrypted result: ";
    // Decrypt results.
    cout << encryptedResult.Decrypt(key) << endl;
    cout << "Decryption done" << endl << endl;
  }
}

int main(int argc, char** argv) { test_fibonacci_number(); }
```

Fig. 7 Transpiler Fibonacci example—testbench code

```
# set up the EVA main code , and define the input vector size - a power of 2
fib = EvaProgram('fib', vec_size=8)
with fib:
  n = Input('number') #sets up the (empty) input dictionary
  #dictionary for the output
  #NB although it is a VECTOR you can't TREAT IT as one
  Output('result', fibonacci_number(n))
#set FHE params - measured in bits, with fixed-point scale for inputs and
#max coefficient range for the outputs
fib.set_input_scales(30)
fib.set_output_ranges(10)
#compile the code for EVA, then run it
print(f'Compiling...')

#set the clock ticking...
start = time.perf_counter()

compiler = CKKSCompiler()
compiled, params, signature = compiler.compile(fib)
#generate pk and sk keys
public_ctx, secret_ctx = generate_keys(params)

#define then encrypt input - here it is simply a number for computing fib
#populate the input dictionary, keyed on 'number'. Note:
#create an initial vector of zeros up to vector size, then set the first one to the value required
inputs = {'number' : [0 for i in range(8)]}
inputs['number'][0] = 10
enc_inputs = public_ctx.encrypt(inputs, signature)
#run FHE
enc_outputs = public_ctx.execute(compiled, enc_inputs)

# stop the clock and report
stop = time.perf_counter()
runtime = stop - start
print(f"compile plus run time: {runtime:0.4f} secs")

#decrypted result
outputs = secret_ctx.decrypt(enc_outputs, signature)

#print out the result(s)
print("Results for computing fibonacci(", inputs['number'][0], "):")
print('decrypted result:', outputs['result'][0])

#compare encrypted computation with clean reference for accuracy (Mean Squared Error)
reference = evaluate(compiled, inputs)
print('reference:', reference['result'][0])
print('MSE', valuation_mse(outputs, reference))
print()
```

Fig. 8 Fibonacci—EVA main code

or both of the compilers could perform better when focused on other problems, and so the next subsection expands upon this to look at some of the basic computations required for many bioinformatics applications.

6.4 *Bioinformatics Applications—Genomic Comparisons*

Background—Edit and Hamming Distance Computations

One of the earliest genomic applications examined by researchers when developing the lower-level FHE libraries was the comparison of genomic sequences, which can be achieved through the calculation of Hamming and edit distance. Hamming (and edit) distance was investigated from an FHE perspective in [33], but that study required a translation of the natural plaintext space into one suitable for FHE (or more precisely in this case, SHE) by developing relevant equality and comparison circuits. In this paper however the aim was simply to develop the Hamming distance code in the usual 'natural' way as an experiment and let the FHE compilers do the heavy lifting in terms of conversion into a suitable plaintext space for FHE. The early work by [34] on Hamming and edit distance was a very useful starting point for using FHE in the genomics arena. The FHE solution was focused on development and optimisation for a plaintext space suited to an FHE library API, rather than for more natural straight DNA sequence datasets, since FHE compilers were not an option at the time. The algorithm developed for Hamming distance, slightly modified in [35], is shown in Fig. 9.

In Fig. 9, h is the Hamming distance, a Variant Call Format (VCF) file is a special file format used in bioinformatics applications for genetic variation datasets, and L is a list of such files indexed by patient. The x and y are the individual records in a VCF file for patients at locations—note that Hamming distance requires the records to be of equal length since it does not measure insertions or deletions (unlike edit distance), and hence contributes a zero count. For edit distance, Cheon et al. presented the algorithm shown in Fig. 10:

```
h ← 0
for i ← 1 to |ℒ|
    if the structural variant given by xᵢ or yᵢ is an insertion or deletion
        hᵢ ← 0
    else if xᵢ ∧ yᵢ = ∅
        hᵢ ← 1
    else if xᵢ and yᵢ reference the same bases but have different alternate alleles
        hᵢ ← 1
    else
        hᵢ ← 0
h = h + hᵢ
return h
```

Fig. 9 Hamming distance [34, 35]

```
for i ← 0 to n
    d(i, 0) ← 0
for j ← 0 to m
    d(0, j) ← 0
for i ← 1 to n
    for j ← 1 to m
        t(i, j) ← (αᵢ = βᵢ)
        d(i, j) ← min{d(i − 1, j − 1) + t(i, j), d(i, j − 1) + 1, d(i − 1, j) + 1}
return d(n, m)
```

Fig. 10 Edit distance algorithm [33]

Table 6 FHE runtime figures—genome applications [27]

Task	Scheme	Dataset	Time (min)	Memory (GB)
Hamming distance	BGV	200 K sequences	8	2.2
Approximate edit distance	BGV	10 K sequences	3	1.3
Genetic testing	BFV	1 query (1 genetic variant) in 50 VCF Files	1	83

In Fig. 10 the edit distance is d(i, j), αi and βj are the specific nucleotides being compared within the DNA sequences α and β, and n and m are the lengths of the two DNA sequences being compared. The function t is used to record a count of each iterative match.

Experimental Results with EVA and the Transpiler The edit and Hamming distance algorithms were the next target routines for the evaluation of the Transpiler and EVA FHE compilers, as the next step up in complexity of application area from straight mathematical examples. Unfortunately the results of Hamming and edit distance experiments on both the Transpiler and EVA were broadly poor and thus inconclusive, but for different reasons.

The Transpiler running times for the implementations of Hamming and edit distance examples could not be ascertained as the executions needed to be terminated manually after failing to complete within two hours. Benchmark figures for these operations, based on the BGV scheme (and thus at the FHE library level), are in Table 6 below and taken from [27]:

Clearly the figures in Table 6 render the current Transpiler version unusable for practical genome comparison purposes. The Transpiler again scored well on expressiveness since the problems could be coded, but again usability made the coding exercises too complex to be included in a transparent way for mainstream bioinformatics software developers.

With respect to EVA, despite several attempts it was again not possible to develop a workaround for the fact that EVA lacks the ability to compare encrypted vector elements, a necessary operation for computing Hamming and edit distance, and this limitation proved fatal to the EVA assessment on both Fibonacci and the genomic examples. From a usability perspective EVA comes out on top again, and the simplicity of coding in python together with encouraging results for performance on standard mathematical computations does suggest potential promise for the future.

6.5 Overall Observations

EVA shows promise as it can be extended [32] and python is a much simpler language to use that C++. Further enhancements would also be needed to allow for a wider set of operations e.g. comparisons. There is no access to vector components and no comparison operations available, which severely limits the programming constructs that can be used—for example it is not possible to compare encrypted vector elements, although with clever mapping this can be encoded into a polynomial plaintext space that uses only XOR and AND operations and will thus preserve security [33]. This is not theoretically controversial since the vector size must be public in any case and also the elements themselves are encrypted. This would make more complex applications, such as edit distance [33], easier to code in EVA since they require access to individual vector element comparisons. The Transpiler however would need to be made more transparent and provide a simpler build and development process. Furthermore the dependency on the google XLS library for an intermediate representation introduces an extra complex component in the encryption chain that widens the scope both for coding errors and also potentially increasing the attack surface of the end-to-end compiler process.

7 Conclusions and Future Directions

There is currently a wide range of code support for FHE solutions. Many of the publicly available FHE libraries provide efficient implementations of well-established FHE schemes and are steadily overcoming earlier performance shortfalls. The major problem that affects all of the FHE library solutions however is that of usability since they all require a significant level of cryptographic expertise and sophistication on the part of the programmer. Researchers in specific domains of application have clearly been able to make significant gains in functionality and applicability, for example via the iDASH program [27], but nonetheless these researchers still required expertise in understanding FHE to translate their respective problems into ones which can be worked on by the FHE libraries. This is clearly being recognised now as a barrier to mainstream adoption, but research is in its early stages. Compiler solutions such as the Transpiler and EVA are a step in the right direction, and although there have been recent successes in using FHE in a bioinformatics and medical setting [24, 28] current work has been very scenario-specific and there is much still to be done to provide a more'general purpose' compiler solution for bioinformatics developers. Beyond performance, the obvious main challenge at present is around appropriate compilation of common language constructs that are data-dependent since, by design, they must operate on encrypted data in an FHE environment. This is likely to be a fruitful area of future research.

Moreover it seems reasonable that FHE compilers could (and perhaps should) be targeted towards specific application areas, so that appropriate focus can be placed on compiler optimisation for domain-specific operations—for example in computational genomics. Then, in principle a TFHE-style approach like CHIMERA [15], which aims to allow FHE scheme switching within an application, could be adopted which can form the basis of an aggregated intermediate plaintext space that facilitates optimised and practical execution.

The complexity of the Transpiler solution, whilst providing much greater expressiveness via the XLS intermediate representation, means that a debugging tool would be of immense help in tracking down whether coding or performance issues reside in the representation itself or are in fact a consequence of parameter setting for a given FHE scheme, and thus FHE compiler debugging is likely to be a fruitful area of investigation. It remains to be seen whether the more recent HEIR initiative [25] will provide further insights. In conclusion the field of FHE continues to grow apace, and the current challenges in compiler research for FHE schemes are beginning to make the promise of FHE more accessible to mainstream software developers.

Acknowledgements The author would like to thank the University of Suffolk for their continued support.

Disclosure of Interests The authors have no competing interests to declare that are relevant to the content of this article.

References

1. Gentry C (2009) Fully homomorphic encryption using ideal lattices. In: Proceedings of the annual ACM symposium on theory of computing, pp 169–178. https://doi.org/10.1145/1536414.1536440
2. Viand A, Jattke P, Hithnawi A (2021) SoK: fully homomorphic encryption compilers. In: Proceedings: IEEE symposium on security and privacy, pp 1092–1108. https://doi.org/10.1109/SP40001.2021.00068
3. Gorantala S, Springer R, Purser-Haskell S, Lam W, Wilson R, Ali A, Astor EP, Zukerman I, Ruth S, Dibak C, Schoppmann P, Kulankhina S, Forget A, Marn D, Tew C, Misoczki R, Guillen B, Ye X, Kraft D, Desfontaines D, Krishnamurthy A, Guevara M, Perera IM, Sushko Y, Gipson B (2021) A general purpose transpiler for fully homomorphic encryption. Cryptology ePrint Archive. Paper 2021/811. https://eprint.iacr.org/2021/811
4. Acar A, Aksu H, Selcuk Uluagac A, Aksu H, Uluagac AS (2018) A survey on homomorphic encryption schemes: theory and implementation. ACM Comput Surv 51(4) (2018). https://doi.org/10.1145/3214303
5. Cheon JH, Kim A, Kim M, Song Y (2017) Homomorphic encryption for arithmetic of approximate numbers. In: Lecture notes in computer science (including subseries Lecture notes in artificial intelligence and Lecture notes in bioinformatics), vol 10624. LNCS, pp 409–437. https://doi.org/10.1007/978-3-319-70694-8_15
6. Chillotti I, Gama N, Georgieva M, Izabachène M (2020) TFHE: fast fully homomorphic encryption over the Torus. J Cryptol 33(1):34–91. https://doi.org/10.1007/S00145-019-09319-X
7. Boneh D, Goh EJ, Nissim K (2005) Evaluating 2-DNF formulas on ciphertexts. Lect Notes Comput Sci 3378:325–341. https://doi.org/10.1007/978-3-540-30576-7_18
8. Sander T, Young A, Yung M (2003) Non-interactive cryptocomputing for NC/sup 1/. In: 40th Annual symposium on foundations of computer science, pp 554–566. https://doi.org/10.1109/SFFCS.1999.814630
9. Ishai Y, Paskin A (2007) Evaluating branching programs on encrypted data. In: Lecture notes in computer science (including subseries Lecture notes in artificial intelligence and Lecture notes in bioinformatics), vol 4392. LNCS, pp 575–594. https://doi.org/10.1007/978-3-540-70936-7_31
10. Brakerski Z, Gentry C, Vaikuntanathan V (2014) (Leveled) Fully homomorphic encryption without bootstrapping. ACM Trans Comput Theory (TOCT) 6(3):1–36. https://doi.org/10.1145/2633600

11. Fan J (2012) Vercauteren: somewhat practical fully homomorphic encryption. Cryptology ePrint Archive. https://eprint.iacr.org/2012/144
12. Regev O (2009) On lattices, learning with errors, random linear codes, and cryptography. J ACM (JACM) 56(6). https://doi.org/10.1145/1568318.1568324
13. Brakerski Z, Vaikuntanathan V (2011) Fully homomorphic encryption from ring LWE and security for key dependent messages. In: Lecture notes in computer science (including subseries Lecture notes in artificial intelligence and Lecture notes in bioinformatics), vol 6841. LNCS, pp 505–524. https://doi.org/10.1007/978-3-642-22972-9_29
14. Chen H, Laine K, Player R, Xia Y (2018) High-precision arithmetic in homomorphic encryption. In: Lecture notes in computer science (including subseries Lecture notes in artificial intelligence and Lecture notes in bioinformatics), vol 10808. LNCS, pp 116–136. https://doi.org/10.1007/978-3-319-76953-0_7
15. Boura C, Gama N, Georgieva M, Jetchev D (2020) CHIMERA: combining ring LWE-based fully homomorphic encryption schemes. J Math Cryptol 14(1):316–338. https://doi.org/10.1515/JMC-2019-0026
16. MarbleHE (2024) MarbleHE—Making FHE more accessible. GitHub. https://github.com/MarbleHE. Last accessed 08 Jun 2024
17. Microsoft (2022) GitHub—Microsoft/EVA: compiler for the SEAL homomorphic encryption library. https://github.com/microsoft/EVA. Last accessed 28 May 2024
18. Halevi S, Shoup V (2014) Algorithms in HElib. In: Lecture notes in computer science (including subseries Lecture notes in artificial intelligence and Lecture notes in bioinformatics), vol 8616. LNCS, pp 554–571. https://link.springer.com/chapter/10.1007/978-3-662-44371-2_31. Last accessed 01 Jul 2024
19. Viand A (2023) Useable fully homomorphic encryption. Ph.D. Thesis, ETH Zurich. https://doi.org/10.3929/ETHZ-B-000613734
20. Gouert C, Mouris D, Georgios Tsoutsos N (2023) SoK: new insights into fully homomorphic encryption libraries via standardized benchmarks. Proc Privacy Enhanc Technol 2023(3):154–172. https://doi.org/10.56553/POPETS-2023-0075
21. Viand A, Jattke P, Haller M, Hithnawi A (2023) HECO: Fully homomorphic encryption compiler. In: 32nd USENIX security symposium (USENIX Security 23). USENIX Association, Anaheim, CA, pp 4715–4732. https://www.usenix.org/conference/usenixsecurity23/presentation/viand
22. Archer DW, Manuel J, Trilla C, Dagit J, Malozemoff AJ, Polyakov Y, Rohloff K, Ryan G, Maloze-Moff AJ (2019) Ramparts: a programmer-friendly system for building homomorphic encryption applications homomorphic encryption; automatic arithmetic circuit generation; lattice-based cryptography; usable security and privacy ACM reference format. In: Proceedings of the 7th ACM workshop on encrypted computing and applied homomorphic cryptography (2019). https://doi.org/10.1145/3338469.3358945
23. Dathathri R, Saarikivi O, Dai W, Laine K, Musuvathi M, Kostova B (2019) EVA: an encrypted vector arithmetic language and compiler for efficient homomorphic computation. In: Proceedings of the 41st ACM SIGPLAN conference on programming language design and implementation, p 16. https://doi.org/10.1145/3385412.3386023
24. de Leon CS, Chua RB (2023) Secure genotype imputation using the hidden Markov model with homomorphic encryption. In: Lecture notes in networks and systems, vol 784. LNNS, pp 181–190. https://doi.org/10.1007/978-3-031-44146-2_18
25. Google (2024) HEIR. https://heir.dev/. Last accessed 12 Aug 2024
26. Microsoft (2022) GitHub—Microsoft/SEAL: Microsoft SEAL is an easy-to-use and powerful homomorphic encryption library. https://github.com/microsoft/SEAL. Last accessed 28 May 2024
27. Kim M (2020) The iDASH competition: progress of homomorphic encryption for genomic privacy. National Institute of General Medical Sciences. https://simons.berkeley.edu/sites/default/files/docs/15663/mirankim043020.pdf. Last accessed 11 Jun 2024
28. Salaga H, Karanikolas N, Kaklamanis C (2023) "SHeMeD": an application on secure computation of medical cloud data based on homomorphic encryption. In: Lecture notes in computer

science (including subseries Lecture notes in artificial intelligence and Lecture notes in bioinformatics), vol 14109. LNCS, pp 116–133. https://doi.org/10.1007/978-3-031-37120-2_8
29. Google (2021) GitHub—google/fully-homomorphic-encryption: an FHE compiler for C++. https://github.com/google/fully-homomorphic-encryption. Last accessed 01 Jun 2024
30. Google (2022) XLS: accelerated HW synthesis. https://google.github.io/xls/. Last accessed 02 Jun 2024
31. Bazel: Using Bazel. https://bazel.build/docs. Last accessed 10 Jun 2024
32. Chowdhary S, Dai W, Laine K, Saarikivi O (2021) EVA improved: compiler and extension library for CKKS. In: WAHC 2021—Proceedings of the 9th workshop on encrypted computing and applied homomorphic cryptography, co-located with CCS 2021, pp 43–54. https://doi.org/10.1145/3474366.3486929
33. Cheon JH, Kim M, Lauter K (2015) Homomorphic computation of edit distance. In: Lecture notes in computer science (including subseries Lecture notes in artificial intelligence and Lecture notes in bioinformatics), vol 8976, pp 194–212. https://doi.org/10.1007/978-3-662-48051-9_15
34. Kim M, Lauter K (2015) Private genome analysis through homomorphic encryption. In: Proceedings of the 4th iDASH privacy workshop. https://doi.org/10.1186/1472-6947-15-S5-S3
35. Wood A, Najarian K, Kahrobaei D (2020) Homomorphic encryption for machine learning in medicine and bioinformatics. ACM Comput Surv (CSUR) 53(4). https://doi.org/10.1145/3394658

Making the Human the Strongest Link in an Integrated Cyber Physical Systems (CPS) Operational Stack Concept for Combined Security, Sustainability, Assurance, and Safety (S2AS)

Amanda C. Goodger, Andrew Jones, and Nicholas H. M. Caldwell

Abstract Organisations of all sizes are increasingly using emerging technologies like Artificial Intelligence (AI) and the Internet of Things (IoT) to enhance effectiveness (Kim and Kumar in Cyber–physical systems: a perspective at the centennial, pp. 1287–1308, 2012; Malatji et al. in Inf Comput Secur 27:233–272, 2019), fostering innovation and economic growth (Zengin et al. in Sustainability 13:2682, 2021; Mourtzis et al. in Energies 15:6276, 2022). However, it is leading to increasing human factor vulnerabilities exposure and now necessitating a move to an integrated human-centric approach involving security, sustainability assurance and safety to transform humans from the perceived weakest link to the strongest (Kaspersky Blog in The human factor in IT security: how employees are making businesses vulnerable from within, 2017; Keidanren (Japan Business Federation) in Toward realization of the new economy and society—reform of the economy and society by the deepening of 'Society 5.0', 2016). Based on Riley's 2014 cyber terrain operational stack (CTOS) (Riley in "Cyber Terrain": a model for increased understanding of cyber activity, 2014), this paper uses case studies to develop and combine an integrated human-centric view with an algorithmic-centric risk management methodology for a hybrid approach in cyber-physical environments. This methodology builds on from the original security triad's evolution (Boyes in Technol Innov Manag Rev 5:28–34, 2015) and aligns with potential new Security, Sustainability, Assurance and Safety (S2AS) risk management atomic attributes, providing an integrated framework for managing risks in cyber-physical ecosystems. The proposed integrated risk management framework potentially can be associated with the UN Sustainable Development Goals (SDGs), linking S2AS risk atomic attributes to identify potential risks and mitigation strategies. This could support fostering inclusive employment, technological

A. C. Goodger · A. Jones · N. H. M. Caldwell (✉)
University of Suffolk, Ipswich, UK
e-mail: n.caldwell@uos.ac.uk

A. C. Goodger
e-mail: a.goodger2@uos.ac.uk

H. Jahankhani and B. Issac (eds.), *Cybersecurity and Human Capabilities Through Symbiotic Artificial Intelligence*, Advanced Sciences and Technologies for Security Applications, https://doi.org/10.1007/978-3-031-82031-1_15

innovation, and responsible consumption via AI and IoT for organisations ((Zengin et al. in Sustainability 13:2682, 2021).

Keywords Cyber-physical environments and ecosystems · Emerging technologies · Security triad

1 Introduction

1.1 Context and Importance Supporting an Integrated Approach

Organisations of all sizes are increasingly using emerging technologies like Artificial Intelligence (AI) and Internet of Things (IoT). These technologies improve operational effectiveness, foster innovation, and drive global economic growth. By integrating AI and IoT, businesses can work more efficiently, cut costs, and create new business models [1–4].

However, relying and focusing more on the technologies, e.g. AI and IoT has potentially obscured the limitations and value of the human factors. While these technologies can automate many tasks, humans are still essential for decisions and planning. Therefore, a strong focus on security, sustainability, assurance, and safety (S2AS) is needed.

1.2 Problem Statement

Mistakes, lack of knowledge, and not enough training can lead to security issues, inefficiencies, and safety risks. It highlights that people are often seen as the weakest link in cybersecurity, cyberresilience and safety [5]. This shows the need to improve both technology and human capabilities to make a combined people and technical approach the primary focus for Cyber-Physical Environments and Ecosystems (CPEEs). The challenge is to create a framework that combines human risk management with technological advances for today's world e.g. Society 5.0 [6]. This paper suggests a new approach, viewing the combined considerations as key assets in ensuring security, sustainability, assurance and safety (S2AS).

1.3 Objective

This paper proposes a holistic approach to the CPEE operational stack, viewing humans as the strongest link. By expanding Riley's 2014 cyber terrain operational

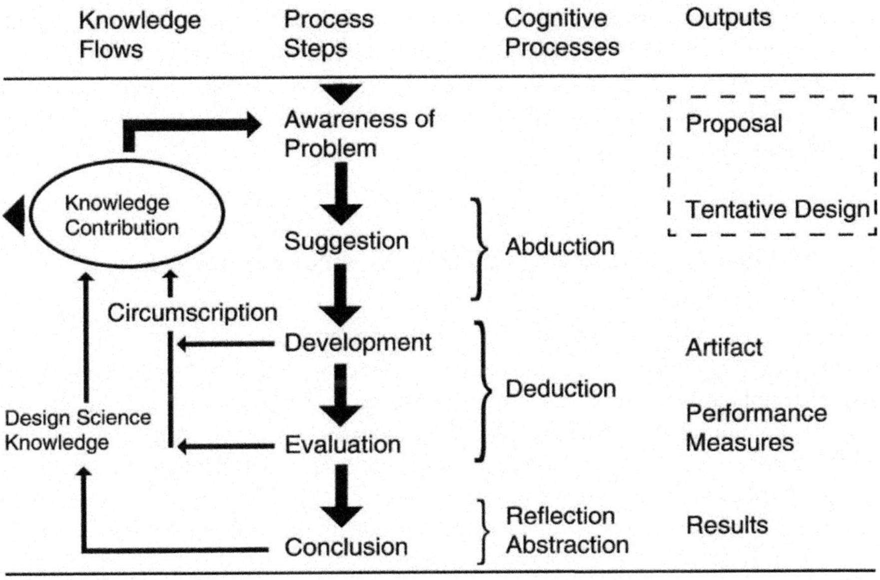

Fig. 1 Processes in the design cycle, adapted from Kuechler and Vaishnavi (2008, 493), based on the original proposal Takeda et al. (1990) [8]

stack (CTOS) [7] and integrating a human-centric physical terrain operational stack (HPTOS), it combines human factors with algorithmic risk management in cyber-physical environments. This methodology suggests for use from various perspectives like top-down and ground-up, aligns with new S2AS attributes. Developed using a metamodel approach within the design cycle (Fig. 1), it enables detailed design, analysis, and evaluation of artefacts and processes [8], discussed later in the paper.

The process involves: awareness of problem phase—outlining the research problem for the proposed model; suggestion phase—providing the overall positioning; development and evaluation phases—using an iterative approach with case studies/action research for flexibility and adaptability; and conclusion—further work and knowledge contribution to evolving the proposed model.

2 Literature Review

This section reviews the existing literature in supporting the proposed approach from the integrated cyber terrain operational stack, human vulnerabilities in integrated systems, the adoption of AI and IoT in SMEs, the evolution of the security triad, and the alignment with the UN Sustainable Development Goals (SDGs).

The term 'ecosystem' was coined in 1930 by Roy Clapham. It was later fine-tuned, by British ecologist Arthur Tansley [9] defining it as an interactive system instituted between a group of living entities and their environment. The main types of ecosystems involve 2 natural ecosystems—terrestrial and aquatic [10], and an artificial ecosystem devised by humans [11]. The Cyber Terrain Operational Stack (CPEE) [7] has been updated in 2023 by Riley to be defined as a cyber-terrain [12]. For the remainder of this paper, it has broadened to the more inclusive to CPEE term.

2.1 Cyber-Physical Versus Digital-Physical Definitions

This paper covers cyber-physical rather than digital-physical environments and ecosystems. Both terms are often used interchangeably, but they have distinct meanings and applications. The CPEE definition for the environment includes the integration of computational (cyber) and physical processes within an artificial ecosystem. In this ecosystem, physical processes are monitored and controlled by computer-based algorithms, often in real-time for seamless integration of cyber and physical components to create systems that can autonomously monitor, control, and optimise physical processes [13]. The Digital Physical Environment and Ecosystem (DPEE) defines the integration of digital technologies with the physical world.

The key differences are: Integration Level—CPEE integrates cyber and physical components for interconnected systems operating in real-time, compared with DPEE where digital tools and representations enhance the physical world; and Technology Level and its application—CPEE include sensors, actuators, networks and real-time computational algorithms for real-time monitoring and control of physical processes.

e.g. autonomous cars, whilst DPEE uses digital twins, augmented reality (AR), virtual reality (VR) and other digital technologies such as in smart cities, digital manufacturing, and digital healthcare, where digital models and simulations are used to optimize physical processes and environments [14] for humans.

2.2 Frameworks and Models Support Cyber-Physical Environments and Ecosystems

Riley's 2014 model stated its provenance is based on DoD framework development [7], but there are references and sources that relate to CPEE.

General Systems Theory (GST) [15–17] emphasizes the holistic nature and interdependence of system components. Living Systems Theory (LST) [18] extends GST to biological and social systems, emphasizing hierarchical organisation from cells to societies [19] and information ecosystems [20] as adopted in the higher layers of Riley's model. Both theories underscore the importance of interactions within complex systems, including Socio-technical systems (STS).

Building on GST, the LST concentrates on the existence of all living biological and social systems, their structure, interaction, behaviour, and development [21].

According to Miller's original conception [22], a 'living system' includes 20 'critical subsystems', which are defined by their functions and visible in numerous systems. Interaction happens through both material/energy and information exchanges. It established the self-organising open systems theory to provide 8 main levels that increased in complexity with potential emergent behaviours (aligning with GST). From levels 1–8 the LST continuum—(1—cells, 2—organs, 3—organisms, 4—groups, 5—organisations, 6—communities, 7—societies and 8—supranational systems). Riley's models [7, 12] covered cyber but did not clearly define the physical aspects for the CPEE.

Built on GST/LST, then Socio-technical systems (STS) underpin CPEE, where socio and technical systems are combined with its contextual situation. Emerging from the establishment of the Tavistock Institute (TI) in 1946 [21], STS reflect the open system GST attribute [15]) through interaction between the STS and the external environments. The Tavistock Institute (TI), through several researchers, evolved it to include resilience on the human aspects through multi-skilling, redundancy, adaptability, and minimum critical specifications to carry out tasks. One of the TI researchers devised the Social-technical 9 design principles (including: Compatibility; Minimal Critical Specification; The Socio-technical Criterion; The Multifunctionality Principle; Boundary Location; Information; Support Congruence; Design and Human Values; and Incompletion). Complementary to CPEE, the STS model for an organisation relates to 4 connected and interrelated variables: task, people, structure (or roles), and technology [23]. Mumford stated there are two primary subsystems [21]. In the organisation (Fig. 2): the social system—individuals and teams, and needs for coordination, control, and boundary management (group level and individual level [24]); and the technical system—includes not only equipment, machines, tools, and technology but also the work organisation (information level and physical level [24]). If STS is not considered effectively in systems design, then it can expand risks that systems will not deliver the intended goals of an organisation [25].

As with CPEE, if STS is not considered involving the 4 interconnected variables, through updating on human dynamics, interdependency issues, technological trends or how organisational developments respond to external influences etc [26] then it increases the system wide risks.

CPEE unites the principles of GST and LST theories, and STSs with complex adaptive systems (CAS) that can be defined as having integrative levels or sets of activities based on pre-existing activities or entities. An early CAS representation is found in Cybernetics as defined by Norbert Wiener [27]. CAS provides system and resilience as an attribute of systems [28] and using many GST/LST attributes [29].

Cyber terrain is central to CPEE based on CAS, involving mapping and understanding the digital landscape, cyberspace, including assets, vulnerabilities, and threats. In 1984, William Gibson coined the word cyberspace in the science-fiction novel Neuromancer. The term 'cyber terrain' was not described until the 2000s [30]. The US Dept. of Defense (DoD) defined cyber terrain based on cyberspace [31], including operating systems or application software, network protocols, computing devices, and even individuals or virtual personas. 8 key areas identified for the

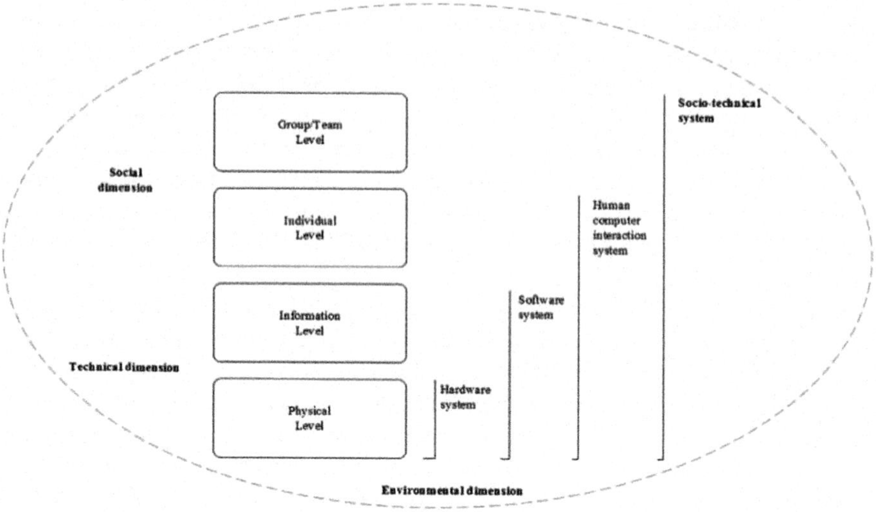

Fig. 2 Information system as a socio-technical system [2]

cyber terrain—including data centres, commercial Internet service providers (ISPs), undersea cables, international standards bodies, basic input/output systems, supply chains, cyber workforce, and innovation for all terrain [32]. Further defined as the 'Key terrain in cyberspace exists at all of the cyberspace planes, which include the geographic, physical, logical, cyber persona, and supervisory planes, but not always linked with physical terrain location'. [33, 34]. It permits users to share information, interact, swap ideas, games, engage in social interactions, and operate business and among many other activities. Plus, it is important to map the cyber terrain i.e., the environment and ecosystems enabling effectiveness of cybersecurity and resiliency [35].

From a military perspective leveraging the Operational Cyber Mission Stack (OCMS) is a framework that enables commanders to visualize, prioritise, and defend cyber elements for mission success. The OCMS does not bridge the gap between physical and logical elements of cyberspace operations [36], which Riley's 2014 model addresses [7]. Riley's model enhances situational awareness and decision-making by integrating these elements [37]

Riley's model [12] has been updated but solely focuses on security. It provides a framework for securing complex, interconnected virtual environments that characterize the metaverse [38], ensuring both user safety and data integrity. It has outlined 7 aspects (including: adaptable flexibility; proactive defence; continuous improvement; collaboration and communication; integration with other frameworks, such as NIST [39], ISO [40, 41], etc.; training and awareness; and strategic planning. Riley now refers to his model as the 'Cyber Terrain Model' (CTM), but for this paper continues to be stated as CPEE [12].

2.3 *Human Vulnerabilities in Cyber-Physical Environments and Ecosystems*

Human vulnerabilities in integrated systems are not a new issue in dynamic interconnected cyber-physical environments [5], the key weakness is poor human decision making. In short: the human is the 'weakest link' in today's environmental resilience and security needs [42]. This is exacerbated and amplified by economic and societal dependence on the information driven environments and ecosystems [43]. Decision making centres on the human capacity to make a definitive choice in each environment, ecosystem, or situation. It can be viewed as a set of options for individual decisions that are affected by a subjective viewpoint, a contextual setting and belief system it emphasizes etc. [44].

Organisational structure vulnerabilities occur via a top-down strategic decision-making and cascaded through a top-down hierarchical structure to lower structural levels/small-scale parts [5]. An option to counteract vulnerabilities is via the GST viability attribute, as seen in the viability system management (VSM) which aids balance (Fig. 3) and supports integrity from short to long-term (Fig. 4):

In Fig. 4, a proposed VSM with 7 functional elements defined as Systems 1–5, 3*, 4*, where System 1 of an organisation or entity comprises of the several elements that perform a specific role. Emerging from its implementation, Systems 2–5 arise from the need to enable each of System 1 operations [45–47].

Other aspects highlighting human vulnerabilities, are the requisite variety impact on the control of systems. It requires the same level of variety as the size of the

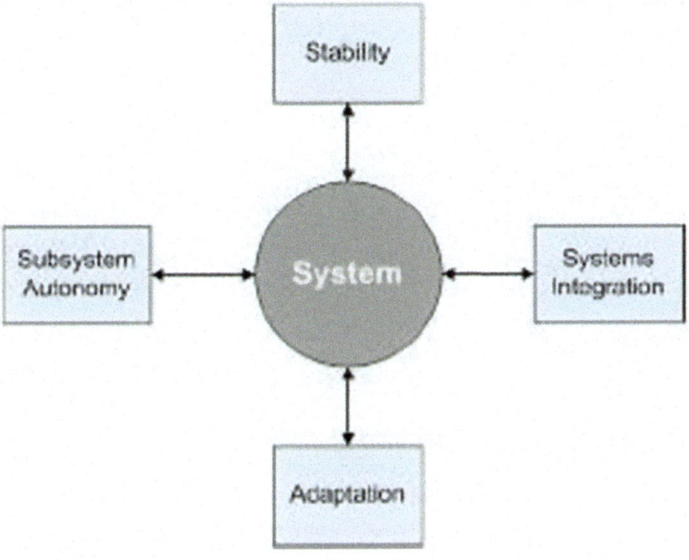

Fig. 3 Viability versus system tension [29]

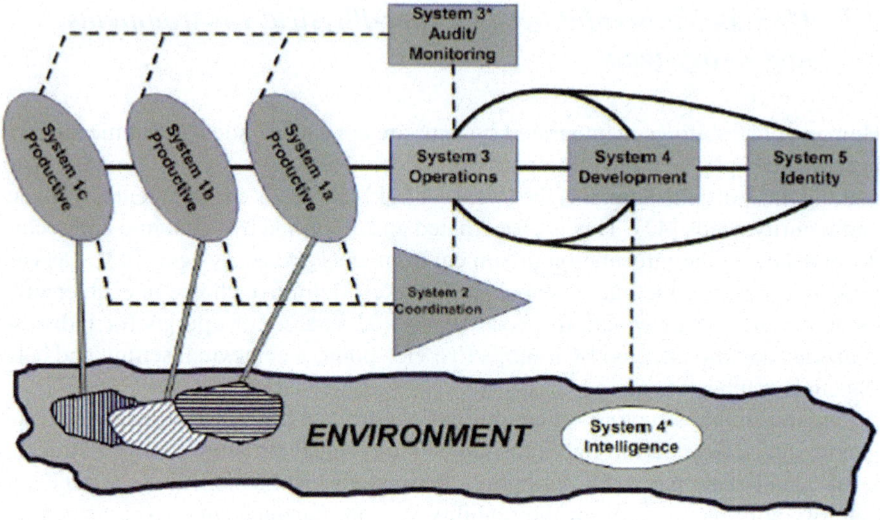

Fig. 4 Viable system model [45, 46]

system to be controlled (Fig. 6). Then if the variety is not effectively managed using a system's amplifiers (e.g., delegation, individual's/group's decisions and commitment, education etc.) and attenuators (e.g., exception reporting, modelling, selective communications etc.) it can cause organisational weaknesses. The ability to regulate and control a system depends on the size of hierarchy, plus feedback and recursion to evaluate and understand the environment, viable system, and management of the system (Fig. 5).

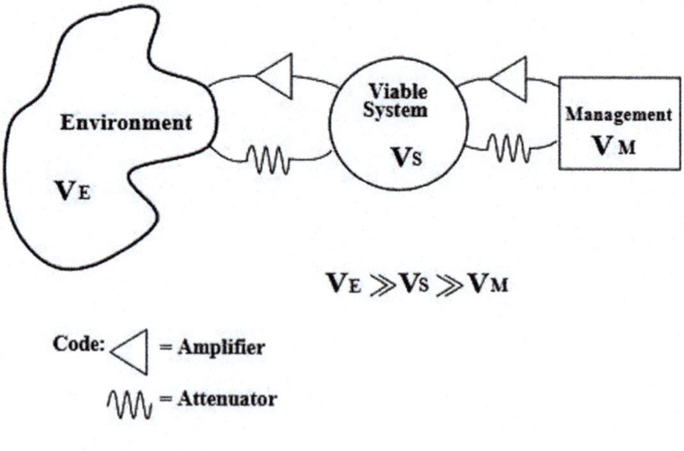

Fig. 5 Managing complexity [48]

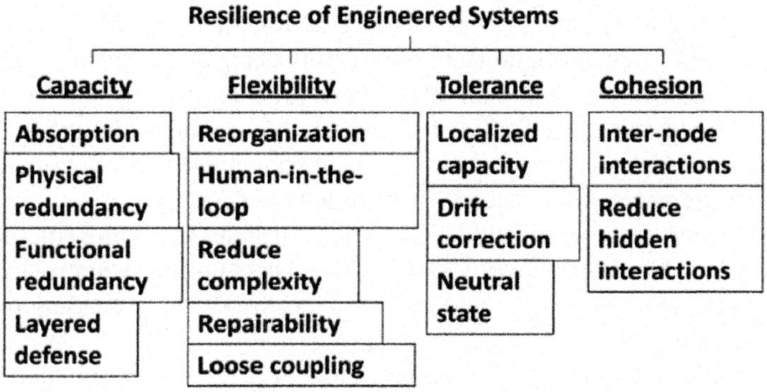

Fig. 6 Resilience principles for engineered systems [49]

To overcome human vulnerabilities in CPEE, then systems engineering (SE) aspects requires resilience as defined by the White House in the US, 'Resilience is: ...the ability to adapt to changing conditions and prepare for, withstand, and rapidly recover from disruption' [49]. Attributes outlined that support systems-based approaches include, capacity (the ability of the system to survive a threat), flexibility (the ability of the system to adapt to the threat), tolerance (the ability of the system to degrade gracefully in the face of a threat), and cohesion for enhancing resilience (the ability of the system to act as a unified whole in the face of a threat.) (Fig. 6). The flexible attribute is aiding humans in the system via a proactive need for human intervention. This principle is subdivided into: the human in control; reduce human error; and the automated function.

The interdependency aspects have altered since these principles were published [49]. Standard methods for humans and technology for usability and utility are required to reduce human error should be employed when there are humans in the loop [50]. In top-down algorithmic dependent societies and context, where there are insufficient algorithmic management capabilities, complexities of measurements and the current limits of existing rationale need more effective human-technology integration [43]. Part of the US-Canada Power System Outage Task Force (2006) recommended robust compliance to reliability governance involving human factors [51].

2.4 Emerging Technologies Cyber-Physical Environments and Ecosystems

As emergent technologies evolve into diverse dynamic environments and ecosystems, they exhibit evolving computational principles that point to digital transformation at all levels across economy and society [52].

In the '21st Century blueprint' across several domains (Fig. 6) considers innovations in biotechnology and nanotechnology, plus expanded forms of computerised intelligence but different to human intelligence [53]. It could be self- learning with increasingly rapid evolution, leading to a cascade in computer intelligence leading to the 'Singularity' [53]. These stem from radical changes in software and associated areas [52] as indicated with increase in AI usage (Fig. 8) [54].

In engineering, complexification uses several attributes including emergence, self-organisation, synergy, non-linearity, non-rigid and decentralised control (s), the existence of power laws, adaptation, and free-scale networks, the prevailing role of non-classical logics etc. [55].

This 'complexification' of engineering sciences (Fig. 9) shows in the way scientists, engineers and philosophers are entering into a living technology' (LT) [55]. It

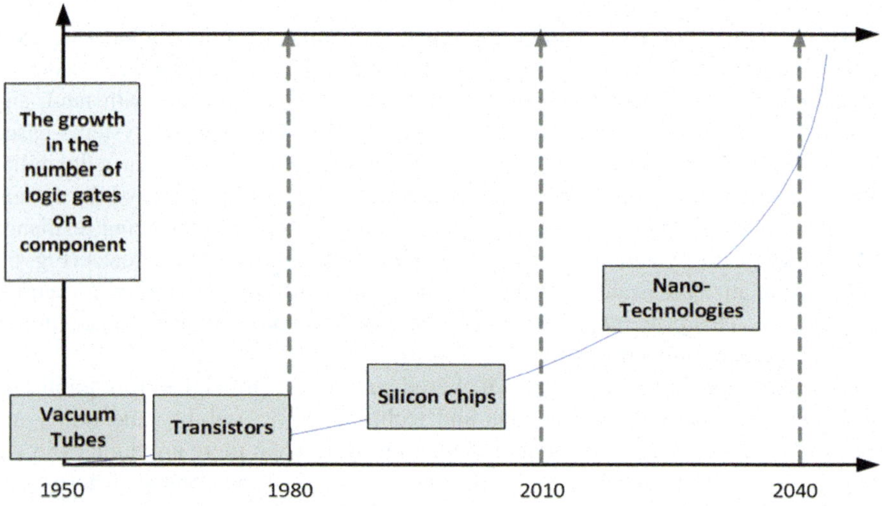

Fig. 7 A long macrotrend [53]

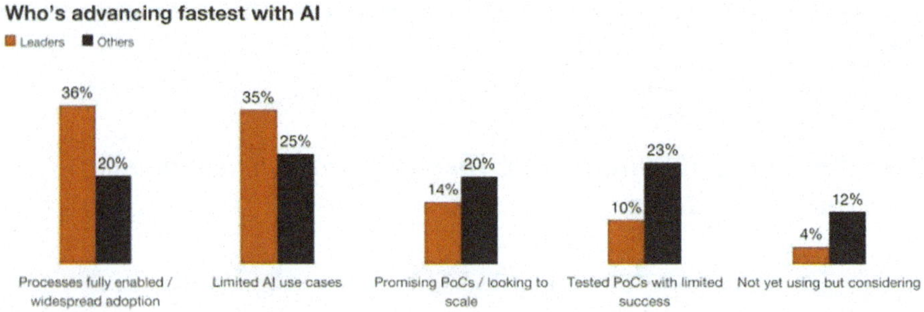

Fig. 8 Q. To what extent is your company looking to integrate AI technologies into its operations? [54]

is based on nature and life using the processes of information and computation that the living systems are carrying on continuously [56]. LT properties consist of adaption, learning, evolution, robustness, autonomous, self-repairing, and self-reproduction. LT is one of 14 themes in artificial life (origins of life, autonomy, self-organisation, adaptation (including evolution, development, and learning), ecology, artificial societies, behaviour, computational biology, artificial chemistries, information, living technology, art, and philosophy). It follows holism and integrates as an interdisciplinary, artificial life.

In Table 1, LT contributions include primary LT consisting of non-living elements, whilst secondary LT comprises living aspects in the components [57], such as governments, societies, and economies [58]. The complexities and dependencies of many cyber-physical systems require adaptability through self-organisation for dynamic systems compared with limited predictability of static systems.

An ongoing issue is how to effectively navigate data and theory-driven know-how [43]. It requires a rethink of the relationship between data and theory, particularly when 'studying algorithmically shaped social phenomena'. Measurement is not static, but now caters for dynamic environments and ecosystems such as with LT mixed networks etc. (Table 1). Predictability capabilities associated with decision-making can influence the outcome, emphasizing the interwoven nature of algorithmically saturated environments and ecosystems [43].

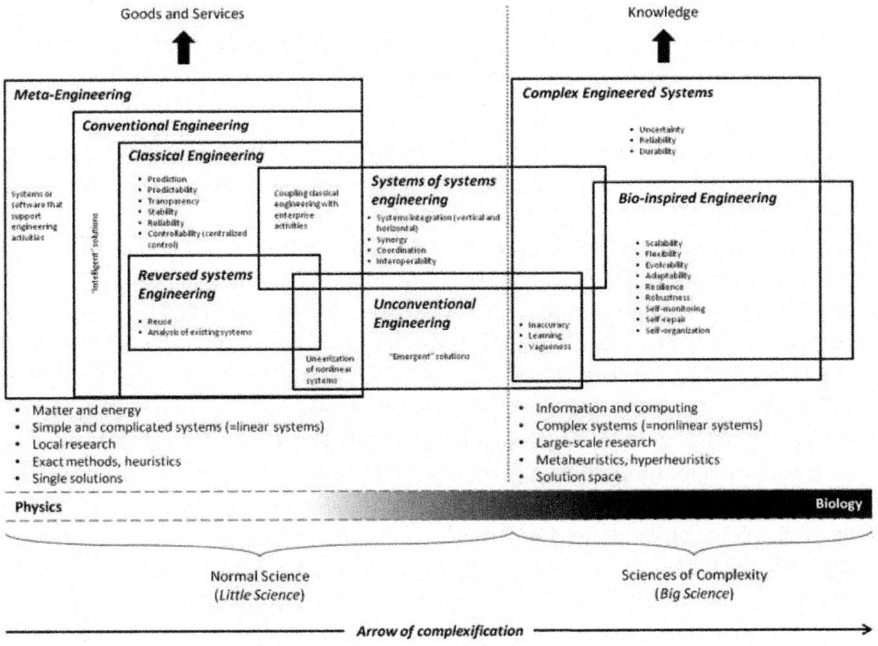

Fig. 9 Complexification of engineering sciences [55]

Table 1 Examples of different kinds of living technology, classified according to synthesis methods and materials [56]

Living technology		
Class	Primary (purely artificial)	Secondary (mixture of artificial and natural)
Wetware	Protocells, self-reproducing and evolving materials	Animal husbandry (horse, wool), yeast in foods (wine, beer, bread), therapeutic use of stem cells, cloning mammals, DNA computers and nanomachines, use of bacteria for drug production, living cells expressing artificial genomes
Hardware	Reproducing and evolving robots, evolving hardware	Electronically controlled insects, optical computation with bacteriorhodopsin, human prosthetic devices, cyborgs
Software	Evolving computational processes (e.g., models of chemical or economic networks)	Autonomous decision support systems composed of evolving (learning) algorithms with human input
Mixed networks of the above	(Subject of current research and development)	Human social networks on Web, human organisations, cities, human economies

All depend on human effort and ingenuity

2.5 *Evolution of the Security Triad in Cyber-Physical Environments and Ecosystems*

With the evolution of emerging technologies like cloud computing, Internet of Things (IoT), and AI, the resilience and security landscape is becoming more complex e.g. Riley's 2023 model [12] (Fig. 10.).

The security triad, initially comprising confidentiality, integrity, and availability (CIA) [59, 60] has evolved since 1980, being included within data regulations, such as the EU's GDPR and potentially a new approach is required [61]. As since 25th May 2018, 'the GDPR Enforcement Tracker lists more than 1600 individual breaches of the GDPR' [62] and the UK Information Commissioners Office (ICO) state that 88% human mistakes accounts for a substantial number of data leak instances [63]. The model that needs to be developed is for all levels and parts of organisations and cyber professionals to counter a growing overwhelming attack/disruption surface, increasing vulnerabilities and sophisticated threats. Several new considerations have been proposed to broaden the CIA triad to improve resilience and security [64].

In 1998, the Parkerian hexad added the atomic attribute of authenticity, possession (ownership), and utility [65]. This expansion addresses broader aspects of information security. The 3 extra attributes provide a more all-inclusive and adaptable approach to resilience and security for the increasing cyber-terrain and ecosystems complexity. "Security by design" in software development suggested the 3 DIE triad

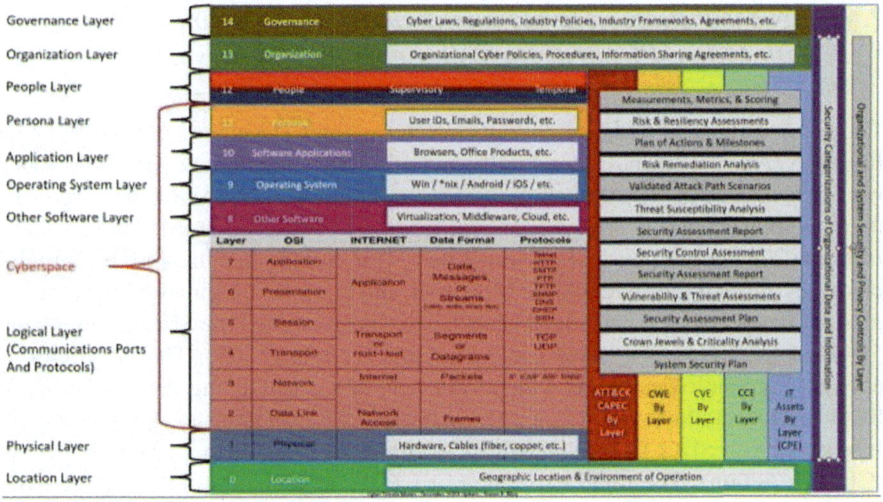

Fig. 10 Updated Riley [7] cyber-terrain model [12]

attributes (distributed, immutable and ephemeral) [66]. Others proposed for information technologies (IT) and operational technologies (OT) include safety, reliability and productivity [37]. This provides for more holistic and sustainable resilience and security including cybersecurity, physical security, and cyber-physical security. The governance model must be the same spanning across IT and OT domains, and sustainability is defined in terms of risks, costs, and processes. This has further extended the attributes to a 'Nonad' model to include safety, resilience, and interoperability [67].

Building on Riley's model [7, 12], S2AS (safety, sustainability, assurance, and security) risk management propose a more comprehensive approach to CPEE. The inclusion of safety and resilience ensures systems can withstand and recover from attacks, while interoperability emphasizes seamless integration across CPEE. Balancing security with sustainability requires a holistic approach. Organisations can integrate security measures into their sustainability initiatives by: adopting green technologies: implementing energy-efficient and eco-friendly security systems reduces environmental impact [68]; risk assessment: conducting thorough risk assessments to identify potential security threats and sustainability challenges [69], cross-functional collaboration: encouraging collaboration between security and sustainability teams to align goals and strategies [70]; continuous monitoring: using IoT and AI for real-time monitoring to ensure both security and sustainability standards are met [69, 70]; and regulatory compliance: maintaining regulations that promote both security and environmental sustainability [69–71].

2.6 UN Sustainable Development Goals Association with Cyber-Physical Environments and Ecosystems

The proposed holistic CPEE aligns with UN Sustainable Development Goals (SDGs), linking S2AS risk attributes to identify and mitigate potential risks in CPEE. SDGs were agreed in 2015 [72], the new industrial revolution, coined by the German government as Industry 4.0 [73], has emerged, leading to Society 5.0 [6]. This evolution from the 2009 cyber-physical environment [74] to Society 5.0 involves overcoming obstacles, termed the 'five walls': Ministries and Agencies, legal system, technologies, human resources, and social acceptance. Japan aligned the SDGs with Society 5.0, establishing an integrated socio-economic system [6, 75]. The integration of emerging technologies aligns with several UN SDGs, including but not limited to SDG 8 (Decent Work and Economic Growth), SDG 9 (Industry, Innovation, and Infrastructure), SDG 12 (Responsible Consumption and Production), and SDG 13 (Climate Action). The rapid expansion of CPEE necessitates addressing potential threats to ensure sustainability, security, and resilience, including adopting AI technologies across multiple economic sectors [76] and for human factors e.g., developing inclusive and sustainable employment opportunities, promoting technological innovation, and advancing responsible consumption practices [3].

3 Model Development Using the Case Study Approach

For the metamodel approach requirements for modelling [77, 78], then a model reflects reality, abstracting key features to study specific questions, while acknowledging its iterative nature and inherent simplifications [79]. Langefors' infological information equation, defined in 1995

$$I = I(SDt)$$

(environmental system's pre-knowledge (S), the data (D) that is decoded by the user, over time (t)), was used in devising the proposed model [20]. A key aspect emphasizes the user's pre-knowledge and experience, making information subjective. Langefors' equation highlights the user collaboration in the proposed model. It introduces (A) intervention level evolved and devised from real-world context (s); (B) object level abstracting for modelling for the proposed model; and (C) meta level outlining the overall capability to analyse overall environments/ecosystems and hierarchical relationships (Fig. 11) [78]. The metamodeling approach highlights the modelling processes and design supportive systems such as for the proposed model.

Developing from metamodeling to reality relationships for the proposed model map to four phases: (1) Identification (user-focused purpose); (2) Specification (requirements like data quality and adaptability); (3) Design (iterative, flexible

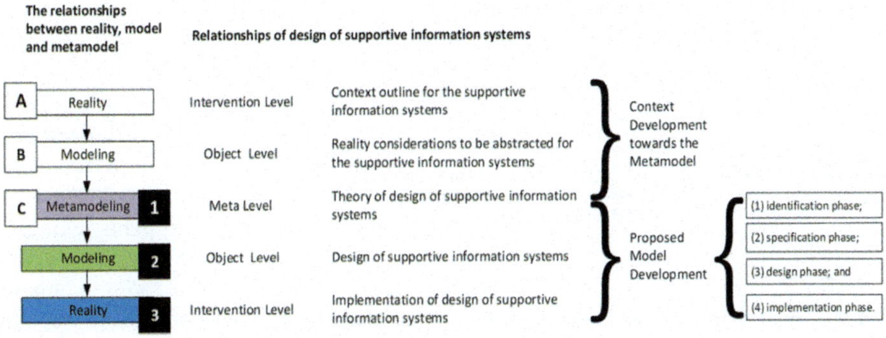

Fig. 11 Adaptation of the metamodel approach [20]

process); and (4) Implementation (adaptable to changing conditions). Each phase ensures the system meets user needs and remains relevant.

A metamodel approach can enable safety and security considerations to be embedded from the beginning, supporting the alignment of diverse methodologies and standards [20]. By promoting a holistic view of system development, this approach ensures that emerging technologies, combined with human factors, are not only innovative but also safe and secure [80]. In Fig. 11, illustrates the use of a proposed unified framework for systematically representing complex systems and their interdependencies, supporting sustainability by assessing environmental impacts and resource efficiency throughout the development lifecycle. Additionally, it ensures consistent documentation and traceability, critical for regulatory compliance, quality control, and early risk identification and mitigation.

3.1 *Iterative Model Development*

The proposed model used the metamodel approach (see Fig. 11) using an iterative approach linking the integrated insights from literature reviews, case studies, and action research through information triangulation. Drawing on abductive reasoning inspired by Riley's 2014 model [7] and corroborated by deductive reasoning through case studies, the model evolved through an iterative method [81]. In Fig. 12, deductive topological attributes [82] were employed, focusing on local/key subjects to refine the model. This theory-building approach combined exploratory and evaluative stages, encompassing single snapshots and diachronic analyses, with parallel and nested case studies involving small and medium-sized organisations to further model development.

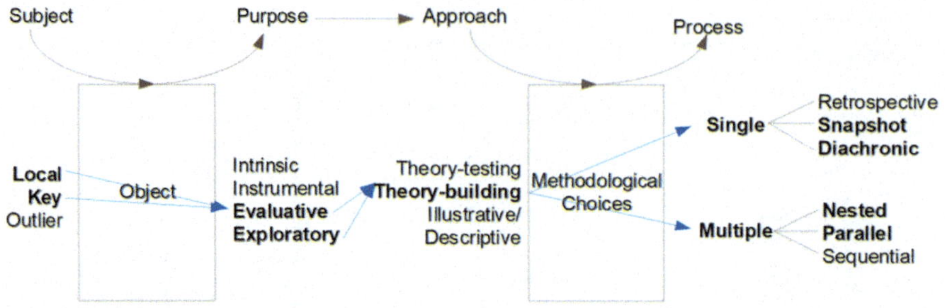

Fig. 12 Short-term outcomes based on small-sized organisations case studies for MM development on main research goal

3.2 Metamodel Development

In Fig. 13 in the paper shows the transition from real-world conditions to the construction of a metamodel, then the ability to evolve the metamodel development across four key phases: (1) Identification—focused on the user's purpose, sourced through steps [A]-[C], drawing on Riley's 2014 model [7] to provide a foundation for the integrated context in step [1] for the proposed model; (2) *Specification*—defining requirements such as data quality and adaptability, outlined in step [2], which begins with the 'Help to Grow' digital audit and includes human factors, algorithm and AI assurance; (3) *Design* —an iterative and flexible process, also developed as part of step [2], emphasizing adaptability in model creation; and (4) *Implementation*—step [3], applied in real-world contexts, ensures the model's adaptability to risk appetite calibration conditions, offering a comprehensive and integrated approach to metamodel development.

In Table 2, the Right-hand proposed model development Steps 1–4 from (Fig. 11) the metamodel development transformed to 3 layers to meet the GAECA Egg Pyramid model (Fig. 13):

3.3 Validation and Verification of the Proposed Model's Development

Using a case study approach (Fig. 12), the proposed model was applied to operational issues in the (a) Large Semiconductor Organisation; (b) Charity trying to digitize to obtain increased fundraising; (c) Charity to meet regulatory and compliance needs; (d) New digital centre within an UK HE institution; (e) Biomedical Startup Organisation; (f) Construction Company; and (g) Post Office's Fujitsu Horizon platform during the purchase process, aimed at saving a local UK sub-Post Office and community shop. The Validation & Verification (V and V) process used the Design Science Cycle (Fig. 1), and the relevant processes and outputs are as follows (Table 3).

Enhancing Cyberbullying Detection with RoBERTa: A Transformer-Based Approach

M. A. Hannan Bin Azhar and Aysha Akter Runa

Abstract This study investigates the effectiveness of transformer-based models, specifically RoBERTa, in detecting cyberbullying on social media platforms compared to traditional machine learning models such as Random Forest (RF) and Long Short-Term Memory (LSTM). Cyberbullying poses significant challenges due to the evolving nature of language and the anonymity provided by digital platforms. The research focuses on fine-tuning RoBERTa for cyberbullying detection and evaluates its performance using a comprehensive real-world dataset comprising approximately 48,000 manually annotated tweets, categorised into various forms of cyberbullying, including explicit and subtle abuses related to ethnicity, age, gender, and other characteristics. Results show that RoBERTa achieved the highest accuracy of 83.9%, outperforming LSTM (77.7%) and RF (79.2%). It excelled in the religion category (accuracy: 96.06%, precision: 96.48%) and ethnicity (accuracy: 98.60%, precision: 98.28%). RF led in the age category (accuracy: 98.31%, precision: 94.54%), with RoBERTa closely behind at 97.53% accuracy. LSTM performed lower, especially in the gender category, with an accuracy of 88.45% and precision of 72.12%. Although the results highlight the effectiveness of RoBERTa in recognising subtle forms of cyberbullying, it faced challenges in real-time applications due to slower inference times and higher computational costs. This research highlights the importance of contextual understanding in cyberbullying detection and the potential of transformer-based models to improve accuracy.

Keywords Cyberbullying · Transformer models · RoBERTa · Natural language processing · Machine learning · LSTM · Random forest · Social media · Deep learning

---

M. A. Hannan Bin Azhar (✉) · A. A. Runa
School of Engineering, Technology and Design, Canterbury Christ Church University, Canterbury, United Kingdom
e-mail: hannan.azhar@canterbury.ac.uk

A. A. Runa
e-mail: ar991@canterbury.ac.uk

H. Jahankhani and B. Issac (eds.), *Cybersecurity and Human Capabilities Through Symbiotic Artificial Intelligence*, Advanced Sciences and Technologies for Security Applications, https://doi.org/10.1007/978-3-031-82031-1_16

1 Introduction

Cyberbullying has emerged as a significant challenge in the digital landscape, primarily facilitated by the rise of social media platforms. Unlike traditional bullying, which occurs in physical spaces, cyberbullying allows perpetrators to harass, intimidate, or insult victims at any time and from any location. This pervasive issue involves the use of digital communication tools to deliberately inflict harm through hostile or abusive content. The anonymity provided by the internet, coupled with the vast reach of social media, exacerbates the problem. Victims often feel powerless as harmful actions can be witnessed and shared with a wide audience, amplifying their impact. A 2019 national survey of middle and high school students by the Cyberbullying Research Center [1] found that 37% of young people have experienced cyberbullying in their lifetime, with 15% admitting to having cyberbullied others, leading to long-term emotional and psychological effects. In severe cases, victims have resorted to self-harm or suicide [2].

With over 5 billion active social media users globally [3], the urgency for effective and scalable cyberbullying detection systems is increasingly evident. The prevalence of cyberbullying is particularly pronounced among younger demographics, as platforms like Twitter, Instagram, TikTok, and YouTube have become popular venues for such behavior due to their extensive user bases and minimal barriers to content posting Current systems prioritise moderating or removing it after the damage has already occurred [4]. The challenge of addressing this issue lies in the dynamic nature of language on these platforms. Slang, memes, and emojis frequently evolve, resulting in comments that may appear harmless but carry harmful undertones in specific contexts. This flexibility complicates the development of automated systems capable of accurately identifying instances of cyberbullying [4].

The primary objective of the research reported in this paper is to enhance cyberbullying detection using transformer-based models, specifically RoBERTa (Robustly Optimised BERT Approach) [5]. This paper aims to fine-tune RoBERTa for cyberbullying detection tasks and compare its performance with traditional machine learning models like Random Forest (RF) [6, 7] and Long Short-Term Memory (LSTM) [8]. Key research goals include developing an optimised preprocessing pipeline to manage the noisy and unstructured nature of social media text, evaluating the models with a real-world dataset from multiple platforms, and analysing their strengths and weaknesses in detecting various forms of cyberbullying, including both explicit and subtle abuses. Additionally, the research will identify challenges related to transformer models, such as the detection of emerging slang, code-switching, and multimedia content.

The remainder of the paper is organised as follows: Sect. 2 provides a comprehensive review of existing literature on cyberbullying detection methods, focusing on both traditional machine learning approaches and recent advancements in transformer-based models. Section 3 outlines the methodology, including details of the dataset, data preprocessing, and model implementation. Section 4 presents the results of the model evaluations and discusses the findings in detail, highlighting the strengths and limitations of each model. Finally, Sect. 5 concludes the paper with a summary of key insights and suggestions for future research.

2 Literature Review

The study of cyberbullying has emerged as a crucial topic within digital communication and internet safety. Researchers are examining its social, psychological, and behavioral aspects, along with developing technological solutions for detecting and reducing its detrimental effects. According to Watts et al. [9], cyberbullying is defined as intentional and repeated harm perpetrated through electronic communication, emphasising both the deliberate nature and persistent occurrence of such behavior. The anonymity offered by digital platforms often encourages individuals to engage in cyberbullying, as they perceive a lower likelihood of immediate repercussions. Studies indicate that victims face a heightened risk of mental health issues, including depression, anxiety, and diminished self-esteem [2]. In light of these findings, researchers are actively working on automated systems designed to identify and combat cyberbullying in real time [7].

2.1 Traditional Approaches

Early detection methods Lexicon-based systems rely on word lists, rule-based approaches apply predefined rules, and mixed-initiative methods combine human reasoning with these techniques. Pérez et al. [10] and Fahrnberger et al. [11] developed instant messaging-based cyberbullying prevention tools using lexicon-based approaches. Pérez et al.'s MISAAC system processes messages through a lexical analyser, assigns an aggressiveness level using a traffic light system, and applies user restrictions based on their color-coded status, which is recalculated after each message. Chen et al. [12] created a Lexical Syntactic Framework (LSF) that uses conversation history, writing style, and syntactic features to compute offensiveness scores for YouTube comments and users. While these systems were effective in identifying explicit bullying, they struggled with implicit cases, such as sarcasm or subtle threats [13]. For instance, a message containing the term "stupid" could be flagged as bullying, regardless of whether it was intended humorously or maliciously. Dinakar et al. [14] integrated common-sense reasoning into their cyberbullying detection system. They created a BullySpace Knowledgebase with over 200 assertions, comparing concepts in messages to detect indirect bullying, such as implicit homophobic remarks. This mixed-initiative approach combines human knowledge with automated detection to catch subtle bullying that traditional systems might miss. Nevertheless, traditional systems often lacked the adaptability required to keep pace with the evolving nature of internet slang, necessitating manual updates that were both time-consuming and error prone.

2.2 Machine Learning Approaches

To enhance detection capabilities, researchers have increasingly turned to machine learning models such as Support Vector Machines (SVM) and RF Naïve Bayes, Logistic Regression, Decision Tree and K-Nearest Neighbour (KNN) for cyberbullying detection tasks [7]. SVM has been widely used in cyberbullying detection studies due to its effectiveness in handling skewed text classification tasks [15]. Early research by Yin et al. [16] demonstrated that SVM, particularly when incorporating sentiment and contextual features, outperformed baseline models, and subsequent studies, like those by Dadvar et al. [17], further improved performance by adding user features.

Tree-based models have been widely used in cyberbullying detection, with Decision Trees achieving high accuracy, in studies like Balakrishnan et al. [18]. RF was favored for its ability to reduce overfitting, with Al-Garadi et al. [19] finding it superior to Naïve Bayes, SVM, and KNN. RF is an ensemble learning method that builds multiple decision trees during training, outputting the most frequent class for classification tasks. It is recognised for its robustness in managing complex feature relationships and mitigating overfitting. In the context of cyberbullying detection, RF's aggregation of results from various trees enhances prediction stability and accuracy.

The Naïve Bayes model is a simple probabilistic approach for classifying cyberbullying events, assuming feature independence, and has achieved approximately 68% accuracy when combined with sentiment analysis [20]. Logistic Regression has been found to outperform Naïve Bayes and SVM in detecting cyberbullying, especially under unbalanced datasets, with studies showing it to be the better model for various contexts and features, including images [21, 22].

Deep learning models consistently outperform traditional machine learning methods in cyberbullying detection, with approaches like Convolutional Neural Network (CNN), hybrid CNN-LSTM, and Bidirectional LSTM (BLSTM) showing superior results [23, 24]. The complexity of neural architecture varies among these models, with CNNs primarily utilised for image and text classification [25, 26], including sentiment analysis [27]. In contrast, LSTM networks excel in learning long-term dependencies due to their internal memory [28]. LSTM is a type of recurrent neural network (RNN), excels at processing sequential data. By incorporating memory cells, LSTM retains information over long sequences, making it adept at tasks where word order and context are vital, such as cyberbullying detection. Research conducted by [28] demonstrates improved performance by LSTM compared to previous machine learning models like SVM, Naïve Bayes, and decision trees. Additionally, BLSTMs enhance input information by encoding it in both forward and backward directions. Although these studies demonstrated machine learning's potential, they underscored the necessity for more advanced models to better capture the nuances of online language [7].

2.3 Transformer-Based Models

The advent of transformer-based models has revolutionised the field of Natural Language Processing (NLP), surpassing previous approaches in various language tasks, such as like topic modeling [29], sentiment analysis [30], recommendation systems [31], and harmful news detection [32]. Initially proposed by Vaswani et al. [33], transformers utilise self-attention mechanisms that allow for parallel text processing and the capture of long-range word dependencies. This marks a significant advancement over earlier models, such as RNNs and CNNs, which often struggled with lengthy sentences and intricate language structures.

One of the most notable transformer models is BERT (Bidirectional Encoder Representations from Transformers) [34], which gained widespread recognition for its ability to analyse text bidirectionally. By considering both the preceding and following words in a sentence, BERT, developed by Google in 2018, provides a deeper understanding of a word's context, improving the model's grasp of meaning and intent. BERT's architecture consists of 12 transformer layers with 768 hidden units and 12 self-attention heads. BERT can be fine-tuned for various NLP tasks, such as question answering and text classification, by adding just one output layer without altering the model's core structure. In cyberbullying detection, Paul and Saha [35] compared BERT with BLSTM, SVM, LR, CNN, and a hybrid of RNN and LSTM across datasets from Formspring [36], Twitter [37], and Wikipedia [38], showing BERT outperforming the other models.

Facebook AI Research (FAIR) addressed the limitations of Google's BERT by introducing RoBERTa in 2019, enhancing training methods to improve performance. Liu et al. [5] identified that BERT was undertrained and made key modifications, such as using dynamic masking, removing next sentence prediction, training on larger datasets with longer sentences, and larger batch sizes. RoBERTa has demonstrated state-of-the-art results in tasks like text classification, sentiment analysis, and question answering, making it particularly effective for applications such as cyberbullying detection. Results from a recent study [39] demonstrated that RoBERTa achieved the best overall performance, particularly in detecting the 'bully' class, outperforming both traditional machine learning and deep learning models like BLSTM with attention. While traditional models also performed well on the balanced dataset, RoBERTa consistently proved to be more effective. These findings confirm that pre-trained language models, such as RoBERTa, are highly competitive and reliable for cyberbullying detection.

3 Methodology

This paper adopts a comprehensive approach to comparing the performance of traditional machine learning models (RF and LSTM) with a transformer-based model (RoBERTa) in detecting cyberbullying. These models were chosen based on their proven effectiveness in handling various aspects of natural language processing and text classification tasks. Each model was implemented using Python and popular deep learning frameworks, including TensorFlow and PyTorch [40]. The methodology involves multiple stages, including data preprocessing, model selection, training, hyperparameter tuning, and evaluation. A key focus of the methodology is ensuring that the models are evaluated on a real-world, cross-platform dataset to enhance the generalisability of the findings.

3.1 Cyberbullying Dataset

Previous studies on cyberbullying detection utilised datasets collected from a variety of social media platforms, including Twitter, Instagram, FormSpring, and YouTube [41]. Numerous datasets have been generated through web scraping these platforms. While most datasets primarily rely on Twitter, with many tweets gathered using the open Twitter API and filtered through offensive hashtags, a significant concern is the class imbalance present in these datasets. The dataset used in this paper was prepared by Wang et al. [42] from six existing studies. The dataset was manually annotated into detailed cyberbullying target classes, including victim's age, ethnicity, gender, religion, other cyberbullying qualities, and non-cyberbullying. To address data imbalance, they augmented the dataset through a semi-supervised approach, resulting in a balanced collection of approximately 48,000 tweets, with about 8,000 tweets randomly sampled from each class, which is publicly accessible [43]. The dataset contains offensive language, including racially charged slurs and expressions of hatred toward political figures. It features harmful gender-related remarks aimed at women and the LGBTQ + community, as well as derogatory statements about various religious groups. Additionally, it highlights age-related bullying experiences, primarily in school settings, reflecting the lasting emotional impact of harassment.

3.2 Data Preprocessing

Given the noisy and unstructured nature of social media text, extensive preprocessing was required to prepare the data for model training. For the LSTM model, extensive preprocessing was applied to prepare the textual data for sequential input. The preprocessing steps included cleaning the text by removing URLs, emojis, mentions, hashtags, and non-alphabetic characters, followed by converting the text to lowercase. Afterward, "stopwords" were removed to eliminate common words, and stemming was performed using the Porter Stemmer to reduce words to their root forms. Finally, tokenisation was used to break down the text into individual words, and the sequences were padded to a maximum length of 42 tokens, based on the mean token length of 19.2 and a standard deviation of 11.4. This ensured that the LSTM model received uniform-length input for its training process.

In the case of the RF model, a simpler preprocessing approach was adopted. Basic text cleaning was performed, similar to LSTM, by removing URLs, emojis, mentions, hashtags, and non-alphabetic characters while converting the text to lowercase. However, "stopwords" were not removed, as RF can benefit from these common words. The key transformation for this model was applying TF-IDF (Term Frequency-Inverse Document Frequency) vectorization [42], which converted the cleaned text into numerical vectors. The vectoriser used a maximum of 5000 features ("max_features" = 5000), ensuring that the most important words were included based on their relevance in the corpus.

For the RoBERTa model, minimal preprocessing was needed due to its advanced pretrained architecture and built-in tokenisation capabilities. The primary text-cleaning tasks involved removing URLs, mentions, and hashtags, but no "stopword" removal or stemming was necessary. RoBERTa's tokeniser was used to convert the cleaned text into tokenised input IDs and attention masks. Each tokenised input had a maximum length of 128 tokens, and these inputs were directly fed into the model. This minimal preprocessing allowed RoBERTa to handle the complexity of language and utilise its pretrained knowledge for text classification in the cyberbullying detection task, ensuring it could process and interpret even subtle language patterns from the dataset.

3.3 Hyperparameter

For LSTM, the key hyperparameters included an embedding dimension of 100, which controlled the size of the word embeddings generated for each token in the input sequence. The model used a single LSTM layer with 100 units to capture the sequential dependencies in the text data. Additionally, the "max_length" of each sequence was set to 42 tokens, based on statistical analysis of the data. The optimiser used for training was Adam (Adaptive Moment Estimation) [44], and the loss function was "sparse_categorical_crossentropy", which is suitable for multiclass classification tasks. The model was trained over 10 epochs with a batch size of 32. Adam is an optimisation algorithm commonly used for training deep learning models. It combines the advantages of two other methods: Stochastic Gradient Descent (SGD) with momentum and RMSProp [44]. Adam keeps track of both the moving average of the gradient (momentum) and the squared gradient (adaptive learning rate), allowing it to adapt the learning rate for each parameter dynamically. This makes Adam well-suited for handling sparse gradients and noisy data.

For the RF model, the main hyperparameter was the number of trees ("n_estimators"), set to 100, meaning the model constructed an ensemble of 100 decision trees to enhance prediction accuracy and reduce overfitting. TF-IDF vectoriser limited to top 5000 words for feature extraction. The model employed the default settings for other parameters, such as "max_depth", which allows trees to grow until all leaves are pure or contain fewer than the minimum samples required to split. The minimum number of samples required to be at a leaf node was also not explicitly defined, thus using the default value of 1.

For the RoBERTa model, the key hyperparameters included the use of the pretrained roberta-base architecture, which had a hidden size of 768. The model's classifier consisted of a fully connected layer with 50 hidden units and a dropout rate of 0.2 to prevent overfitting. The learning rate for fine-tuning RoBERTa was set to 2e-5, and the AdamW optimiser with eps = 1e-8 was used. AdamW is a variant of the Adam optimiser that incorporates weight decay directly into the optimisation algorithm, rather than applying it as a separate step [44]. This adjustment helps mitigate overfitting by encouraging smaller weights during training. The learning rate scheduler was set up with a linear warmup, and the model was trained for 4 epochs, which was determined to be sufficient for convergence without overfitting.

3.4 Evaluation Metrics

The performance of the models was evaluated using various metrics commonly applied in multi-class classification tasks, including accuracy, precision, recall, F1-score, and the confusion matrix. Accuracy represents the proportion of correctly classified instances relative to the total number of instances, serving as a general metric for assessing model performance. Precision measures the proportion of positive predictions that are correct, which is particularly important in scenarios where minimising false positives is crucial. Recall, also known as True Positive Rate (TPR), evaluates the model's ability to correctly identify actual positive instances, making it essential when the focus is on reducing false negatives. Additionally, the F1-score combines precision and recall, providing a balanced measure of performance that is especially advantageous in handling imbalanced datasets. The confusion matrix offers a comprehensive view of the model's performance by detailing true positives, true negatives, false positives, and false negatives for each class, allowing for a deeper understanding of where the model is succeeding or struggling. Furthermore, false positive rate and false negative rate are critical for understanding the model's performance in specific contexts, helping to identify areas where improvements can be made. The model was validated using an 80/20 train-test split, ensuring it was trained on 80% of the data and evaluated on 20% of unseen data. This method provided a robust way to assess generalisation performance.

4 Results and Discussions

The performance of the three models—RF, LSTM, and RoBERTa—was evaluated on the dataset, with results summarised in Tables 1 and 2. Their effectiveness in classifying cyberbullying tweets reveals varying capabilities and limitations in this complex text classification task. RF achieved an overall accuracy of 79.2%. It excelled in specific categories, such as "religion" with a precision of 94.98% and "age" with a precision of 94.54%. However, it struggled with the "not_cyberbullying" class, achieving a precision of 49.47%, and faced challenges in classifying the "other_cyberbullying" class with a precision of 51.57%. This indicates that while RF effectively captures well-defined classes, it encounters difficulties with less distinctive language cues, leading to misclassifications in overlapping categories.

LSTM, which incorporates advanced sequential processing, achieved an overall accuracy of 77.7%. It mirrored some strengths and weaknesses of RF, particularly in recognising "age" and "ethnicity" with precisions of 94.06% and 98.31%, respectively. However, it faced similar issues with the "not_cyberbullying" class (precision: 50.77%) and "other_cyberbullying" (precision: 51.54%). The LSTM model's ability to capture context and sequential dependencies in text is notable, yet it still struggles in categories where features are less distinct, or language is ambiguous.

RoBERTa, a transformer-based model fine-tuned for language understanding, outperformed both previous models with an impressive accuracy of 83.9%. It excelled in classifying nuanced categories such as "ethnicity" (precision: 98.28%) and "age" (precision: 97.65%), indicating its ability to understand contextual subtleties crucial for accurate classification. RoBERTa also improved the identification of the "not_cyberbullying" class (precision: 64.67%) and "other_cyberbullying" (precision:

Table 1 Overall performance of different models

Model	Accuracy (%)	Precision	Recall	F1-Score
LSTM	77.7	0.767	0.772	0.767
RF	79.2	0.797	0.792	0.794
RoBERTs	83.9	0.836	0.836	0.834

Table 2 Comparison of performances on different types of cyberbullying

Cyberbullying type	Model	Accuracy (%)	Precision	Recall
Religion	LSTM	93.17	0.932	0.931
	RF	94.50	0.949	0.944
	RoBERTa	96.06	0.964	0.960
Age	LSTM	96.73	0.940	0.967
	RF	98.31	0.945	0.983
	RoBERTa	97.53	0.976	0.975
Gender	LSTM	88.45	0.721	0.884
	RF	79.71	0.892	0.797
	RoBERTa	87.01	0.852	0.870
Ethnicity	LSTM	92.67	0.983	0.926
	RF	98.09	0.983	0.981
	RoBERTa	98.60	0.982	0.986

59.16%), showcasing its effectiveness in handling complex text structures. This superior performance can be attributed to its robust architecture, which captures long-range dependencies and subtle meanings in text data, allowing it to perform well across all categories.

Confusion matrices were generated (Fig. 1) for each model to further analyse their performance. The LSTM model demonstrated strong TPRs of 0.97 for "age" and 0.93 for "ethnicity", indicating good effectiveness in identifying these categories. However, the TPR for "not_cyberbullying" was lower at 0.46, reflecting challenges in detecting this category. Similarly, LSTM struggled with "other_cyberbullying", achieving only a 0.46 TPR, suggesting areas for improvement. Detection of "gender" was relatively better, with a TPR of 0.89.

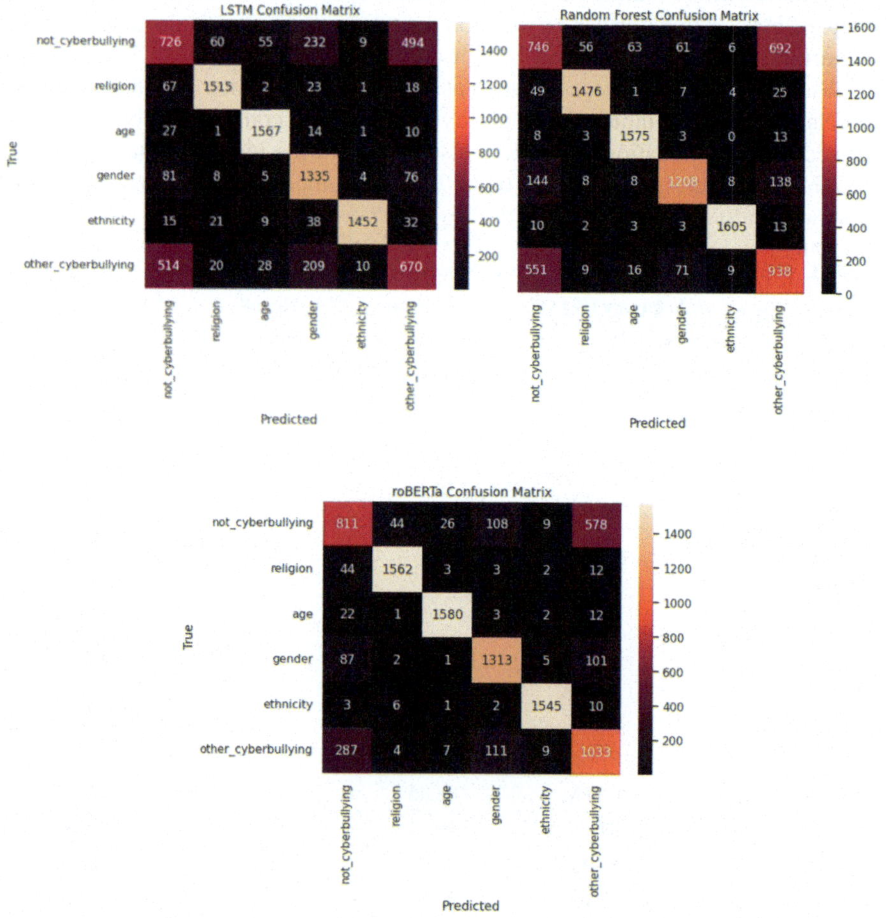

Fig. 1 Confusion matrix of LSTM (top-left), RF (top-right), RoBERTa (bottom)

In contrast, the RF model performed consistently across several categories, achieving a TPR of 0.98 for both "age" and "ethnicity." However, it also faced challenges in detecting "not_cyberbullying", with a TPR of 0.46, similar to LSTM. For "gender", RF achieved a TPR of 0.80, reflecting moderate success, while it attained a TPR of 0.59 for "other_cyberbullying", which was better than LSTM but still indicates room for improvement.

The RoBERTa model exhibited the most consistent performance across categories, achieving a TPR of 0.52 for "not_cyberbullying", 0.96 for "religion", and 0.98 for "ethnicity". RoBERTa also maintained high performance in detecting "age" with a TPR of 0.98, and for "other_cyberbullying", it achieved a TPR of 0.71, outperforming both LSTM and RF in this category.

Overall, while all models excelled in detecting "age" and "ethnicity", RoBERTa stood out for its balance between precision and consistency. However, the results suggest that all models, including RoBERTa, could benefit from further refinement to improve detection in categories like "not_cyberbullying" and "other_cyberbullying".

While RoBERTa outperformed the other models in terms of accuracy, it was also the most computationally intensive. RoBERTa required approximately 3.5 h to train, compared to 0.1 h for RF and 0.23 h for LSTM. In terms of inference time, RoBERTa took 200 ms per sample, while RF and LSTM were significantly faster at 10 ms and 25 ms per sample, respectively. The trade-off between accuracy and computational efficiency is an important consideration for real-world applications. In scenarios where real-time inference is required, such as on social media platforms where millions of comments are generated every minute, the slower inference time of RoBERTa may be a limitation. On the other hand, if accuracy is the primary concern and computational resources are available, RoBERTa may be the better choice.

The RF model achieved an accuracy of 79.2%, primarily due to limitations in feature representation and the inherent complexities of the cyberbullying text data. This model uses TF-IDF vectorisation for feature extraction, which, while effective, may not fully capture the semantic relationships and contextual nuances present in the text. Additionally, the dataset may contain noise and overlapping categories, which complicate classification tasks. To enhance performance, it is recommended to implement feature engineering techniques, such as creating new features that capture sentiment or context, and to adopt more advanced text representation methods like Word2Vec, GloVe, or BERT embeddings [42], which could lead to better outcomes.

Similarly, the LSTM model also achieved an accuracy of 77.7%. Its performance was likely affected by various factors, including data preprocessing, model architecture, and training dynamics. The preprocessing steps—such as removing URLs, emojis, and "stopwords", along with stemming—while simplifying the dataset, may have resulted in the loss of important contextual information. The use of a single LSTM layer with 100 units seems too simplistic for capturing the complexities of the data, and training for only 10 epochs might not provide adequate learning opportunities. To improve LSTM performance, strategies like data augmentation—through techniques such as synonym replacement or back-translation—could introduce additional training examples. Furthermore, employing advanced preprocessing methods like lemmatisation and utilising context-sensitive embeddings could enhance semantic understanding. Exploring deeper LSTM architectures, incorporating dropout layers for regularisation, and fine-tuning hyperparameters using techniques like Grid Search [45] could also improve model convergence and accuracy.

In contrast, the RoBERTa model achieved a higher accuracy of 83.9%, but its performance was still hindered by specific implementation choices and the characteristics of the dataset. While RoBERTa has a robust architecture, setting the parameter "freeze_bert" to False, which allows for weight fine-tuning during training, may lead to overfitting, especially given the limited number of training epochs and a relatively small dataset. The choice of a standard learning rate of 2e-5 may not have been optimal for this task, potentially resulting in inadequate convergence. Additionally, the simplicity of the classification layer—comprising a linear layer followed by ReLU activation and dropout—might not sufficiently capture the details in the data.

5 Conclusions

The evaluation of RF, LSTM, and RoBERTa models for cyberbullying detection highlighted varying strengths across different categories. RoBERTa emerged as the most accurate, with an overall accuracy of 83.9%, particularly excelling in identifying subtle forms of cyberbullying such as those related to ethnicity and age. It outperformed RF, which achieved an accuracy of 79.2%, and LSTM, with an accuracy of 77.7%, in handling context-dependent language and complex text structures. However, RoBERTa's performance was hindered by slower inference times and higher computational demands, making it less suitable for real-time applications. Both LSTM and RF demonstrated faster inference times but struggled with ambiguity in classes such as "not_cyberbullying" and "other_cyberbullying", which are more challenging to classify due to less distinct language features.

Despite these promising results, challenges remain. All models exhibited weaknesses in detecting gender-based cyberbullying, suggesting that further refinement is necessary in this area. Additionally, the heavy reliance on textual data limits the ability to detect cyberbullying that incorporates multimedia content, such as images and videos, which is becoming increasingly common on social media. Future research should focus on optimising model architectures, such as deeper LSTM layers and more efficient RoBERTa tuning, to improve performance in detecting subtle cyberbullying. Addressing class imbalance through data augmentation and exploring multimodal systems that analyse both text and multimedia content are key priorities. Additionally, improving real-time detection with model compression and enhancing dataset diversity for better generalisation will help build more scalable, accurate solutions. Hybrid models combining traditional machine learning with deep learning approaches also warrant exploration for more robust cyberbullying detection.

References

1. Patchin JW, Hinduja S (2020) 2019 Cyberbullying data. Cyberbullying research center. https://cyberbullying.org/2019-cyberbullying-data. Last accessed 23 Oct 2024
2. John A, Glendenning AC, Marchant A, Montgomery P, Stewart A, Wood S, Lloyd K (2018) Hawton K self-harm, suicidal behaviours, and cyberbullying in children and young people: systematic review. J Med Internet Res 20(4):e129. https://doi.org/10.2196/jmir.9044
3. Dixon S (2023) Number of social media users worldwide from 2017 to 2028. statista., Available at: https://www.statista.com/statistics/278414/number-of-worldwide-social-network-users/. Last accessed 23 Oct 2024
4. Gongane VU, Munot MV, Anuse AD (2022) Detection and moderation of detrimental content on social media platforms: current status and future directions. Soc Netw Anal Min 12(1):129. https://doi.org/10.1007/s13278-022-00951-3
5. Liu Y, Ott M, Goyal N, Du J, Joshi M, Chen D, Levy O, Lewis M, Zettlemoyer L, Stoyanov V (2019) RoBERTa: a robustly optimized bert pretraining approach. ArXiv, abs/1907.11692
6. Al-Garadi MA, Varathan KD, Ravana SD (2016) Cybercrime detection in online communications: The experimental case of cyberbullying detection in the Twitter network. Comput Hum Behav 63:433–443
7. Teng TH, Varathan KD, Crestani F (2023) A comprehensive review of cyberbullying-related content classification in online social media. Expert Syst Appl 244:122644
8. Eronen J, Ptaszynski M, Masui F (2022) Comparing performance of different linguistically-backed word embeddings for cyberbullying detection- arXiv preprint arXiv:2206.01950
9. Watts LK, Wagner J, Velasquez BJ, Behrens PI (2017) Cyberbullying in higher education: A literature review. Comput Hum Behav 69:268–274
10. Pérez PJC, Valdez CJL, Ortiz MDGC, Barrera JPS, Pérez PF (2024) MISAAC: instant messaging tool for cyberbullying detection [online]. Available from http://worldcomp-proceedings.com/proc/p2012/ICA7994.pdf. Last accessed 23 Oct 2024
11. Fahrnberger G, Nayak D, Martha VS, Ramaswamy S (2014) SafeChat: a tool to shield children's communication from explicit messages. IN: international conference on innovations for community services (I4CS). 14th. Reims. 4–6, 2014, 80 – 86
12. Chen Y, Zhou Y, Zhu S, Xu H (2012) Detecting offensive language in social media to protect adolescent online safety, international conference on privacy, security, risk and trust (PASSAT) and social computing (SocialCom). Amsterdam, 3–5:2012
13. Miljana Mladenović, Vera Ošmjanski, and Staša Vujičić Stanković. 2021. Cyber-aggression, Cyberbullying, and Cyber-grooming: A Survey and Research Challenges. ACM Comput. Surv. 54, 1, Article 1 (2022), 42 pages. https://doi.org/10.1145/3424246
14. Dinakar K, Jones B, Lieberman H, Picard R, Rose C, Thoman M, Reichart R (2012b) You too?! Mixed-initiative LDA story matching to help teens in distress, International AAAI Conference on Weblogs and Social Media (ICWSM 2012). 6th. Dublin. 4–7, 2012, 74–81
15. Van Hee C, Jacobs G, Emmery C, Desmet B, Lefever E, Verhoeven B et al (2018) Automatic detection of cyberbullying in social media text. PLoS ONE 13(10):e0203794. https://doi.org/10.1371/journal.pone.0203794
16. Yin D, Xue Z, Hong L, Davison BD, Kontostathis A, Edwards L (2009) Detection of harassment on web 2.0. Proceedings of the content analysis in the WEB, 2(0), Madrid, Spain, 1–7
17. Dadvar M, Trieschnigg D, Ordelman R, de Jong, F. (2013). Improving cyberbullying detection with user context. In Proceedings of the 35th European conference on Advances in Information Retrieval (ECIR'13). Springer-Verlag, Berlin, Heidelberg, 693–696. https://doi.org/10.1007/978-3-642-36973-5_62
18. Balakrishnan V. Khan S, Arabnia H.R (2020) Improving cyberbullying detection using Twitter users' psychological features and machine learning computers and Security, 90, Article 101710. https://doi.org/10.1016/j.cose.2019.101710
19. Al-Garadi MA, Varathan KD, Ravana SD (2016) Cybercrime detection in online communications: the experimental case of cyberbullying detection in the Twitter network computers in human behavior, 63:433–443. https://doi.org/10.1016/j.chb.2016.05.051

20. Sanchez H, Kumar S (2011) Twitter bullying detection. ser. NSDI 12(2011):15
21. Mangaonkar A, Hayrapetian A, Raje R (2015) Collaborative detection of cyberbullying behavior in Twitter data. 2015 IEEE International Conference on Electro/Information Technology (EIT), Dekalb, IL, USA, 611–616. https://doi.org/10.1109/EIT.2015.7293405
22. Hosseinmardi H, Rafiq RI, Han R, Lv Q, Mishra,S (2016) Prediction of cyberbullying incidents in a media-based social network. 2016 IEEE/ACM International Conference on Advances in social networks analysis and mining (ASONAM), San Francisco, CA, USA, 186–192, https://doi.org/10.1109/ASONAM.2016.7752233
23. Hani J, Nashaat M, Ahmed M, Emad Z, Amer E, Mohammed A (2019) Social media cyberbullying detection using machine learning International J. Adv. Comput. Sci. Appl. 10(5):703–707. https://doi.org/10.14569/IJACSA.2019.0100587
24. Agrawal S, Awekar A (2018) Deep learning for detecting cyberbullying across multiple social media platforms. In Pasi G, Piwowarski B, Azzopardi L, Hanbury A. Advances in information retrieval European conference in information retrieval (ECIR) 2018
25. Simonyan K, Zisserman A (2015) Very deep convolutional networks for large-scale image recognition. Int. Conf. Learn. Represent. 1–14
26. Chen T, Xu R, He Y, Wang X (2017) Improving sentiment analysis via sentence type classification using BiLSTM-CRF and CNN. Expert Syst Appl 72:221–230
27. Kim Y (2014) Convolutional neural networks for sentence classification. Proceedings of the 2014 Conference on empirical methods in natural language processing (EMNLP). Doha, 1746–1751. arXiv: 1408.5882. https://doi.org/10.3115/v1/D14-1181
28. Dass A, Daniel DK (2022) Cyberbullying detection on social networks using LSTM Model, 2022 International Conference on Innovations in Science and Technology for Sustainable Development (ICISTSD), Kollam, India, pp 293–296. https://doi.org/10.1109/ICISTSD55159.2022.10010559
29. Ogunleye B, Maswera T, Hirsch L, Gaudoin J, Brunsdon T (2023) Comparison of topic modelling approaches in the banking context. Appl. Sci. 13:797
30. Zhao A, Yu Y (2021) Knowledge-enabled BERT for aspect-based sentiment analysis. Knowl.-Based Syst. 227:107220
31. Yang N, Jo J, Jeon M, Kim W, Kang J (2022) Semantic and explainable research-related recommendation system based on semi-supervised methodology using BERT and LDA models. Expert Syst Appl 190:116209
32. Lin SY, Kung YC, Leu FY (2022) Predictive intelligence in harmful news identification by BERT-based ensemble learning model with text sentiment analysis. Inf Process Manag 59:102872
33. Vaswani A, Shazeer N, Parmar N, Uszkoreit J, Jones L, Gomez AN, Kaiser Ł, Polosukhin I (2017) Attention is all you need. Adv Neural Inf Process Syst 30:6000–6010
34. Devlin J, Chang MW, Lee K, Toutanova, K (2018) Bert: pre-training of deep bidirectional transformers for language understanding. arXiv, arXiv:1810.04805
35. Paul S, Saha S (2022) CyberBERT: BERT for cyberbullying identification: BERT for cyberbullying identification. Multimed Syst 28:1897–1904
36. Reynolds K, Kontostathis A, Edwards L (2011) Using machine learning to detect cyberbullying. In Proceedings of the 10th International Conference on machine learning and applications and workshops, Honolulu, HI, USA, 18–21 2011; IEEE: Piscataway, NJ, USA, 2:241–244
37. Waseem Z, Hovy D (2016) Hateful symbols or hateful people? Predictive features for hate speech detection on twitter. In Proceedings of the NAACL Student Research Workshop, San Diego, CA, USA, 13–15, pp 88–93
38. Wulczyn E, Thain, N, Dixon L (2017) Ex machina: Personal attacks seen at scale. In Proceedings of the 26th International Conference on World Wide Web, Perth, Australia, 3–7:1391–1399
39. Ogunleye B, Dharmaraj B (2023) The use of a large language model for cyberbullying detection. Analytics. 2(3):694–707. https://doi.org/10.3390/analytics2030038
40. Facebook AI, RoBerta, https://pytorch.org/hub/pytorch_fairseq_roberta/, Last accessed 23 Oct 2024

41. Ejaz N, Razi F, Choudhury S (2024) Towards comprehensive cyberbullying detection: a dataset incorporating aggressive texts, repetition, peerness, and intent to harm, computers in human behavior, vol. 153. https://doi.org/10.1016/j.chb.2023.108123
42. Wang J, Fu K, Lu CT (2020) Sosnet: a graph convolutional network approach to fine-grained cyberbullying detection. In Proceedings of the 2020 IEEE International Conference on Big Data (Big Data). 1699–1708
43. Cyberbullying Dataset, Available online: https://drive.google.com/drive/folders/1oB2fan6GVGG83Eog66Ad4wK2ZoOjwu3F, Last accessed 23 Oct 2024
44. Loshchilov I, Hutter F (2019) Decoupled weight decay regularization. International Conference on Learning Representations (ICLR 2019). https://doi.org/10.48550/arXiv.1711.05101
45. Marco R, Ahmad SSS, Ahmad S (2023) An improving long short term memory-grid search based deep learning neural network for software effort estimation. Int. J. Intell. Eng. Syst. 16:164–180. https://doi.org/10.22266/ijies2023.0831.14

Rethinking the Human Factor in Information Security. From an Information-Centric Perspective to One Based on Organizational Culture

Jeimy J. Cano M.

Abstract Modern organizations know that information becomes a strategic asset to develop opportunities and initiatives that allow them to position themselves in a specific sector, create competitive advantages and innovative experiences for their customers. In this sense, the treatment of information by the people of a company, in the scenario of its key processes and objectives, is essential to protect itself from adverse events that compromise its promise of value. Therefore, this paper proposes a review of the current perspective of information security of people based on compliance, awareness and punishment, based on know-how and associated with training and standard practices, to then propose a complementary view based on behaviors, conditions and consequences of unforeseen events, which allows to recognize biases, detect cognitive barriers and place information security in the culture of the organization, where individuals understand the risks, actively want to do the right thing and are provided with all the support to do so.

Keywords Information security · Information security awareness · Human factor · Organizational culture

1 Introduction

Security and cybersecurity challenges continue to advance as the digital surface enabled by increased connectivity and new digital initiatives available to organizations and nations increases. This trend is not going to change in the coming years and a special vigilant posture will have to be maintained in order to warn of new vulnerabilities and impacts that can be exploited by insider attackers or threat groups (usually external) either to realize their own motivations or particular agendas with their allies respectively [1].

Jeimy J. Cano M. (✉)
Universidad de los Andes, Law Faculty, Cra 1E No.18A-10, Bogotá, Colombia
e-mail: jcano@uniandes.edu.co

H. Jahankhani and B. Issac (eds.), *Cybersecurity and Human Capabilities Through Symbiotic Artificial Intelligence*, Advanced Sciences and Technologies for Security Applications, https://doi.org/10.1007/978-3-031-82031-1_17

Much of the literature available to date focuses on technical actions and geopolitical instability conditions that are exploited by threat groups that end up with cyber operations and relevant impacts on both organizations and companies [2]. However, the internal attacker or the challenges of the organizational culture of information security (OCIS), is little reviewed in depth and generally ends up located in training and awareness issues that to date are maintained in many organizations with limited success or often as a compliance requirement that must be met [3].

In this sense, multiple academic and industry reports confirm that the most important security breaches recorded are due to human behaviors [4, 5], and it is important to emphasize the word behaviors which require a different review and attention, and not just talk about training, tests or practices, when what is at stake is situational awareness and the elements of the environment [6], as well as its contents to motivate a concrete action against a risk scenario, which can be defined as a threat environment (an adversary), the vulnerabilities of the organization and the person, as well as the impacts that can be derived from the exploitation of the latter [3].

To date, when the subject of people's safety is reviewed, the answer ends up referring to training courses (classroom and virtual) that are generally carried out to meet a requirement, without having the expected appropriation effect. In this way, this particularly strategic process for today's companies has been distorted in the midst of the vortex of information and targeted messages (marketing, political and social) that end up influencing or confusing people regarding the protection of the companies' information assets [7].

In this sense, this document proposes a review of the current practices of people's security based on awareness, compliance and punishments (disciplinary and social), which have shown little significant progress in the transformation of people's behavior in the treatment of information [8], to introduce a complementary proposal based on behaviors, conditions and consequences, allowing an understanding of the actions of people in a specific context that enables the appropriation of the risks involved in the protection of the information assets of the companies.

This article begins with a conceptual review of the human factor in information security, followed by a review of the current perspective of people's information security, which is generally based on "know-how to do". It then presents the transformation of people's information security framed in the interconnected dynamics of organizations and individuals' behaviors, which privileges a culture of coherence and appropriation, to finally raise some conclusions that synthesize the reflections raised.

2 The Human Factor in Information Security. Some Conceptual Elements

Security has been necessary at the governmental level to protect national interests, ensure the safeguarding of citizens and preserve order. In the organizational context, protection has been based on safeguarding trade secrets, as well as key information about their operations and customers from events such as loss, disclosure or unauthorized use [9].

Although organizations in general recognize that information is a key element in the development of their activities, little attention or priority is given to its protection by their executives [10]. This reality reveals the need to redefine the companies' collective understanding of information security, in order to tune in to information protection as a minimum requirement that everyone in an organization must have when dealing with it.

It follows from the above that achieving a reliable environment for corporate information assets requires the permanent and sustainable application of security and control practices that become part of the organizational culture, i.e., part of the company's theory in use that influences the behavior of employees with respect to information protection [11].

The culture of information security in the context of the companies of the 20th century privileged the development of competences related to know-how (in the field of application and performance); therefore, it seeks to insist on the application of training courses oriented to the protection of information, and therefore emphasizes the validation of concrete actions to safeguard information. This culture seeks to make the established procedures repeatable and reliable and, therefore, the sanctions for those who, although aware of the procedures, do not comply with the competencies declared by the company with regard to the treatment of information.

Now, an information security culture based on the considerations of a 21st century organization requires knowledge of the context, of the current conditions of the company, in order to define not only the access strategies, but also the guidelines for its use, in order to establish the scenario where it will be deployed and the impacts that a possible inappropriate use may materialize. This implies not only individual behaviors and practices, but also an understanding of the context itself and the way in which its treatment may affect the company's interests.

Thus, different investigations insist that achieving a reliable information protection environment, adjusted to the dynamics of the organization, must place the human being as the center of the reflections and practices, in order to build from there a vigilant and resistant posture that not only responds to the regulatory or compliance requirements established for the company, but to the conditions and characteristics of the company's processes that allow individuals to understand and appropriate what they do to protect information assets, and how the defined controls are not limitations to "do", but enablers to achieve the company's objectives [12, 13].

3 Current Perspective on Individual's Information Security: Awareness, Compliance and Punishments

In particular, today's organizations, when talking about people's information security practices, generally situated as an operational issue, focus on "ACP": awareness, compliance and punishment, which privilege the know-how that demands a particular knowledge imparted by the organization and that must be accepted and executed in association with the company's current policies and procedures.

Awareness is understood as the exercise that makes the protection of information assets relevant and important in organizations. In this exercise, people recognize the information that is sensitive and associated with the company's promise of value, which allows increasing the level of importance and cognitive attention of the person with respect to this topic. Unfortunately, this valuable and necessary effort ends up relegated to a presentation with interesting slides that are important for the organization, not for the concrete and situated exercise of people in their work processes in the organization. At the end of the awareness session, everyone signs the attendance sheet and the requirement is fulfilled, which will be presented to the verification and control entities, and the organization has the evidence that it has complied with the mandate to train its employees on this topic.

Compliance is based on ensuring the use and due attention to the policies, standards and procedures applied to the protection of information, which respond to the key risks identified by the organization in the face of possible internal or external events that may compromise the security and control measures in place to prevent unauthorized access, leaks or deterioration of the integrity of the information. Notwithstanding the above, people usually end up complying with the rules and procedures for reasons such as ensuring a specific guideline, without knowing or knowing why they should do so or what risk is being covered. This situation leads to an application without awareness of what is to be secured and protected with regard to the company's strategic information assets.

When people are not consistent in their behavior with regard to the treatment of information expected by the organization, disciplinary punishments from the compliance (and administrative) areas and social sanctions, generally recriminations and criticisms for not passing the expected tests, are more likely to frustrate and discourage them than to inspire them to improve [14]. Punishment is associated with not repeating and applying the knowledge previously imparted about the risk for which they were trained, the error as a result, which usually does not take into account the particular context and conditions in which it occurs, nor the elements of human dynamics that may be present such as distractions, emotions and concerns that may affect the good performance of the person in the face of the adverse event.

In summary, we could say that the current security agenda for individuals is associated with the following elements:

- Focused on information
- The human factor is the challenge to overcome
- Based on awareness, compliance and punishments (social and disciplinary)

- Ensure training, testing and practice.

This agenda of many organizations places the security of people in an eminently operational and low priority level, since the confidence in the protection of information is placed in the available technological security mechanisms, as well as the applications that materialize the preventive controls implemented in the different IT platforms. This confirms the illusion that the greater the investment in technological security, the greater the protection available to the organization, creating a false sense of security that is strengthened and confirmed by the absence of adverse events, creating greater blind spots in the organization and the dynamics of its processes. Therefore, it is important to keep in mind that "the absence of evidence is not evidence of absence".

4 Transforming Individual's Information Security: Behaviors, Conditions and Consequences

Transforming the current state around people practices in information handling requires initially having a strategic and convergent purpose for security that links both the physical and the digital or cyber environment. The purpose of people security is to understand and manage insider risk, that is, risk associated with a person exploiting, or intending to exploit, their legitimate access to an organization's assets for unauthorized purposes. To this end, the people security system must be designed to detect early weak signals of possible unusual behavior and limit its materialization into specific abuses of information processing [15].

For this, as in the current model "ACP" must be considered: behaviors, conditions and consequences, which must be contextualized and motivated by a particular content that is attractive and situated in the dynamics of the individual.

The behaviors of people respond to a multidimensional view of the human being that, seen from the perspective of an iceberg, the tip is the visible action, underneath are identified the patterns and changes that occur in the individual, the next level are the thought structures that influence the patterns, and finally the mental models that underlie the attitudes, beliefs, customs, expectations, values or culture that enable the thought structures [16]. Talking about behaviors is not an eminently educational action, it is an exercise of cultural transformation that requires a situated understanding of the person and his integral perspective to understand his actions and the motivations behind them.

The conditions are the elements of the environment where the person develops and acts, such as the social and work dynamics, the perception of the environment, his personal relationships, his logic and rationality, elements that situate people's behaviors and give meaning to their concrete actions in the companies. Each of these elements define the triggers of human actions, which when interacting with people's own mental models create an inner motion that translates into an interaction, which according to the alchemy founded on the conditions and their thought structures, creates patterns to end up with concrete actions that determine their social dynamics [7].

Consequences are the natural effects of behaviors situated in the conditions of the environment, which correspond to a specific and particular scenario for each person. That is, people in similar environments may act differently and therefore create unforeseen effects, which may challenge the security and control measures beyond their capabilities, revealing turning points in the reading of the organizational culture of information security, which translate into strategies other than punishments or social recriminations, such as support for people facing their personal and work problems, recognizing mistakes as a learning window that reveals blind spots, illustrating the analysis of behaviors in the different information systems as a personal and corporate vigilant posture and a deterrence communication at three levels, those who are in the process of joining the organization, those who are already employees, and external agents who would like to recruit a current collaborator of the company [15].

All of the above, behavior, conditions and consequences are immersed in a context from which the security area cannot abstract in order to understand the dynamics of the risks in the treatment of information, under penalty of making an inadequate reading that ends in the punishment of the event per se, an inaccurate understanding of what leads to the behavior and why the installed controls were not able to identify it. In this scenario, it is necessary to motivate an organizational transformation that overcomes a culture of compliance and motivates a culture of coherence.

While the compliance culture emphasizes checking and enforcing people's adherence to prescriptive standards through management oversight and auditing, the culture of coherence seeks to ensure that people understand the risks, actively want to do the right thing and are helped to do it, i.e., take ownership of the risks and be aware of the consequences of not applying the established safety and control measures. In this way, the culture of regulatory compliance, which often ends up being I comply and then I lie, due to the pressure to complete a task, the lack of understanding of the rules or the belief that the rules do not make sense, is broken with a situated and coherent perspective for a person in their process based on doing the right things, with meaning and aligned with the challenges of the organization and its customers [15].

To summarize, we can say that the new security agenda for individuals must be aligned with the following elements:

- Focused on information security culture
- Human error is the challenge to understand
- Based on consequences, conditions and behaviors
- Recognize biases, detect cognitive barriers and place information security in the culture of the organization.

This transformation places the security of people in an eminently strategic plane and focused on the culture of the organization, since the confidence of information protection is based on the understanding of the dynamics of people and human groups of a company, as well as the company's own challenges and how the treatment of information makes the difference not only in the processes but for customers. This confirms the need to understand information security in twenty-first century organizations from the need for permanent change and renewal where information flows within the organization at the key points that are required, so that those who are authorized to use it create and develop the concepts and proposals more in line with the conditions of the environment and the strategic objectives of the company. It is no longer the access control conditions that prevail, but the guidelines for the use of information in a particular context and conditions that make the difference in each process of the company.

5 The Phishing Case: Beyond Training

Today, organizations seek to reduce unsafe behaviors by following strategies of punishments, penalties and extinction that do not favor the recognition of the context and consequences of such behaviors, which ends up creating a spiral of irregular behaviors and harmful actions in the future, ending up with unexpected situations, without a substantial improvement towards the construction of a OCIS adjusted to the dynamics of the organization.

On the other hand, efforts are noted that increase behaviors associated with negative reinforcements and recovery, which implies ethical distinctions and a culture of non-reporting that ends up generating more opaque points in the operation and treatment of information, which will possibly go unnoticed or hidden incubating possible security breaches, where their background and particular dynamics are not known, given that there are other different elements at stake such as incentives or recognitions at the corporate level.

Thus, if we analyze the behavior "Clicking on suspicious links or links from unknown sources", following ABC Model (Antecedents, Behaviors, Consequences) [17], we could have the following analysis:

Type of consequence:

- Negative for the individual—Inhibits behavior.
- When the consequence follows the behavior:
- Immediately while the behavior occurs or immediately after.
- Probability that the consequence will occur:
- With certainty the consequence will occur almost or 100%.

Given that the consequence of the behavior is perceived when the action takes place and with certainty that the effects are had and perceived, there is a greater influence to proceed with the behavior change.

However, in reality what happens is that although the behavior has a negative consequence, its effects are not seen in the immediate future and the consequence only occurs once or a few times, in line with the above, people will choose not to change the behavior and maintain the unsafe condition despite having been trained and exposed to simulations of similar events.

The challenge is therefore to bring the individual closer to the consequences of his behaviors and to reveal in a situated way, where he carries out his activities, how his results will be impacted and in a systemic way, the results of the organization, given that the effects will propagate given the high interaction and coupling of the systems in the dynamics of the corporate business.

This means connecting behaviors, conditions and consequences in a situated manner, clearly detailing the impacts and consequences of the actions taken, which allow the person to recognize their own cognitive barriers, understand the proximate effects of their behaviors on the organization and take ownership of information security risks in their daily practice.

6 Conclusions

Anyone trying to address the challenge of information security in people will need answers to three key questions [15].

- What problem are we trying to solve? What is the nature and magnitude of the internal risk we face?
- What tools do we need to solve the problem? What kind of information security systems or tools for people are needed?
- How are we doing? How effective is the people information security program?

Faced with the above questions, the executive teams demand answers that translate into maturity models to reduce the uncertainty that this type of initiative generates and try to frame the decisions required for the people's information security program [18]. In this sense, the maturity of this type of program means that it contains the necessary components articulated with the dynamics of the organization and its environment, and that these are working properly, and not that the system has reached a level where it is still adequate, nor that the organization can relax [15]. That is, maturity is a trend that is confirmed by a sustained effort over time that responds to the dynamics of the environment (changes in risks) and to the adaptation of the behaviors of the threat actors to the defensive measures available to the organization.

Thus, the security of people's information as the system of protection measures through which an organization understands and manages internal risk, that is, protecting organizations from the people who work for them [15], implies taking a holistic view of the challenge of protecting information and their strategic assets, overcoming the exclusively compliance view and advancing in a situational awareness that creates psychologically safe environments, where that which does not go as planned becomes a challenge to understand, where the human being is the center and source of both the threats and the defenses of current and future organizations.

Therefore, given the current dynamics of a highly interconnected and digital environment people's information security must be advanced in an uncertain territory posed by the dynamics of business and the conditions of individuals, a space to explore new strategies to understand human error and create spaces for reflection and response to reveal biases and cognitive barriers in the treatment of information, as well as better technological capabilities that reveal emerging and unusual patterns of behavior that found strategic advantages for the organizational culture of information security.

References

1. Zeijlemaker S (2022) Managing the dynamic nature of cyber security. A future-proof strategy, this is how it works. Netherlands. Disem Institute
2. Barker, J.: Hacked. The secrets behind cyber attacks. London, UK: Kogan Page Limited (2024).
3. Martin P (2019) The rules of security. Staying safe in a risky world. Oxford, UK: Oxford Press
4. Munich Re (2024): Munich Re global cyber risk and insurance survey 2024. https://www.munichre.com/en/insights/cyber/global-cyber-risk-and-insurance-survey.html
5. Vishwanath A (2022) The weakest link: how to diagnose, detect, and defend from phishing. Cambridge, MA. USA: MIT Press
6. Renaud K, Ophoff J (2021) A cyber situational awareness model to predict the implementation of cyber security controls and precautions by SMEs. Organizational Cybersecurity J. Practice, Process and People. https://doi.org/10.1108/OCJ-03-2021-0004. https://www.emerald.com/insight/content/doi/https://doi.org/10.1108/OCJ-03-2021-0004/full/html
7. Cano J (2024) Phishing and human error. Beyond training and simulations. In Rocha A, Adeli H, Dzemyda G, Moreira F, Colla V (eds) Information systems and technologies. WorldCIST 2023. Lecture Notes in Networks and Systems, vol 799. Springer, Cham. https://doi.org/10.1007/978-3-031-45642-8_12
8. Alnatheer M (2012) Understanding and measuring information security culture in developing countries: case of Saudi Arabia (Doctoral Thesis). Queensland University of Technology, Australia. https://core.ac.uk/download/pdf/18312508.pdf
9. Hurlburst G, Bojanova I, Sobel A, Crosby K (2014) Security or privacy? A matter of perspective. IEEE Computer. 47:94–98. https://doi.org/10.1109/MC.2014.327
10. Cano J (2023) Boards and cyber risk: a grieving process for executives and board members? 2023 IEEE Colombian Caribbean Conference (C3), Barranquilla, Colombia. 1–6. https://doi.org/10.1109/C358072.2023.10436197
11. AlHogail A, Mirza A (2014) Information security culture: a definition and a literature review. World congress on computer applications and information systems (WCCAIS). 17–19. IEEE. Hammamet, Tunisia. https://doi.org/10.1109/WCCAIS.2014.6916579
12. Michalos M (2021) The contribution of fostering a cyber security culture in organizations' cyber resilience. Master Thesis. School of Social Science. Hellenic Open University. https://www.researchgate.net/publication/351710901_The_contribution_of_fostering_a_cyber_security_culture_in_organizations%27_cyber_resilience
13. Cano J (2021) Organizational culture for information security: a systemic perspective on the articulation of human, cultural and social systems. ISACA J. 3. https://www.isaca.org/resources/isaca-journal/issues/2021/volume-3/organizational-culture-for-information-security
14. McAlmont S (2024) Positive approaches to cybersecurity training. Information week. https://www.informationweek.com/cyber-resilience/positive-approaches-to-cybersecurity-training
15. Martin P (2024) Insider risk and personnel security. An introduction. Oxon, UK: Routledge
16. Goodman M (1997) Systems thinking: What, why, when, where, and how? The systems thinker. 8(2). From: https://www.appliedsystemsthinking.com/supporting_documents/Intro4WsandHow.pdf

17. Agnew J, Daniles A (2010) Safe by accident? Take the luck out of safety. Leadership practices that build a sustainable safety culture. Atlanta, GA. USA: Performance Management Publication
18. Proctor P (2022) What Cybersecurity Metrics Should I Report to My Board? Gartner Blog. https://blogs.gartner.com/paul-proctor/2022/04/18/what-cybersecurity-metrics-should-i-report-to-my-board

AI-Driven Risk Assessments: Advancing Cybersecurity and Sustainability

Onyebuchi Sunday Onwuajuese, Husnain Rafiq, Sarah McHale, Princess Obianuju Ugochukwu, and Shirley Hunter-Barnett

Abstract In an increasingly interconnected world, the complexity of global security, safety, and sustainability challenges is escalating. Digital transformation, cyber threats, and environmental risks demand dynamic and comprehensive risk management approaches. This paper explores the potential of AI-driven risk assessments in addressing these challenges, focusing on how AI enhances both cybersecurity and sustainability efforts. AI technologies enable real-time threat detection, predictive analytics, and proactive risk mitigation, offering significant improvements over traditional risk management methods. Furthermore, the introduction of quantum computing in AI-driven risk assessments provides unprecedented computational power, enhancing the accuracy and speed of risk identification and mitigation strategies. However, the integration of AI and quantum computing also introduces ethical concerns related to bias, transparency, and accountability. This study critically examines the effectiveness of AI-driven models in various sectors, including cybersecurity, critical infrastructure, and environmental management, highlighting both their advantages and limitations. By addressing emerging risks through the convergence of AI, privacy, and security frameworks, this paper emphasises the need for collaboration between governments, industries, and academia to ensure the ethical and effective application of AI and quantum technologies in risk management. The findings present a pathway for balancing innovation with responsible AI deployment in an increasingly volatile global landscape.

Keywords AI-driven risk assessment · Cybersecurity · Sustainability · Predictive analytics · Quantum computing · Ethical AI · Critical infrastructure · Risk management

O. S. Onwuajuese · H. Rafiq (✉) · S. McHale · P. O. Ugochukwu · S. Hunter-Barnett
Department of Computer Science, Edge Hill University, Ormskirk, UK
e-mail: husnain.rafiq@edgehill.ac.uk

H. Jahankhani and B. Issac (eds.), *Cybersecurity and Human Capabilities Through Symbiotic Artificial Intelligence*, Advanced Sciences and Technologies for Security Applications, https://doi.org/10.1007/978-3-031-82031-1_18

1 Introduction

In today's hyperconnected world, the complexity of global security, safety, and sustainability challenges is intensifying. Digital transformation and geopolitical instability have exposed critical infrastructure and supply chains to cyber threats such as ransomware and data breaches, while climate change heightens environmental risks like extreme weather and resource scarcity [1, 2]. The rise of technologies such as the Internet of Things (IoT) and artificial intelligence (AI) has further complicated this landscape, creating both opportunities and vulnerabilities. Traditional risk management approaches are increasingly inadequate for addressing these evolving challenges. Static, reactive models fail to keep pace with the dynamic nature of modern threats, making integrated, proactive risk assessment essential [3, 4]. Risk assessment is pivotal in managing modern security and sustainability challenges by identifying and prioritising threats. Most risk assessment methods consider elements of three types: assets, vulnerabilities, and threats [5]. Combining these elements, the risk can be expressed as the product of them (Fig. 1). In cybersecurity, it helps organisations mitigate vulnerabilities, while in environmental contexts, it informs resource allocation and resilience planning [6, 7]. Moreover, comprehensive risk assessment anticipates how risks may evolve and interact, addressing their far-reaching, interconnected impacts [8]. With real-time insights and dynamic adaptation, modern risk assessments enable organisations to respond swiftly to emerging threats and enhance their resilience [9].

AI has revolutionised risk assessment, enabling real-time threat detection and predictive analytics across sectors. In cybersecurity, AI models detect anomalies and predict attacks, while in environmental management, they forecast disasters and optimise resource use [10]. AI's ability to process vast datasets offers unparalleled precision in risk prediction. However, AI's reliance on data quality raises concerns about bias, and its increasing role in decision-making introduces ethical challenges regarding accountability and transparency [11]. As AI reshapes risk management, balancing its benefits with these risks is crucial.

2 The Evolution of Risk Assessment in a Hyperconnected World

2.1 *Traditional Versus AI-Driven Risk Assessment*

Traditional risk assessment methods are often manual, involving structured steps to identify, analyse, and manage known risks based on historical data. These methods typically rely on deterministic models, focusing on static threats that evolve slowly or are relatively predictable [12]. The process is time-consuming, requiring extensive

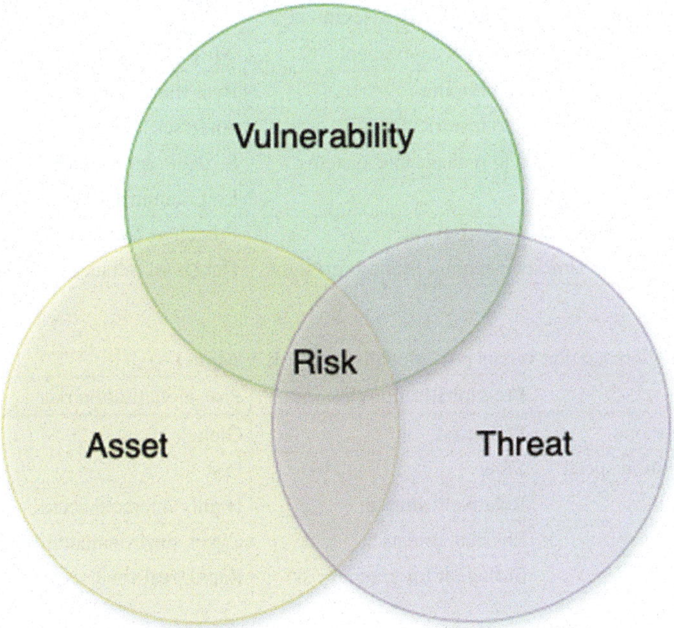

Fig. 1 The components of the basic risk assessment method

human input and is generally reactive, addressing risks only after they have manifested. This approach has limitations, especially in highly dynamic environments like cybersecurity, where threats can emerge rapidly.

In contrast, AI-driven risk assessments automate much of the analysis and enable real-time, predictive insights. Machine learning algorithms can sift through vast datasets, identifying patterns and potential risks that would be difficult to detect manually [4]. This shift from reactive to proactive risk management allows organisations to anticipate threats and take preventive action. AI also enhances scalability, making it possible to assess risks in large, complex systems without the same level of human intervention required in traditional methods. Despite the benefits, AI introduces challenges, such as algorithm bias and reliance on quality data inputs, which can affect the accuracy of predictions [11]. Table 1 summarises the key differences between traditional and AI-driven risk assessments, emphasising AI's ability to enhance speed, scalability, and proactivity.

Table 1 Traditional versus AI-driven risk assessment

Criteria	Traditional risk assessment	AI-driven risk assessment
Speed	Slow and periodic	Real-time, continuous
Data dependency	Limited to historical data	Analyses large, real-time datasets
Scalability	Limited to manageable systems	Scalable across complex networks
Human input	High	Low, automated processes
Proactivity	Reactive, post-event	Proactive, anticipatory
Challenges	Time-consuming, limited foresight	Data bias, reliance on quality inputs

Table 2 Pre-globalisation versus post-globalisation risk dynamics

Criteria	Pre-globalisation risk	Post-globalisation risk
Geographic scope	Localised	Global
Risk propagation speed	Slow	Fast
Complexity	Relatively simple	Highly interconnected
Main risks	Physical threats	Cyber, environmental, supply chain
Response time	Sufficient for gradual risks	Rapid, real-time response required

2.2 *Impact of Globalisation and Digital Transformation on Risk Dynamics*

Globalisation and digital transformation have significantly altered risk dynamics. The rise of global supply chains and interconnected digital infrastructures has created new vulnerabilities. For instance, the widespread adoption of cloud computing and IoT has expanded the attack surface, making it easier for cyber-attacks to affect multiple systems simultaneously [13]. In a hyperconnected world, risks that were once localised can now propagate globally, creating cascading failures across industries.

Digital transformation has also accelerated the pace of risk emergence. In traditional models, risks evolved slowly, giving organisations time to respond. However, in today's digital environment, risks can appear suddenly and spread rapidly, leaving little time for conventional risk management strategies to adapt [14]. Table 2 illustrates the key shifts in risk dynamics, highlighting how globalisation and digital transformation have influenced the speed and complexity of modern risks.

2.3 *The Need for Continuous and Real-Time Risk Assessment*

The rapid evolution of risks in a hyperconnected world highlights the inadequacy of periodic risk assessments. Traditionally, organisations conducted risk assessments quarterly or annually, often using static data. However, modern risks, particularly in areas like cybersecurity and supply chain management, demand real-time monitoring

Table 3 Periodic versus real-time risk assessment

Criteria	Periodic risk assessment	Real-time risk assessment
Frequency	Quarterly or annually	Continuous
Data source	Historical data	Real-time data
Response speed	Slow	Immediate
Risk awareness	Limited to assessment periods	Ongoing, dynamic
Monitoring tools	Manual, scheduled audits	Automated, AI-driven monitoring

to detect emerging threats [15]. For instance, a cyber-attack can occur and spread across networks in seconds, making continuous risk assessment essential for timely interventions.

AI enables this shift to continuous, real-time risk assessment by automating data collection and analysis. AI tools can continuously monitor system performance, flagging deviations from normal behaviour that might indicate a risk [4]. Real-time assessments provide decision-makers with up-to-date insights and allow for immediate adjustments to security measures. Table 3 compares periodic and real-time risk assessments, highlighting the advantages of AI-driven approaches.

3 AI and Machine Learning in Risk Identification and Prediction

AI and machine learning (ML) have significantly improved the detection of emerging risks. Traditional methods often rely on static models and past data, limiting their effectiveness in identifying rapidly evolving threats. In contrast, AI systems can process large datasets in real time, recognising patterns and anomalies that may indicate risks [4]. This ability to detect deviations from normal activity is particularly valuable in cybersecurity, where AI tools can identify potential attacks before they fully materialise. Predictive analytics powered by AI allows organisations to forecast risks based on trends, enabling them to address potential threats before they cause harm [16].

In cybersecurity, AI has proven effective in identifying and mitigating threats. For example, improved algorithms and machine learning approaches enable AI to assist telecom operators in detecting anomalous patterns, predicting potential vulnerabilities, and proactively adjusting defensive strategies. Secondly, it examines AI-based methods for dynamic risk evaluation and adaptive cybersecurity measures [17]. Similarly, AI's application in critical infrastructure protection has enhanced the resilience of essential services. A notable example is AI's use in water management systems, where predictive models help prevent equipment failures by analysing real-time data from sensors [18]. These examples illustrate how AI improves both the detection of risks and the ability to act proactively in mitigating potential impacts.

AI's role extends to predicting environmental and societal risks. For example, AI models have been used to forecast extreme weather events, such as hurricanes, based on satellite data and climate trends [19]. In addition, AI has been applied to public health by predicting the spread of diseases, allowing governments to allocate resources efficiently during pandemics [20]. These models have proved crucial in helping organisations and governments prepare for and mitigate the impacts of environmental and societal risks.

4 Quantum Computing and AI in Risk Assessment

Quantum computing, when combined with AI, is poised to revolutionise risk assessment by dramatically increasing the speed and accuracy of data analysis and decision-making processes. Traditional AI models, while effective, are often limited by the computational power required to process vast datasets in real-time. Quantum computing, with its ability to perform complex calculations exponentially faster than classical computers, could significantly enhance AI's predictive capabilities, enabling more precise risk identification and mitigation strategies [21]. In financial services, for example, quantum AI could analyse market trends and portfolio risks at unprecedented speeds, allowing for near-instantaneous assessments of market volatility and investment risks [22]. Furthermore, in cybersecurity, quantum-enhanced AI could detect and respond to threats in real-time, minimising vulnerabilities by quickly analysing and adapting to new attack vectors [23]. However, while the integration of quantum computing with AI promises transformative benefits, it also presents challenges related to the ethical and regulatory frameworks required to govern such powerful technologies [24].

5 AI-Driven Risk Assessments in Cybersecurity

AI enhances the identification of emerging cyber threats by analysing vast amounts of data to detect anomalies in real time. Traditional systems, which rely on known threat signatures, are often limited in spotting novel attacks. In contrast, AI can learn from patterns in data, identifying previously unknown vulnerabilities [3]. This capability is especially crucial for detecting zero-day attacks and other threats that evolve rapidly. AI's adaptability ensures it remains effective as attackers develop new techniques [4].

Ransomware, insider threats, and advanced persistent threats (APTs) pose significant risks to organisations. AI's ability to monitor systems continuously makes it an effective tool in mitigating these threats. For ransomware, AI can detect early signs of an attack, such as abnormal encryption activity, and prevent large-scale damage [10]. Insider threats, which often bypass traditional security measures, can also be identified by AI systems that track unusual behaviour or access attempts [3]. Moreover, AI is effective against APTs by recognising long-term attack patterns that might be missed by human analysts [25].

AI-powered tools enable continuous monitoring, providing real-time insights into system activities. Unlike traditional security measures that rely on periodic checks, AI systems offer uninterrupted surveillance, ensuring that emerging threats are detected as soon as they occur [3]. In terms of incident response, AI can automate actions such as isolating compromised devices or blocking malicious IP addresses, reducing response times and limiting damage [10]. This capability allows organisations to manage incidents more efficiently, reducing the impact of attacks.

6 Sustainability and Risk Assessment: AI's Role in Environmental Risk Management

AI has become an essential tool in predicting climate and environmental risks, offering more precision than traditional models. AI can process large datasets from satellite imagery and sensors to forecast events such as floods and droughts [26]. This enables governments and organisations to anticipate environmental challenges and take preventive measures. AI models, unlike static systems, can identify patterns in climate data that might otherwise go unnoticed, providing early warnings for natural disasters like hurricanes and wildfires [27]. This improved predictive accuracy helps mitigate the impact of climate-related risks.

AI is increasingly being used to improve sustainability reporting and manage environmental risks for businesses. With growing pressure to meet environmental standards, companies rely on AI to track energy usage, carbon emissions, and resource consumption in real time [28]. AI tools can also forecast potential environmental risks to supply chains, allowing businesses to prepare for disruptions caused by extreme weather or resource shortages. This proactive approach helps organisations align their operations with sustainability goals while managing climate-related risks more effectively [19].

A key example of AI's role in environmental risk management is its use in natural disaster forecasting. In the US, AI models have been used to predict hurricanes by analysing real-time meteorological data and historical storm patterns [29]. This has led to more accurate predictions and earlier warnings, allowing for timely evacuations and better allocation of resources. In Japan, AI has been deployed to predict earthquakes by analysing seismic data, helping authorities issue early warnings and reduce casualties [30]. Post-disaster recovery efforts are also enhanced by AI-powered tools that assess damage through satellite imagery and drones, speeding up recovery processes.

7 The Convergence of Privacy, Cybersecurity, and Risk Assessment

Data Privacy Impact Assessments (PIA) and cybersecurity risk management have traditionally operated as separate entities, with PIAs focusing on compliance and privacy risks, and cybersecurity primarily addressing system protection. However, with increasing data breaches, integrating the two has become essential [31]. A combined approach allows organisations to address privacy and security risks holistically, identifying vulnerabilities related to both data handling and system security. This integration helps ensure that data protection measures align with broader cybersecurity strategies, improving overall resilience.

AI plays a crucial role in helping organisations comply with global privacy regulations like the GDPR. Automated AI tools can track how data is collected, stored, and used across various systems, ensuring that practices adhere to legal requirements [32]. For example, AI can automatically identify potential privacy risks, such as unauthorised access or improper data sharing, and alert organisations to take corrective action. These tools also streamline compliance processes, reducing the manual effort needed to meet complex privacy standards.

The future of privacy and security risk management lies in their continued convergence, with AI driving this shift. As regulations tighten and cybersecurity threats become more sophisticated, organisations will need integrated frameworks that address both privacy and security risks simultaneously [33]. AI will be essential in automating the continuous monitoring of data and security protocols, ensuring that privacy policies are upheld while defending against cyber threats. This convergence will not only improve compliance but also enhance trust in digital systems by providing a more cohesive approach to risk management.

8 Challenges and Ethical Considerations in AI-Driven Risk Assessment

Bias in AI-driven risk assessments remains a significant challenge. If the data used to train AI systems is not representative, it can lead to biased outcomes, impacting decisions in areas such as credit scoring or hiring [34]. To address this, it is crucial to ensure that datasets are diverse and inclusive. Another key issue is transparency. AI models, especially complex machine learning systems, often operate in a "black box," making it difficult to understand how decisions are made [35]. This lack of transparency raises concerns about accountability, as it becomes unclear who is responsible when an AI system makes an incorrect or biased decision.

AI systems require vast amounts of data, often including sensitive personal information. This dependence on data introduces risks related to privacy and security. Ensuring that AI systems comply with data protection regulations like GDPR is a major challenge, as these systems process and analyse personal data at scale [32].

Furthermore, AI systems themselves can be vulnerable to cyberattacks, where malicious actors attempt to manipulate data inputs or outputs, leading to flawed risk assessments [3]. Techniques such as data anonymisation and encryption can help manage these risks, but balancing privacy and data utility remains difficult.

The legal landscape surrounding AI-driven risk assessments is still evolving. Current regulations, such as GDPR, impose strict requirements on data handling but often lack clarity on how AI systems should be regulated [36]. One key challenge is determining liability when AI systems make errors or produce biased outcomes [34]. As AI becomes more integrated into decision-making processes, there is a growing need for clearer legal frameworks that address issues of transparency, fairness, and accountability. Future regulations will likely focus on ensuring that AI systems are used responsibly, with clear guidelines on how data is processed and how decisions are made.

9 Conclusion

AI has revolutionised risk assessment by enhancing real-time threat detection, predictive analytics, and decision-making in cybersecurity and environmental management. While AI improves the identification of risks like ransomware and natural disasters, ethical challenges such as bias and transparency remain. Moving forward, AI's role will expand, but this growth must be paired with strong regulation to protect privacy and ensure ethical use. Effective risk management requires collaboration between AI developers, cybersecurity experts, and sustainability professionals. By working together, they can build systems that are not only technically robust but also socially responsible. Combining expertise ensures that AI models address privacy concerns while enhancing resilience in managing both cyber and environmental risks. Governments must create regulatory frameworks that guide the ethical use of AI in risk assessment, while industry leaders should invest in transparent, responsible AI systems. Academia's role is to drive research and innovation, ensuring future professionals are equipped to handle emerging risks. Together, these sectors can shape a future where AI is used ethically and effectively to address global challenges.

References

1. Adebimpe Bolatito Ige (2024) Eseoghene Kupa, and Oluwatosin Ilori, "Aligning sustainable development goals with cybersecurity strategies: ensuring a secure and sustainable future." GSC Adv. Res. Rev. 19(3):344–360. https://doi.org/10.30574/gscarr.2024.19.3.0236
2. Karuturi Satish SRV (2024) Cybersecurity in a hyperconnected world: challenges and solutions. Redshine Archive. 11:4. https://doi.org/10.25215/9358099984.06
3. Cheimonidis P, Rantos K (2023) Dynamic risk assessment in cybersecurity: a systematic literature review. 01, multidisciplinary digital publishing institute (MDPI). https://doi.org/10.3390/fi15100324

4. Ibrahim Adedeji Adeniran (2024) Christianah Pelumi Efunniyi, Olajide Soji Osundare, and Angela Omozele Abhulimen, "Enhancing security and risk management with predictive analytics: A proactive approach." Int. J. Sch. Res. Eng. Technol. 4(1):032–040. https://doi.org/10.56781/ijsret.2024.4.1.0021
5. Baybulatov AA, Promyslov VG (2022) A metric for the IACS availability risk assessment. In: 2022 International Russian Automation Conference (RusAutoCon), IEEE, pp 750–754. https://doi.org/10.1109/RusAutoCon54946.2022.9896250
6. Tsiodra M, Panda S, Chronopoulos M, Panaousis E (2023) Cyber risk assessment and optimization: a small business case study. IEEE Access 11:44467–44481. https://doi.org/10.1109/ACCESS.2023.3272670
7. Strupczewski G (2021) Defining cyber risk. Saf Sci 135:105143. https://doi.org/10.1016/j.ssci.2020.105143
8. Raveendran A, Renjith VR, Madhu G (2022) A comprehensive review on dynamic risk analysis methodologies. J Loss Prev Process Ind 76:104734. https://doi.org/10.1016/j.jlp.2022.104734
9. Olaniyi OO, Olabanji SO, Abalaka AI (2023) Navigating risk in the modern business landscape: strategies and insights for enterprise risk management implementation. J Sci Res Rep 29(9):103–109. https://doi.org/10.9734/jsrr/2023/v29i91789
10. Rane N, Choudhary S, Rane J (2024) Artificial intelligence for enhancing resilience. J. Appl. Artif. Intell. 5(2):1–33. https://doi.org/10.48185/jaai.v5i2.1053
11. Whang SE, Roh Y, Song H, Lee J-G (2023) Data collection and quality challenges in deep learning: a data-centric AI perspective. VLDB J 32(4):791–813. https://doi.org/10.1007/s00778-022-00775-9
12. Maidana RG, Parhizkar T, Gomola A, Utne IB, Mosleh A (2023) Supervised dynamic probabilistic risk assessment: Review and comparison of methods. Reliab Eng Syst Saf 230:108889. https://doi.org/10.1016/j.ress.2022.108889
13. Affia AO, Nolte A, Matulevičius R (2023) IoT security risk management: a framework and teaching approach. Inform. Educ. 22(4):555–588. https://doi.org/10.15388/infedu.2023.30
14. Rodríguez-Espíndola O, Chowdhury S, Dey PK, Albores P, Emrouznejad A (2022) Analysis of the adoption of emergent technologies for risk management in the era of digital manufacturing. Technol Forecast Soc Change 178:121562. https://doi.org/10.1016/j.techfore.2022.121562
15. Kumar S, Mallipeddi RR (2022) Impact of cybersecurity on operations and supply chain management: Emerging trends and future research directions. Prod Oper Manag 31(12):4488–4500. https://doi.org/10.1111/poms.13859
16. Xu H et al. (2024) Leveraging artificial intelligence for enhanced risk management in financial services: current applications and future prospects. Academic J. Sociol. Manag. J. Home: ajsm.suaspress.org | CODEN: AJSMAS, vol. 2(5). https://doi.org/10.5281/zenodo.13765819
17. Philip Olaseni Shoetan (2024) Olukunle Oladipupo Amoo, Enyinaya Stefano Okafor, and Oluwabukunmi Latifat Olorunfemi, "Synthesizing AI'S impact on cybersecurity in telecommunications: a conceptual framework." Comput. Sci. IT Res. J. 5(3):594–605. https://doi.org/10.51594/csitrj.v5i3.908
18. Lowe M, Qin R, Mao X (2022) A review on machine learning, artificial intelligence, and smart technology in water treatment and monitoring. Water (Switzerland), 14:9. https://doi.org/10.3390/w14091384
19. Gupta S, Modgil S, Kumar A, Sivarajah U, Irani Z (2022) Artificial intelligence and cloud-based collaborative platforms for managing disaster, extreme weather and emergency operations. Int J Prod Econ 254:108642. https://doi.org/10.1016/j.ijpe.2022.108642
20. Chintala SK (2022) AI In Public Health: Modelling Disease Spread And Management Strategies. NeuroQuantology, 20(8):10830–10838. https://doi.org/10.48047/nq.2022.20.8.nq221111
21. Gill SS et al (2022) Quantum computing: a taxonomy, systematic review and future directions. Softw Pract Exp 52(1):66–114. https://doi.org/10.1002/spe.3039
22. Zeguendry A, Jarir Z, Quafafou M (2023) Quantum Machine Learning: a review and case studies. Entropy 25(2):287. https://doi.org/10.3390/e25020287

23. Portmann C, Renner R (2022) Security in quantum cryptography. Rev Mod Phys 94(2):025008. https://doi.org/10.1103/RevModPhys.94.025008
24. Atik J, Jeutner V (2021) Quantum computing and computational law. Law Innov Technol 13(2):302–324. https://doi.org/10.1080/17579961.2021.1977216
25. Hasan MdM, Islam MU, Uddin J (2023) Advanced persistent threat identification with boosting and explainable AI. SN Comput Sci 4(3):271. https://doi.org/10.1007/s42979-023-01744-x
26. Cowls J, Tsamados A, Taddeo M, Floridi L (2023) The AI gambit: leveraging artificial intelligence to combat climate change—opportunities, challenges, and recommendations. AI Soc 38(1):283–307. https://doi.org/10.1007/s00146-021-01294-x
27. Chen L et al (2023) Artificial intelligence-based solutions for climate change: a review. Environ Chem Lett 21(5):2525–2557. https://doi.org/10.1007/s10311-023-01617-y
28. Gaur L, Afaq A, Arora GK, Khan N (2023) Artificial intelligence for carbon emissions using system of systems theory. Ecol Inform 76:102165. https://doi.org/10.1016/j.ecoinf.2023.102165
29. Liu J, Lee J, Zhou R (2023) Review of big-data and AI application in typhoon-related disaster risk early warning in Typhoon Committee region. Trop. Cyclone Res. Rev. 12(4):341–353. https://doi.org/10.1016/j.tcrr.2023.12.004
30. Pwavodi J, Ibrahim AU, Pwavodi PC, Al-Turjman F, Mohand-Said A (2024) The role of artificial intelligence and IoT in prediction of earthquakes: Review. Artificial Intelligence in Geosciences 5:100075. https://doi.org/10.1016/j.aiig.2024.100075
31. Wairimu S, Iwaya LH, Fritsch L, Lindskog S (2024) On the evaluation of privacy impact assessment and privacy risk assessment methodologies: a systematic literature review. IEEE Access 12:19625–19650. https://doi.org/10.1109/ACCESS.2024.3360864
32. van Bekkum M, Zuiderveen Borgesius F (2023) Using sensitive data to prevent discrimination by artificial intelligence: Does the GDPR need a new exception?, Computer law and security review. 48:105770. https://doi.org/10.1016/j.clsr.2022.105770
33. Chen Y, Esmaeilzadeh P (2024) Generative AI in medical practice: in-depth exploration of privacy and security challenges. J Med Internet Res 26(1):e53008. https://doi.org/10.2196/53008
34. Felländer A, Rebane J, Larsson S, Wiggberg M, Heintz F (2022) Achieving a data-driven risk assessment methodology for ethical AI. Digit. Soc. 1(2):13. https://doi.org/10.1007/s44206-022-00016-0
35. Osasona F (2024) Olukunle Oladipupo Amoo, Akoh Atadoga, Temitayo Oluwaseun Abrahams, Oluwatoyin Ajoke Farayola, and Benjamin Samson Ayinla, "Reviewing the ethical implications of ai in decision making processes." Int. J. Manag. Entrep. Res. 6(2):322–335. https://doi.org/10.51594/ijmer.v6i2.773
36. Timan T, Mann Z (2021) Data protection in the era of Artificial Intelligence: trends, existing solutions and recommendations for privacy-preserving technologies. In: the elements of big data value, cham: Springer International Publishing, pp 153–175. https://doi.org/10.1007/978-3-030-68176-0_7

Power of Social Media Meta Data: Enhancing Cyberbullying Detection

Zhou Zhou and Tasmina Islam

Abstract Cyberbullying poses a significant challenge in today's digital society due to its widespread reach and persistent effects. This paper investigates the role of metadata in enhancing cyberbullying detection across three datasets from YouTube, Twitter, and Sina Weibo. Using Correlation-Based Feature Selection and machine learning classifiers (Naive Bayes, Random Forest, and Bagging), the impact of user and social media meta-features on detection performance is analysed. The results show that while user metadata had minimal influence on detection accuracy, social media metadata, particularly, Retweets, Favourites, and SenderFollowers, significantly improved detection performance, enhancing accuracy by up to 9.7%.

Keywords Cyberbullying detection · YouTube · Metadata · Twitter · Sina Weibo

1 Introduction

With the increasing digital interconnectedness of society, pluralistic values continue to spark intense clashes and conflicts in cyberspace. Cyberbullying remains a prevalent issue, with incidents becoming more complex and widespread, creating a persistent wave of online harassment. As people's daily lives are now deeply intertwined with the internet, the psychological impact of cyberbullying cannot be overlooked. Although the COVID-19 pandemic heightened stress levels and increased digital reliance, the ongoing effects of digital overexposure, combined with lingering social isolation in some contexts, continue to elevate the risk of cyberbullying [1].

Cyberbullying is typically the sending or posting of harmful or cruel texts or images over the internet [2]. It can take various forms, including deception, harassment, cyberstalking, slander, degradation, xenophobia, and disguise [2]. Compared to traditional bullying, cyberbullying is not limited by time or place and is often characterised by anonymity and lack of supervision [3], which makes it more prevalent

Z. Zhou · T. Islam (✉)
Department of Informatics, King's College London, London, UK
e-mail: tasmina.islam@kcl.ac.uk

H. Jahankhani and B. Issac (eds.), *Cybersecurity and Human Capabilities Through Symbiotic Artificial Intelligence*, Advanced Sciences and Technologies for Security Applications, https://doi.org/10.1007/978-3-031-82031-1_19

than traditional bullying. To make matters worse, cyberbullying may not disappear over time; it can remain online and continue to torment users [4].

Content-based cyberbullying detection has become a key area of cyberbullying research. Current state-of-the-art methods for cyberbullying detection combine contextual and sentiment features, such as, cursed word dictionaries, history of user activity, grammatical features and sentiment indicators derived from user generated content with text mining methods [5]. Moreover, integrating sentiment analysis with linguistic models has proven effective in enhancing the detection of specific forms of cyberbullying, such as racism and sexism, on platforms like Twitter [6]. At the same time, a diverse array of metadata is available within cyberbullying datasets, encompassing user-based metadata such as age, gender, pro-social behavior, social media membership duration, political affiliation, occupation, education level, religion, and geographic ethnicity. Additionally, content-based metadata can be classified into categories including linguistic features, popularity metrics, temporal features, and category classifications [7]. This paper aims to analyse the role of metadata in cyberbullying detection, with a specific focus on comparing the performance of various machine learning algorithms.

According to a German survey, some 32.8% of young people experience cyberbullying at least once every two months, while 9% say they are harassed to varying degrees on a weekly or even daily basis [8]. Another study by BBC News also showed that over half of young people experience bullying [9]. Similar studies are emerging, and these have brought much thought to how to curb cyberbullying, verbal attacks and other vicious behaviour that have become a hot topic of discussion in society. In terms of the effects of cyberbullying, victims of cyberbullying often face psychiatric and psychosomatic illnesses. Cyberbullying can cause severe psychological trauma and stress to victims, often leading to depression, anxiety, insomnia and other emotions, and may even lead to suicidal behaviour. Therefore, creating a safe and clean Internet environment through cyberspace governance has become a public outcry. This also suggests that in-depth research on cyberbullying is extremely crucial. In recent years, how to detect cyberbullying and how to prevent it has become a key part of research, but relatively little analysis has been done on cyberbullying metadata. By analysing cyberbullying textual metadata, we can firstly examine the impact of metadata on cyberbullying for the purpose of better detecting and preventing cyberbullying. Secondly, in subsequent research, the metadata can be used to classify different subtypes of cyberbullying and to study more complex cyberbullying and its related measures.

The remainder of the paper is organised as follows: Sect. 2 describes literature reviews on recent development of cyberbullying detection systems. Section 3 discusses the methodology and experimental set up including data processing and model training. Section 4 analyses the results. Finally, Sect. 5 concludes the paper.

2 Literature Review

This section reviews existing studies on cyberbullying detection, the effectiveness of various algorithms, and the role of metadata in cyberbullying detection models.

An evaluation of performance results on the YouTube dataset showed that using a Naive Bayes classifier to identify cyberbullying and offensive videos, and by mining video metadata to identify users on YouTube, the method had an accuracy of over 67% [7]. As cyberbullying is more commonly conveyed through text, this method offers potential for future exploration in textual metadata. It has been suggested that certain LIWC (Linguistic Inquiry and Word Count) categories, such as death, appearance, religion, and sexuality, serve as indicators of cyberbullying. Research combining metadata (e.g., user features and image content) with textual features found a higher frequency of negative emotions and a lower incidence of positive emotions in cyberbullying posts [10].

Other studies have demonstrated that the use of LIWC measurement values can enhance cyberbullying detection models. The study reported by Nand et al. showed that using LIWC values outperformed traditional algorithms that rely on filtered and unfiltered word features, with an F-score improvement of up to 10% [11]. Similarly, Cheng et al. utilised LIWC to extract psychological features by counting words associated with emotions, personality traits, and psychological motivations, which resulted in improved model performance [12].

In addition, a meta-analysis of gender differences in cyberbullying showed that men were more likely to be cyberbullies than women; however, this gender difference was influenced by age. It was found that females were more likely to use cyberbullying when they were younger, while males were more likely to use cyberbullying when they were older than females [13]. However, studies have also shown that gender and age lack associations with cyberbullying victimisation [14, 15]. Differences in the samples surveyed may lead to varying findings. For example, a positive association between age and the frequency of victimisation was found in one study of 11–14- and 10–15-year-olds [16], while another study involving a sample of 12–20-year-olds showed a negative association between age and bullying [17].

In previous research, language features, context metadata features, etc. can be used to automatically identify cyberbullying. The maximum likelihood probability of considering words belonging to a particular category showed an accuracy of 0.67, an F-score of 0.59 and a precision of 0.44 [18]. Similarly, in the cyberbullying detection for YouTube comments, research shows that the accuracy of naive bayes is approximately 63% [19]. Nandakumar et al. also carried out an analysis based on precision and running time complexity and showed that the Naive Bayes algorithm obtained nearly 50% of the precision and had a running time of about 2 ms [20]. The analysis using the Naive Bayes classifier algorithm can be done easily, while the accuracy of the results is improving somewhat in the course of continuous improvement research.

Studies have shown that stringing together bullying features and semantics and feeding them into a linear SVM classifier can give good performance results [4].

Dadvar et al. used a supervised learning method to train an online harassment detection classifier and used the support vector machine (SVM) model in WEKA as a classification tool. The results were expressed by precision (0.31), recall (0.15), and F-measure (0.2) [21]. In the cyberbullying detection study for League of Legends Forum and World of Warcraft Forum, SVM demonstrated good performance results with accuracy of 98% and F-score of 49% [22]. One study analysed performance based on Receiver operating characteristic (ROC) performance and true positive rate under three models (textual feature model, social feature model and composite model), with the SVM classification algorithm, the ROC value and True Positive Rate were best at 0.703 and 0.733 respectively [23]. Cheng et al. used 80% of the dataset for training and the rest for testing and analysed the results using F-Score (0.665) and AUC (0.714) [24].

In the study by Bozyiğit et al., RF obtained the best recall score (0.909) in an experiment that included additional social media features [25]. Chatzakou et al. used Twitter-based features to distinguish between bullies and attackers. The results show that among tree classifiers, random forests can obtain the best accuracy (73.45%) [26]. Balakrishnan et al. stated in the report that random forest is the most popular technique used in multiple cyberbullying detection studies because its performance in processing multiple features may be correlated [27].

In the study of Huang et al., the bagging classification algorithm combined with textual features resulted in a ROC of 0.587 and a TP of 0.067, and the classification result of combined social features was a ROC of 0.749 and a TP of 0.374 [23]. If all features are considered comprehensively, ROC is 0.700 and TP is 0.211 [23].

Talpur and O'Sullivan Through the extracted features and applied to the decision tree algorithm, only the best results are shown in SMOTE, and the kappa statistic is 0.479 [28]. Huang et al. used decision trees to study the relationship between social network characteristics (for example, the number of friends) and users, and the results showed that the accuracy rate was 62.8% [23].

Based on the data collected by Akshi Kumar and others from comments on YouTube videos, the results are analysed according to indicators such as precision, accuracy, recall, and F-score. The results were assessed using four algorithms (Random Forest, k-Nearest Neighbour, Sequential Machine Optimisation, and Naive Bayes), with k-nearest neighbour achieving the highest accuracy of about 83% [29]. In the study by Bozyiğit et al., the results obtained by the KNN classification algorithm in the two experiments (with and without additional social media features) were very close, with accuracy of 0.802 and 0.814, and precision of 0.876 and 0.872, respectively [25]. While these studies provide valuable insights, they reveal a research gap in understanding the impact of metadata on cyberbullying detection. This paper will examine how metadata can be effectively utilised to enhance the classification of cyberbullying.

3 Methodology and Experimental Set-Up

Various machine learning models, such as Naive Bayes, Random Forest, Support Vector Machine, Bagging, Decision Tree, and K-Nearest Neighbor, have been utilised through the classification tool WEKA to investigate the role of metadata in cyberbullying detection. The performance score of the experiment is primarily calculated using four indicators: accuracy, precision, recall, and F-score (also called F1-score). Following the preprocessing of the dataset and the extraction of features, the performance scores of the three datasets are analysed across different classification algorithms.

3.1 Datasets

Three datasets were used in this study.

Dataset 1 consists of cyberbullying and non-cyberbullying labelled text data from the YouTube social media platform [30]. Its metadata includes the number of comments, membership duration, number of uploads, profanity in UserID, and age.
Dataset 2 is a Turkish cyberbullying dataset that includes social media features [25]. The data is labelled as True (cyberbullying), while non-cyberbullying entries are labelled as False. Metadata features include Retweets, Favourites, Hashtags, Media, SenderAccountYears, SenderFollowings, SenderFavourites, and Emojis. Since the text is in Turkish, it was translated into English for analysis.
Dataset 3 is a Chinese dataset for sex discrimination and hate speech detection from Sina Weibo [31]. Metadata includes age and likes. The text, originally in Chinese, was translated into English for the purpose of this study. From the original 8970 data points, a random sample of 2000 was selected. Table 1 provides the data distribution.

Table 1 Datasets

	Total	Cyberbullying	Not cyberbullying	Source	Original language	Test language
Dataset1	3464	3047	417	YouTube	English	English
Dataset2	5000	2500	2500	Twitter	Turkish	English
Dataset3	2000	1000	1000	Sina Weibo	Chinese	English

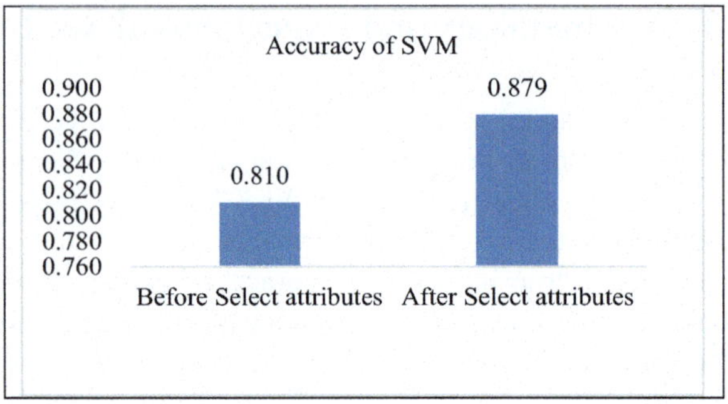

Fig. 1 Dataset1 accuracy of SVM before/after feature selection

3.2 Word Vector and Feature Selection

The datasets mentioned above contain textual data, primarily comments or tweets, which are extensive and complex. Initially, the text is converted to a string format using the NominalToString filter, and then the StringToWordVector filter transforms the string into a set of numeric features representing word occurrence in the text [32]. Given the large number of word vectors generated for each dataset, feature selection is necessary.

Feature selection was performed using Correlation-Based Feature Selection (CFS) with a threshold set at approximately 0.05. Learner-Based Feature Selection selected only a small number of features, which was insufficient for the subsequent performance calculations. Comparatively, Information Gain-Based Feature Selection and Correlation-Based Feature Selection yielded both overlapping and distinct results. After evaluating the performance of the Random Forest classifier, Correlation-Based Feature Selection demonstrated better overall performance.

Following feature selection, Dataset 1 resulted in 63 features, including 5 meta-features and 58 word-vectors. Dataset 2 had 67 features, including 8 meta-features and 60 word-vectors. Dataset 3 contained 76 features, comprising 2 meta-features and 74 word-vectors.

After feature selection, the performance of cyberbullying detection has improved significantly. Using dataset 1 as an example, the accuracy of cyberbullying detection was improved by 6.9% after feature selection compared to no feature selection, as calculated by the support vector machine classification algorithm, as shown in Fig. 1.

3.3 Evaluation Metrics

Four commonly used performance metrics are applied to evaluate the effectiveness of cyberbullying detection models. Table 2 presents the formulas for each metric, where true positives (TP), false positives (FP), true negatives (TN), and false negatives (FN) are denoted accordingly [6].

Table 2 Performance metrics

Metrics	Formula
Accuracy	$\frac{TP+TN}{TP+TN+FP+FN}$
Precision	$\frac{TP}{TP+FP}$
Recall	$\frac{TP}{TP+FN}$
F-score	$\frac{2*Precision*Recall}{Precision+Recall}$

Accuracy is the simplest evaluationz metric, representing the ratio of correctly predicted instances to the total number of samples. It has been widely adopted in cyberbullying detection studies.
Precision, also referred to as Positive Predictive Value, reflects the proportion of true positive instances among all samples predicted as positive.
Recall, also called True Positive Rate (TPR), represents the percentage of actual positive instances that were correctly predicted.
F-score: Ideally, both precision and recall are important, but there can be trade-offs between the two. The F-score is a weighted harmonic mean of precision and recall. When both are equally important, the F-score is used.

Since higher precision and recall are desired, a larger F value indicates better performance. In addition to accuracy, precision, recall, and F-score, **True Positive Rate (TPR)** and **False Positive Rate (FPR)** were also measured as additional evaluation metrics. Relying solely on accuracy can sometimes be misleading, especially when the sample is imbalanced, so other metrics are necessary to provide a more comprehensive assessment of performance [33].

3.4 Training and Testing

In this study, tenfold cross-validation is employed for model evaluation. In this method, the dataset is divided into ten equal parts. For each iteration, nine parts are used for training, and the remaining one is used for testing. This process is repeated ten times, with each part serving as the test set once. The accuracy (or error rate) is recorded for each fold, and the overall performance of the model is obtained by averaging the accuracy (or error rate) across all 10 folds.

To further enhance the reliability of the performance estimate, it is common to perform multiple iterations of tenfold cross-validation (e.g., 10 repetitions of tenfold cross-validation) and then average the results. This approach provides a more robust estimate of the algorithm's accuracy. A key advantage of tenfold cross-validation is that every observation is used for both training and validation, and each observation serves as a test case exactly once.

3.5 Parameter Setting

For the six machine learning algorithms mentioned at the beginning of Sect. 3, the default parameter settings were retained, with only specific modifications explained below.

Naive Bayes: Naive Bayes is a family of simple "probabilistic classifiers" based on applying Bayes' Theorem. It assumes that the features are conditionally independent of each other. The general settings were kept as default, except for the 'useSupervisedDiscretisation' parameter, which was modified to 'True' (the default is False). This change converts numeric features into nominal ones, meaning that word vector values and meta-features were discretised. Previous studies suggest that setting 'useSupervisedDiscretisation' to True can improve classification performance and enhance feature visualisation [34, 35]. As shown in Fig. 2, this modification improved the model's accuracy from 0.762 to 0.814.

K-Nearest Neighbor (KNN): The number of neighbours (K parameter) is crucial for KNN's performance. The default value K is 1, with other common values include 3, 7, 11, and 21 [36]. In Dataset 1, different K values influenced accuracy, with the lowest accuracy observed at K = 1. Choosing K = 5 or 7 increased accuracy by 1.6% (Fig. 2). However, in Datasets 2 and 3, maximum accuracy was achieved with K = 1, and accuracy decreased as K increased.

Bagging: The base classifier used in the Bagging ensemble method is typically a Decision Tree (DT) by default. After testing, it was observed that accuracy results differed by approximately 3% when using various tree-based classifiers (Fig. 3). Among these, Random Forest demonstrated the highest accuracy; however, it also had the longest runtime, as expected due to its ensemble nature.

Random Forest: This algorithm builds multiple decision trees and combines their outputs for improved accuracy. Modifying its parameters resulted in only minor improvements, so the default settings were maintained.

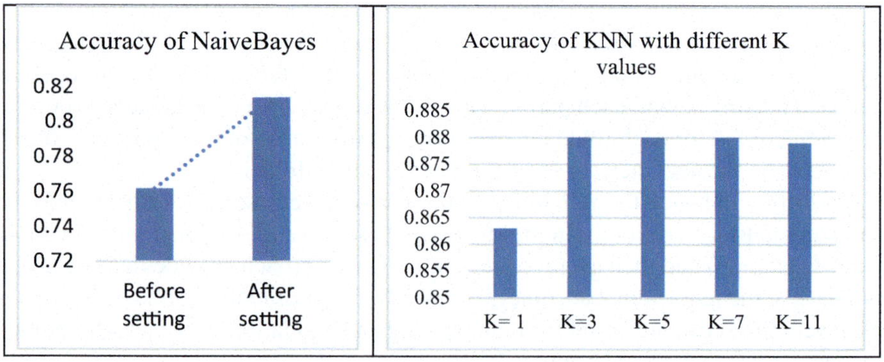

Fig. 2 Dataset1 accuracy of using different parameter settings (left) Naïve Bayes an (right) KNN

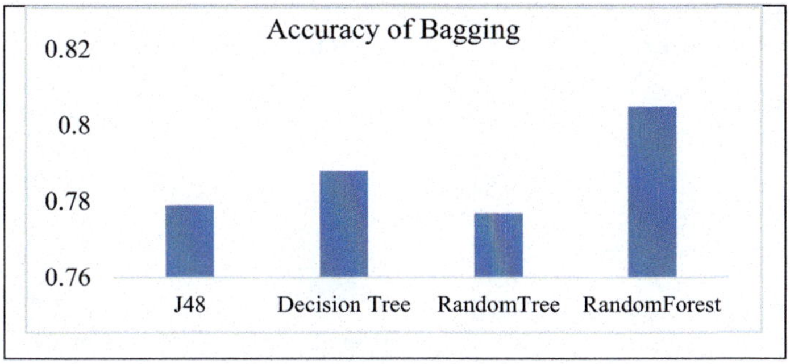

Fig. 3 Dataset2 accuracy of using different base classifiers

Support Vector Machine (SVM): Attempts to modify parameters showed minimal improvement in performance metrics, and in some cases, performance decreased. Therefore, default settings were used for subsequent testing.

Decision Tree (DT): Using the default settings yielded the best performance during testing.

3.6 Comparison of Algorithms

The performance of different classification algorithms, especially when utilising full features (including meta-features), varies significantly in terms of accuracy. Table 3 presents the accuracy of these algorithms across three datasets.

The results shown in Table 3 indicate that Naive Bayes produced the lowest accuracy, while KNN achieved the highest, with a margin of 6.6% between them. This disparity is likely due to the class imbalance in Dataset 1, which contains 3047 samples labelled as non-cyberbullying and only 417 labelled as cyberbullying.

To investigate Naive Bayes' sensitivity to class imbalance, we applied the Synthetic Minority Over-sampling Technique (SMOTE) was applied using WEKA's built-in filter [37]. SMOTE artificially generates new minority class samples to balance the dataset. This technique was used exclusively to understand Naive Bayes' performance on imbalanced data; it was not applied in subsequent experiments.

Table 3 Accuracy of different algorithms

	Dataset1	Dataset2	Dataset3
Naive Bayes	0.814	0.802	0.944
Random Forest	0.875	0.797	0.952
SVM	0.879	0.747	0.94
Bagging	0.878	0.805	0.946
DT	0.875	0.783	0.952
KNN	0.88	0.767	0.946

Fig. 4 Dataset1 accuracy of Naïve Bayes before/after SMOTE

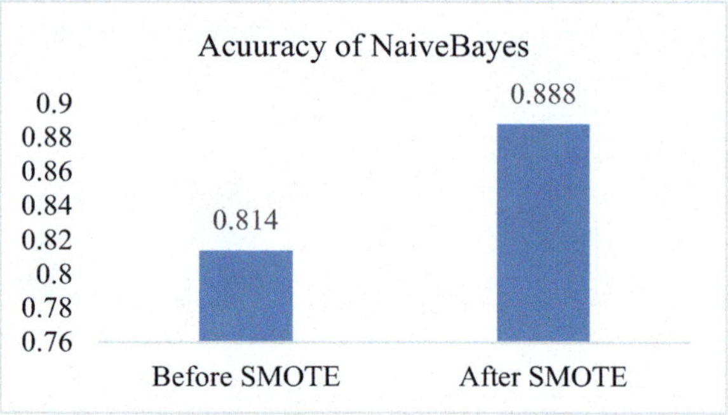

After applying SMOTE, Naive Bayes' accuracy improved significantly, as shown in Fig. 4. However, the performance of other algorithms remained mostly unaffected, with Random Forest and Bagging achieving minimal variation in accuracy (0.889 and 0.875, respectively).

In Dataset 2, Bagging achieved the highest accuracy (0.805), while SVM (Support Vector Machine) had the lowest (0.747), as shown in Table 3. Naive Bayes and Random Forest also demonstrated strong performance.

For Dataset 3, Random Forest and Decision Tree both achieved an accuracy of 0.952, with SVM slightly lower at 0.940. Overall, the performance difference between the six classification algorithms was minimal in Dataset 3, where all algorithms performed well.

Based on the comparative analysis, Random Forest and Bagging consistently exhibited superior performance across all datasets. Naive Bayes, as a commonly used classification method, was also selected for further testing. Consequently, the next set of experiments will focus on these three algorithms: Naive Bayes, Random Forest, and Bagging.

4 Result and Evaluation

In this section, three classification algorithms (Naïve Bayes, Random Forest, and Bagging) as mentioned in 3.4.2 are used on the datasets that include one or more meta-features to analyse how these meta-features impact the performance of cyberbullying detection systems. Dataset 1 primarily consists of user meta-features, Dataset 2 contains mostly social media meta-features, and Dataset 3 includes both user and social media meta-features.

4.1 User Meta-Features Evaluation

Dataset1 was collected from YouTube, with metadata including age, number of comments, number of uploads, membership duration, and profanity in the user ID, as shown in Table 4. To evaluate the impact of meta-features on cyberbullying detection, experiments were conducted using a base dataset without any metadata, six datasets with a single meta-feature, and one dataset combining two meta-features. The two selected meta-features (number of comments and number of uploads) were chosen based on their ranking among the top 20 features identified by Correlation-Based Feature Selection. Descriptions of the test datasets are shown in Tables 5 and 6.

Table 4 Dataset 1 meta features

Feature	Description
Age	Age of posters or commenters
Number of comments	Number of comments on posts
Membership duration	The membership duration of the sender on YouTube
Number of uploads	Number of uploads of posts
Profanity in UserID	Whether the UserID contains any obscene words

Table 5 Description of Dataset 1 test data with feature sets

Dataset	Description
featureSet1	Only word vectors of dataset 1
featureSet1 + metafeature	Word vectors and all meta features of dataset 1
featureSet1 + Age	Word vectors and a meta feature Age of dataset 1
featureSet1 + Number of Comments	Word vectors and a meta feature Number of Comments of dataset 1
featureSet1 + Membership Duration	Word vectors and a meta feature Membership Duration of dataset 1
featureSet1 + Number of Uploads	Word vectors and a meta feature Number of Uploads of dataset 1
featureSet1 + Profanity in UserID	Word vectors and a meta feature Profanity in UserID of dataset 1
featureSet1 + Number of Comments + Number of Uploads	Word vectors and two meta features Number of Comments and Number of Uploads of dataset 1

Table 6 Dataset 3 test dataset description

Dataset	Description
featureSet3	Only word vectors of dataset 3
featureSet3 + age	Word vectors and a meta feature age of dataset 3
featureSet3 + likes	Word vectors and a meta feature likes of dataset 3

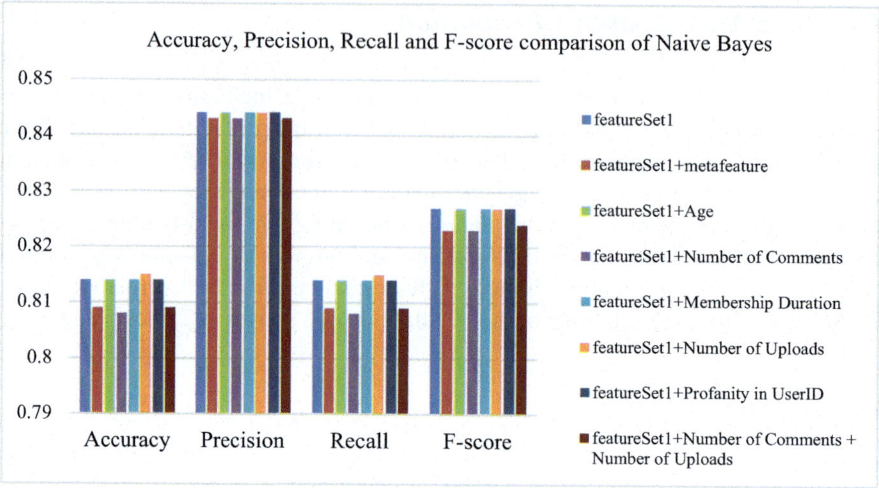

Fig. 5 Dataset1 accuracy, precision, recall and F-score comparison of Naïve Bayes

Naïve Bayes: As shown in Fig. 5, the accuracy of Naïve Bayes remained consistent (0.814) across most datasets, regardless of whether meta-features such as age, member-ship duration, or the number of uploads were included. The minimal variation (0.5–0.6%) observed suggests that user meta-features do not have a substantial impact on cyberbullying detection using this algorithm. This indicates that, from a real-world perspective, user-specific details like age or membership duration may not provide significant predictive value in identifying abusive behavior. Instead, these features could introduce unnecessary complexity without improving detection rates.

Radom Forest: With Random Forest, the inclusion of meta-features resulted in a small but consistent improvement in accuracy and recall (0.3% to 0.6% as shown in Fig. 6). This suggests that Random Forest can capture subtle patterns better than Naïve Bayes, but the improvements remain marginal. In practical applications, this means that while user metadata may slightly enhance detection, it is unlikely to justify the additional computational resources required for real-time systems.

Bagging: The results from the Bagging classification algorithm showed only a minimal difference in accuracy 0.1% across datasets with varying meta-features, indicating that meta-features had little to no impact on cyberbullying detection (Fig. 7).

The results from Dataset 3 (as shown in Figs. 8, 9 and 10) indicate that the age meta-feature has minimal effect on cyberbullying detection performance. Combined with the findings from dataset 1, it can be concluded that the user meta-feature Age has a negligible impact on cyberbullying detection, suggesting that age is not a determining factor in whether someone engages in cyberbullying.

Additionally, other user meta-features from dataset 1, such as Membership Duration and Profanity in UserID, also show no significant impact on detection performance. However, social media meta-features like Number of Comments and Number

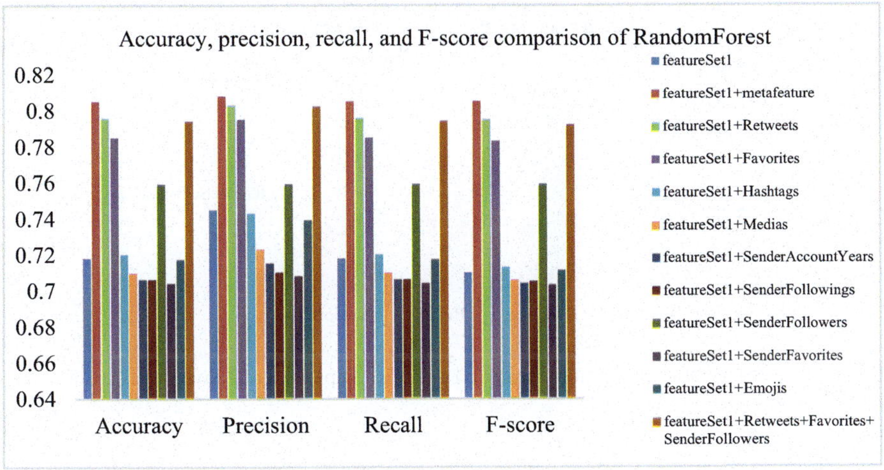

Fig. 6 Dataset1 accuracy, precision, recall and F-score comparison of Radom Forest

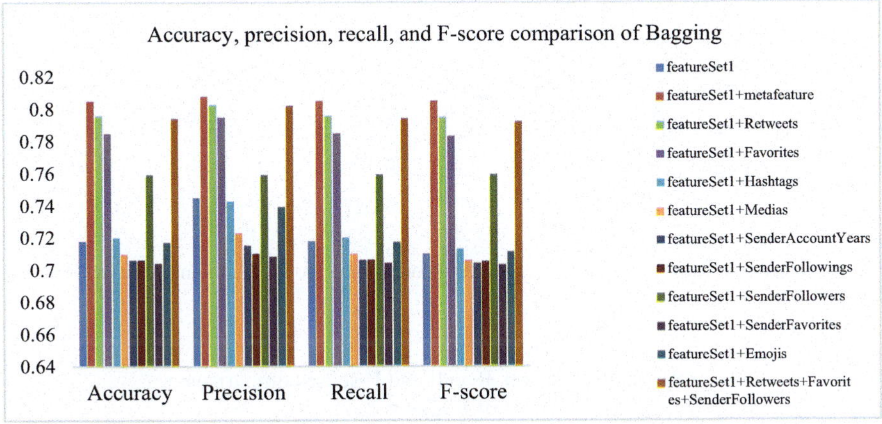

Fig. 7 Dataset1 accuracy, precision, recall and F-score comparison of Bagging

of Uploads, though not influential in this case, warrant further investigation, which will be explored in the next tests using Dataset 2.

4.2 Social Media Meta-Features Evaluation

In Dataset 2, eight social media meta features were selected, as described in Table 7. The dataset was expanded into 12 datasets, each incorporating different combinations of meta-features, as outlined in Table 8, to assess their impact on cyberbullying

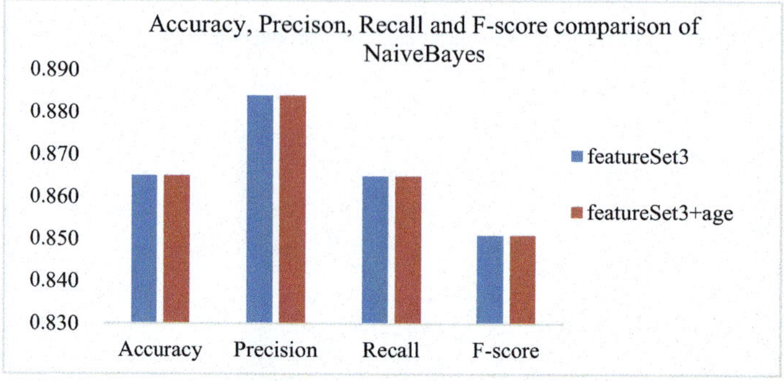

Fig. 8 Dataset3 accuracy, precision, recall and F-score comparison of Naïve Bayes (with/without age)

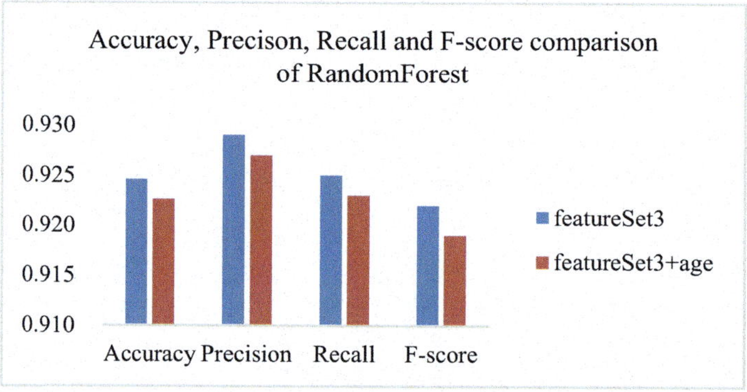

Fig. 9 Dataset3 accuracy, precision, recall and F-score comparison of Random Forest (with/without age)

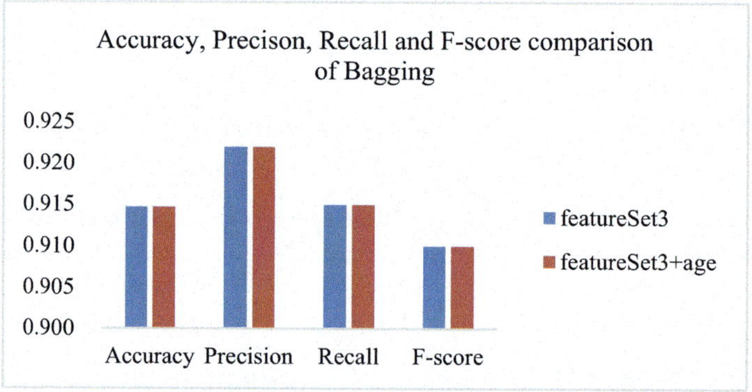

Fig. 10 Dataset3 accuracy, precision, recall and F-score comparison of Bagging (with/without age)

detection. The study in [25] demonstrated that adding these meta-features alongside traditional word vectors significantly improved performance. In this paper, we employ various data preprocessing and feature selection techniques, combined with machine learning algorithms, to conduct the tests and evaluate these improvements.

Naïve Bayes: As shown in Fig. 11, incorporating meta-features with word vectors significantly enhances detection performance. The dataset (featureSet2 + metafeature) improved accuracy by 9.7%, precision by 7.9%, recall by 9.7%, and the F-score by 10.3%. These results indicate that metadata can complement text-based analysis by providing context about user behavior and engagement, which is crucial in detecting nuanced cases of cyberbullying. For example, the number of retweets or Favourites can help identify viral tweets that are more likely to spread bullying content.

However, the effects of individual meta-features vary. Retweets, for instance, boosted accuracy by 9.3%, indicating that this feature strongly correlates with the propagation of bullying content. While Favourites improved accuracy by 8.6%, SenderFollowings had a marginal or even negative impact, suggesting that the mere act of following others does not reveal much about the likelihood of a tweet being abusive. This implies that real-world applications should prioritise metadata that captures user engagement (such as retweets and likes) rather than social network metrics like followings when developing cyberbullying detection systems.

On the other hand, the SenderFollowers meta-feature showed moderate improvements, with a 4.9% increase in accuracy, 3.1% in precision, 4.9% in recall, and 5.5% in F-score. While these results indicate some positive impact, the effect is less pronounced compared to Retweets and Favourites.

The remaining meta-features, including Hashtags, Medias, SenderAccountYears, SenderFavourites, and Emojis, had minimal impact on cyberbullying detection using Naive Bayes. Correlation-Based Feature Selection (CFS) identified Retweets, Favourites, and SenderFollowers as the top-performing meta-features. When these were incorporated into featureSet2 for performance testing, the results closely matched those of featureSet2 + meta-feature, further confirming their importance.

Table 7 Social media meat features

Feature	Description
Retweets	The number of times a tweet was shared by other users
Favourites	The number of users who liked a tweet
Hashtags	The number of topics (special tags) in a tweet
Medias	The number of photos, videos, etc. in a tweet
SenderAccountYears	The membership duration of the sender on Twitter
SenderFollowers	The number of users following the sender of a tweet
SenderFollowings	The number of users followed by the sender of the tweet
SenderFavouritess	The number of times the sender liked other tweets

Table 8 Dataset 2 test dataset description

Dataset	Description
featureSet2	Datasets that include only word vectors of dataset 2
featureSet2 + metafeature	Dataset with word vectors and all meta features of dataset 2
featureSet2 + Retweets	Dataset including word vectors and a meta feature Retweets of dataset 2
featureSet2 + Favourites	Dataset including word vectors and a meta feature Favourites of dataset 2
featureSet2 + Hashtags	Dataset including word vectors and a meta feature Hashtags of dataset 2
featureSet2 + Medias	Dataset including word vectors and a meta feature Medias of dataset 2
featureSet2 + SenderAccountYears	Dataset including word vectors and a meta feature SenderAccountYears of dataset 2
featureSet2 + SenderFollowings	Dataset including word vectors and a meta feature SenderFollowings of dataset 2
featureSet2 + SenderFollowers	Dataset including word vectors and a meta feature SenderFollowers of dataset 2
featureSet2 + SenderFavourites	Dataset including word vectors and a meta feature SenderFavourites of dataset 2
featureSet2 + Emojis	Dataset including word vectors and a meta feature Emojis of dataset 2
featureSet2 + Retweets + Favourites + SenderFollowers	Dataset including word vectors and three meta features Retweets, Favourites, and SenderFollowers of dataset 2

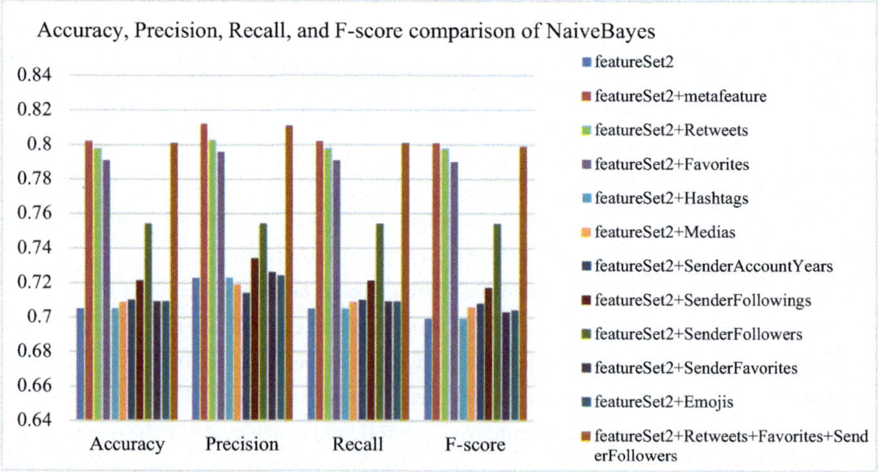

Fig. 11 Dataset2 accuracy, precision, recall and F-score comparison of Naïve Bayes

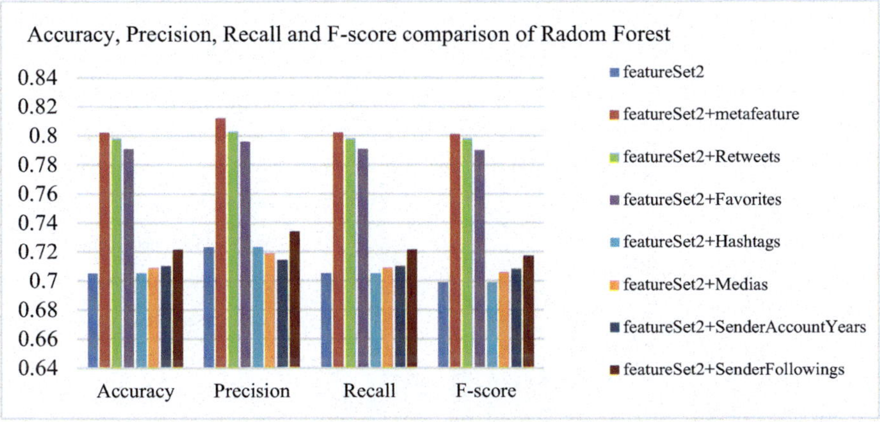

Fig. 12 Dataset2 accuracy, precision, recall and F-score comparison of Random Forest

Random Forest: Fig. 12 illustrates that while Random Forest slightly underperforms compared to Naïve Bayes, the overall trend remains similar. Retweets, Favourites, and SenderFollowers emerge as the most impactful meta-features, enhancing both accuracy and recall by 8.2%. In terms of performance metrics like accuracy, recall, and F-score, datasets containing these features consistently outperformed others. Notably, the dataset with all meta-features, as well as the one combining Retweets, Favourites, and SenderFollowers, demonstrated substantial improvements: accuracy increased by 8.2 and 7.6%, precision improved by 5.8 and 5.3%, recall by 8.2 and 7.6%, and F-score by 9%. Random Forest's strength in handling non-linear relationships likely enables cyberbullying detection systems to capture more intricate patterns of abusive behavior.

Among the individual meta-features, Retweets had the most significant impact, boosting accuracy by 6.7%, followed by Favourites with a 6% increase, and SenderFollowers, which saw a more modest 1.5% improvement. In contrast, datasets that included SenderFollowings and SenderFavourites experienced a noticeable decline in accuracy, dropping by 3.8% and 4.9%, respectively. Datasets featuring Hashtags, Medias, SenderAccountYears, and Emojis showed little to no effect, remaining largely stagnant.

These findings suggest that certain features, particularly those related to user engagement, Like Retweets and Favourites, are far more valuable for detecting cyberbullying than others. Consequently, real-world systems should prioritise engagement related metadata to improve detection performance, ensuring resources are allocated to the features that offer the greatest benefit.

Bagging: As with the previous classifiers, Bagging also benefits from the inclusion of metadata, as shown in Fig. 13. The most significant improvement was observed when all meta-features were combined, leading to an 8% boost across all performance metrics. Retweets and Favourites were once again key contributors, with retweets increasing accuracy by 7.8% and the F-score by 8.5%. The consistency of these

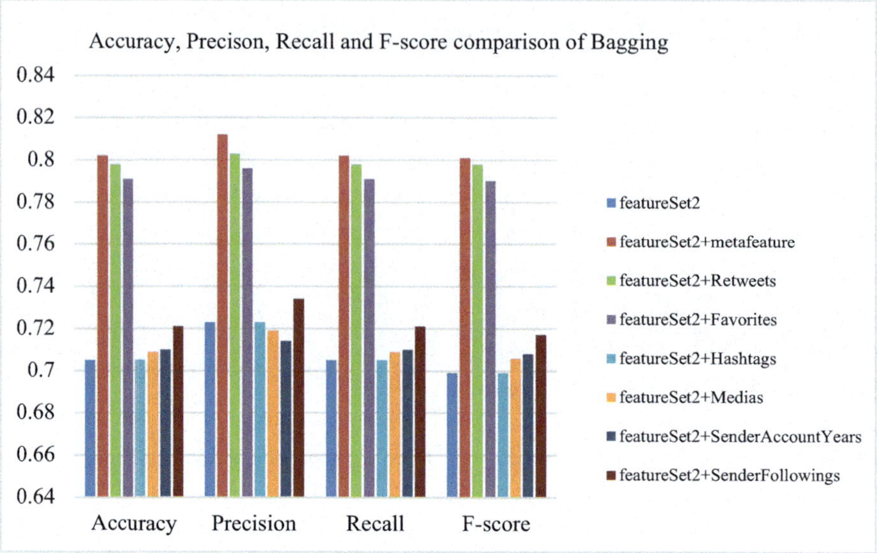

Fig. 13 Dataset2 accuracy, precision, recall and F-score comparison of Bagging

results across different classifiers further strengthens the argument that social media engagement data should be central in any real-world cyberbullying detection system.

Additionally, Dataset 3 was evaluated to verify the role of the "Likes" feature, which corresponds to the Favourites meta-feature in Dataset 2. Figures 14, 15, and 16 show that across all classifiers, the inclusion of likes improved accuracy, precision, recall, and F-scores. The impact of likes on cyberbullying detection is evident, particularly in Naïve Bayes, where F-score increased by 9.2%. This indicates that user engagement features such as likes can be pivotal in real-world applications, as they provide measurable indicators of how content resonates with the audience, an important aspect when identifying harmful content that spreads widely.

As shown in Figs. 14, 15, and 16, regardless of the classification algorithm used, the dataset containing "likes" demonstrated better performance, with improvements across all four performance metrices. Under Naive Bayes, accuracy increased by 7.9%, precision by 6.1%, recall by 7.9%, and F-score by 9.2%. For Random Forest, accuracy, precision, recall, and F-score improved by 2.7%, 2.5%, 2.7%, and 3%, respectively. With Bagging, the improvements were 3.1%, 2.7%, 3.1%, and 3.5%, respectively. Among the three algorithms, Random Forest achieved the best overall performance, with an accuracy of 0.952. These results indicate that social media metadata, such as "likes," positively impacts cyberbullying detection, enhancing its overall performance.

The results from all three classifiers demonstrate that incorporating social media metadata (especially Retweets, Favourites/Likes, and SenderFollowers) significantly enhances cyberbullying detection. In real-world applications, these findings can guide the design of more robust cyberbullying detection systems. For example, by

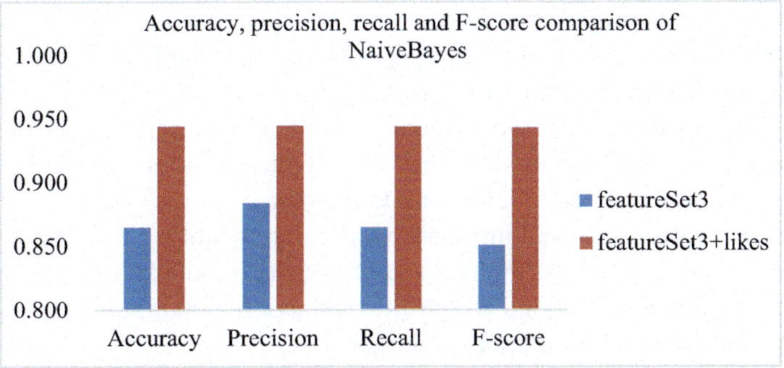

Fig. 14 Dataset3 accuracy, precision, recall and F-score comparison of Naïve Bayes (with/without likes)

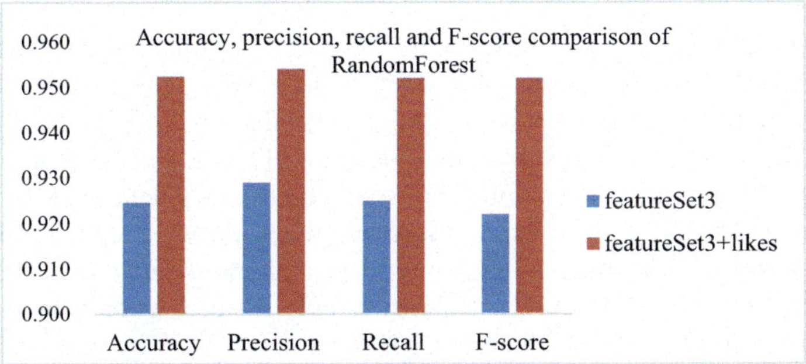

Fig. 15 Dataset3 accuracy, precision, recall and F-score comparison of Random Forest (with/without likes)

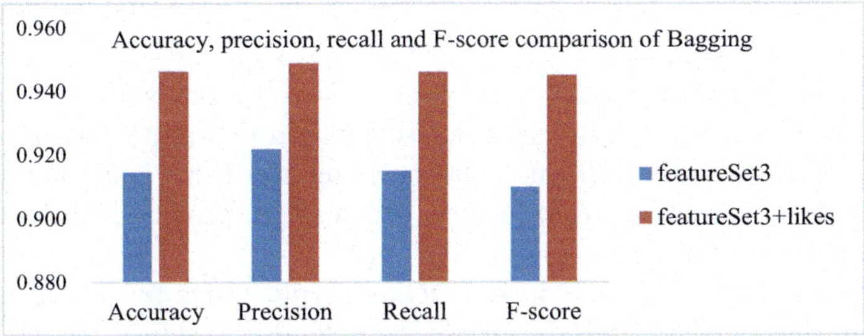

Fig. 16 Dataset3 accuracy, precision, recall and F-score comparison of Bagging (with/without likes)

prioritizing content with high retweets and likes, systems can quickly identify harmful posts that are gaining visibility and address them before they spread widely. Detection systems can minimise false positives by concentrating efforts on content with high engagement metrics, which are more likely to be harmful and have a broader impact. As engagement patterns shift over time, systems can adapt by continually prioritizing the features that most strongly correlate with bullying behavior, making them more resilient to changing user interactions.

In conclusion, integrating social media meta-features, particularly those tied to user engagement, significantly enhances the effectiveness of cyberbullying detection. These features provide crucial context that can improve both the accuracy and timeliness of intervention, making them vital components for developing scalable, real-world detection systems. Platforms that prioritise metadata-driven approaches can better identify and respond to harmful behavior, potentially leading to more effective prevention and mitigation strategies.

5 Conclusion

In today's digital society, cyberbullying has become a significant issue, necessitating effective detection mechanisms. This study explored the impact of various metadata types on the detection of cyberbullying across datasets from YouTube, Twitter, and Sina Weibo, employing Naive Bayes, Random Forest, and Bagging classifiers. The results highlighted the importance of social media metadata, such as Retweets, Favourites, and SenderFollowers, in improving the accuracy, recall, and F-score of detection systems. In particular, the Twitter dataset demonstrated substantial performance improvements when these social media meta-features were included, with the Bagging algorithm showing an increase in F-score of up to 9.5% compared to models using only word vectors. Similarly, metadata such as "likes" in the Sina Weibo dataset contributed to notable gains in detection metrics, underscoring the utility of social media metadata in enhancing cyberbullying detection.

In contrast, user metadata, such as age and account duration—had minimal impact on detection performance in both the YouTube and Sina Weibo datasets. Although the immediate influence of user metadata appears limited, it remains a promising area for further research. Combining user metadata with other specialised feature sets or applying it to different datasets may reveal more significant effects. This could be particularly relevant when analysing user behaviour patterns, such as posting frequency or engagement history, which might provide additional insights into cyberbullying tendencies.

The findings from this study have practical implications for social media platforms aiming to improve their cyberbullying detection systems. By incorporating social media metadata, such as retweets, likes, and follower information, platforms can significantly enhance the precision of their automated tools, creating safer online environments. Although user metadata alone may not have shown a substantial impact in this study, its potential to improve detection performance when combined with behavioural and social data should not be overlooked. Such combinations could lead to more personalised monitoring systems, enabling platforms to better understand user behaviour patterns and develop evolving risk profiles to help mitigate harmful activities. Additionally, exploring how specific social media features like Retweets or Likes correlate with cyberbullying, especially within groups such as high-engagement users or influencers, could provide deeper insights into behaviour-driven detection strategies.

Future research should also focus on developing more sophisticated detection tools that can process metadata and word vectors in multiple languages. Since this study relied on translating non-English datasets into English, creating models that directly analyse multilingual data would greatly enhance the effectiveness of cyberbullying detection across different linguistic and cultural contexts. These advancements could further refine the practical application of metadata-driven models, making detection systems more adaptable and effective across global social media platforms.

In conclusion, this study highlights the significant role of social media metadata in advancing cyberbullying detection, while also pointing to the untapped potential of user metadata for more specialised applications. These findings pave the way for future research to explore metadata-driven models in greater depth, particularly in multilingual and cross-cultural contexts, ultimately contributing to more robust and adaptable cyberbullying detection systems that support both prevention and intervention efforts across diverse online environments.

References

1. Yang F (2021) Coping strategies, cyberbullying behaviors, and depression among Chinese netizens during the COVID-19 pandemic: a web-based nationwide survey. J Affect Disord 281:138–144
2. Feinberg T, Robey N (2009) Cyberbullying. Educ Dig 74(7):26
3. Tokunaga RS (2010) Following you home from school: a critical review and synthesis of research on cyberbullying victimization. Comput Hum Behav 26(3):277–287
4. Zhao R, Zhou A, Mao K (2016) Automatic detection of cyberbullying on social networks based on bullying features. In: Proceedings of the 17th international conference on distributed computing and networking. pp. 1–6
5. Van Bruwaene D, Huang Q, Inkpen D (2020) A multi-platform dataset for detecting cyberbullying in social media. Lang Resour Eval 54(4):851–874
6. Wang L, Islam T (2023) Automatic detection of cyberbullying: racism and sexism on Twitter. In: Cybersecurity in the Age of Smart Societies: Proceedings of the 14th international conference on global security, safety and sustainability. Springer, Cham, pp 105–122
7. Marathe S, Shirsat KP (2015) Approaches for mining youtube videos metadata in cyber bullying detection. Int J Eng Res Technol 4:680–684
8. Bergmann MC, Baier D (2018) Prevalence and correlates of cyberbullying perpetration. Findings from a German representative student survey. Int J Environ Res Public Health 15(2):274
9. BBC News (2019) Over half of young people have experienced bullying. Bbc.co.uk/newsround/48923195. Last accessed: Oct 2024
10. Hosseinmardi H, Mattson SA, Rafiq RI, Han R, Lv Q, Mishra S (2015) Detection of cyberbullying incidents on the instagram social network. arXiv preprint arXiv:1503.03909
11. Nand P, Perera R, Kasture A (2016) How bullying is this message?: A psychometric thermometer for bullying. In: Proceedings of COLING 2016, the 26th international conference on computational linguistics: technical papers. pp 695–706

12. Cheng L, Guo R, Silva Y, Hall D, Liu H (2019) Hierarchical attention networks for cyberbullying detection on the Instagram social network. In: Proceedings of the 2019 SIAM international conference on data mining. Society for Industrial and Applied Mathematics, pp 235–243
13. Barlett C, Coyne SM (2014) A meta-analysis of sex differences in cyber-bullying behavior: the moderating role of age. Aggressive Behav 40(5):474–488
14. Hinduja S, Patchin JW (2008) Cyberbullying: an exploratory analysis of factors related to offending and victimization. Deviant Behav 29(2):129–156
15. Betts LR, Spenser KA, Baguley T (2022) Describing disclosure of cybervictimization in adolescents from the United Kingdom: the role of age, gender, involvement in cyberbullying, and time spent online. J Genet Psychol 183(1):40–53
16. Kowalski RM, Limber SP (2007) Electronic bullying among middle school students. J Adolesc Health 41(6):S22–S30
17. Dehue F, Bolman C, Völlink T (2008) Cyberbullying: youngsters' experiences and parental perception. Cyberpsychol Behav 11(2):217–223
18. Marathe SS, Shirsat KP (2015) Contextual features based naïve Bayes classifier for cyberbullying detection on YouTube. Int J Sci Eng Res 6(11):1109–1114
19. Dinakar K, Reichart R, Lieberman H (2011) Modeling the detection of textual cyberbullying. In: Fifth international AAAI conference on weblogs and social media
20. Nandakumar V, Kovoor BC, Sreeja MU (2018) Cyberbullying revelation in Twitter data using Naïve Bayes classifier algorithm. Int J Adv Res Comput Sci 9(1)
21. Dadvar M, De Jong F (2012) Cyberbullying detection: a step toward a safer internet yard. In: Proceedings of the 21st international conference on World Wide Web. pp 121–126
22. Vo HHP, Tran HT, Luu ST (2021) Automatically detecting cyberbullying comments on online game forums. In: RIVF international conference on computing and communication technologies (RIVF). IEEE, pp 1–5
23. Huang Q, Singh VK, Atrey PK (2014) Cyber bullying detection using social and textual analysis. In: Proceedings of the 3rd international workshop on socially-aware multimedia. pp 3–6
24. Cheng L, Li J, Silva Y, Hall D, Liu H (2019) PI-bully: personalized cyberbullying detection with peer influence. In: The 28th international joint conference on artificial intelligence (IJCAI)
25. Bozyiğit A, Utku S, Nasibov E (2021) Cyberbullying detection: utilizing social media features. Expert Syst Appl 179:115001
26. Chatzakou D, Kourtellis N, Blackburn J, De Cristofaro E, Stringhini G, Vakali A (2017) Mean birds: detecting aggression and bullying on twitter. In: Proceedings of the 2017 ACM on web science conference. pp 13–22
27. Balakrishnan V, Khan S, Arabnia HR (2020) Improving cyberbullying detection using Twitter users' psychological features and machine learning. Comput Secur 90:101710
28. Talpur BA, O'Sullivan D (2020) Cyberbullying severity detection: a machine learning approach. PLoS ONE 15(10):e0240924
29. Kumar A, Nayak S, Chandra N (2019) Empirical analysis of supervised machine learning techniques for Cyberbullying detection. In: International conference on innovative computing and communications. Springer, Singapore, pp 223–230
30. Elsafoury F (2020) Cyberbullying datasets, Mendeley Data, V1. Available at: https://www.kaggle.com/datasets/saurabhshahane/cyberbullying-dataset
31. Jiang A, Yang X, Liu Y, Zubiaga A (2022) SWSR: a Chinese dataset and lexicon for online sexism detection. Online Soc Networks Media 27:100182
32. StringToWordVector (weka-dev 3.9.5 API) last accessed Oct 2024. https: // weka.sourceforge.io/doc.dev/weka/filters/unsupervised/ feature/StringToWordVector.html
33. Leonard LC (2017) Web-based behavioral modeling for continuous user authentication (CUA). In: Advances in computers, vol 105. Elsevier, pp 1–44
34. Yamak ZR (2018) Multiple identities detection in online social media. Doctoral dissertation, Normandie Université
35. Witten IH, Frank E, Trigg LE, Hall MA, Holmes G, Cunningham SJ (1999) Weka: practical machine learning tools and techniques with Java implementations

36. Brownlee J (2016) Machine learning mastery with Weka: analyze data, develop models, and work through projects. Machine Learning Mastery
37. Frank E (2019, January 30) Oversampling and undersampling. Last accessed: Oct 2024, from https://waikato.github.io/weka-blog/posts/2019-01-30-sampling/

Cybercrime as a Threat to the Commercial Banking Sectors in Bangladesh: A Critical Analysis

Kudrat-E.-Khuda Babu, Nuntira Hongsrisuwan, A. M. Jahid, Md. Ahsan Ullah, Subrata Debnath Shuvo, and Mahmudul Hasan Shobuj

Abstract The main objective of this paper is to shed light on the negative impact of cybercrime on the economy of Bangladesh and also to find out the prevailing typologies of cybercrime in the banking sector. The literature review of this study attempts to explore the relationship between technological development and cybercrime in a global context. Moreover, the patterns and trends of cyber-crime in the Bangladeshi banking sector have been reviewed. At the same time, how the Bangladeshi banking sector has consistently been victimized by cybercrime, the prevalence of cybercrime in various commercial banks and what are the systemic flaws in dealing with cyber-crime are discussed in the investigative part of this paper. The recommendations section of this paper provides some guidance on how to combat cybercrime and build a strong cyber security framework.

Keywords Bangladesh · Cyber security · Cybercrimes · The ICT Act 2006 · The Digital Security Act, 2018

K.-E.-K. Babu (✉) · Md. A. Ullah
Department of Law, Daffodil International University, Dhaka, Bangladesh
e-mail: kekbabu@gmail.com

N. Hongsrisuwan
Faculty of Engineering, Shinawatra University, Pathum Thani, Thailand

A. M. Jahid
Graduate Student, Dept. of Communication Studies, Texas Tech University, Lubbock, TX, USA
URL: https://www.thedailystar.net

S. D. Shuvo
MSc International Criminal Justice and Intelligence, University of Portsmouth, Portsmouth, UK

M. H. Shobuj
LLM, University of South Wales, Newport, UK

H. Jahankhani and B. Issac (eds.), *Cybersecurity and Human Capabilities Through Symbiotic Artificial Intelligence*, Advanced Sciences and Technologies for Security Applications, https://doi.org/10.1007/978-3-031-82031-1_20

1 Introduction

Bangladesh has set a target to transform the country into completely cashless by 2031, meaning that all money transactions, such as payments, will be made solely online, according to the Bangladesh Bank, the central bank of Bangladesh. To achieve the goal, the country targets achieving 30 percent of transactions digitally by 2025. Despite the highly positive trajectory, there is a growing concern over the cybersecurity as it will be very critical to ensure the safe and smooth transition to a cashless society. Whether it is national or any corporate level, cybersecurity is a fundamental component of any security concepts [70]. Ensuring appropriate cybersecurity measures is paramount in combating any form of cybercrime [14]. Cybercrime refers to any illegal activity carried out using or in relation to a computer system or network, encompassing crimes such as unlawful entry and the transmission or dissemination of information through a computer system or network [45]. However, Halder and Jaishankar [28], explain cybercrime as criminal acts perpetrated against individuals or groups with the deliberate intention of damaging their reputation or causing physical or mental injury or loss, either directly or indirectly, through digital communication networks such as the internet (including chat rooms, emails, notice boards, and groups) and mobile phones. They defined cybersecurity as the use of technology to inflict physical, social, psychological, or financial damage upon a target, who may be individuals, groups, or organizations [28]. Since the advent of e-banking, or digital banking, Bangladesh's banking sector has been going through a transitional period. The banks have had difficulties in addition to changes in their daily operations [80]. According to Dhaka Bank Chairman Abdul Hai Sarker, cashless and paperless banking is a risky business if it is seen it from the security aspect. "Therefore, we have to ensure security if we want to sustain digital transformation," said Sarker. Since IT has taken center stage, cybercrimes are now viewed as a greater threat to our financial industry. Any system can be accessible through digital technologies and the internet, therefore a nation's reserves and data could be subject to unauthorized access. The threat is growing daily due to our increased ability to move digitally in cyberspace [59]. Given that the e-banking system has emerged as a major target for criminal behavior in financial institutions, ensuring strong cyber security is essential to safeguarding the reserves and maintaining the confidentiality of customer information. If not, it might become a serious threat in the future and cause emerging nations like Bangladesh to suffer even more losses [1]. Finding reactive solutions to the cybercrime problem is a difficult undertaking, particularly for poor countries like Bangladesh because cybercrime typically involves international elements [73].

2 Literature Review

2.1 Cybercrime and Technology

Cybercrime is considered as one of the significant phenomena in today's world [79].

Bangladesh has encountered various facets of cybercrime, particularly following the first decade of the twenty-first century, with the banking sector being notably affected as it experiences more challenges than other sectors [60]. Bangladesh's banking sector has to navigate a challenging trajectory unless it intensifies its cyber security efforts [75]. According to Kharouni [43], the tools and infrastructure used to conduct cybercrime are computers and networks as a channel, source, device, target or location of the crime. Goel [26] stated that cyber-crime boundaries have expanded into the digital space due to the proliferation of digital commerce and e-transactions. Along with other parts of the world, cybercrime has also increased in Bangladesh, especially the country's banking sector has become a vulnerable area for such crimes in recent years [76]. Thanks to the evolving nature, technology, especially the internet, has become the part and parcel of people from all walk of life. Internet has also connected people to people from all over the globe simultaneously. The information superhighway, sometimes known as "the web," makes it possible to connect to the rest of the world through the internet, among other things. Because of this, everyone has access to their place of employment, bank, and electricity provider, as well as risky travel options and cybercriminals who are just waiting for naive internet users to become their next victim [40]. According to Joveda et al. [39], cyberspace has turned into an excellent medium for sharing information and data in the e-banking industry. As the dependency of people on digital horizon has increased, banks are bringing about a fundamental change and jumping on cashless banking as it is convenient for both bankers and customers. However, on the flip side, information technology misuse has increased in risk within Bangladesh's banking sector [39].

It is very important for ensuring cybersecurity and be aware of crimes being committed in the digital world. Cashless and paperless banking is a risky business if we see it from the security aspect. Therefore, security needs to be insured if the country wants to sustain digital transformation. In 2023, there were 2365 cyberattacks worldwide, affecting 343.3 million victims, as reported by Forbes. According to the Cybersecurity Ventures 2023 Official Cybercrime Report, the cumulative cost of damages resulting from cybercrime is projected to attain $10.5 trillion by 2025. Dhaka Tribune, an English language daily, in a study in February 5, 2018, revealed that the banking industry in Bangladesh is in the serious risk of being attacked by hackers. A Tunisian hacker who introduced as "Human Mind Cracker" penetrated the website of Islami Bank Bangladesh, the largest private Shriah-based bank in Bangladesh, on January 6, 2013. The county's largest state-run bank Sonali Bank came under attack on December 2, 2015, when a hacker with the name, KinGnCa, breached the lender's network security and took control of the website. Later in February 2016, a large number of ATMs of Dutch-Bangla Bank, City Bank, and

Trust Bank, the leading private commercial banks in Bangladesh, were came under the attack of hackers [37].

2.2 *Trends of Cybercrime*

Ozarslan [57] thinks that although every firm have the risk of being victimized by cybercrimes, cybercriminals mostly set their targets considering two issues—revenue and the largest effect. This why the cyber attackers usually targets financial institutions, such as banks or non-bank financial institutions (NBFIs), financial service providers, including mobile financial service providers (MFSs) as those institutions have the two criteria. In financial sector especially, the banks deal and possess the highest amount of money than any other entities or companies and manage extremely valuable data virtually—ranging from details of bank cards, accounts and customers and other sensitive data. Moreover, their ongoing digital transformation initiatives, intricate regulatory landscape, complex supply chain dynamics, and COVID-19-induced hybrid work habits increase the probability of hackers acquiring and using such data [8]. Consequently, the financial industry is becoming a target for cybercriminals. Even though security flaws have historically alarmed financial institutions, their concerns have grown significantly as a result of the far bigger functional and reputational consequences of cyber hazards [65]. Financial services have become more automated and seamless to meet consumer demand for faster and easier transactions. And because of this, the fear of various types of fraud and insecurity has increased. Due to its importance compared to other industries, the financial sector of a country is more vulnerable to cyberattack and the possibility of damage is high. Regular business operations in a country or region could be severely disrupted by an attack on that nation's or region's data security [67]. Banks' aging infrastructure and high profit margins make them more attractive to cybercriminals [46]. Small-scale money withdrawals from a bank network are simple, while larger-scale money theft requires a little more effort [22].

2.3 *Cybercrime and Banking Industry*

According to Kuepper [44], the prevalence of cyberattacks has reached such a level that most banks have experienced some kind of attack and such incidents is increasing. The trend of these attacks is increasing at a significant rate in financial institutions. A significant cyberattack that could destroy a financial institution is not unthinkable [55]. For having interconnection among banks, a cyberattack on one bank has the potential to spread throughout the banking system, threatening the stability of the financial institution. Moreover, US banks are particularly vulnerable to state-sponsored cyberattacks [51]. Cyberattack intensified in the beginning of the COVID-19 pandemic in 2020 [29]. According to a research report of Packetlabs [58], maintaining data in the banking industry is very expensive. Due to the huge value of resulting data, the banking industry has been identified as particularly vulnerable to cyberattacks. By hacking financial data in various organizations, they are sold for millions of dollars. The potential attack picture for banks is similar to their digital presence [20]. Moreover, the digital evolution of the industry has resulted in mobile apps, electronic transfer systems and other online services that companies provide, also increasing cyber security risks [52]. Although digital banking systems have made banking services easier and increased customer access, security flaws have made it easy for scammers to attack. To protect data, banking and cyber security must be given importance simultaneously [2].

Financial institutions, such as banks, are more vulnerable to cyberattacks than businesses in other industries as found in a research literature of Archon [3]. As a result of data security breaches and other security-related incidents, banks experience revenue losses, operational disruptions, and brand and customer reputation damage. In recent years, the number of cyberattacks on such financial institutions has risen due to the increasing sophistication of attackers and the expansion of potential targets and entry points [66]. Both the number of people using banking services and the quantities of newer forms of banking technologies have increased largely. Thus, the banking sector is especially susceptible to cybersecurity breaches, the frequency and sophistication of which are increasing across all industries [3]. In a research of Center for Strategic and International Studies [13], financial institutions have been discovered to be the most prevalent victims of cyber-attacks. Banks are where money is stored, and cybercriminals are using them for extortion, theft, and fraud, while nation-states and hacktivists use the financial sector for political and ideological influence. Regulators are taking notice and implementing new cyber risk mitigation measures to combat the growing danger to the institutions under their oversight [18]. The Strategic Technologies Program (STP) looks at the rise of cyber risks to the monetary system, as well as legal and regulatory efforts to solidify its defenses [21].

Archon's [3] research finds that other financial institutions, including banks, are more vulnerable to cyber-attacks than businesses in any other industry found in the literature. As a result of cyber-attacks and other security-related incidents, banks suffer revenue loss as well as operational disruptions and damage to brand and customer reputations due to financial institutions' data security breaches. Meanwhile, over the years, the types of cyber-attacks have become more virulent and the scope and efficiency of their attacks have increased. Hence the number of cyber-attacks in this sector has increased [66]. As the number of people using banking services and the amount of new banking technologies have increased dramatically, so has the frequency and sophistication of cyber security breaches in the banking sector [3].

According to a study by the Center for Strategic and International Studies [13], financial institutions are the most vulnerable to cyber-attacks. A bank is an institution where money is stored, and criminals use cyberattacks to steal and defraud that money. And if nation-states and hacktivists use the financial sector for political and ideological influence, the situation becomes even direr. As the scale and severity of cyberattacks continue to increase, regulators of financial and private institutions are increasingly scrambling to protect their institutions and adopting new cyber risk mitigation measures [18]. For example, the Strategic Technologies Program (STP) looks at legal and regulatory efforts to strengthen their defenses against the risk of cyberattacks on their financial systems [21]. According to Collier [16], majority of the financial institutions come under attack across the world for stealing money. Those entities are targeted mainly because of economic links. Adequate knowledge on the types of cyberattacks, such as espionage or ransomware, and strategies of attackers help effectively fight against cyber-attacks. Relying on tactical inspections and restoration of high-impact cyber-attacks globally [31]. Mandiant's most recent M-Trends 2022 research has found that three area—industrial and professional Services and financial sector—were attacked by cyber criminals mostly in 2021. The same industries come under attack across the globe every year. The patterns of cyberattacks expose the overall situation of security aspects. The more the internal computers of the financial are inter connected with networks, the more they are in risk of attacks from outside [7]. The cyberattacks include illegal money transfer, extortion, taking ransom after cyberattack. However, to carry out a ransomware, a hacker had to send huge times [9]. The cyber attackers visit the victim's network again and again before launching the attack. They try to be familiar with the victim's network and try to find out the critical and weak zones that are very sensitive for the survival of the organizations [23].

3 Cybercrime as a Global Phenomena

According to International Monetary Monetary Fund & Capital Markets Department [35], banking industry is now in a challenging situation considering the geopolitical and macroeconomic factors across the globe. So to assess and mitigate the impending risks, the scrutiny of the ongoing practices in the banking industry if very crucial. So a technology-based methods has been developed to manage the risks [77]. Now the financial services such as mobile financial services and banking services are now at the doorsteps of ordinary people, thanks to the spread of mobile networks and communication and information technology in daily life [30]. But while technology has made financial services more affordable and accessible, it has also increased the risks of being the target of a cyberattack [34]. In addition to stealing money, criminals can now access sensitive data and spy on companies, affecting bank's bottom line [25]. To tackle these kinds of cybercrimes, the banking sector needs to collaborate with the authorities concern in order to develop a strategy [47]. Asian countries have emerged as the most vulnerable region in the banking sector. Not only Bangladesh, financial establishments in Japan, the Philippines, Taiwan, Thailand, and Vietnam have also come under cyber-attacks, with a loss of billions of dollars throughout the area [78]. The incidents of stealing from remittance are recurring due to frequent attacks on electronic infrastructures of the financial institutions [69]. The cyberattacks in most cases become successful in Asian region compared to other parts of the world due to weak cybersecurity measures, a lack of knowledge and engagement in cybercrime protection [15]. However, although this region has now turned into a safe haven for cybercrime, the financial institutions like banks come under cyberattacks on only in the Asian region but also across the globe [38].

4 Research Methodology

This study used a qualitative research methodology in conjunction with key informant interviews (KIIs) and a number of case studies drawn from the interviewee's experience as well as secondary sources in order to associate the findings with the particular topic. Qualitative approaches facilitate understanding in a manner that is challenging to do with a closed-ended survey [11]. The data collection tool employed an open-ended questionnaire to ascertain the views, beliefs, and current state of affairs of the pertinent banks. For this study, a sample of 28 bank employees from the IT department—particularly those from the Cyber Security Unit (CSU) of multiple banks—were questioned. Senior Principal Officers of Pubali Bank PLC, Sonali Bank PLC, Eastern Bank PLC, Janata Bank PLC, Brac Bank PLC, and the national bank of Bangladesh were among the assistant programmers present. Key informant interviews (KIIs) were conducted concurrently with the experts. There were 25–50 min interview times. Interviews were conducted via audio recordings, which were then transcribed for analysis and discussion. Throughout the interview, brief notes were collected. Every interview was open-ended and rather loosely structured. The purposeful sampling strategy was employed in this study's data collection. The entire study period, from August 13, 2022 to November 15, 2022, was undertaken.

Data was collected between August 19, 2022, and October 29, 2022. Certain obstacles had to be overcome during the study period in order to complete the research activity in line with the goals. Studies on cybercrime and cybersecurity in the banking sector, especially when compared, are extremely rare in Bangladesh [42]. Due to numerous banks' refusal to allow the study to interview them about their cyber security and systems, the sample size is tiny. Participants in many interviews opted to remain anonymous while knowing that the competent development authorities were ineffective, since they did not want to criticize the authorities.

5 Findings and Data Analysis

5.1 Types of Cybercrimes in the Banking Sector

All banks employ the Core Banking Solution (CBS) to centrally manage the flow of funds and information. Every bank, such as Bangladesh Bank, Pubali Bank, and so forth, has an autonomous and distinct CBS [41]. Typically, CBS is bought from any supplier. Bangladesh Bank has purchased its CBS from the India's Tata Company [6].

Fig. 1 Bangladeshi Banks are at high risk of cyber-attacks. Photo Credit: *Dhaka Tribune*

However, Pubali Bank, a private commercial bank in Bangladesh for the first time has created its own CBS [74]. Initially, CBS was purchased and launched, allowing branch users to log in and provide information about their everyday activities and bank transfers [24]. Prior to the advent of CBS, banks could destroy physical or hard copies. However, cybercrimes started to occur in the form of digital fraud or money theft through CBS. Bangladesh Bank Heist of 2016 is an example of such attacks or data leaks (Fig. 1).

Currently, a national effort is underway to establish uniformed CBS for each bank in Bangladesh as this sector has faced different kinds of cybercrimes. A Brac Bank officer Mr. Shahin claims that cybercrimes in Bangladesh's banking sector are growing in complexity and variety. He said the Bangladesh Bank heist, in which money is taken via false transfers, took place as a result of unauthorized access to client accounts. Ransomware, spyware, hacking, and encryption are just a few of the techniques cybercriminals employ to get into our networks and steal personal information. People are regularly tricked into divulging their login passwords by phishing emails and other social engineering tactics. Data manipulation and breaches pose a serious threat to the integrity of our operations. These attacks are increasingly targeting foreign exchange management and debit/credit card reconciliation. Additionally, DOS and DDOS attacks interfere with our services and expose personal information in some cases, sophisticated techniques such as man-in-the-middle attacks have also been observed. Attacks happen all the time, whether they are effective or not Ransomware attacks on BRAC Bank and Social Islami Bank Limited and bank loot attacks on Bangladesh Bank. Recently, Bangladesh Krishi Bank's CBS also got hacked. Financial institutions are frequently the target of fruitless attacks, particularly banks. The bank needs to continuously monitor in order to stop these [19]. The Bangladesh Computer Council (BCC) maintains the sensors that are used to identify attacks. Antivirus software is essential, as it helps detect and prevent cyberattacks [74]. Via online, illicit funds obtained through cybercrime are moved, creating a massive circle. The same risks apply to internal (banking official) rule and regulatory violations as they do to exterior attacks [17]. Bangladesh Bank's Cyber Security Unit (CSU) upholds the regulations in line with this.

6 Evolution of Cybercrime in the Banking Sector

Cybercrime in the banking sector has progressed in emerging nations like Bangladesh because of the growing internet penetration and quick digitization of financial institutions, as well as the consequent rise in the number of skilled cybercriminals. The Banking Regulation & Policy Department (BRPD) of Bangladesh Bank issues some directives for all the banks from 2011 to 13 to this end. The rise of the dark web in 2004–05 was also a major reason behind the evolution cybercrimes in the developing countries [53]. Cybercriminals target developing countries mostly considering their comparatively weak security measures. The incidents of cybercrimes increased massively after 2010 [56]. However, people learnt about cybercrimes after 2012 and they have a clear understanding 2016. According to Motaleb (alias), an IT officer of Bangladesh Bank, said that he is an eyewitness of how the financial industry of Bangladesh is becoming a victim of cybercrime. Even before the 2016 Bangladesh Bank money heist, the Central Banker website had been subjected to several cyberattacks. Those attacks were launched from several neighboring countries, including Sri Lanka. As a result, Bangladesh Bank started taking steps like system updates including penetration testing. Bangladesh Computer Council is one of the organizations that started monitoring cyberattacks in Bangladesh at the beginning. This council emphasizes the importance of software and hardware security. Even before 2016, commercial banks were not in any position regarding cyber security. Later they began to bring about a major change in collaboration with security agencies and experts. Security related upgrades in the banking sector have increased exponentially. From 2019 to 20, strong measures against cyber threats began. The impact and damage caused by cyberattacks in the banking sector is more severe than other sectors, as banks lose both information and money due to attacks. In some cases, information is more valuable than money. So those confidential information and data of certain banks are as vulnerable as money [5]. Cyber attackers often track specific IPs using VPNs. But the current situation is better than previous years due to greater awareness. The situation is somewhat under control after the Bangladesh bank heist, although there are instances of not changing passwords for a long time due to lack of awareness [48]. Recently banks are focusing on standards and compliance as part of effective measures to ensure cyber security [50].

7 Risk Scale of the Banking Systems

Bangladesh currently has 61 scheduled banks. Each of these banks has a separate CBS [41]. Analyzing past cyber-attacks, banks in this country are always at risk of cyber-attacks [1]. As a result, cyber security systems are not completely secure. This CBS needs to be kept up to date with software and discard the use of outdated technology and software [32]. There are some cybercrime related cases which are actually like test cases and they are not effective enough yet finding out and monitoring these cases is not an easy and trivial task. Private commercial bank official Imran (alias) said since the theft from Bangladesh Bank in 2016, the country's banking system has come under strict control and supervision. Still, all the bank officials, especially the senior officials, are still less aware of cyber security. The situation has become more vulnerable due to banks' diverse customer base and lack of IT know-how. So awareness campaigns and training programs are needed to eliminate these problems. However, cyberattacks have decreased somewhat due to IT audits and recommendations regarding security. It is very important to protect banking institutions through awareness as well as careful monitoring and compliance with appropriate regulations. Among the private commercial banks, BRAC Bank is an example of highly advanced third generation banks thanks to its advanced technologically and compliance management. This bank is among the top banks in term of cyber security aspects. Mr Rahim (alias), an officer of Brac Bank, however said this bank still is not out of concern of cyberattacks despite of having technological advancements and maintaining security compliance. They have also come under ransomware attacks, he said. According to Rahim, any bank can be victimized of small or massive cyberattacks. In term of the cyberattack aspects, usually banks are categorized into two parts—one, the banks who have already leant that they has come under cyberattack and the second one, those who are yet to ascertain whether they were attacked or not [75]. Considering the cyber security aspects, Pubali Bank, another private commercial bank, is now in leadership position technologically [74]. However, the biggest challenge to lessen the cybersecurity risks in this sector is because the middle aged and elderly people have no digital literacy due to their backdated in the mindset of technology [27].

8 Limitations of the Banking Sector Regarding Cyber Security

Banks have different financial concerns about cyber security depending on whether they are government or private [49]. Cybersecurity regulations are frequently viewed as additional costs, and a major contributing cause is the general lack of interest in this topic, according to Mr. Akash (alias), an employee of Bangladesh Krishi Bank employee. According to Akash, since cyber security requires huge investment and does not bring any financial benefits, the management of banks does not want to

Fig. 2 Cybersecurity constraints in banking in Bangladesh. Photo Credit: https://divobd.com/

spend money in this sector. But it is very important to understand that spending resources on cyber security is an investment of sorts because it is an investment to protect the investment, the outcome of which is not readily available. However, setting priorities for technology advancements is also challenging because employees of any bank come from diverse backgrounds. So there are conceptual differences about their technology. Lack of technical skills especially among the management members of a bank is significant. Crossing this divide between technology and management is therefore not an easy task. Naturally, security is somewhat different from other IT operations, such as systems, databases, etc., due to security risks. But proactiveness is often overlooked by security personnel. Therefore, this Human Resource (HR) gap prevents any production overall [68] (Fig. 2)

So harmony between people and process ae very crucial although it a big challenge to combination it together despite having adequate technological support. Pubali Bank officer Mr. Rajib Ahmed (Alias), said the major obstacles in ensuring a strong cyber security are using old technology and not updating it, employing outdated software, lack of proper training channels, not having enough skilled personnel and not being aware of security issues. Banks usually prioritize and focus on their business. Managers in banks are actually the leaders of the sector and are less focused on IT and cyber security, so many banks don't pay much attention to buying new tools like management tools or SIM privileges. After all, properly maintaining a bank's cyber security is very difficult. After purchasing a particular CBS, banks have to maintain the specific service level agreement (SLA) of that CBS, which is a constant source of concerns for banks. After the Bangladesh Bank heist in 2016, the central bank took several initiatives, including training on cyber security issues, but these initiatives were later halted due to the Covid-19 pandemic. Nowadays the entire world is concerned about cyber security, and for the same reasons banks are contributing significantly to security solutions by developing policies, IT audits and other measures [72]. Another major way to ensure cyber security is to develop skilled human resources with appropriate use of technology. Because cyber-attacks are often undetectable and there is no guarantee that an attack can occur at any time and place [68].

9 Probable Ways of Dealing with Cybercrime

There is an important security triad called PPT which consists of People, Process and Technology. The first here is people, who work regarding cyber security. The second is the process, how people work in cyber security [71]. Here is a line of difference between ordinary people and technical people and they act differently according to different levels, which is the process. The last one is technology, what and how the technology is used to tackle cybercrime, like machine learning (ML), artificial intelligence (AI) etc. [71]. In order to effectively combat cybercrime, one must make an effort to address each of the triad's components, such as raising public awareness, developing and refining processes, and employing cutting-edge technology [61]. Mr. Ashraf (aka) an official of Sonali Bank Plc said that the most important task in dealing with cybercrime is to increase awareness. Another important task here is to be vocal about both internal and external threats. This requires technology Banks must maintain security guidelines and be aware of systems such as internal and external audits of foreign companies. In that case the security standards will be focused and the authorities will have sufficient sincerity. In that case, not only the management authority, but other big officials should be more attentive. After the 2016 Bangladesh Bank cyber-attack, cyber security is ensured through a different IT department like Bangladesh Bank CSU. Similarly every bank has a separate security department. Since banks are providing digital based services, they are now always vulnerable to cyber-attacks [33]. Sonali Bank has a Computer Incident Response Team (CIRT) working round-the-clock to detect cybercrime and alert the bank immediately in case of any cyber incident. BRAC Bank established the first CSU in Bangladesh in 2009–10. Currently other banks have also established CSU. Mrs Neetu (alias), an executive of Pubali Bank, said the advantage of the Security Operation Center (SOC) is that it monitors the system round the clock. When a cyberattack occurs, one system tries to attack another from an IP address that is displayed on a map. In this case, the severity of the attack is determined by the risk scale. Now not only leading banks but also private banks have introduced CSU. Here, the SIM solution acts as the base and the alarm will be generated automatically. While some programs could meet the needs of people and processes but not technology, Pubali Bank is concentrating more on the latter. Even though Pubali Bank currently lacks personnel and procedures, they are still in the lead when it comes to these two. However, when it comes to technology and other digital equipment, Brac Bank is without a doubt in the lead [54].

10 Discussion

With time, Bangladesh becomes more and more susceptible to cybercrime [12]. A variety of tactics are used by cybercriminals in the banking sector, including hacking, malware, ransomware, encryption, social engineering techniques like phishing emails, data manipulation and breaching, foreign exchange management, DOS and DDOS attacks, compromise of debit or credit cards, data leakage of passwords, credit card information, etc., and man-the-middle attacks [47]. Despite improvements in cybersecurity aspects, cybercrime still affects the banking sector, and causes a loss of both money and data [5]. The anomie theory holds that societal standards and values have an impact on people's propensity to engage in deviant behavior. This theory emphasizes how important it is to encourage moral behavior and honesty in the banking sector in order to lessen the likelihood of internal threats and collaboration with outside hackers within the context of cybercrime [10]. According to the social learning theory, people will imitate criminal behavior if they witness it and are frequently encouraged to do so. This suggests that imitation and common standards could allow cybercrime to spread throughout the banking sector. This concept highlights the significance of motivating stakeholders and bank employees to embrace a culture of moral conduct and cybersecurity awareness [64]. Whether successful or not, attacks happen all the time. In particular, unsuccessful attacks tend to occur in various institutions, thus banks must constantly monitor to prevent them [75]. As seen during an attack, antivirus software is essential and the first line of defense to break into a machine. Types of cybercrime in Bangladesh include illegal money transfer and violation of rules and regulations by both foreign and domestic sources [74]. When compared to the rich countries of the world, not only Bangladesh, but all the countries of the growing South Asia lag behind in terms of security and concern. Bangladesh naturally lags behind industrialized countries in policy and strategy [78]. Due to differences in economic capabilities, banks in Bangladesh, like underdeveloped countries, have much lower financial budgets for cyber security than developed countries [36]. In developing countries, most of the annual budget is spent on infrastructure. As well as incurring expenses related to construction and operation of data centers and purchase of hardware. But not enough budget is kept in the field of security. Developing countries like ours are fortunate to have access to the necessary equipment. As a result Bangladesh has to depend on third countries for any IT solution. Hence geographical location, dependence and foreign policy become important factors here [36].

Routine activity theory states that cybercrime is more likely when there are no acceptable targets or capable guardians. This theory emphasizes the necessity of national cybersecurity capabilities and training programs in Bangladesh in order to effectively combat cyberthreats [62, 63]. Resources have to be built up and an IT hub is essential to be grown in the part of our side of the globe. Authorities must take steps regarding this, to improvise our cyber security fields. IT experts of Bangladesh are not specified to an extent, e. g. security, database etc. Most of the IT officials rather know a bit about hybridization of all components like linux, windows, security etc. and thus, they are not specialized enough which is seen almost in every developed country. All kinds of security systems are developed by giant countries, countries like Bangladesh do nothing but maintain them. This lack of knowledge is a prime factor here, as well as the lack of proper training on the basis of security [76]. Every bank should have internal tools to detect the presence of various types of malware such as phishing mails. Cyber-criminals steal money from certain banks' CBS with access after the attack. Dealing with such attacks is a major challenge for commercial banks. In terms of capacity, private banks are ahead of government banks. But in terms of risk, private banks are more risky than public sector banks. In order to combat cybercrime in banks with ties to both the public and private sectors, it is important to promote moral decision-making and accountability. The Moral Disengagement Model elucidates the psychological mechanisms that enable people to rationalize immoral behavior [4]. As stated by Unit21 [77], digital visibility increases the likelihood of attacks, even private ones. It's crucial to remember, though, that some banks—like Brac and EBL—are now technologically capable of handling any form of cybercrime. The private commercial banks have established adequate infrastructure and operating policies, and they have achieved notable strides in the field of cyber security. Due to their strict regulations, private banks have no losses whereas certain institutions are still trying to reduce their losses. So that they can work harder and spend the money on things like cyber security management and business policy. The risk of cybercrime in the banking sector of Bangladesh and different types of cybercrime are discussed above. Apart from this, the necessity and necessity of ensuring proper cyber security system in both private and public banks of the country was also discussed. Apart from this, a comparative picture of Bangladeshi banks with developed countries including Asian countries is presented in dealing with cybercrime. Not only the government but the authorities of the banks should always be well prepared to fight in cyberspace. Cyber-attacks on banks are actually aimed at stealing and destroying a country's foreign exchange reserves and assets. Adequate cyber security policy must be ensured to be a viable solution to cybercrime prevention. At the same time the concerned authorities should be made aware and create harmony with the general public.

11 Recommendations

The following recommendations should be considered by the government of Bangladesh and the concerned authorities in order to solve the existing and upcoming cybersecurity-related problems in the Baking sectors of Bangladesh.

(i) In Bangladesh, there is no other option for transferring money than the banking industry. Each individual must be a client of a specific bank. In light of this, protecting the security of the banking industry is crucial in comparison to other industries. People don't give a damn about the details of their bank transactions—such as the amount, fees, or deductions from the primary account—and are less concerned about them. People actually engage with a binary transaction, such as "I did the work, now pay me," and have very little concept of cyber security. It doesn't matter if they pay me slightly less or a little more.

(ii) There is a Near Field Communication (NFC) Regulation of Bangladesh Bank which is a wireless medium by using card service. The Department of Banking Inspections (DBI) is a new unit which has been formed with 7 sub-units in Bangladesh Bank for security issues. A Customer Interest Protection Center (CIPC) has been formed for customer satisfaction in Bangladesh Bank. But it is a matter of sorrow that a large number of people do not even know about these services, which should be met up as soon as possible. Cyber awareness is a must and basic cyber security must be introduced for all people as well as a certain chapter in schools. Thus, enhancing awareness, especially, security awareness and granting more funds in cybersecurity issues can be effective enough to prevent cybercrime.

(iii) There are many more things to do with the PPT, e.g. compliance management, increase the experts and cyber security resources, making people aware and raising processes etc. The position of Chief Technology Officer (CTO) is now filled a bit, but still to be fulfilled. The position of Chief Information Security Officer (CISO) is still empty on a large scale. But their presence is a thing of hope in various banks, which was absent even a few years ago. Pubali Bank is in a process with recruitment of CISO and enlarges the position of visibility. They are in a better position now along with Brac Bank and this could be followed by other banks.

(iv) Resources must be built up enough, as IT hubs are essential to be grown in the part of our side of the globe. Authorities must take steps regarding this, to improvise our cyber security fields as well as other regulations regarding the cyber world.

(v) Everyone is under the security threat as every single thing is trackable in this era, e. g. private photos, information, password, identity proof etc. Every phone data is used in investigation and the large scale of digital footprint has enhanced the risk. Users should be more aware regarding their pin, information, card etc. especially, while making transactions via mobile banking like Bkash, Rocket, Nagad etc.

(vi) Many times information is accessed from the mobile banking account via fake one-time password (OTP) and through successful ransomware attacks on the network, so people must be aware of suspicious transactions. People can be impacted not only personally but also organizationally.

(vii) For banks, transaction monitoring, vulnerability assessment, penetration testing, updating software, using antivirus, arranging proper training sessions both for IT and non-IT professionals and using security event management software are some essential procedures to be met up in order to build a proper secured banking system.

12 Conclusion

Bangladesh is a developing nation with fast-growing banks as per the The World Bank Report, 2023. Recent banking industry cyber-attacks demonstrate the urgent need for Bangladesh to boost cyber security. The main reason for cybercrime is digital footprint mobility. Bangladesh's digital presence is growing. As the country's reserve grows, digital security is more important than ever. Financial institution cybersecurity prioritizes user resources. More people are ditching cash, hence more transactions are online. Digital payments like credit and debit cards must be safe. This type of cybercrime can steal money or access confidential data. Criminals target banks because they store a lot of data and money. Financial institutions will feel more prepared to handle rising threats when they switch from reactive to proactive security. Poor manpower placement is a key issue. Especially the government. Private Banks are ahead of operated banks in cyber security and cybercrime prevention. Bangladesh is performing better than other surrounding countries, but it needs to do more to catch up to Asia's technologically secure and proactive nations and the rest of the world. Since everything is trackable and digital footprints are vast, everyone is at danger of security breaches. The world is becoming a community, and no one can stay healthy by avoiding technology. Security is crucial since it safeguards our sensitive data and money. Thus, social and cyber security awareness is essential to combat banking sector cybercrime with financial institutions and respected state agencies. After all, banking should abandon reactive cybercrime response and implement proactive security measures. It's crucial to protect the state's reserves, citizens' money and data, and banks' digital transactions.

References

1. Afrin S (2023) Cybersecurity challenges in Bangladesh's financial sector. The Daily Star. https://www.thedailystar.net/supplements/digital-transformation-bangladeshs. Accessed 10 May 2024
2. Ali M (2023) Exactly why is cybersecurity crucial for the banking industry? VaporVM. https://vaporvm.com/exactly-why-iscybersecurity-crucial-for-the-banking-industry. Accessed 12 Jul 2024
3. Archon (2022) Cyber threats in the Banking Industry. https://www.archonsecure.com/blog/banking-industry-cyber-threats. Accessed 20 May 2024
4. Bandura A, Barbaranelli C, Caprara G, Pastorelli C (1996) Mechanisms of moral disengagement in the exercise of moral agency. J Pers Soc Psychol 71:364–374
5. Bangladesh: data protection in the financial sector (2022) Data Guidance. https://www.dataguidance.com/opinion/bangladesh-data-protection-financial-sector. Accessed 12 Feb 2024
6. Bangladesh Bank (2023) Core Banking Solution (CBS) features and controls. Bangladesh Bank. https://www.bb.org.bd/aboutus/regulationguideline/dfim/cbs_nbfi_sep2023.pdf. Accessed 11 Jan 2024
7. Barney N (2022) What is network security? Definition, importance and types. TechTarget. https://www.techtarget.com/searchnetworking/definition/network-security. Accessed 18 Mar 2024

8. Bellucco A (2023) Financial technology (Fintech): its uses and impact on our lives. Investopedia. https://www.investopedia.com/terms/f/fintech.asp. Accessed 31 Jan 2024
9. Benny A (2023) 17 most common types of cyber attacks and examples (2024). https://www.aura.com/learn/types-of-cyber-attacks. Accessed 15 Oct 2024
10. Bernburg JG (2019) Anomie theory. In: Criminology and criminal justice. https://doi.org/10.1093/acrefore/9780190264079.013.244. Accessed 21 Feb 2024
11. Busetto L, Wick W, Gumbinger C (2020, May 27) How to use and assess qualitative research methods. Neurol Res Pract. https://doi.org/10.1186/s42466-020-00059-z. Accessed 30 Mar 2024
12. Byron RK, Zaman A (2015) Bangladesh vulnerable to cyber-attacks. The Daily Star. https://www.thedailystar.net/business/bangladesh-vulnerable-cyber-attacks-73048. Accessed 15 Jul 2024
13. Center for Strategic and International Studies (2022) financial sector cybersecurity. https://www.csis.org/programs/strategic-technologies-program/archives/cybersecurity-and-governance/financial-sector. Accessed 23 Sep 2023
14. Chen S, Hao M, Ding F et al (2023) Exploring the global geography of cybercrime and its driving forces. Humanit Soc Sci Commun 10:71. https://doi.org/10.1057/s41599-023-01560-x. Accessed 14 Feb 2024
15. Choudhury SR (2016, October 18) Asia a target for hackers due to large victim pool, low awareness, lack of disclosure norms and use of unlicensed technology. CNBC. https://www.cnbc.com/2016/10/18/asia-a-target-for-hackers-due-to-large-victim-pool-low-awareness-lack-of-disclosure-norms-and-use-of-unlicensed-technology.html. Accessed 21 Aug 2024
16. Collier J (2022) Insights: the financial sector remains a popular target for cybercriminals. Gulf Business. https://gulfbusiness.com/insights-the-financial-sector-remains-a-popular-target-for-cybercriminals. Accessed 12 June 2024
17. Comply Advantage (2021, February 24) How are cybercrime and money laundering related? ComplyAdvantage. https://complyadvantage.com/insights/cybercrime-money-laundering/. Accessed 5 Mar 2023
18. Cossin D, Lu AH (2021) Board oversight of cyber risks and cybersecurity. IMD Business School. https://www.imd.org/research-knowledge/corporate-governance/articles/board-oversight-cyber-risks-cybersecurity/. Accessed 18 Oct 2023.
19. Dhaka Tribune (2021, April 2) Cyber-attacks hit over 200 organizations including Bangladesh Bank, BTRC. Dhaka Tribune. https://www.dhakatribune.com/bangladesh/242875/cyber-attacks-hit-over-200organizations. Accessed 28 Jun 2023
20. Digby A (2023, February 13) Attack surface management for financial services. Informer.io. https://informer.io/resources/attack-surfacemanagement-financial-services. Accessed 31 Jan 2024
21. Elliott J, Jenkinson N (2020, December 7) Cyber risk is the new threat to financial stability. International Monetary Fund. https://www.imf.org/en/Blogs/Articles/2020/12/07/blog-cyber-risk-is-the-new-threat-to-financial-stability. Accessed 15 Apr 2024
22. Enhalo (2021) Cybercrime is a major threat to the financial sector. https://enhalo.co/360-security/cybercrime-major-threat-to-financial-sector-Eriksson. Accessed 12 May 2024
23. Eriksson J, Giacomello G (2022) Cyberspace in space: fragmentation, vulnerability, and uncertainty. https://doi.org/10.4324/9781003110224-8/cyberspace-space-johan-eriksson-giampiero-giacomello. Accessed 17 Oct 2024
24. FYNDNA TechCorp (2022, March 28) Evolution of CBS. LinkedIn. https://www.linkedin.com/pulse/evolution-cbs-fyndna. Accessed 21 Jul 2024
25. Fatoki J (2023) The influence of cyber security on financial fraud in the Nigerian banking industry. Int J Sci Res Arch 9:503–515. https://doi.org/10.30574/ijsra.2023.9.2.0609
26. Goel S (2016) Cyber-crime: a growing threat to Indian Banking Sector. Int J Sci Technol Manage 5(12)
27. Gregory S (2023, May 4) Challenges faced by the banking industry in the digital age. Mondo. https://mondo.com/insights/challenges-facing-banking-industry-digital-age/. Accessed 16 Jan 2024

28. Halder D, Jaishankar K (2011) Cyber crime and the victimization of women: laws, rights, and regulations. IGI Global, Hershey, PA, USA. ISBN 978-1-60960-830-9
29. Hopkins J (2021) Cyber security in the age of COVID-19: a timeline and analysis of cyber-crime and cyber-attacks during the pandemic. NCBI. https://www.ncbi.nlm.nih.gov/pmc/articles/PMC9755115/. Accessed 18 Aug 2024
30. Hossain M, Islam M, Sattar Z, Choudhury A, Dasgupta P, Elahi SM (2020) Role of mobile phones in household welfare, women's empowerment. The Financial Express. https://thefinancialexpress.com.bd/views/role-of-mobile-phones-in-household-welfare-womens-empowerment-1603111521. Accessed 12 Jul 2023
31. IP Specialist off (2022, March 15) Prevention against ransomware attack. LinkedIn. https://www.linkedin.com/pulse/prevention-against-ransomwareattack-ipspecialistofficial. Accessed 28 Feb 2024
32. IT Convergence (2023, April 6) Risks of using outdated operating system. IT Convergence. https://www.itconvergence.com/blog/risks-of-using-outdated-operating-system/. Accessed 11 Jan 2024
33. Imeson M (2017) How banks are combating the cyber threat. The Banker. https://www.thebanker.com/How-banks-are-combating-the-cyber-threat-1506931215. Accessed 21 Sep 2024
34. Imperva (2023, December 7) Financial services cybersecurity threats and solutions Imperva. Imperva, Inc. https://www.imperva.com/learn/data-security/financial-services-cybersecurity/. Accessed 12 May 2024
35. International Monetary Monetary Fund, & Capital Markets Department (2023) Global financial stability report April 2023: safeguarding financial stability amid high inflation and geopolitical risks. International Monetary Fund. https://www.imf.org/en/Publications/GFSR/Issues/2023/04/11/global-financial-stability-report-april-2023. Accessed 15 Oct 2024
36. Investopedia (2024) Importance and components of the financial services sector. Investopedia. https://www.investopedia.com/ask/answers/030315/what-financial-services-sector.asp. Accessed 23 Jun 2024
37. Islam S (2018) Majority of banks still vulnerable to cyber-attacks. https://www.dhakatribune.com/opinion/special/2018/02/05/majority-banks-still-vulnerablecyber-attacks. Accessed 19 Sep 2024
38. Jones A (2016) Cybercrime: the growing threat to global banking. https://internationalbanker.com/banking/cybercrime-growing-threat-global-banking/. Accessed 21 Jul 2024
39. Joveda N, Khan MT, Pathak A (2019) Cyber laundering: a threat to banking industries in Bangladesh: in quest of effective legal framework and cyber security of financial information. Int J Econ Finan 11(10)
40. Kabir R (2018) Cyber crime in Bangladesh: a growing threat in Banking Business. https://www.academia.edu/25668378/Cyber_crime_in_Bangladesh_A_growing_threat_in_Banking_Business. Accessed 12 May 2024
41. Kashem A, Hossain S, Abadin Z (2020, September 25) State-owned commercial banks seek to develop core banking solution software. The Business Standard. https://www.tbsnews.net/economy/banking/state-owned-commercial-banks-seek-develop-core-banking-solution-software-137425. Accessed 21 Jun 2024
42. Khan TA, Mahmood M, Elahi SM (2021) Cyber security and the role of Bangladesh Bank. The Financial Express. https://thefinancialexpress.com.bd/views/reviews/cyber-security-and-the-role-of-bangladesh-bank-1612015431. Accessed 12 May 2024
43. Kharouni L (2012) Automating online banking fraud automatic transfer system: the latest cybercrime toolkit feature (Rep.)
44. Kuepper J (2022) Cyberattacks and the risk of bank failures: how major cyberattacks can impact banks. Investopedia. https://www.investopedia.com/articles/personal-finance/012117/cyber-attacks-and-bank-failures-risks-you-should-know.asp. Accessed 27 May 2024
45. Kundu S, Islam K, Jui T (2018) Cyber-crime trend in Bangladesh, an analysis and ways out to combat the threat. 1–2. https://doi.org/10.23919/ICACT.2018.8323799
46. Lagarde C (2018, June 22) Estimating cyber risk for the financial sector. International Monetary Fund. https://www.imf.org/en/Blogs/Articles/2018/06/22/blog-estimating-cyber-risk-for-thefinancial-sector. Accessed 23 Sep 2024

47. Lodha H (2022) An overview of cyber crimes in banking sector. https://www.legalserviceindia.com/legal/article-7694-an-overview-of-cyber-crimes-in-banking-sector.html. Accessed 12 May 2024
48. Luthra P, John P (2023, May 23) Financial cyber-crimes high due to lack of awareness, vulnerabilities in system, says expert. CNBC TV18. https://www.cnbctv18.com/technology/financial-cyber-crimes-high-due-to-lack-of-awareness-vulnerabilities-in-system-says-expert-16742171.htm. Accessed 22 Sep 2024
49. Mahmood M, Amir SR, Rahman M, Khan RR, Anjum SN, Elahi SM (2019) The current state of the banking industry in Bangladesh. The Financial Express. https://thefinancialexpress.com.bd/views/views/the-current-state-of-the-banking-industry-in-bangladesh-1556380055. Accessed 23 Feb 2024
50. Martinez P (2021, November 14) Cybersecurity compliance: the key to cross-border banking. LinkedIn. https://www.linkedin.com/pulse/cybersecurity-compliance-key-cross-border-banking-martinez-cissp. Accessed 12 May 2024
51. Mee P, Schuermann T (2018, September 14) How a cyber attack could cause the next financial crisis. Harvard Business Review. https://hbr.org/2018/09/how-a-cyber-attack-could-cause-the-next-financial-crisis. Accessed 21 Jun 2024
52. Najjar A (2023, July 25) Digital transformation and cybersecurity challenges for businesses resilience: issues and recommendations. NCBI. https://www.ncbi.nlm.nih.gov/pmc/articles/PMC10422504/. Accessed 27 Aug 2024
53. Nazah S, Huda S, Abawajy J, Hassan M (2020) Evolution of dark web threat analysis and detection: a systematic approach. IEEE 171796–171819.https://doi.org/10.1109/ACCESS.2020.3024198
54. Nazmul AK, Hassan F, Hossain MM (2019) Impact of stress on employees performances: a case study on Pubali Bank Limited. AIUB J Bus Econ 16(1):163–181
55. Nelson A (2021) The global cyber threat to financial systems—IMF F & D. International Monetary Fund. https://www.imf.org/external/pubs/ft/fandd/2021/03/global-cyber-threat-to-financial-systems-maurer.htm. Accessed 18 Mar 2024
56. Otieno D (2020) Cyber security challenges: the case of developing countries. https://www.researchgate.net/publication/346485466_Cyber_security_challenges_The_Case_of_Developing_Countries. Accessed 15 Oct 2024
57. Ozarslan S (2022) Ransomware risk mitigation strategies for Financial Institutions. Picus. https://www.picussecurity.com/key-threats-and-cyber-risks-facing-financial-services-and-banking-firms-in-2022. Accessed 23 July 2024
58. Packetlabs (2021) Cybersecurity threats in the Banking Sector. https://www.packetlabs.net/posts/banking-and-cybersecurity. Accessed 23 Aug 2024
59. Pandey P (2020, May 28) Massive shift to digital space increases the need for cyber Security experts. Times of India. https://timesofindia.indiatimes.com/education/news/massive-shift-to-digital-space-increases-the-need-for-cyber-security-experts/articleshow/76072041.cms. Accessed 28 Feb 2024
60. Paul S (2022, September 19) Bangladesh is at serious risk of cyber-crimes. Dhaka Tribune. https://www.dhakatribune.com/opinion/oped/294341/bangladesh-is-at-serious-risk-of-cybercrimes. Accessed 12 May 2024
61. Plutora (2020, September 8) What is operational intelligence? A detailed introduction. Plutora. https://www.plutora.com/blog/people-processtechnology-ppt-framework-explained. Accessed 31 Jan 2024
62. Ramirez ER (n.d.) Introduction to cybercrime—computers and criminal justice. College of DuPage Digital Press. https://cod.pressbooks.pub/crimj1165/chapter/module-1/. Accessed 28 Jan 2024
63. Ramirez ER (2024) Criminological explanations of cybercrime – computers and criminal justice. College of DuPage Digital Press. https://cod.pressbooks.pub/crimj1165/chapter/module-3/. Accessed 19 May 2024
64. Rogers MK (2006) The development of a meaningful hacker taxonomy: a two dimensional approach. Digital Invest 97–102

65. Sen K (2023, August 21) Why is the finance sector a target for cyber attacks? UpGuard. https://www.upguard.com/blog/finance-sector-cyber-attacks. Accessed 4 Aug 2024
66. SentinelOne (2023, August 22) Cyber attacks on Financial Institutions: why banks are caught in the crosshairs. SentinelOne. https://www.sentinelone.com/blog/a-cyberwar-on-financial-institutions-why-banks-are-caught-in-the-crosshairs/. Accessed 12 May 2024
67. Sergi S (2023) Cybersecurity threat trends. Deloitte. https://www2.deloitte.com/uk/en/insights/topics/cyber-risk/global-cybersecurity-threat-trends.html. Accessed 19 Jul 2024
68. Sharma V (2024, January 17) Cybersecurity in Banking Sector: importance, threats, challenges. KnowledgeHut. https://www.knowledgehut.com/blog/security/cyber-security-in-banking. Accessed 17 Aug 2024
69. Sharma J, Verma S (2023) Mounting cases of cyber-attacks and digital payment
70. Shea S, Guide S (2024) What is cybersecurity? Definition from TechTarget. TechTarget. https://www.techtarget.com/searchsecurity/definition/cybersecurity. Accessed 18 Jul 2024
71. Simon B (2019) Complete guide to the PPT framework. Smartsheet. https://www.smartsheet.com/content/people-process-technology. Accessed 18 mar 2024
72. Srinivasulu K (2023, August 1) Cybersecurity threats in digital banking sector. ET CIO. https://cio.economictimes.indiatimes.com/news/digital-security/cybersecurity-threats-in-digital-banking-sector/102298156. Accessed 22 May 2023
73. Stanciu V, Tinca A (2017) Exploring cybercrime—realities and challenges. J Account Manage Inf Syst 16:610–632. https://doi.org/10.24818/jamis.2017.04009. Accessed 21 Jul 2024
74. The Daily Star (2023, November 22) Pubali Bank prioritizes robust infrastructure and security for successful digitalization. The Daily Star. https://www.thedailystar.net/supplements/digital-transformation-bangladeshs-financialindustry-2023/news/pubali-bank-prioritizes-robust-infrastructure-and-security-successful. Accessed 20 Jun 2023
75. The Digital Banker (2023, September 4) Poor cyber security puts banks in Bangladesh at peril. The Digital Banker. https://thedigitalbanker.com/poor-cyber-security-puts-banks-in-bangladesh-at-peril/. Accessed 22 May 2024
76. Ullah M (2020) Banking sector the biggest risk to Bangladesh economy: survey. The Business Standard. https://www.tbsnews.net/economy/banking/banking-sector-biggest-risk-bangladesh-economy-survey-45535. Accessed 12 May 2024
77. Unit21 (2023, May 10) Risk management in banking: types + best practices for mitigation blog. Unit21. https://www.unit21.ai/blog/risk-management-in-banking. Accessed 23 Jul 2024
78. Vo DH (2020, July 4) Financial inclusion and stability in the Asian region using bank-level data. NCBI. https://www.ncbi.nlm.nih.gov/pmc/articles/PMC7334937/. Accessed 12 May 2024
79. Welance (2023, April 28) Impact of cybercrime on individuals, businesses, and society. LinkedIn. https://www.linkedin.com/pulse/impact-cybercrime-individuals-businesses-society-join-welance. Accessed 24 May 2023
80. Zaman A (2023, November 22) Evolution of digitalization in Bangladesh's financial sector. The Daily Star. https://www.thedailystar.net/supplements/digital-transformation-bangladeshs-financial-industry-2023/news/evolution-digitalization-bangladeshsfinancial-sector-3475836. Accessed 10 Oct 2024

Comprehensive Analysis of mSpy Within Covert Operations

Sohail Mohammed and Shahrzad Zargari

Abstract The rapid development of mobile technology is creating new opportunities for crime targeting smart devices and wireless networks, posing major challenges to public safety. Law enforcement must skilfully collect and present digital evidence in court to maintain its reliability and accuracy. The main topics of this research are the characteristics, applications, and effects of mobile spyware in law enforcement and intelligence operations. Through in-depth research, we explore the characteristics of mobile spyware and how they affect the integrity and recovery of forensic evidence. This study examines the latest advances in spyware technology and methods for collecting real-time statistics on calls, messages, images, social media interactions, and location data. Previous research shows that sophisticated mobile spyware often evades detection, making it easier to gather evidence. Such sophisticated spyware can challenge established forensic techniques by removing or altering important information. However, this study shows how effective cell phone surveillance and detective programs like mSpy are in tracking crimes and improving the quality of evidence collection. The results show how using the latest tools can lead to better detection and analysis. In addition, the research emphasizes the importance of approaches to solve the growing challenges of mobile devices. The study then provides law enforcement organisations and covert agents with useful information about the effectiveness of mobile spy apps and programs like mSpy in crime prevention. By using this cutting-edge technology, law enforcement agencies can greatly improve the efficiency of their investigations and ensure the integrity of digital evidence used in court.

Keywords MSpy · Digital forensics · UMobix · The Seizure of digital evidence · Mobile operating system

---

S. Mohammed · S. Zargari (✉)
Sheffield Hallam University, Sheffield, UK
e-mail: s.zargari@shu.ac.uk

H. Jahankhani and B. Issac (eds.), *Cybersecurity and Human Capabilities Through Symbiotic Artificial Intelligence*, Advanced Sciences and Technologies for Security Applications, https://doi.org/10.1007/978-3-031-82031-1_21

1 Introduction

1.1 Background of Study

Mobile devices have shaped both personal communication and organisational operation over the years as they've become invaluable tools that facilitate communication to business. Evidenced by Silver (2019), showing social connections between users have exceeded to 5 billion during 2019. However, technological advances in mobile technology over the years showed an increase in shadow of abuse and exploitations, posing a significant threat to individuals' privacy within cyber security.

1.2 Recent Events in the Abuse of Technology

Two surveys, 2013 and 2015, emphasised the risks connected to technology, especially when it comes to manipulation and misuse. 62% of respondents to the 2013 poll reported horrifying instances of internet stalking, and 98% of participants reported having experienced abuse, mostly through text messages [1]. The 2015 poll also highlighted the dangers that exist on social media, 25% of participants had experiences with extortion and coercion related to the disclosure of personal photos or videos [2]. Based on the results of both polls, people experienced severe abuse and danger, with spyware playing a substantial role.

The Pegasus programme from the NSO Group is a well-known instance of this type of spyware. Pegasus is designed to stealthily enter systems, evading antivirus software and enabling its operators to operate it from a distance. It may access personal information, listen in on conversations, and even turn on cameras and microphones without the user's awareness [3]. The Pegasus Project investigation by the guardian revealed widespread abuse of this technology by governments across the globe to target political opponents, activists, and journalists [4]. Also, Facebook and WhatsApp filed lawsuits against the NSO Group in 2019, claiming that the illegal use of their servers for the distribution of Pegasus spyware violated multiple laws, including the Computer Fraud and Abuse Act (CFAA). The NSO Group filed appeals, but the Supreme Court finally denied them to have the case dismissed on the grounds of foreign official immunity and derivative sovereign immunity [5].

1.3 Advantages of Spyware and Case Relation

When utilised properly, spyware can be a powerful tool for national security and law enforcement. According to The Guardian's 2022 analysis, they show recent instances from 2018 to 2020 highlighting how successful it is at keeping an eye on illegal activity on mobile devices. In 11 investigating cases, surveillance technology and spyware were used by Canadian police to obtain vital data, including emails, calls, and text messages [6]. Law enforcement can obtain evidence lawfully with the help of mobile spyware programmes like Flexi Spy, mSpy, and UMobix, which contribute to both criminal prevention and public safety upkeeping. Spyware improves the investigative powers of law enforcement when used responsibly and under supervision.

The primary scope of this study is to identify and analyse the performance of mSpy for covert operations. This involves identifying real-time data and extraction with the comparison of forensic tools for evaluating the overall performance. Existing studies have shown the challenges faced with mobile devices that could cause a huge problem with the performance, reliability, and accuracy of data. However, there is a lack of research in spyware that contributes with covert operations, in an experimented approach.

The goal of this paper is to address the gap in knowledge and provide a comprehensive analysis on mSpy application for covert and law enforcement companies seeking to evaluate the suitability of mSpy in terms of detection and prevention of cybercrimes.

2 Literature Review

2.1 Challenges of Mobile Forensics

Due to the rapid growth of technology, there are several obstacles in the field of mobile forensics such as: Advancement of technology, bypassing device locks, WhatsApp deleted messages and battery life are some of the major areas of difficulty that have been uncovered via comprehensive research in this area.

2.1.1 Advancement of Technology

The research study conducted by Montasari and Hill [7], offers a thorough examination of the difficulties brought on by the quick development of mobile device technology and its consequences for the field of mobile forensics. In the dynamic and ever-changing field of information and communication technology (ICT), mobile technology is very essential to government, trade, and communication among other areas. Since mobile devices are used by society for a wider range of functions, communication methods have become more complex than they were in the past. Forensic investigators face a significant obstacle because they must constantly adjust to keep up with the technological improvement. Because mobile technology is developing so quickly, forensic investigators need to regularly update their policies and procedure to handle the new capabilities and complexity that come with mobile devices. Additionally, Humphries et al. [8] study highlights the difficulties presented by cutting-edge methods like security hardening, the lack of procedural standard, management of large volumes of data and limited availability of datasets for the mobile devices. This study emphasises the ways in which technical developments add complexity to forensic investigations, underscoring the necessity of continuous innovation and adaptability in the area [8].

2.1.2 Bypassing Device Lock

The difficulties of mobile forensics are clarified by Fukami et al. [9], especially regarding encrypted and locked user data. Since modern gadgets incorporate cutting-edge security features, they draw attention to how difficult it is to unlock devices without the correct user authentication credentials. Zhang et al. [10] provide examples of how thieves can circumvent device locks by creating false fingerprints using silica gel, jeopardising commonly used fingerprint identification methods. Furthermore, the influence of fingerprint scanners on mobile devices is examined by Yu et al. [11], using the Vivo X80 Pro as an example, that has a mounted fingerprint scanner. Fingerprint presses made by accident have the potential to lock the device, which presents a big problem for forensic experts when trying to recover evidence. These research highlights the dynamic field of mobile forensics, and the difficulties investigators face when obtaining important evidence.

2.1.3 Battery Life

As it comes to mobile forensic investigation, it is critical to carefully preserve electronic devices when they are seized. Goel and Kumar [12] provides light on the intricacies involved in this stage, emphasising a disadvantage related to the use of blocking material or turning on aeroplane mode to protect devices from third party or remote sources. According to their research, doing so could hasten the device battery depletion and cause it to malfunction. This result could hinder the forensic procedure and jeopardise the integrity of the information contained within. Furthermore, the study conducted by Sharma et al. [13] provides additional insights into this issue. They propose that removing the battery of the device, when its necessary for identify verification may cause it to lock the sim card. In some cases, this procedure might even cause the data kept in volatile memory to be totally erased. These results from both authors illustrates one of the biggest challenges that forensic investigators find when maintaining the integrity of the evidence and guaranteeing the device operability. To maintain the accuracy and dependability of the forensic procedure, efforts to address these issues must be made precisely and thoughtfully.

2.1.4 Burner Phones

The study conducted by McTurk [14] illuminates the complex relationship that exists between terrorists and mobile technology. It highlights the widespread usage of low-cost burner phones, such as the Nokia 110 4G, for anonymous communication during criminal actions. These throwaway gadgets offer a fleeting form of communication, making it more difficult for law enforcement to follow movements. According to McTurk, forensic investigators encounter difficulties when trying to recover data from these disposable devices because of the anonymity offered by prepaid plans and carriers. Furthermore, cloud technology facilitates cyberterrorism activities like hacking and money laundering, providing terror groups with a channel of communication [15].

Zinkus et al. [16] highlight data extraction difficulties caused by inaccessible APIs and disabled data ports, which emphasise the difficulties prepaid burner phones present for mobile forensics. Standard classification system extracts are available, however, memory space allotment constraints require skilled decryption to successfully navigate the forensic inquiry process.

2.1.5 The Seizure of Digital Evidence

The study conducted by Karagiannis and Vergidis [17] highlights that law enforcement faces a significant obstacle when it comes to seizing digital evidence. One major obstacle is that there isn't a widely accepted definition of cybercrime, which makes it difficult to create standardised protocols for obtaining evidence. In addition, accessing digital evidence and acquiring the power required to seize it are challenges that investigators frequently face. Multinational investigations involving cybercrime offenders become more complex because of the disparities in techniques used by various nations and organisations. Karagiannis and Vergidis [17], recommend creating forums where law enforcement agencies can exchange best practices and experiences in investigating cybercrimes to overcome these obstacles, promote cooperation, and improve their overall ability to apprehend digital evidence. This cooperative strategy highlights how crucial it is for law enforcement organisations to share expertise and learn from one another to effectively tackle cybercrime.

2.1.6 Mobile Operating Systems (OS)

Hewawitharana [18] highlights where desktop/laptop operating systems and mobile operating systems (OS) vary significantly technologically. Mobile Operating Systems, in contrast, only support one programme at a time and do not have true multitasking features. Long-running background processes or memory limitations may cause interruptions to background programmes, which are suspended until the user actively requests them to resume. Qiao et al. [19], on the other hand, clarified the complex issues surrounding mobile augmented reality (AR), highlighting the difficulties associated with navigating various computing platforms, operating systems, and data types. Compatibility problems linger, hindering the development of web augmented reality even with the growing popularity of hybrid mobile apps. Adding more complexity is the need to support several operating systems, display devices, and sensors. The problem of integrating virtual content from various systems is another. To overcome these issues, academic institutions, corporate entities, and standards organisations must work together extensively. This highlights some of the major obstacles that forensic investigators must overcome to access and evaluate data across various mobile platforms and technologies.

2.1.7 Encryption

Modern cell phones prioritise data security with strong encryption settings including password locks, encrypted memory cards, and biometric authentication, providing serious hurdles for forensic investigators, according to Krishnan et al. [20]. With the rising complexity of decryption and password recovery, techniques like chip-off extraction, Flasher box, and JTAG become important. Investigations are made more difficult by anti-theft techniques including automatic data destruction following unsuccessful efforts. Fukami et al. [9] draw attention to the significant challenges that modern mobile forensic methods face because of sophisticated security measures. They bring up the FBI-Apple debate from 2015, which highlighted issues with encryption laws and government access to encrypted devices. Considering increasing encryption methods, Fukami et al. emphasise the significance of continuing research to preserve the integrity of investigative procedures. This is significant since existing approaches frequently call for intrusive modifications and have trouble working with devices that are securely locked down. As a result, it is critical to create new approaches that can accommodate these advanced security protocols without sacrificing forensic integrity.

2.1.8 Evidence Extractions

Fernando [21] study, emphasises the need for open-source tools to retrieve evidence from a variety of devices, including iOS and Android, and draws attention to the financial constraints in mobile forensics. To extract artefacts at a lower cost than commercial offerings, tools such as BitPim, iPhone Backup Analyzer, and Linux Memory Extractor (LIME) offer alternatives. However, Krishnan et al. [22] concentrate on the difficulty of obtaining evidence from various portable devices, including tablets, PDAs, wearable technology, GPS units, and smartphones. For thorough evidence extraction using both physical and logical methods, specialised technologies such as Oxygen and Cellebrite are essential. The significance of modifying forensic methods to stay up with the quickly changing mobile technology is emphasised by both studies. To guarantee effective evidence extraction, forensic experts need to keep up with the most recent developments and comprehend a variety of operating systems and device architectures. These revelations highlight the ever-changing landscape of mobile forensics and the necessity for adaptable, modern tools and methods to efficiently handle evidence extraction across various platforms.

2.1.9 Malicious Application

According to Sharma et al. [23] research, the spread of harmful programmes presents a serious risk to users of Android and iOS in modern digital environment. Even with the strict screening procedure of reliable stores like Google Play and Apple's App Store, the scammers sometimes find a way in and trick unwary consumers. Cybercriminals utilise this appearance of validity as cover to covertly install malware on consumers mobile devices. Due to this, consumers may believe that this was just an update of the application on the mobile phone, but they are unaware of the malicious operation taking place in the background. The attacker is then able to use spyware that is deployed to gather private information, endangering the confidentiality, integrity, and availability (CIA) of user's data.

2.1.10 Android-Based Spyware

Research study by Naser et al. [24] highlights the frequency of Android-based spyware, which becomes a focus of attention. The reason for this increased focus is that Android is quite popular and has an open ecosystem, which makes it easy for bad actors to take advantage of. However, further research must be done to take a through approach and look at other platforms, such as iOS and IoT devices. Ignoring vulnerabilities on other platforms could unintentionally increase security concerns and draw attention away from other platforms, making investigators just focusing on Android when spyware attacks have happened on other platforms. To preserve integrity of all digital devices and guarantee a comprehensive understanding of emerging dangers, they must be some mitigation steps taken place.

2.1.11 WhatsApp Messages

Salamh et al. [45] study emphasises the challenges forensic investigators encounter while trying to retrieve deleted messages and data from volatile memory (RAM). The study concentrated on WhatsApp's forensic artefacts in RAM on mobile devices, observing that since erased messages are not stored in the database, volatile data is prone to deletion, particularly when anti-forensics or power outages occur. Furthermore, Griggio et al. [25] noted that WhatsApp keeps messages on its servers until the receiver views them, with a 30-day maximum retention duration if the recipient is not online. Due to the short window of time for data retrieval, this temporary server storage poses difficulties for forensic investigations. Investigators might find it difficult to get this data if WhatsApp doesn't cooperate promptly. According to

Heath et al. [26], to overcome these difficulties in recovering digital evidence from messaging platforms, forensic investigators must work with service providers and apply creative problem-solving strategies.

2.2 Research Areas and Gap's

In Table 1 are both the research areas gathered from literature review and the gaps that were identified within the research areas:

2.3 Comparison of Spyware Applications

This is the comparison of spyware applications to explain all advantages and disadvantages (as seen in Table 2). This will then prove which spyware application best suits this study.

3 Research Method

3.1 Introduction

As explained in the recent chapter, the lack of research regarding mobile spyware such as of mSpy in effect to covert investigators have been limited. Although crimes have increased hugely in the cyber dynamics; there is minimal research on how effective Spyware can be conducted in the use of it for law enforcement perspective that can prevent crimes. This is especially important were serious crimes like murder, drug dealing, kid napping, and more could be prevented. This chapter will outline each step that was taken to successfully achieve the aim of the project; to better understand mSpy in the use of covert investigators, specifically on its features like real-time on social media such as WhatsApp, Snapchat, and Instagram. As suggested by Rozgonjuk et al. [29], the social media application mentioned above are very popular and nearly every user on a mobile device uses them in today's generation.

3.2 Research Design

The experiment will consist of 3 phases, the first phase focuses on installing the application covertly, the second phase relates to collect the data from the targeted phone, and finally the third phase focuses on Transmission of data. Each phase will

Table 1 Both the research areas and gaps identified

Research areas: Topic—challenges of mobile forensics	**Gaps identified** from the research area
Advancement of technology	• Inadequate protocols in place to manage the latest technological developments in mobile devices • Policies and procedures need to be revised to keep up with the rapid advancement of mobile technologies • Restricted mobile device dataset accessibility hinders the work of forensic investigators
Bypassing device lock	• Unable to unlock encrypted devices in the absence of valid user authentication credentials • Issues with data extraction from gadgets with sophisticated security mechanisms, such as fingerprint sensors
WhatsApp deleted messages	• Due to the messages being temporarily stored on WhatsApp servers, there is a limited time frame for retrieving deleted data • Difficulties in retrieving data from volatile memory and decoding deleted messages
Battery life	• Absence of regulations to maintain device integrity during seizures without accelerating battery discharge • Risks of taking out a device's battery include the possibility of data loss and SIM card lockout
Burner phones	• Data recovery on prepaid burner phones is challenging since data ports are disabled and APIs are not accessible • The anonymity that prepaid plans and phone providers offer makes it difficult to connect specific persons with pieces of information
The Seizure of digital evidence	• The absence of a commonly agreed-upon definition of cybercrime makes it difficult to standardise procedures for gathering evidence • Accessing digital evidence can be complicated when dealing with many organisations and jurisdictions
Mobile operating system (OS)	• The inability of mobile operating systems to multitask effectively hinders forensic investigations • Compatibility problems in utilising several operating systems, data kinds, and computer platforms

(continued)

Table 1 (continued)

Research areas: Topic—challenges of mobile forensics	**Gaps identified** from the research area
Encryption	• As encryption technology advances, it becomes harder to crack passwords and decrypt device data • Obstacles to data recovery from devices equipped with anti-theft features like automated device erasure
Evidence extractions	• Budgetary constraints prevent the purchase of commercial forensic tools, hence open-source alternatives must be used • Issues with obtaining evidence via PDAs, wearable technology, and other portable electronics
Malicious application	• Possibility of malware infiltration even with the strictest app store screening practices
Android-based spyware	• Malware primarily targeting Android platforms ignores weaknesses on iOS and internet of things devices • Thorough study is required on new threats on all digital channels

outline each step to go further or step back from the experiment, as shown in the research design. Figure 1 shows the Research Design flowchart for each phase:

3.3 *Phases*

3.3.1 Installation Phase

When installing mSpy application, there was few steps to follow for registration that allowed the installation to take place on the target device. After the installation, mSpy will run quietly in stealth mode in the background so that the user of the device is unaware of it. It gathers a lot of information covertly, such as browsing histories, call logs, text messages, emails, and location data. Only authorised people can access the distant server where this collected data is safely transferred. It is possible to develop well-informed and convincing conclusions by carefully examining the information acquired using data from experiments to extract insightful information.

3.3.2 Data Collection Phase

During this phase, Real-time data was collected from the different features that mSpy offers such as WhatsApp, Instagram, Snapchat, keylogging, call logs, Geo location,

Table 2 Advantages and disadvantages of different spyware

Spyware applications	Advantages	Disadvantages
mSpy	**User friendly interface**: With its user-friendly design, mSpy's interface is easy to use for users with varying technical background. A flawless experience is guaranteed for all users, regardless of skill level with technology **Extensive monitoring in real-time**: It offers a host of monitoring functions, such as social media monitoring, GPS position tracking, text messages tracking, Geo-location and Geo-fenced tracking, and more **Stealth mode**: Monitor activities with mSpy without anyone noticing. It operates silently in the background of the intended device, allowing to watch covertly and not alert the user **Compatibility**: The user will value that mSpy runs well on both IOS and Android devices, giving them the freedom to choose the ones that most closely matches their tastes. It all comes down to providing users with flexibility and convenience to suit their demands **Customer service**: If consumers have any problems or inquiries, mSpy's customer service is available and quick to respond	**Limited features**: mSpy's features is less extensive than that of Flexi SPY, which could negatively affect user experience **Cost**: mSpy's subscriptions option can be seen somewhat expensive, especially for scenarios involved prolonged or multiple device use. This could present a financial challenge for customers **Detection risks**: It is possible that the person utilising the tracked device will find evidence of mSpy, especially if they frequently run malware or spyware scan **Privacy concerns**: Because mSpy monitors personal data without the owner's express consent, it is acceptable to have privacy concerns while using it **Reliability problems**: A few users might experience problems with mSpy's accuracy and dependability of data it provides, such as incomplete monitoring reports or lags in, which could reduce the program's efficacy

(continued)

Table 2 (continued)

Spyware applications	Advantages	Disadvantages
Flexi Spy	**Improved capabilities**: With a variety of cutting-edge capabilities, including phone interception, ambient recording, and remote camera activation. Flexi SPY elevates monitoring to a new level and gives customer a wide range of surveillance alternatives **Customised settings**: Flexi SPY gives customer the ability to change the monitoring options to fit their own requirement and tastes, making for a more personalised monitoring experience **Real time monitoring**: Aswell as mSpy, Flexi SPY provides a user with instant access to live data streams from the target device, improving the capacity to remain informed and flexible **Discrete operations**: Flexi SPY operates from the background of the target device, much like mSpy, to ensure covert surveillance without notifying the targets device	**User friendly**: Those who are not tech-savvy may find it difficult to sort through Flexi SPY's complex features and customisation possibilities as they may feel a little daunting **Pricing**: If users want to extend the period of time or are interested in the premium services, the cost of a subscription may be greater than average **Detection risks**: If the target user checks for malware or spyware, there's a chance they may find Flexi SPY **Legal considerations**: Due to advance monitoring tools like call interception, it may breach privacy regulation in some areas, it is important to be aware of any potential legal ramification before utilising them
UMobix	**Monitoring social media and dating apps**: Keeps an eye on a variety of social networking sites and well-known apps **Keystrokes**: Similarly to mSpy it can capture keystrokes, including unsent messages. Ensuring that no typed content is not overlooked **Real time location tracking**: It can quickly exploit target's location in real time **Remote control of devices**: Takes control of smartphone remotely. So, it can uninstall unwanted apps and keep an eye on how it is used **Appropriate application use and screen time management**: Monitors targets apps usage and screen time	**Geo fencing**: Not available feature on UMobix but available on both Flexi SPY **Installation application**: Extra caution may be required while setting up apps on iOS devices to guarantee smooth setup procedure

Note: The Spyware comparison is referenced from two different authors: **mSpy**: [27]. **UMobix and Flexi SPY**: [28]

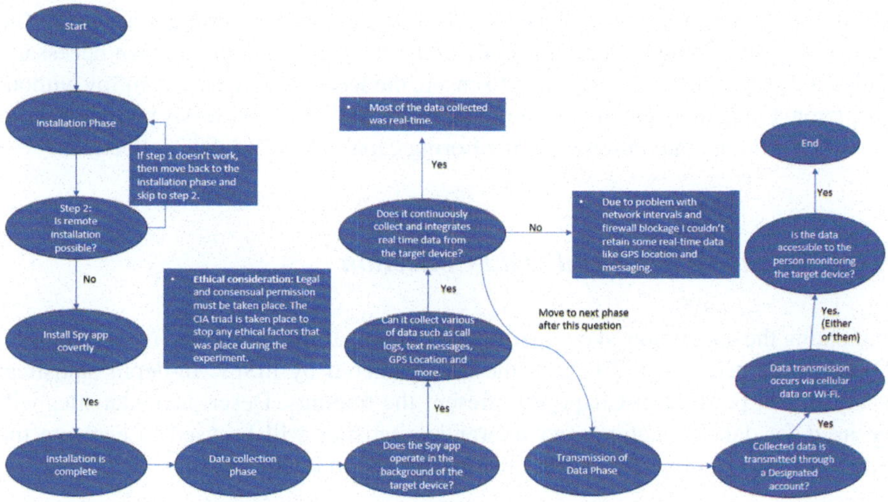

Fig. 1 The research design flowchart for each phase

Geo fencing etc. The data collected was then used to be compared between both incoming and outgoing features from mSpy with forensic tools for validation testing to see if the accuracy and reliability of mSpy is more capable than forensic tools.

3.4 Transmission of Data Phase

The transmission of data is where the collected data is from a specific web server designed from mSpy. When logging in to the account the user can investigate each data collected but the target device user must have either Wi-fi or Cellular connection i.e. this will allow the incoming and outgoing features to be real-time. If not transmission of data will not be real-time, and data will be stored locally rather than sending it through to the server straight-away. When the target device is connected to an internet connection the data stored locally will sent a packet out to the global interface. This communication between the client and server is called a 'Three-Way-Handshake'.

3.5 Implementation of mSpy

When implementing mSpy, research went through a careful process to seamlessly integrate it into covert operations. The initial steps involved extensive preparation and familiarization to ensure smooth integration of mSpy with pre-configuring and hiding itself in an application called "update service". The target device OPPO X3 Find Lite was carefully selected for its compatibility and sensitivity of mSpy functions. During the installation phase as mentioned in the research design, physical access to the device was often relied upon, or social engineering tactics applied when necessary. This careful approach was necessary to ensure the secret deployment of mSpy without suspicion. Using these calculated methods, mSpy has been seamlessly integrated with covert operations, maximizing its monitoring capabilities and data collections.

3.6 Validating Data and Data Collection

Validating the data required comparing the outputs of forensic tools from Cellebrite, Oxygen Forensics, and IEF with the data gathered by mSpy. In-depth statistical analysis will be used in Chap. 4 to present the findings. Descriptive statistics will be used for data collection, and inferential statistics will be used to evaluate the effectiveness of mSpy and its implications for covert operations.

3.7 Experimental Setup

The experiment conducted in each phase will take place on cloned Digital Forensics Windows 22 64-bit machine. However, after following these phases, the data collected will be compared with multiple forensic tools. Different forensic tool will have their own VM machine but using the same cloned used by mSpy. This will be concurrent throughout the project, to allow an accurate investigation. Figure 2 below is the VM'S experimental setup design:

3.8 Experiment Testing

In this section, Experiment testing was done to analyse mSpy comparison to text message, call logs, emails, Geo location, geo fencing and social media platforms for real-time data. To gain real-time data, the features will be compared from server to the original communication made from the target device or identifying a location that matches a communication from the server to the target device. Furthermore, incoming, and outgoing features was compared from mSpy to forensic tools to suggest the accuracy and reliability between them. Finally, refer to 3.8.1 all the problems and Solution are suggested from the experiment testing.

3.8.1 Problems and Solutions During Experiment Testing

Table 3 below presents all the problems and solutions that were identified during the experiment testing.

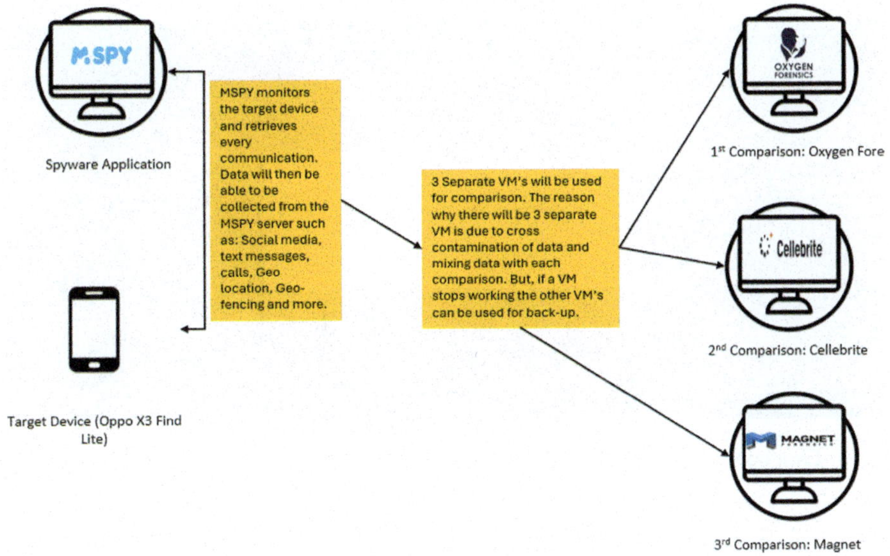

Fig. 2 Experimental setup from the VM design

Table 3 The problems and solutions identified during the experiment testing

Problems	Solution
(P1): Real-time functionality like geolocation isn't working properly because of network interval problem	**(S1)**: Changing the JavaScript code itself is the best solution, but it isn't doable without administrative rights
(P2): Any potentially hazardous or spyware apps that are executing on the computers are successfully blocked by the network firewall	**(S2)**: We were able to solve most of P1 and P2 by creating a private network connection inside the virtual machine (VM)
(P3): Images are not loading on the mSpy server	**(S3)**: To fix this, turn off real-time protection and the windows firewall

Compatibility Problem

During the trial testing, significant problems were found, particularly in relation to compatibility problems with Android devices. Due to the device's restrictions, such as lack of root access, these issues hindered physical extractions using forensic tools. A Cellebrite example is shown in the following Fig. 3:

Problem:

Physical extraction, a useful technique for obtaining a lot of user data, including information from social media, encountered difficulties on the mobile device. The reason it failed was that root access was not available on the device, which is required for these kinds of extractions. Because root access was limited to earlier models, things became much more complicated with the device's most recent model

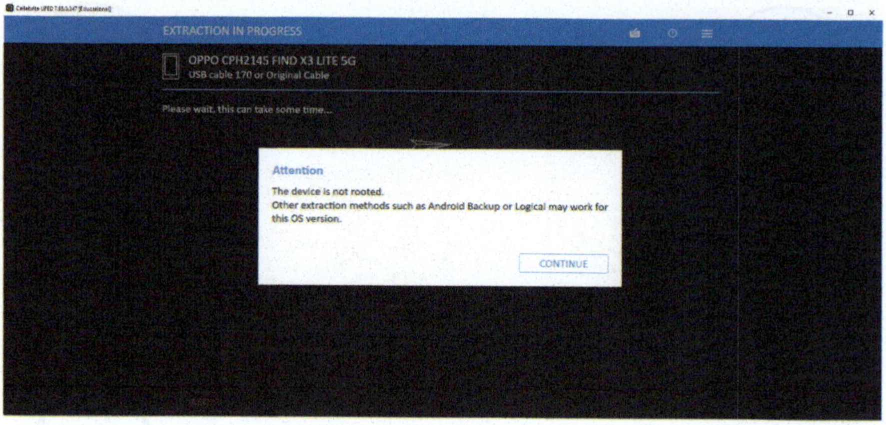

Fig. 3 Error occurred with Cellebrite's physical extraction

3.8.2 Scenario-Based Experiment

The scenario for this experiment, was that the covert operator strategically selects mSpy as a potential tool for enhancing surveillance capabilities in covert operations and law enforcement endeavours. Collaborating with a participant, we construct a scenario where they pose as a married couple, providing a realistic context for the evaluation. With the participant's consent, the owner installs mSpy on the target's mobile device to assess its performance in providing real-time data and intelligence. The participant, who is residing at Cantor Building, Sheffield Hallam, serves as the backdrop for the investigation. Throughout the experiment, ethical and legal boundaries are meticulously observed to ensure participant rights and privacy are protected. The findings from the evaluation aims to inform the practical application of mSpy in covert operations and law enforcement, while emphasising ethical considerations and adhering to legal frameworks.

3.9 Legal and Ethical Considerations

There are several difficulties and moral ramifications associated with the use of mSpy. Different jurisdictions have different legal standards for using these kinds of surveillance tools and they frequently walk a narrow line between legitimate intelligence collection and invasions of privacy. It is impossible to overlook the ethical issues raised by the possibility of power abuse and individual rights violations. Some sources from articles and journals must be identified such as: RIPA [30], Covert Surveillance code of practise [31] and spy laws [32]. But, due to the experiment implemented by a scenario based there was not concerns of legal and ethical considerations in place. However, during the experimented testing, CIA triad must be followed throughout

Fig. 4 CIA triad diagram for ethical and legal consideration by [44]

the process because of communications can be 'viewed as risk factor' as said from Alkhudhayr et al. (2019). For the experiment to proceed, permission and authorisation from the participant were granted. Figure 4 below shows a CIA triad diagram for ethical and legal considerations.

3.10 Conclusion

In summary, this study's multifaceted quantitative method with a deductive approach, which centred on experimental data, offers a solid framework for assessing how the mSpy application is used in covert operations. The approach's focus on statistical analysis and controlled experiment to provide insightful information on the implications and efficacy of mSpy, advancing knowledge of its use in covert surveillance. Overall, the experiment that was conducted had shown how effective mSpy compared to other forensic tools could be for law enforcement and covert operators in collecting evidence.

4 Results and Discussion

4.1 Introduction

The findings and analysis of the experiment described in Chap. 3 are presented in this chapter. It accomplishes Objectives 1, 2, and 3 by assessing the mSpy tool's reliability and effectiveness in covert operations and by comparing it to other forensic tools to verify the collected data. Each of the three experimental phases is represented by a section in this chapter. In the first phase, mSpy's ability to install covertly is assessed to make sure the target is unaware of its presence. The second phase concentrates on gathering data in real-time and validating all incoming and outgoing communications by contrasting mSpy's data with that from forensic tools like Oxygen Forensics, Cellebrite PA, and Magnet. The third phase evaluates how well the three-way handshake between the client and server sent data. The evaluation of mSpy's value in accomplishing the study's aims, objectives, and research question ends in this chapter. Additionally, it contrasts the results with the literature reviewed in.

4.2 Results

4.2.1 Phase 1—Installation of mSpy

Physical Access and Configuration

To install the mSpy programme secretly, the study needed direct access to the target phone during the first phase of physical access and configuration. Several configurations were refined to improve the program's stealth and performance. These included turning off notifications, turning on accessibility features, and turning off battery optimisation to keep the app from shutting down to save energy. As seen in the Fig. 5 below, after mSpy was installed, it functioned covertly in the background on the target device under the guise of "**Update Service**" without being noticed. This allowed complete access to the functions and data of the device while guaranteeing continuous monitoring without the user's knowledge. MSpy's operation throughout this phase was covert, which fit in well with the study's focus on discretion and stealth. mSpy reduced the possibility that the device owner would discover it by hiding alerts and seeming like a regular system application. This stealth strategy had several benefits. It decreased the possibility that the spyware would be found and eliminated while enabling continuous surveillance and giving instant access to phone logs, message histories, and location data. Moreover, the capacity for covert operations permitted extended monitoring, which made it possible to gather comprehensive data necessary for taking further steps or making decisions. In summary, this stage of the inquiry established the foundation for efficient gathering and examination of data, guaranteeing the accomplishment of the study's goal.

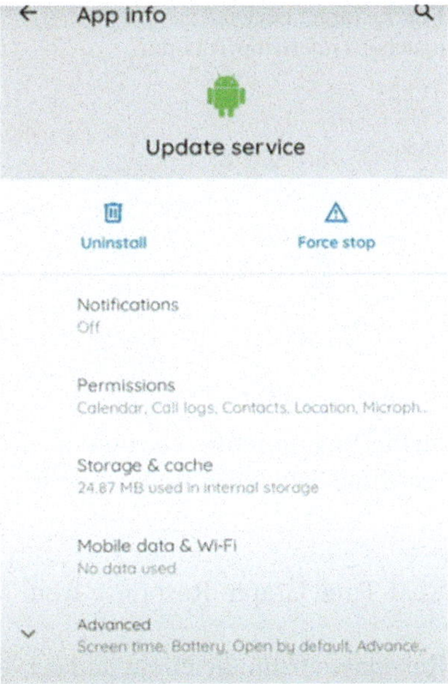

Fig. 5 A preconfigured application called "Update Service"

4.2.2 Phase 2—Data Collection

Real Time Collection of Data

Using the mSpy tool, real-time data gathering was carried out throughout this phase. This procedure provided insight into the effects of the many functionalities that mSpy has to offer. The Synchronised timestamps between the client and server provide an example of real-time data collecting. The timestamps at 7:44 pm on February 16th match, as seen in Fig. 6 (mSpy web server) Fig. 7 (target device), suggesting that the real-time data collection process was successful. Throughout the course of the inquiry, these outcomes remained constant.

The linked timestamps show how well mSpy works at delivering real-time information. This feature ensures that the information gathered from the intended device is instantly accessible on the mSpy website, enabling prompt observation and examination. The dependability of mSpy's real-time data collection features is confirmed by the consistent matching of timestamps during the inquiry. In summary, mSpy's real-time data collecting verified to be dependable and efficient. Timestamped synchronisation between the mSpy web server and the target device guaranteed precise and timely data capture. Applications that need instant access to data, such social media interactions, geolocation monitoring, and other real-time operations, depend heavily

Fig. 6 mSpy web server—Timestamp 7:44 pm

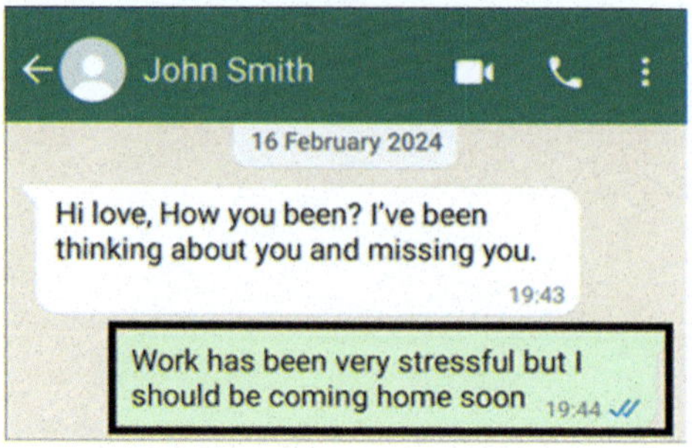

Fig. 7 Target device match—Timestamp 7:44 pm

on this functionality. The outcomes of this phase demonstrate the resilience of mSpy's real-time data collection capabilities.

Real-Time Graph: Response from Mobile Device and mSpy Server

Responses from the client and server were computed for real-time capturing to validate the functionalities of the Real-Time server. This is shown in the graph Fig. 8 below:

The graph data indicates there is continuous direct communication between the mobile device and other users, with consistent communication answers at 100%. On the other hand, beginning on February 17, mSpy server data shows a 25% decrease from the predicted slope, reaching 75%. This decrease points to the discontinuation

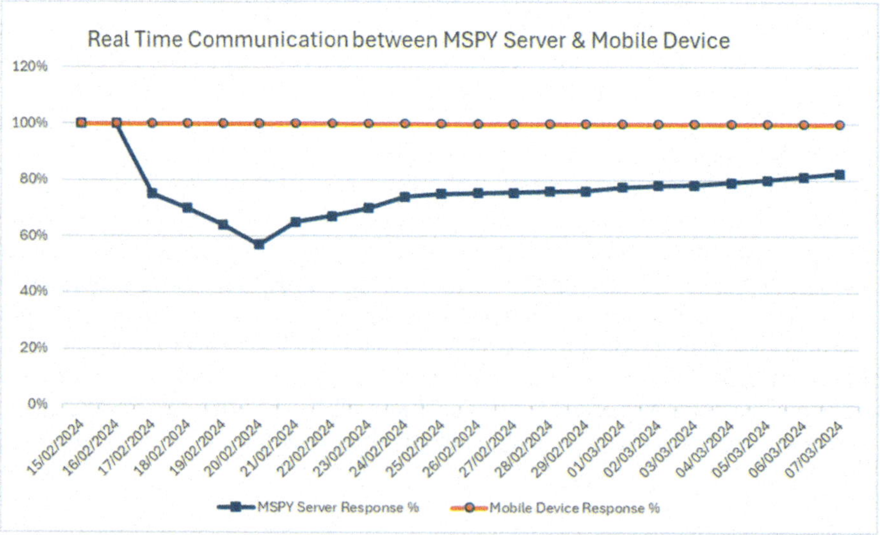

Fig. 8 Real Time communication from mSpy server and mobile device

Table 4 All concordant results from Real-Time graph

Dates	Percentage (both mobile and target response %)
15/02/2024	100
16/02/2024	100
25/02/2024	75
26/02/2024	75
27/02/2024	76
28/02/2024	76

of real-time data provision for geolocation and WhatsApp-like services. The network's discovery of a malicious application on the web server's firewall was the main reason, supporting earlier worries (P1 and P2). To solve these problems, several trial-and-error techniques were used, including the solutions suggested in the experiment testing (S1 and S2). Concordant results proved the correctness and dependability of real-time data despite these difficulties, both before and after issue identification (see Concordant Results from Real-Time Graph).

Concordant Results from Real-Time Graph

Concordant results indicate that the data provided by the mSpy server, and the mobile device is consistent, which is an important finding from the Real-Time graph analysis. When the Mobile Device Response % and the mSpy Server Response % percentage values line up, the findings are considered congruent. All concordant results from the Real-Time graph is shown in the Table 4 below:

These findings show that, on days, data transfer between the mobile device and the mSpy server performs consistently. The readings that stay at 100% on February 15 and 16 indicate that the two entities are communicating continuously and in sync with each other. On the other hand, modest variations are noted for the next two dates, when the percentages fall to 75% on February 25 and 26 and slightly rise to 76% on February 27 and 28.

4.2.3 Comparison Analysis of mSpy with Forensic Tools (Oxygen Forensic, Cellebrite PA and Magnet) Both Incoming and Outgoing Feature Communications

Incoming and outgoing communication features are compared between mSpy's performance and that of three competitive forensic tools: Oxygen Forensic, Cellebrite PA, and Magnet. By means of methodical data gathering and enhancement, perceptions regarding the efficiency of mSpy in retrieving significant data are contrasted with the potential of proven forensic tools. This comparison provides important

Table 5 All incoming communication from mSpy in comparison to forensic tools

Categorised features	Total incoming from mSpy	Total incoming from Cellebrite PATotal incoming from Cellebrite	Total incoming oxygen forensic	Total incoming from Magnet
Text messages	37	51	47	51
Calls	1	2	1	2
Images	6	7	5	12
Geo-fencing	20	0	0	0
Geo-location	522	0	0	0
WhatsApp	31	0	38	0
Instagram	6	0	12	0
Snapchat	8	0	5	0

insights for forensic investigations by illuminating the relative advantages and disadvantages of each tool in the collection and examination of incoming and outgoing communication data.

Incoming Features Communication Results

All incoming communications were included in the data that mSpy gathered throughout the monitoring period, reducing any latency that might be caused by firewall blocks or network intervals (P1: network interval, P2: firewall block, and Windows real-time protection concerns serve as the major exceptions), however the solution for both were suggested in. Table 5 below provides an overview of the comparison between the incoming communication features from mSpy, Oxygen Forensics, Magnet, and Cellebrite PA:

The graph below provides a visual representation of comparing incoming communication results across various classified features from Oxygen Forensics, Cellebrite PA, mSpy, and Magnet. Because each feature group is represented by a different colour, it is easy to discern how incoming communications are allocated. The graph provides a concise overview of the differences in performance between the forensic tools, highlighting mSpy's effectiveness in collecting communication, including WhatsApp chat, as compared to the forensic tools. There are also noticeable differences in the identification of calls, texts, and images, which offers crucial data for contrasting the features of incoming communications (Fig. 9).

Based on the comparison shown in the table and graph, mSpy performed better than other forensic tools when it came to recording incoming communications. It was especially good at recording WhatsApp conversations. According to the data, mSpy recorded 31 WhatsApp conversations, whereas Magnet and Cellebrite did not produce any results for this feature. Regarding text messages, mSpy found 37 that came in, which is marginally less than the 51 that Cellebrite, Magnet, and

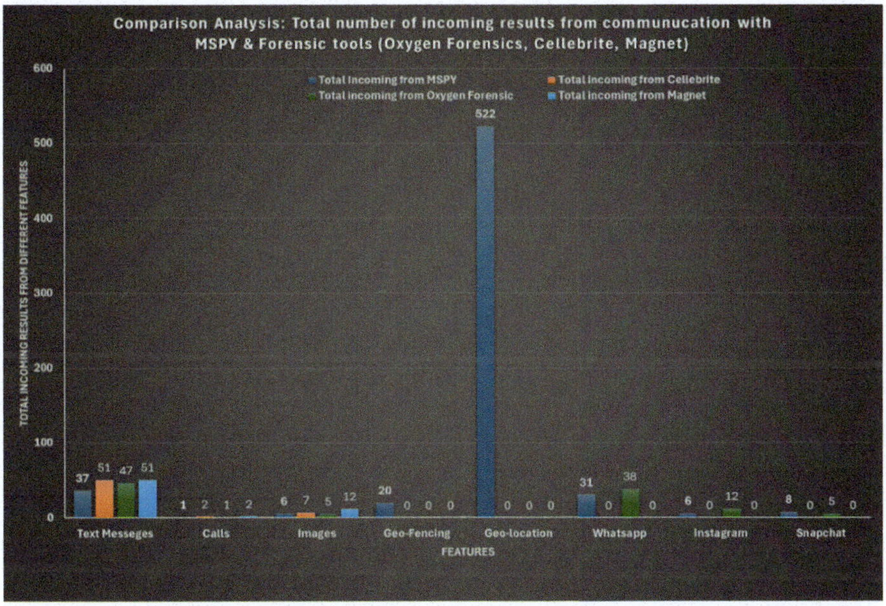

Fig. 9 Incoming communication results graph

Oxygen Forensics found and the 47 that Oxygen Forensics found. Nevertheless, mSpy captured 10 fewer messages than Oxygen Forensics, although 51 was the same number of incoming communications for Magnet and Cellebrite. Furthermore, with 522 geolocation records and 20 geofencing results, mSpy showed a strong advantage over the forensic tools in real-time geolocation and geofencing functionalities.

However, there are several reasons why mSpy's data collecting differs from that of other forensic programmes. Unlike forensic technologies, mSpy's real-time monitoring approach may be hindered by network outages or firewall blocks, which could result in removing some results. Furthermore, mSpy's ability to gather thorough data without the need for device modification like rooting or jailbreaking is hampered by the restrictions placed by more recent handset models, including the Oppo × 3 Find Lite, which prevent low-level access to data. Although mSpy has advantages in real-time monitoring, its effectiveness can be degraded in situations where network or device restrictions inhibit data collecting.

Outgoing Feature Communication Results

Table 6 below summarises the findings of mSpy's analysis of outgoing communication data in comparison to Cellebrite, Oxygen Forensics, and Magnet.

Table 6 All outgoing communication results from mSpy in comparison to forensic tools

Categorised features	Total outgoing from mSpy	Total outgoing from Cellebrite	Total outgoing from oxygen forensics	Total outgoing from magnet
Text messages	15	15	15	15
Calls	14	2	1	2
Images	10	10	13	22
Geo-fencing	2	0	0	0
WhatsApp	15	0	29	0
Instagram	5	0	8	0
Snapchat	4	0	3	0

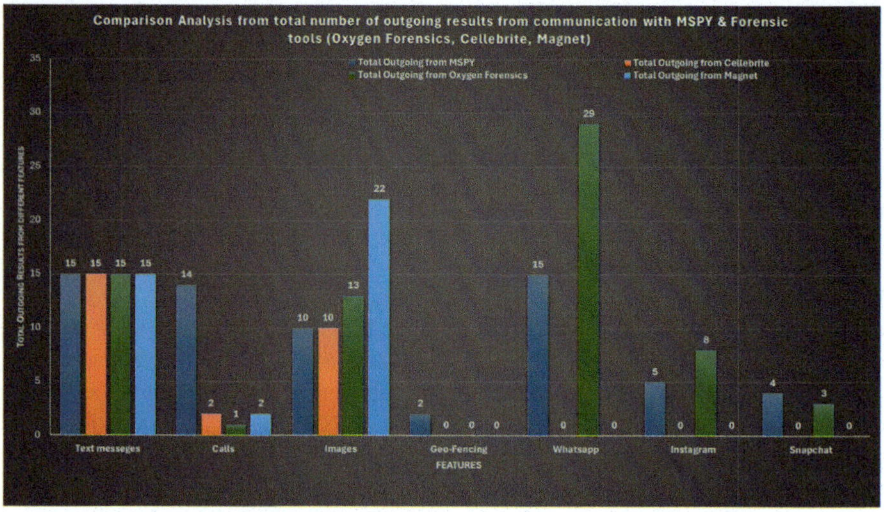

Fig. 10 Outgoing communications results graph

The graph below shows the outgoing communication findings from Oxygen Forensics, Cellebrite PA, mSpy, and Magnet across a variety of categorised criteria. It is easy to compare the distribution of outgoing communications because each feature group is identified by a distinct colour. This graphic comparison demonstrates how well mSpy records several kinds of outgoing communications, including as calls, WhatsApp messages, and geo-fencing information. mSpy is particularly good at a few areas, such geo-fencing and Snapchat, where other programmes perform poorly or not at all. To compare the characteristics and efficiency of different forensic instruments in gathering outgoing communications, the graph provides essential data (Fig. 10).

According to the outgoing communication results from table and graph, mSpy did well across several categories. Like Cellebrite, Oxygen Forensics, and Magnet, mSpy gathered 15 text messages, demonstrating consistent text message data gathering. In terms of call data, mSpy performed noticeably better than the other tools, recording 14 calls as opposed to 2 from Cellebrite, 1 from Oxygen Forensics, and 2 from Magnet, indicating its edge in this feature. When it came to image data collection, Oxygen Forensics and Magnet gathered slightly more than mSpy, with mSpy recording 10 outgoing images that matched Cellebrite, while Oxygen Forensics collected 13 and Magnet collected 22. mSpy demonstrated its exceptional ability in the Geo-Fencing area by capturing two records in a row, whereas the other tools yielded no results.

In an overall conclusion deduced from this phase of the study, it can be claimed that the approach of the primary tool was successful. The comparison data gathered from all incoming and outgoing features were compared. The fact that mSpy was able to gather 15 outgoing WhatsApp messages and Oxygen Forensics captured 29, but Cellebrite and Magnet produced no results, highlights mSpy's capability to obtain WhatsApp data. Cellebrite and Magnet produced no results for Instagram, but mSpy recorded 5 outgoing messages compared to Oxygen Forensics' 8 records. While Oxygen Forensics only recorded three outgoing messages on Snapchat, mSpy recorded four, and Cellebrite and Magnet produced no results. In comparison to other forensic tools, mSpy's outgoing communication findings show that it can capture accurate and succinct data in a variety of categories. This points to mSpy's dependability as a tool for maintaining a track on and recording a broad variety of outgoing communications, enhancing its potent performance in gathering incoming communications and adding to its value for forensic investigations.

Although mSpy outperformed forensic tools in this phase in terms of accuracy and performance, there could be other reasons for the discrepancy, such as extraction method constraints (physical extraction is not available), and unresolved parameter p1 concerns. Nevertheless, using the suggested approach, parameter p2 was successfully determined and put into practice. Most difficulties were successfully resolved, as explained in Sect. 3.8.1. However, a significant challenge remained regarding network intervals because administrator privileges and Java code accesses were unavailable. This resulted in parameter p1 remaining partially unresolved, posing a substantial difficulty for the inquiry, while parameter p2 was entirely resolved.

4.2.4 Phase 3—Transmission of Data Phase

TCP Three-Way Handshake

At the last stage of communication, data is transferred from the client (the mobile device that is the target) to the mSpy server through a TCP three-way handshake. Within a layered network architecture, the Transmission Control Protocol (TCP) is an end-to-end connection-oriented protocol that supports applications running on several networks. Between applications operating on hosts interacting via an IP network, TCP guarantees the dependable, ordered, and error-checked delivery of a stream of bytes. To track the sequence of arriving packets, each device in a TCP connection keeps note of its current state and associated sequence number. To indicate that the data has been successfully received and that the device is prepared

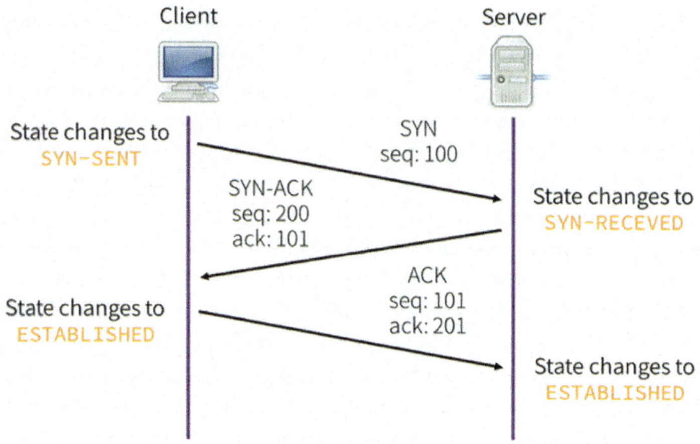

Fig. 11 TCP 3-way-Hanshake in connection to the client (target phone) and the server (mSpy)

for more incoming packets, the recipient device sends back an acknowledgement (ACK) packet with an acknowledgement number after receiving a new packet.

The TCP three-way handshake procedure, which consists of these three crucial phases, is required to establish a TCP connection between the client and server. These steps are the following:

- **SYN Packet**: To establish a connection, the client (target device) sends the mSpy server a synchronise (SYN) packet. This packet establishes the connection by containing a random sequence number.
- The synchronize-acknowledgement **(SYN-ACK)** packet is the server's response to receiving the SYN packet. This packet indicates that the server has received the client's SYN packet. It also contains an acknowledgement number that matches the client's sequence number. The packet has its own random sequence number.
- **ACK Packet**: The client then replies to the server with an ACK packet. This packet verifies that the client has received the server's SYN-ACK packet by acknowledging the sequence number of the server.

The following Fig. 11 shows the TCP 3-way handshake in connection to the Client (Target phone) and the server (mSpy):

Following the completion of this handshake, the sequence numbers on both sides synchronise, creating an encrypted connection. The client and server can send and receive data independently because of this synchronisation. The figure below shows all of this in action towards the 3-way handshake and this was in referenced by a study from Hsu et al. [33] which they explain in detail about TCP three-way handshake for a successful connection.

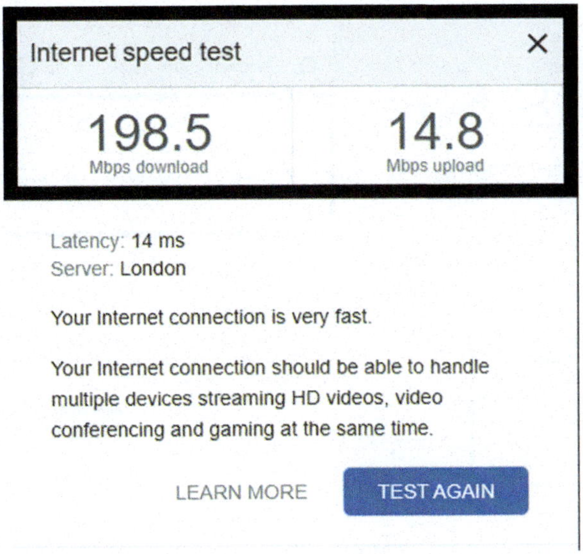

Fig. 12 The upload speed is 14.8 Mbps, while the download speed is 198.5 Mbps for Sheffield Hallam University Wi-Fi

Wi-Fi/Mobile Data Connectivity

Wi-Fi/ Mobile data connectivity is essential during the data transmission phase since every packet transmitted from the client (the target mobile device) must transit through both an upload and download mechanism. The accuracy of data transmission and speedy uploads are made possible by dependable Wi-Fi connectivity, which has a substantial impact on the transmission process between the mobile device and the server. Figure 12 shows Sheffield Hallam University's reliable and fast Wi-Fi connections, which make uploading and downloading simple in regulated lab environments. Figure 8 illustrates the notable decreases in upload and download speeds that occurred when the target device was linked to external internet networks (Fig. 13).

There were considerable decreases in speed at locations where package limits made the internet connections slower. The target device's connection and communication with the mSpy server may be hampered by these variations in internet speed, underscoring the significance of dependable internet access for efficient data transfer.

Although mobile data consumption presented its own set of issues, Wi-Fi access was essential for data transmission. The participant encountered limitations imposed by their mobile data provider, which restricted the use of mSpy. Moreover, poor mobile signal reception remained an issue even with access to 4G and 5G networks. These problems imply that mSpy's operation could be affected by several fundamental factors other than network connectivity.

Table 7 below presents the key problems faced for Mobile data.

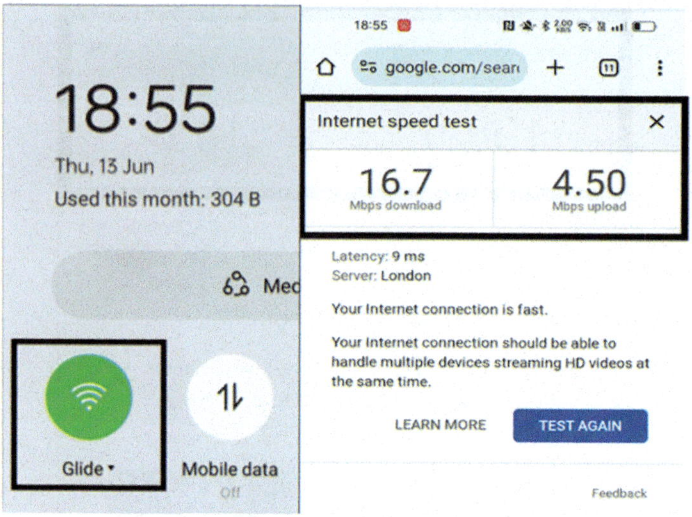

Fig. 13 The download speed of 16.7 Mbps and upload speed of 4.50 Mbps for the external Wi-Fi connection "Glide" respectively, indicate a notable decline from Fig. 12. Restrictions pertaining to the Wi-Fi package are the cause of this decrease

Table 7 The several key problems with mobile data

Category	Potential issue	Description
Network restriction	Restrictions from carrier	Mobile data usage by mSpy may be restricted by device settings or mSpy itself. Make sure mobile data usage is enabled by checking the app's settings
	Network Sype	Weak mobile signals continued to cause problems with data transfer even with access to both 4G and 5G networks, suggesting that mSpy's functionality may be impacted by other underlying variables
App settings	Data usage settings	mSpy's use of mobile data may be limited by the device's settings. Due to restricted access to the carrier's administrative settings, these restrictions could not be changed
	Battery optimization	The target device used for the study has battery optimisation settings, however they were disabled. Because of this, certain apps like 'mSpy' were able to use background data even when they weren't linked to Wi-Fi. Frequently, this function is turned on automatically to save battery life
Device settings	Mobile data limits	It is possible to reduce background data usage by configuring data usage settings. Nevertheless, during the investigations, it was confirmed that this setting was disabled
	Data saver mode	When the battery is low, data saver mode, which was initially turned off, could be turned on automatically and possibly impact data transmission

4.2.5 Discussion

The objectives of the study were effectively met, and the findings confirmed the usefulness of mSpy in covert operations by offering real-time information to support the research question.

Objectives Achieved

Objective 1) Evaluate the performance of forensic tools regarding data extraction on mSpy application

The objective of this research was to assess forensic tools information service capabilities, with a focus on the mSpy tool specifically. By analysing and contrasting popular forensic tools like Oxygen Forensics, Cellebrite PA, and Magnet, the research proved that mSpy is dependable for capturing both inbound and outbound calls. MSpy fared better in real-time features including social media monitoring, geolocation, and geofencing, despite physical extraction techniques on forensic tools being problematic because of the device's model. Since mSpy did not require a jailbreak to perform a full/physical extraction, unlike all other forensic tools, it can be used for covert operations without requiring any device modifications.

The study accomplished this objective and made a substantial contribution to the field of digital forensics by demonstrating mSpy's skill in real-time data collecting. Moreover, results in line with earlier studies, including those of Swaminathan and Danks [34], emphasised the significance of thorough evaluation frameworks for determining the effectiveness of monitoring systems. The varied uses of surveillance technology were further emphasised by findings from studies conducted by organisations like the ACLU from Slobogin and Brayne [35], with a focus on the importance of CCTV cameras in preventing crime. The study added to the body of information already available about the importance of CCTV cameras and made the case for using mSpy to speed up criminal investigations because of its real-time geolocation tracking features.

Although there were several extraction techniques investigated with forensic tools like Cellebrite, Magnet, and Oxygen, such as image extraction and sophisticated logical/backup approaches, operational problems like Cellebrite freezing during APK extraction pointed to areas that needed improvement. These difficulties highlighted how important it is to keep developing forensic tools to improve their dependability and intuitiveness.

Overall, the study's thorough examination of the benefits and drawbacks of forensic instruments enabled it to accomplish this objective and offer a solid basis for deliberations regarding digital forensics. Its conclusions have had a big influence on the sector, guaranteeing effective administration of justice by pinpointing problem areas and providing useful and efficient information.

Objective 2) Evaluation of forensic tools in analysis of mSpy application

The investigation accomplished this objective through multiple phases, the first of which involved setting up and gaining physical access to the target phone so that mSpy could be covertly placed on it. This configuration prevented the device user from detecting the programme by changing it to "Update Service," ensuring persistence and stealth.

mSpy proved its adeptness in obtaining real-time data during the data collection phase. Accurate and timely data capture was confirmed by means of synchronised timestamps between the client and server, hence validating the tool's capabilities.

The comparative analysis that examined both incoming and outgoing communication aspects showed that mSpy outperformed Magnet, Oxygen Forensics, and Cellebrite PA. Even with network and firewall issues that made it difficult to gather data, mSpy performed remarkably well in gathering real-time geolocation information and WhatsApp discussions. This comparison demonstrated the benefit of mSpy in real-time monitoring and data retrieval without the need for device change.

Analysis of outgoing communication data confirmed mSpy efficacy, especially regarding call and WhatsApp message recording. MSpy dependability was highlighted by the comparison with other forensic tools, which highlighted its correctness and consistency in obtaining outgoing communication data. Also, The TCP three-way handshake protocol made assurance that the client and the mSpy server were sending data securely during the transmission phase. Accurate data transfer required steady Wi-Fi connectivity, underscoring the significance of network stability in forensic investigations.

Overall, the thorough assessment of the study validated the accomplishment of the goal and showed that mSpy is more effective than other forensic technologies. MSpy demonstrated its worth as a useful tool for digital forensic investigations despite obstacles like network outages, providing dependable and effective real-time data collecting and communication analysis. These results make a substantial contribution to the focus by guaranteeing the prompt administration of justice and offering perceptions into the areas where forensic technologies could be more useful.

Objective 3) A comprehensive analysis of mSpy application regarding to be used by law enforcement

The study achieved this objective by closely evaluating mSpy's effectiveness and acceptability for usage by law enforcement. An extensive analysis demonstrated mSpy's ability to operate covertly and dependably on target devices. mSpy demonstrated how to switch off alerts, navigate battery optimisation options, and hide as a system programme during installation. This ensured seamless, continuous operation, which is essential for prolonged, covert law enforcement monitoring. Harkin et al. [36] state that such use must be ethically acceptable and require agreement from both parties.

MSpy's ability to collect real-time data, such as geolocation, geofencing, and social media interactions, was tested throughout the data collection phase. The synchronised timestamps between the target device and the mSpy server verified the precise and reliable real-time data transmission, which is essential for swift law enforcement activities. MSpy's consistent performance under varied network conditions confirmed its dependability and robustness. Slobogin and Brayne [35] suggests that digital monitoring techniquescan help the government and security forces tackle serious crimes, even in the absence of statistics from the government.

Moreover, mSpy outperformed other forensic tools, such as Oxygen Forensics, Cellebrite PA, and Magnet, in certain domains, most notably the acquisition of geolocation data and WhatsApp messages. Comparing mSpy to the other technologies, it was clear that its extensive surveillance capabilities were superior. MSpy's practicality and convenience of use for law enforcement operations, where prompt and dependable deployment is crucial, are further enhanced by its capacity to gather extensive data without requiring device changes, such as rooting or jailbreaking. mSpy provides a less intrusive option than jailbreaking, which law enforcement agencies occasionally use to decrypt devices during investigations, according to Grimmelmann [37].

The study also considered network limitations and mobile data usage settings as potential data transfer obstacles. Despite these difficulties, mSpy continuously showed dependable data transfer via mobile and Wi-Fi networks, made possible by the TCP three-way handshake protocol. This in-depth examination, which includes real-world testing and comparisons with other forensic tools, validates mSpy's efficacy as a solid, dependable, and stealthy tool for law enforcement organisations that can gather copious amounts of precise data for inquiries.

Research Question Achieved

It is evident from the objectives mentioned above that the research issue was effectively addressed. The study demonstrated the dependability and efficiency of mSpy in real-time data collection and offered thorough insights into how it can be used for covert operations. This was accomplished by thorough testing and comparison study with well-known forensic instruments, which identified areas that needed to be improved to guarantee the timely administration of justice.

The results we obtained are supported by Zhang et al. [38] study on real-time data collecting for mobile applications. Their investigation confirms mSpy's strong real-time data collection capabilities even under varied network conditions by addressing the issues and potential solutions for dependable data transfer. 39 also examine how surveillance technologies are developing, highlighting how they affect security and privacy. They support mSpy's commitment to public safety, which makes it possible for law enforcement to fight crime more successfully. Their results are consistent with more general developments in the field of surveillance technologies related to the gathering of covert data.

Furthermore, Kaur, Singh, and Kumar's comparative review of mobile device forensic techniques (2021) highlight the significance of thorough evaluation frameworks. Their results support our methodology of comparing mSpy's capabilities against those of Oxygen Forensics, Cellebrite PA, and Magnet, highlighting mSpy's better real-time data capture capabilities.

These findings support the success and usefulness of mSpy and validate our research on its suitability for covert operations as well as its important role in the field of digital forensics. Because of this all-encompassing approach, mSpy is not only a helpful tool for covert operations and law enforcement agencies to obtain data in real-time, but it also stays up to date with the most recent advancements and technological challenges in surveillance systems.

Key Literature Comparison to This Study

They are 6 different key literature comparison that was found to provide the similar/different attributes from different authors towards this study to fill the research gaps. Table 8 below summarises the following.

5 Conclusions

Overall, the study was successful because it closely followed the main aim of the study, which was to evaluate the usefulness of mSpy to aid covert operations. We have shown how real-time data analysis with mSpy improves tracking capabilities and makes it easier to stop illegal activity. On the other hand, the design of the project imposed internal limitations as seen below in Sec. 5.1.

5.1 Limitations

One major drawback was the time limits, which made it difficult to gather and evaluate all the data, especially for the mobile device of the covert agent. Understanding the dynamics of covert communication may be hampered by this constraint. Prioritising quicker networks is necessary because slower networks also resulted in transmission problems. The validity of the results would have been improved by a comparison study of other spyware programmes. To gain further insight into the behaviours of covert communication, future studies should overcome these constraints as seen in 5.2.

Table 8 6 key literatures for the comparison to the study

Literature topic	Authors and year	Key findings from literature	Comparison to this study from literature (i.e. similarities, differences, and contradictions)	Result
Advancement of technology	**(1)**: Montasari and Hill [7], **(2)**: Humphries et al. [8]	According to **(1)**: Montasari and Hill [7], it is crucial to constantly adapt to stay up with technological changes Likewise, **(2)**: Humphries et al. [8] highlight other difficulties, such as data management's intricacies and the significance of strengthening security protocols. These observations highlight how crucial it is for forensic processes to demonstrate inventiveness and adaptability in the face of changing technology environments	**(1)**: Montasari and Hill [7] identified challenges in data extraction, exemplified by devices such as the 'Vivo X3 Find Lite' used in this study. They suggested that the device posed significant hurdles to extraction, including limitations in obtaining physical and full image data using forensic tools. This finding supports the study's observation of difficulties encountered during data extraction from modern mobile devices Furthermore, **(2)**: Humphries et al. [8] highlighted the decreased availability of extensive datasets, attributing this limitation to constraints in forensic extraction methods. Their study aligns with the findings of the current research, reinforcing the notion that forensic investigators must continually adapt and refine their policies and procedures to overcome evolving challenges in data extraction and ensure the accuracy and reliability of forensic analyses	Similar

(continued)

Table 8 (continued)

Literature topic	Authors and year	Key findings from literature	Comparison to this study from literature (i.e. similarities, differences, and contradictions)	Result
Seizure of digital evidence	(3): Karagiannis and Vergidis [17]	The authors **(3)**: Karagiannis and Vergidis [17] emphasised the difficulty in defining cybercrime because there is no agreed-upon definition. This has consequences for the creation of uniform protocols for the acquisition of evidence. In tackling this issue, they underlined how crucial it is to have international cooperation and to follow best practices	**(3)**: Karagiannis and Vergidis [17] support the *significance of standardising protocols for* digital evidence seizure that is emphasised in this study. They also underscore the critical role that international cooperation plays in tackling the legal and procedural obstacles related to evidence validation, especially in situations like the ones covered in this study, like mSpy, where evidence needs to be examined by law enforcement to be admissible in court	Similar
Mobile operating systems	(4): Hewawitharana [18], (5): Qiao et al. [19]	According to **(4)**: Hewawitharana [18], mobile operating systems are very different from desktop and laptop operating systems, especially in that they aren't as good at multitasking In the study by **(5)**: Qiao et al. [19], there are compatibility problems with mobile AR across various OS, platforms, and data kinds, which calls for close cooperation to resolve	This research is supported by a study **by (4)**: Hewawitharana [18], which found that the target device's restricted multitasking capabilities led to operational issues, like crashes when many apps were open, which hampered prompt communication The results of this study are consistent with those of **(5)**: Qiao et al. [19], who found compatibility problems in mobile augmented reality across several operating systems. While Qiao et al. concentrated on augmented reality, this study found compatibility issues with app extractions that resulted in freezes and downgrades, underscoring the larger challenge of mobile app compatibility in forensic investigations	Similar/ Different

(continued)

Table 8 (continued)

Literature topic	Authors and year	Key findings from literature	Comparison to this study from literature (i.e. similarities, differences, and contradictions)	Result
Battery life	(6): Goel and Kumar [12], (7): Sharma et al. [13]	When conducting mobile forensic investigations, **(6)**: Goel and Kumar [12] draw attention to the possibility of increased battery drain and device malfunction when utilising blockage materials or aeroplane mode. According to **(7)**: Sharma et al. [13], taking out the device's battery may cause problems including SIM card lockout and data loss in volatile memory. These problems affect the operation of the gadget and the integrity of the evidence, their resolution calls for accuracy and caution	**(6)**: Goel and Kumar [12], bolster their research by emphasising problems with using the smartphone in aeroplane mode. In this investigation, background programmes that were running prevented data from reaching the server, and although using aeroplane mode preserved battery life, it also jeopardised data integrity by possibly causing data loss However, **(7)**: Sharma et al. [13] departs from this work as the investigation just collected mSpy data and did not entail physically removing the battery or collecting data from SIM cards	Similar/ Different
Malicious application	(8): Sharma et al. [23]	According to **(8)**: Sharma et al. [23], there is a considerable risk of infection on iOS and Android smartphones (2022). Malware that poses as updates is installed by cybercriminals to take advantage of vulnerabilities, even despite stringent app store screening. Due to users' ignorance of these hostile behaviours, data availability, confidentiality, and integrity are compromised	This study is substantially supported by **(8)**: Sharma et al. [23], since mSpy also secretly acquired information from the targeted user without their awareness. Users can lessen this by installing antivirus software on their mobile devices, which will identify and prevent spyware and other invasions	Huge similarity

(continued)

Table 8 (continued)

Literature topic	Authors and year	Key findings from literature	Comparison to this study from literature (i.e. similarities, differences, and contradictions)	Result
Android based spyware	(9): Naser et al. [24]	According to **(9)**: Naser et al. [24], the open nature and extensive usage of Android-based malware make it a more appealing target for hackers. To guarantee complete prevention strategies, they advise extending study to encompass iOS and IoT gadgets. Disregarding vulnerabilities on other platforms may elevate the danger of security breaches and deflect focus from threats that extend beyond Android. To properly handle new threats and preserve the integrity of all digital platforms, a comprehensive approach is necessary	As the major platform for analysing mSpy installation and data gathering was Android, the study by **(9)**: Naser et al. [24] provides significant support for this investigation. It highlights how much simpler it is to download malware on Android as opposed to iOS and other platforms because of its open nature. Given that concentrating only on Android may cause important vulnerabilities in other systems to go unnoticed, it is recommended that research be extended beyond Android to other platforms	Huge similarity

Other Literature studies:
Although not essential to the study, the additional literature provides valuable perspectives. McTurk [14] and Zinkus et al. [16] emphasize the importance of emergency phones for security and privacy. Krishnan et al. [20], Fukami et al. [9] focuses on encrypted techniques for cyber defence. Although slightly different from spyware, Fernando [21] explores evidence extraction, which is essential for digital forensics. The understanding of mobile security is tangentially enhanced by discussions of WhatsApp communication and digital forensics by Salamh et al. [45], Griggio et al. [25] and Heath et al. [26]. Zhang et al. [10] reveal the illegal use of fake fingerprints, while Fukami et al. [9] points out the problems of unlocking devices. The study by Yu et al. [11] on devices such as the Vivo X80 Pro highlight the challenges of forensic mining, although these studies provide a broad overview of mobile security and forensics, although not directly related to spyware detection and mitigation

5.2 Recommendation

Future research should create action plans that ensure adherence to deadlines and milestones to overcome time constraints. This strategy supports maintaining project momentum and completing tasks on time. In addition, we recommend that future research focus on evaluating the effectiveness of reliable and high-speed networks, including 5 GHz Wi-Fi, to improve the reliability of data transmission. Anon [40] notes that EE 4G and 5G mobile networks offer speeds from 1 Mbps to a few hundred, and 5 GHz Wi-Fi can reach speeds of up to 1300 Mbps [41], which is much faster than the usual 5-100 Mbps of 2.4 GHz Wi-Fi speed [42, 46]. It shows how 5 GHz Wi-Fi networks can provide more reliable and faster data transmission, increasing the efficiency of process coverage and tracking functions. Finally, future researchers should consider conducting experiments with different spyware to ensure comprehensive verification of the results.

References

1. Bonar EE, Cunningham RM, Collins RL, Cranford JA, Chermack ST, Zimmerman MA, Blow FC, Walton MA (2019). Feasibility and acceptability of text messaging to assess daily substance use and sexual behaviours among urban emerging adults. NIH PMC5889069. https://doi.org/10.1080/16066359.2017.1310205
2. Maddocks S (2018) From non-consensual pornography to image-based sexual abuse: charting the course of a problem with many names. Aust Fem Stud 33(97):345–361. https://doi.org/10.1080/08164649.2018.1542592
3. Greengard S (2024, May 3) Pegasus (spyware)|description, origins, spying, and controversies. Encyclopaedia Britannica. https://www.britannica.com/topic/Pegasus-spyware
4. Church KW, Chandrasekar R (2023) Emerging trends: risks 3.0 and proliferation of spyware to 50,000 cell phones. Nat Lang Eng 29(3):824–841. https://doi.org/10.1017/s1351324923000141
5. Penney J, Schneier B (2022) Platforms, encryption, and the CFAA: the case of WhatsApp v NSO Group. Schulich Law Scholars. https://digitalcommons.schulichlaw.dal.ca/cgi/viewcontent.cgi?article=2788&context=scholarly_works
6. Cecco L (2022, July 8) 'Asleep at the wheel': Canada police's spyware admission raises alarm. The Guardian. https://www.theguardian.com/world/2022/jul/07/canada-police-spyware-admission-surveillance-experts
7. Montasari R, Hill R (2019, January 1) Next-generation digital forensics: challenges and future Paradigms. IEEE conference Publication|IEEE Xplore. https://ieeexplore.ieee.org/abstract/document/8688020/authors
8. Humphries G, Nordvik R, Manifavas H, Cobley P, Sorell M (2021) Law enforcement educational challenges for mobile forensics. Forensic Sci Int Digital Invest 38:301129. https://doi.org/10.1016/j.fsidi.2021.301129
9. Fukami A, Stoykova R, Geradts Z (2021) A new model for forensic data extraction from encrypted mobile devices. Forensic Sci Int Digital Invest 38:301169. https://doi.org/10.1016/j.fsidi.2021.301169
10. Zhang P, Feng P, Li Q, Chen Z (2019, January 24) Private data acquisition method based on system-level data migration and volatile memory forensics for android applications. IEEE. Retrieved 3 January 12 C.E., from https://ieeexplore.ieee.org/stamp/stamp.jsp?arnumber=8625420

11. Yu Y, Niu Q, Li X, Xue J, Liu W, Lin D (2023) A review of Fingerprint sensors: mechanism, characteristics, and applications. NIH, PMC10305017. https://doi.org/10.3390/mi14061253
12. Goel M, Kumar V (2019) Layered framework for mobile Forensics analysis. Soc Sci Res Netw. https://doi.org/10.2139/ssrn.3351029
13. Sharma Y, Shagufta Noval S, Jain A, Sabitha B, Ramya T (2022, December 14) Forensics-as-a-service: a review of mobile forensics. IEEE conference Publication|IEEE Xplore. https://ieeexplore.ieee.org/abstract/document/10072726
14. McTurk B (2019, March) Forensic professionals' views on the lack of standards in the digital forensic field: a generic qualitative inquiry—ProQuest. ProQuest. Retrieved 17 Jan 2024, from https://www.proquest.com/docview/2206331024?pq-origsite=gscholar&fromopenview=true&sourcetype=Dissertations%20&%20Theses
15. Cartwright B, Weir GRS, Frank R (2019, April 29) Cyberterrorism in the cloud: through a glass darkly. University of Strathclyde. https://pureportal.strath.ac.uk/en/publications/cyberterrorism-in-the-cloud-through-a-glass-darkly
16. Zinkus M, Jois TM, Green M (2021, May 26) Data security on mobile devices: current state of the art, open problems, and proposed solutions. arXiv.org. https://arxiv.org/abs/2105.12613
17. Karagiannis C, Vergidis K (2021) Digital evidence and cloud forensics: contemporary legal challenges and the power of disposal. Information 12(5):181. https://doi.org/10.3390/info12050181
18. Hewawitharana C (2022) Challenges in mobile OS (Android, IOS, Windows)—a comparative analysis. ResearchGate. https://www.researchgate.net/publication/364955512_Challenges_in_Mobile_OS_Android_iOS_Windows_-A_Comparative_Analysis
19. Qiao X, Ren P, Dustdar S, Liu L, Ma H, Chen J (2019, April 1) Web AR: a promising future for mobile augmented reality—state of the art, challenges, and insights. IEEE J Mag|IEEE Xplore. https://ieeexplore.ieee.org/document/8643424
20. Krishnan S, Zhou B, Kyun An M (2021, December 9) Smartphone forensic challenges. Retrieved 2 Feb 2024, from https://shsu-ir.tdl.org/server/api/core/bitstreams/89c23f94-9d9e-4fdc-beb4-eb8e236d251d/content
21. Fernando V (2021, April 19) Cyber forensics tools: a review on mechanism and emerging challenges. IEEE conference Publication|IEEE Xplore. https://ieeexplore.ieee.org/abstract/document/9432641/authors#authors
22. Krishnan S, Zhou B, An MK (2019) Smartphone forensic challenges. Int J Comput Sci Secur 13(5):183–184. https://www.cscjournals.org/manuscript/Journals/IJCSS/Volume13/Issue5/IJCSS-1501.pdf
23. Sharma BK, Yadav V, Purba MK, Sharma Y, Kumar V, Ishant, Mehta P (2022) Challenges, tools, and future of mobile phone forensics. Journalppw. Retrieved 5 Mar 2024, from https://journalppw.com/index.php/jpsp/article/view/4132/2704
24. Naser M, Albazar H, Abdel-Jaber H (2023) Mobile spyware identification and categorization: a systematic review. Informatica 47(8). https://doi.org/10.31449/inf.v47i8.4881
25. Griggio CF, Nouwens M, Klokmose CN (2022) Caught in the network: the impact of WhatsApp's 2021 privacy policy update on users' messaging app ecosystems. In: CHI conference on human factors in computing systems. https://doi.org/10.1145/3491102.3502032
26. Heath H, MacDermott Á, Akinbi A (2023) Forensic analysis of ephemeral messaging applications: disappearing messages or evidential data? Forensic Sci Int Digital Invest 46:301585. https://doi.org/10.1016/j.fsidi.2023.301585
27. Bang D (2024, March 8) MSpy reviews [2024] is MSpy the best spy App? https://www.linkedin.com/pulse/mspy-reviews-2024-best-spy-app-ditto-bang-63mdc
28. Wise J (2023, July 20) Umobix versus Flexispy comparison 2024: which is better? EarthWeb. Retrieved 25 Mar 2024, from https://earthweb.com/umobix-vs-flexispy/
29. Rozgonjuk D, Sindermann C, Elhai JD, Montag C (2021) Comparing smartphone, WhatsApp, Facebook, Instagram, and Snapchat: which platform elicits the greatest use disorder symptoms? Cyberpsychol Behav Soc Netw 24(2):129–134. https://doi.org/10.1089/cyber.2020.0156
30. Office H (2023) Regulation of investigatory powers act 2000 (RIPA). GOV.UK. https://www.gov.uk/government/publications/regulation-of-investigatory-powers-act-2000-ripa/regulation-of-investigatory-powers-act-2000-ripa

31. Office H (2018) Covert surveillance and property interference revised code of practice. https://assets.publishing.service.gov.uk/media/5ba37401e5274a55cdb89bce/201800802_CSPI_code.pdf
32. Anon (2014) Spy laws in the UK|Spy Equipment UK. Spy Equipment UK. https://www.spyequipmentuk.co.uk/spy-laws/
33. Hsu FH, Hwang YL, Tsai CY, Cai WT, Lee CH, Chang KW (2016) TRAP: a three-way handshake server for TCP connection establishment. Appl Sci 6(11):358. https://doi.org/10.3390/app6110358
34. Swaminathan N, Danks D (2024, March 22) Application of the NIST AI risk management framework to surveillance technology. arXiv.org. https://arxiv.org/abs/2403.15646
35. Slobogin C, Brayne S (2022) Surveillance technologies and constitutional law. NIH 6:219–240. 38074421. https://doi.org/10.1146/annurev-criminol-030421-035102
36. Harkin D, Molnar A, Vowles E (2019) The commodification of mobile phone surveillance: an analysis of the consumer spyware industry. Crime Media Cult 16(1):33–60. https://doi.org/10.1177/1741659018820562
37. Grimmelmann J (2020) Spyware versus spyware: software conflicts and user autonomy. Heinonline 16.1. https://scholarship.law.cornell.edu/cgi/viewcontent.cgi?article=2797&context=facpub
38. Zhang X, Cao Z, Dong W (2020, July 31) Overview of edge computing in the agricultural Internet of Things: key technologies, applications, challenges. IEEE J Mag|IEEE Xplore. Retrieved 13 Apr 2024, from https://ieeexplore.ieee.org/abstract/document/9153160
39. Zhang W, Min Ong L, Findlay M (2023) Regulating AI and big data in emerging economies. Google Books. Retrieved 10 Mar 2024, from https://books.google.co.uk/books?hl=en&lr=&id=HuDoEAAAQBAJ&oi=fnd&pg=PA158&dq=Zhang,+W.+(2023).+technology+developments+in+surveillance+and+emphasise+how+they+affect+privacy+and+security&ots=KmjMbT_OQ7&sig=x1-4laAH8zSrW1ZVAYh1BmBjkeA
40. Anon (2023, September 6) What mobile speed. EE Community. https://community.ee.co.uk/t5/Mobile-Broadband/What-mobile-speed/td-p/1291553#:~:text=On%20EE%204G%20Network%20your,Speed%20depending%20on%20the%20Location
41. Silver, L. (2019). Smartphone Ownership Is Growing Rapidly Around the World, but Not Always Equally. [online] Pew Research Center's Global Attitudes Project. Available at: https://www.pewresearch.org/global/2019/02/05/smartphone-ownership-is-growing-rapidly-around-the-world-but-not-always-equally/.
42. Slobogin, C., & Brayne, S. (2022). Surveillance Technologies and Constitutional Law. Annual Review of Criminology, 6(1). https://doi.org/10.1146/annurev-criminol-030421-035102.
43. Alkhudhayr, F., Alfarraj, S., Aljameeli, B., & Elkhdiri, S. (2019). Information Security: A Review of Information Security Issues and Techniques. 2019 2nd International Conference on Computer Applications & Information Security (ICCAIS). https://doi.org/10.1109/cais.2019.8769504.
44. Coleman D (2020, May 11) COVID-19 cybersecurity framework: confidentiality, integrity, availability. Roebuck Technologies. https://www.roebucktech.com/it-blog/covid-19-cybersecurity-framework-confidentiality-integrity-availability/.
45. Salamh FE, Karabiyik U, Rogers MK (2020, October 20) Asynchronous forensic investigative approach to recover deleted data from instant messaging applications. In: IEEE conference Publication |IEEE Xplore. https://ieeexplore.ieee.org/abstract/document/9297227?casa_token=vpvhYyj-QOkAAAAA:uBxuBbhbWjeKNsF3PVKk80J33rIM-Fv5R99-FDt-Ca6eGl9RsDJygbum_s6wJRjbLrz0pe8ztpg.
46. Wi-Fi S (2020, September 13) What's the difference between 2.4 and 5 GHz WiFi? Social WiFi. https://socialwifi.com/knowledge-base/wifi-technology/difference-between-24-and-5-ghz-wifi/#:~:text=The%20main%20difference%20is%20speed,support%20up%20to%201300%20Mbps. **This is reference 42.**

Automated Penetration Testing Using Generative Artificial

Veerababu Gullinkala and Sina Pournouri

Abstract This research explores the integration of ChatGPT, a Large Language Model (LLM), with traditional penetration testing tools such as Nmap, Metasploit, Whois, and Msfvenom. By leveraging ChatGPT's natural language processing abilities, the study introduces a novel approach to penetration testing that simplifies tool usage and improves task efficiency. The model assists in understanding complex queries, generating targeted attack scripts, and analyzing data. This fusion enhances both the accuracy and speed of penetration testing while making it more accessible to newcomers. The study also emphasizes the importance of ethical AI use in cybersecurity. The findings suggest that combining language models with technical tools fosters better communication among security experts and opens new possibilities for AI-assisted cybersecurity practices. Ultimately, the research paves the way for smarter, more efficient penetration testing methods as organizations strive to protect their digital assets.

Keywords Penetration testing · Generative artificial · Chatgpt · Large language model · LLM · AI-assisted cybersecurity · Ethics

1 Introduction

1.1 Penetration Testing

Malicious network assaults have escalated in severity in recent years, posing a serious risk to the data security of people, companies, and even whole nations [1]. The goal of penetration testing is to identify security holes in a system by simulating malicious cyberattacks in an authorized manner. Penetration testing is a tried and true method of testing the safety of digital assets by proactively searching for and taking advantage

V. Gullinkala · S. Pournouri (✉)
Sheffield Hallam University, Sheffield, UK
e-mail: S.Pournouri@shu.ac.uk

H. Jahankhani and B. Issac (eds.), *Cybersecurity and Human Capabilities Through Symbiotic Artificial Intelligence*, Advanced Sciences and Technologies for Security Applications, https://doi.org/10.1007/978-3-031-82031-1_22

of flaws [2]. This technique mimics hacker assaults without really compromising the uptime of the systems being tested. In other words, the goal of penetration testing is not to compromise a system or steal data, but to make it more secure [1].

By simulating an attack against the system, penetration testing can determine how well its security safeguards are performing. On the other side, real-world examples of security threats may be used to raise awareness of these concerns among the appropriate staff. Penetration testing, as contrast to hacking, is performed after a contract has been signed with a company or corporation and results in a report. Penetration testing identifies potential security holes, which are promptly patched thereafter, successfully averting actual breaches.

1.2 Traditional Penetration Testing

Traditional penetration testing is a hands-on method of gauging a system's or network's defenses. The process often entails a group of cybersecurity experts simulating an attack on the target system in order to locate potential weak points. These experts use a wide variety of resources and methods to identify areas of weakness, break into systems without authorization, and evaluate the potential damage of an assault.

1.3 Artificial Intelligence Based Penetration Testing

To automate steps in the penetration testing process, AI-based penetration testing makes use of AI and machine learning technologies. Scanning for security holes, modelling threats, and simulating attacks are all possible with the help of AI. Algorithms trained with machine learning can sift through massive amount of data in search of trends, security holes, and potential exploits. Effectiveness, accuracy, and the discovery of previously unknown, complicated vulnerabilities are all aided by AI-powered penetration testing tools.

1.4 Penetration Testing with ChatGPT and Google Bard

The utilization of Chat GPT and Google as natural language processing technologies has the potential to automate a range of tasks within the realm of cybersecurity.

The utilization of ChatGPT and Google bard for automating security tasks can offer significant advantages to security testing. This includes enhanced accuracy of results, as well as time and effort savings. Chat bots have the capability to automate security testing tasks, including vulnerability scanning and penetration testing, which are often repetitive and time-consuming. The utilization of this approach can enhance the precision and effectiveness of security testing by furnishing comprehensive data and perspectives on susceptibilities in software systems. Furthermore, the utilization of ChatGPT and Google Bard can facilitate the process of detecting novel vulnerabilities and formulating more resilient security testing scenarios.

This task can be automated using ChatGPT and Google BARD. These Chat bots have the capability to undergo training in order to detect and leverage weaknesses within software systems. This can provide valuable insights to security testers regarding the behavior of actual attackers. This can provide organizations with a more comprehensive comprehension of their software systems' security posture and aid them in enhancing it.

1.5 Related Work

The study "Evaluating the Impact of ChatGPT on Exercises of a Software Security Course" [3] bears significant similarity to the present research, which focuses on the application of large language models (LLMs) such as ChatGPT and Google Bard in penetration testing. This investigation examines ChatGPT's efficacy in identifying and remediating code vulnerabilities, aligning closely with the emphasis on leveraging LLMs for penetration testing purposes.

Similar to tasks associated with penetration testing, the findings indicate that ChatGPT can effectively detect security issues, automate testing methodologies, and propose remediation strategies [3]. Furthermore, the study highlights critical limitations, such as challenges related to cross-file vulnerability identification, which are essential considerations for the current research.

Additionally, the paper's exploration of the educational impact of LLMs contributes to a broader understanding of the implications of integrating these models into security operations, underscoring the transformative potential of LLMs in enhancing both security practices and educational outcomes within the field.

Software applications, scripts, and algorithms can automate parts of the penetration testing process, as noted by Chu [1], with the goal of making the process more productive, reliable, and scalable. While automated penetration testing is still in its early stages, most of the literature frames penetration testing as a planning problem, often using attack trees, graphs, or planning domain languages. Some studies have attempted to translate vulnerability scanner findings to exploitation tools, though they frequently struggle with limited information, ambiguity, and scalability issues [1].

Chu's [1] study aims to achieve real-time automation through the belief-desire-intention (BDI) paradigm, contributing to the field of automated penetration testing. The thesis also focuses on creating an ontology for penetration testing using the Semantic Web Rule Language (SWRL) to improve knowledge reuse and reasoning capabilities, with an evaluation in an IoT context. This work adds to the automation of penetration testing, particularly for IoT.

In comparison to the proposed research, although both aim to improve penetration testing, automated testing and integrating large language models (LLMs) like ChatGPT and Google Bard take different approaches. This article provides a useful overview of the AI penetration testing landscape, highlighting current applications and methods.

The study also addresses the challenges of keeping pace with the growth of digital services, which could be relevant to the proposed investigation of incorporating LLMs into penetration testing. While the article does not directly address ChatGPT's role, its findings and methods can inform future research and offer insights into the pros and cons of integrating LLMs into penetration testing.

Strengths of Chu's Thesis include real-time automation, which could enhance the efficiency of penetration testing by reducing manual labor and enabling faster vulnerability detection and resolution. The use of the BDI model offers a structured approach to decision-making, potentially improving accuracy and reliability in automated testing. The ontology based on SWRL supports knowledge reuse and enhances reasoning, which may improve the identification and exploitation of vulnerabilities. Lastly, the emphasis on experimentation and evaluation shows a commitment to empirical review, offering insights into the performance and limitations of the automated penetration testing model.

1.6 Aim

This research will examine the feasibility and efficacy of real-time implementation of penetration testing using large language models, notably ChatGPT and Google BARD. Its goal is to determine whether such an implementation is possible, and if so, how effective it would be. The purpose of this research is to evaluate the capabilities of these language models in simulating attacks, conducting social engineering, identifying web application vulnerabilities, automating vulnerability assessments, and enhancing red teaming exercises. This project aims to establish the potential benefits and limitations of leveraging large language models to assess the security of computer systems and networks. This will be accomplished by exploring the practicality of incorporating these models into the penetration testing process. To achieve the aim of this research, the following objectives are defined:

- Determine whether and how large language models may be used to effectively manage the complicated nature of actual penetration testing scenarios.
- Examining the strengths and weaknesses of LLM's such as ChatGPT and Google BARD as vulnerability generators for real-world scenarios.
- The goal of this study is to compare and contrast the capabilities of ChatGPT and Google BARD with regard to automating the various phases of a penetration test.
- The purpose of these studies is to confirm that ChatGPT and Google BARD are useful and effective for doing real-world penetration tests.

- To give guidelines and suggestions for using LLMs successfully in real-time penetration testing, taking into account possible obstacles and factors.

2 Experimental Methodology

2.1 Experimental Design

The experimental design consists of multiple rounds, each involving a set of test cases to assess the compatibility of ChatGPT with various penetration testing tools. The experiments are conducted in a carefully controlled environment to ensure consistent conditions and accurate results.

During this stage, the proper integration of ChatGPT into the penetration testing process is explored. The goal is to enhance communication between the user and penetration testing tools by utilizing ChatGPT API features.

Each iteration focuses on a specific penetration testing activity, such as network scanning, exploit testing, or report generation. Test cases are designed to cover a range of potential scenarios and complexity levels relevant to the task. The test cases involve creating a chatbot using the ChatGPT API. When the user asks a penetration testing-related question (based on their specific requirements), the artificial intelligence executes the test and delivers the results to the user via the chatbot.

2.2 Recognizing the Value of ChatGPT Integration

During this stage, the intention is to improve the communication between the user and the penetration testing tools by using chatgpt API features.

2.2.1 ChatGPT Functionality

Chatgpt features are thoroughly investigated so that its strengths and weaknesses may be grasped. To do so, it is necessary to review the OpenAI-supplied API documentation, code samples, and best practices. Researchers may efficiently use ChatGPT for penetration testing if they familiarize themselves with its fundamental ideas and procedures.

2.2.2 Integration Architecture

The integration architecture is made in such a way that the user, the chatbot interface, and the penetration testing tools can all talk to one another without any hitches. Integration may be made easier with the use of APIs, middleware, and other routes of communication that may be developed as part of the design. Chatgpt is built with regard for how it will communicate with the chosen tools and respond to user inquiries.

2.3 Mapping User Questions to Penetration Testing Actions

Integrating the penetration testing tools requires translating user queries to relevant actions inside those tools. Researchers need to learn all aspects of the chosen tools' command structures and syntaxes and figure out how questions posed by users might be translated into actual commands.

User queries are analyzed to determine their purpose, and then mapped to the features of the penetration testing tools. For example, understanding that a user query like "What are the open ports?" can be mapped to an Nmap scan command to retrieve information about open ports on a target system.

2.4 Training ChatGPT

Training In order to train the language model to create correct and contextually appropriate tool commands for penetration testing activities, ChatGPT makes use of the gathered dataset of prompt-command pairs. ChatGPT must be trained to learn how to translate user questions into actionable tool instructions, therefore this phase of development is critical.

2.5 Experimental Setup and Test Cases

In order to conduct the experiments and assess the performance of the integrated system, a specialized environment must be configured as part of the experimental setup. It entails preparing the necessary hardware and software for the integration and checking for compatibility and good operation.

The following items may make up the experimental setup:

Hardware: A computer system that meets the system requirements of the integrated system and the penetration testing tools of choice. The performance, memory, and storage capacity of the hardware must be enough for proper functioning.

Software: The OS, programming languages, libraries, and frameworks needed to include Chatgpt, and the penetration testing tools are all considered essential software components. In order to prevent incompatibilities and conflicts, it is important that all of the software versions work well together.

Tools for Penetration Testing: The experimental environment should include the chosen penetration testing tools, such as Nmap, Metasploit, and Nessus, installed and set properly. The most recent versions of the tools should be used, and they should all work together seamlessly inside the chatbot environment.

ChatGPT, the chatbot interface, and the penetration testing tools all work together as part of an integrated system, which must be set up and configured correctly. This involves laying the groundwork for component-to-component communication, maintaining consistent data flow, and facilitating effortless user-to-system engagement.

Example tests cases:

To determine how well the integrated system performs under different penetration testing conditions, a collection of test cases has been developed. The capabilities and limits of the system can only be accurately evaluated if the test cases include a wide variety of use scenarios.

Examples of possible test scenarios are:

Scanning Networks: The processes of scanning a single host, scanning multiple hosts, scanning individual ports, and detecting services serve as critical test cases for evaluating the overall efficacy of the system in mapping a network. Through the implementation of these methodologies, it is possible to assess the system's performance and accuracy in network mapping activities.

During exploit testing, the integrated system looks for vulnerabilities in target systems and creates Metasploit instructions to attack them. The produced commands and exploits' success rates are evaluated.

2.6 Metrics for Evaluation

Metrics are used to evaluate how well the integrated system is functioning. Metrics used for analysis should be relevant to the study's goals and provide light on the system's viability and efficiency.

Metrics such as these may be used often while evaluating the integrated system:

Accuracy of Commands, how closely a set of created tool commands matches the intended user queries or tasks. For each test scenario, the produced commands are compared to the intended instructions to establish command accuracy.

An evaluation of the findings generated by the integrated system, when compared to those obtained through the manual execution of identical penetration testing procedures, allows for a comprehensive assessment of the quality and effectiveness of the developed instructions. This comparative analysis facilitates the identification of any inconsistencies or deviations, thereby enhancing the understanding of the system's performance relative to traditional methods.

Metrics used for evaluation should be consistent with the study's goals and provide a holistic picture of the integrated system's efficacy, precision, and usability.

2.7 Ethical Consideration

Research involving penetration testing and the incorporation of AI must give careful attention to ethical problems. Ethical considerations and safeguards for participants' rights and confidentiality must be included into every stage of the study process.

- **Data Privacy and Security**: To ensure the safety and confidentiality of sensitive information collected during testing, it is essential to implement appropriate data protection measures. This includes the application of data anonymization techniques, such as encryption, and the enforcement of strict access controls. Furthermore, compliance with all relevant data privacy regulations must be rigorously maintained throughout the process
- **Confidentiality**: Safeguard the confidentiality of collected data and ensure that it is not disclosed to unauthorized individuals or entities. Stringent access control measures should be applied to prevent any breaches of information security.
- **Legal Compliance**: Ensure that the study fully complies with all applicable privacy and data protection laws, as well as relevant regulations governing penetration testing. Additionally, adherence to the highest ethical standards must be maintained throughout the research process.
- **Risk Mitigation**: The potential risks associated with the integrated system, including unforeseen outcomes, security vulnerabilities, or misuse of the technology, must be thoroughly identified and addressed. Proactive measures should be taken to minimize the impact of these risks, ensuring the system's integrity and secure operation.

3 Implementation

In this part, the specifics of combining ChatGPT with various penetration testing tools are presented. By providing a chatbot interface that is driven by ChatGPT, the solution seeks to improve both the effectiveness and the quality throughout the penetration testing process.

3.1 Integration Architecture

The following elements make up the various components that make up the integration architecture for the implementation of ChatGPT with penetration testing tools:

1. Tools for Penetration Testing At the heart of the integration is a collection of commonly used penetration testing tools, such as nmap, Metasploit, and Sublist3r, together with a variety of additional information collecting programs. These tools provide users the ability to scan networks, take advantage of vulnerabilities, collect information, and carry out a variety of security evaluations.

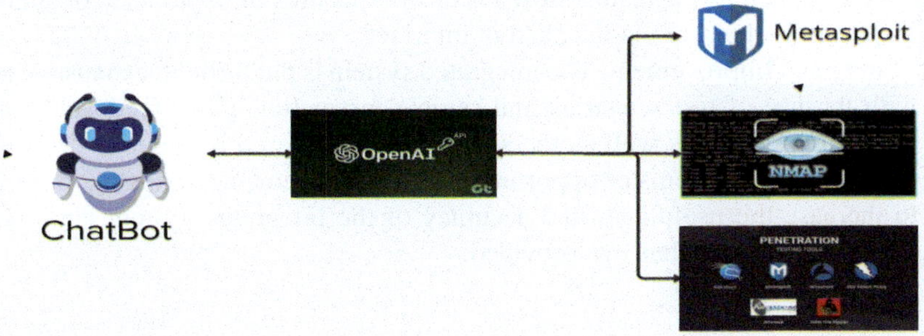

Fig. 1 Chatpentest tool

2. The Python programming language is used to construct a chatbot interface, and the OpenAI API is used in order to include ChatGPT. Users are able to communicate with the penetration testing tools using a conversational interface that is provided by the chatbot. It offers a conversational experience that is based on natural language and enables users to submit instructions, ask questions, and get replies from the system.
3. OpenAI API: The integration of ChatGPT into the chatbot interface is made easier by the OpenAI API. It makes it possible to get model-generated replies based on user inputs and makes it possible for the chatbot and ChatGPT to communicate with one another in a smooth manner (Fig. 1).

3.2 Implementations Steps

Installation and Configuration: All of the essential tools for penetration testing, as well as Python and any other relevant libraries, have been installed and setup on the system. The OpenAI Application Programming Interface (API) is configured, and the authentication particulars, such as the API key, are acquired.

Constructing a Chatbot: The Python programming language is used throughout the construction of a chatbot. For the purpose of managing API requests and answers, libraries like as requests, OpenAI, and python-nmap library for the nmap tool are used. The OpenAI API is queried with the results of the user inputs that are recorded, processed, and transmitted. After that, the answer from the API is retrieved and shown to the user.

Integration with Penetration Testing Tools: The chatbot is integrated with the penetration testing tools by executing the relevant instructions depending on user inputs. This allows for seamless communication between the chatbot and the tools. The user's instructions are then converted by the chatbot into the appropriate

command-line arguments or API requests for the related tools. The output from the tools, such as the results of scans or the specifics of exploits, is obtained and displayed to the user via the chatbot interface.
Testing and Improvement: The integrated system is put through exhaustive testing with the use of test scenarios and sample instructions. The feedback from the penetration tester as well as the supervisor is gathered in order to determine if there are any problems or opportunities for improvement. Refinements are made to increase the usefulness and accuracy of the integrated system, and they are based on the input that was provided.

3.3 Technical Considerations

- **Security**: Sophisticated security protocols have been put into place to protect both the authenticity and the privacy of user instructions and the results they produce. In order to safeguard sensitive information, encrypted communication protocols and access control mechanisms were implemented.
- **Scalability**: The integration was meant to be scalable, which means that it can accommodate a high number of simultaneous user interactions and help facilitate the execution of several tool commands in simultaneously. Scalability was one of the design goals for the integration.
- **Compatibility**: In order to enable a smooth integration with already established penetration testing settings, it was important to verify that the system was compatible with a wide variety of operating systems, tool versions, and network configurations.

4 Experimental Results of ChatGPT Integration to Penetration Testing Tools

4.1 Introduction

The functionality of the chatgpt model is evaluated and discussed in this chapter within the context of a real-world setting. When compared to manual Penetration testing, the chatgpt model was much more efficient. The chatgpt model was stronger to the other techniques in terms of ease of use, automation, real-time, and uncertainty. Moreover, the chatgpt model was better to the other approaches.

4.2 *Experiments*

The suggested integration of chatgpt model to penetration testing tools was put through its paces in a real environment on one of the hackthebox capture the flag blue machines so that the performance study and assessment could be carried out. The experiment had a parrot machine serving as the attacking agent and a blue machine serving as the target.

4.3 *Example-1: Experimental Results of Chatgpt Integration to Nmap*

The combination of ChatGPT and Nmap makes the process of penetration testing much simpler. This is achieved via the automation of the creation of Nmap commands in response to user inquiries and the subsequent execution of Nmap scans on the IP addresses that the user specifies. The procedure and its results are outlined below in the following steps:

An AI-powered penetration testing assistant may be created by integrating the OpenAI API, Python, and Nmap programming languages together. This combines the capability of natural language processing with the capabilities of network scanning. Because of this connection, users are now able to communicate with the system by utilizing natural language questions. The artificial intelligence language model, which is given by the OpenAI API, then creates Nmap instructions depending on the user's input. After that, the subprocess module in Python is used to carry out the instructions in order to do network scans and collect information about the target systems' security posture.

The following is a rundown of the important components and the responsibilities that they play in the integration:

4.3.1 OpenAI API

OpenAI API gives users access to cutting-edge language models such as GPT-3, which are able to interpret natural language queries and provide results that are eerily similar to those produced by humans [4]. In this particular integration, the API key is what allows the connection to the language model and the interactions with it to take place.

The API can be accessed by sending HTTP requests in any programming language; however, official bindings are provided for the Python programming language, along with an official library for Node.js. In this context, programming is conducted using Python [5].

To install the official Python bindings for OpenAI, simply enter the following command in the terminal: pip install openai && pip3 install openai

4.3.2 Openai Authencation

Authentication within the OpenAI API is managed through the use of API keys. The necessary API key for requests can be obtained by accessing the API Key management page.

It is essential to consider the API key as a confidential identifier. Therefore, it must not be shared with others or exposed in client-side applications, such as browsers or software interfaces. For production environments, all requests should be routed through a backend server, where the API key can be securely loaded from an environment variable or a key management service, in accordance with security best practices.

The API key should be included in the Authorization HTTP header for all API requests, formatted as follows:

Import openai

openai.api_key = "API_KEY"

4.3.3 Using Python

Python is a flexible programming language that has extensive libraries and modules, all of which work together to make data processing, scripting, and automation much easier. Python is used throughout this integration as the main language for programming in order to construct the penetration testing assistant and to orchestrate the communication that takes place between the user, the AI model, and Nmap.

4.3.4 Integration Workflow

Interaction with the User during the Integration of the OpenAI API, Python, and Nmap:

The user interaction is a crucial component of the integration since it helps to bridge the gap between the user's intent and the execution of Nmap scans by making use of natural language queries. The Python script may either provide a command-line interface (also known as CLI) or a graphical user interface (also known as GUI), both of which can use by the user to interact with the system. But here I used only command line interface CLI. The process of user contact may be broken down into the following steps:

Instructions

When the user runs the penetration testing helper, they are greeted with a welcome message that provides a quick overview of the capabilities of the tool as well as instructions on how to make use of it. It's possible that the instructions will provide some instances of legitimate queries as well as the syntax for naming target IP addresses.

User Queries

The user is presented with a popup to submit their natural language inquiry about Nmap scanning while using the system. The user is able to indicate the intended scanning steps or goals by typing their inquiry in a conversational way.

By providing ChatGPT with appropriate training focused on Nmap and other pertinent subjects, it can be equipped to address a variety of inquiries related to Nmap output. This includes offering insights on Common Vulnerabilities and Exposures (CVEs) associated with identified vulnerabilities.

ChatGPT has the potential to become proficient at understanding Nmap commands, analyzing their output, and delivering useful information on vulnerabilities, open ports, services, and other topics with the right amount of training data and some fine-tuning. Additionally, it may aid in the generation of Nmap instructions that are specifically customized to the requirements of the scanning.

Example inputs:

User input: Find open ports on the target system.
User input: can you perform a tcp scan (we can see in above Fig. 2).
User input: Are there any advanced scanning techniques to detect os and version? (we see in below Fig. 3).

Natural Language Processing (NLP)

The HTTP request library in Python is used to communicate with the OpenAI API and provide the user's inquiry. The API's language model analyses the input via natural language processing (NLP) methods, therefore gaining a grasp of the user's purpose and producing a suitable Nmap command that is in line with the query.

Nmap Command Generator

After the language model has finished generating the Nmap command, it is sent back to the Python script where it was originally stored. After that, the command that was produced is shown to the user so that they may examine it before the real scan is

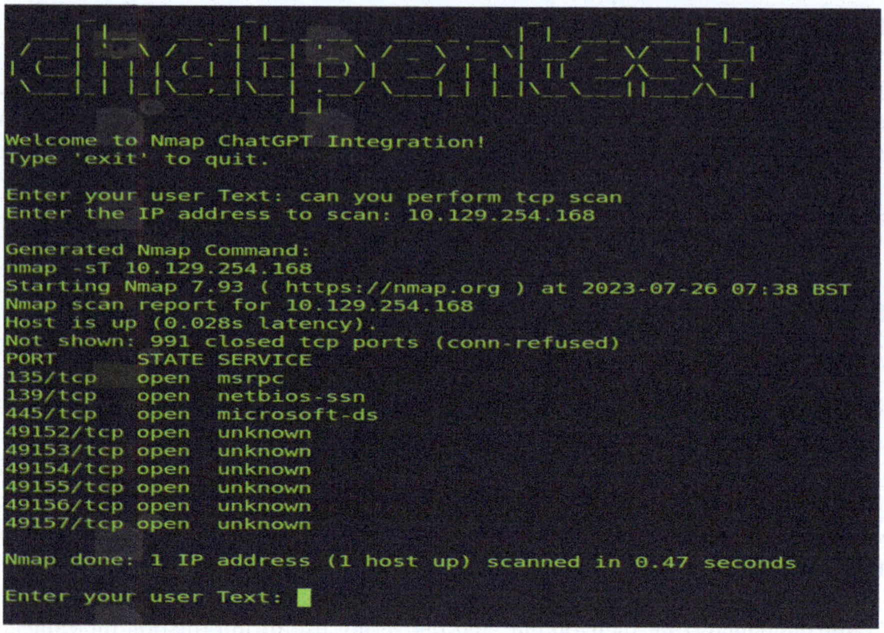

Fig. 2 Example-1

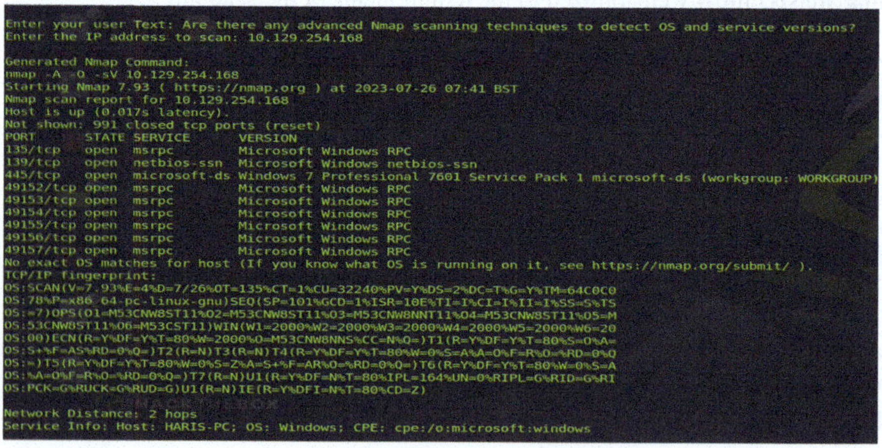

Fig. 3 Example-2

carried out. The Python script makes certain that the created command accurately represents the user's purpose and is pertinent to the IP address that was supplied as the destination.

Review and Confirmation by the User

The created Nmap command is delivered to the user by the Python script, and the user reviews and confirms the command. They have the ability to verify the command and make any alterations or revisions that may be required. The review stage is necessary because it ensures that the produced command is in accordance with the user's aims and that it appropriately represents the scan that was intended.

Execution of the Nmap Scan

The Python script makes use of the subprocess module in order to carry out the Nmap scan as soon as the user has confirmed the Nmap command. The Nmap command is executed in the shell environment by the subprocess module, and the output, which provides information about the results of the network scan, is captured.

Presentation of the Output

When the Nmap scan has finished, the Python script will display the results of the scan to the user in a manner that is easy to understand. Information such as open ports, running services, information on the operating system, and possible vulnerabilities discovered during the scan are often included in the output of the process.

User Interaction Loop

The user interaction loop gives the user the option to continue using the penetration testing assistance by either inputting further inquiries or leaving the programme at any time. If the user desires to conduct another scan or investigate other scanning possibilities, they are able to enter a new query, at which point the procedure resumes with user queries as the starting point.

4.4 Experimental Analysis of Chatgpt Integration to Nmap

4.5 Advantages

4.5.1 The Benefits of Automation and Efficiencies

Because the integration automates the process of constructing Nmap instructions, it considerably reduces the mental strain that is placed on users. Even users with a little amount of knowledge with Nmap are able to provide meaningful and accurate scan instructions with the help of ChatGPT natural language processing capabilities, which are used in this process.

4.5.2 Ability to Use and Accessibility

A user-friendly interface is provided by the system, which abstracts the complexity of Nmap's command-line interface. Because users may communicate with the tool using normal language, penetration testing is made more approachable to a larger audience. This audience includes security experts as well as security enthusiasts.

4.5.3 Customization and Relevance

Through the usage of the is relevant function, the integration gives users the ability to set their own personalized relevance criterion for Nmap commands. Because of its adaptability, the system may be adapted by users to unique penetration testing situations, guaranteeing that the produced instructions are in accordance with the particular objectives and prerequisites of each individual user.

4.6 Challenges

4.6.1 Contextual Understanding

Even though ChatGPT has amazing language processing skills, it may on occasion provide instructions that are theoretically correct but are not appropriate for the user's particular circumstances. Validation and evaluation by a human are essential steps in the process of determining whether or not the produced instructions are relevant to and suitable for the target system.

4.6.2 Security and Validation

Since Nmap is a powerful tool that is able to carry out penetrating scans, there is a possibility that it may be abused by scanning unauthorized computer systems or computer networks. It is very necessary to put in place reliable validation methods, as well as to educate users on how to utilise them, in order to avoid the possibility of abuse and to adhere to ethical penetration testing practices.

4.6.3 Error Handling

The integrated system has to have sophisticated error-handling methods built into it so that it can deal with instances in which Nmap instructions are produced incorrectly or problems occur while the system is being scanned. Providing users with error messages that are not only understandable but also instructive can improve their overall experience and make it easier for them to detect and resolve problems.

4.6.4 API Key Management

Because the application must have an API key in order to use ChatGPT, it is vital to establish proper security measures in order to protect the key and prevent unauthorized access. This will ensure that the integration maintains its integrity and confidentiality.

4.7 Example-2: Experimental Results of ChatGPT Integration to WHOIS Tool

Natural language queries are used throughout the domain information retrieval process thanks to the integration of ChatGPT with WHOIS capabilities, which makes the penetration testing procedure more efficient (Fig. 4).

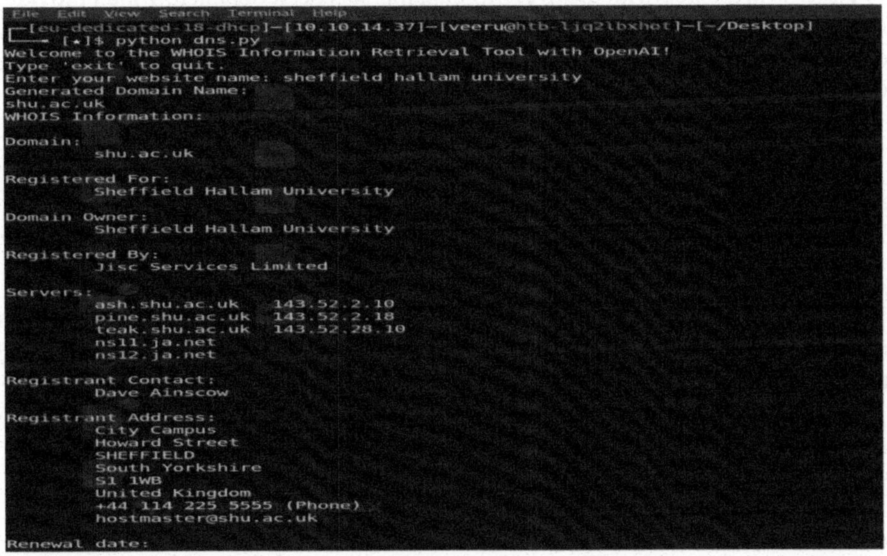

Fig. 4 WHOIS integration

4.7.1 User Interaction

Entering a website name (no need URL or Ip) is the first step in the WHOIS information retrieval process taken by users. Users are greeted by the integration, given instructions on how to use the tool, and provided with a straightforward user interface via which they may input questions.

4.7.2 Domain Generation

Using OpenAI language model, ChatGPT will construct a domain name for the user depending on the query they provide. The formed domain is shown to the user, and it is checked to ensure that it does not include any prefixes such as "www," "http://," or "https://," so that the user may provide more accurate input for WHOIS lookups.

4.7.3 Whois Data Retrieval

Using the Python subprocess module, the system asks the WHOIS database for information about the created domain. Details about a domain's registration, ownership, when it expires, and its nameservers are included inside the WHOIS data.

4.7.4 Output

The user is presented with the WHOIS data that has been obtained, which provides information on the registration and ownership details of the domain. Users may examine this data to determine the authenticity of the domain as well as any possible dangers that may be related with it.

4.8 Experimental Analysis of ChatGPT Integration to WHOIS Tool

The combination of ChatGPT with WHOIS functionality presents the penetration testing process with a number of noteworthy benefits as well as challenges including the following.

4.8.1 Advantages

- Because of the connection, users are now able to engage with WHOIS lookup using natural language queries, which makes the process more straightforward and user-friendly. Users with varied degrees of technical skill will find the procedure easier to navigate if it is approached in this way.

- The creation of domain names in ChatGPT automates the input process, which reduces the need for users to manually prepare domain names for WHOIS inquiries. Consequently, ChatGPT saves time. The process of retrieving information is made more efficient as a result of this automation.
- The integration makes it easier for users to acquire pertinent WHOIS information for domains, which helps users check the integrity of websites and locate potentially malicious domains.

4.8.2 Challenges

- There is a possibility that WHOIS lookups may provide inaccurate results if the user's intended domain does not always completely correspond with the domain that ChatGPT generates for them. Users are encouraged to double check the created domain to confirm that it is accurate.
- There is a wide range of quality and comprehensiveness in the WHOIS data provided by various registrars and domains. The accuracy of the data that was obtained may be compromised in some circumstances due to the omission or incorrect representation of vital elements.
- There is a possibility for abuse, and there may also be privacy problems, since the WHOIS information might expose important facts about who owns a domain. When making use of this integration, users are encouraged to take care and adhere to ethical best practices.

4.9 Example-3 Experimental Results of ChatGPT Integration to Msfvenom

This section will explore the outcomes of integrating ChatGPT with the msfvenom tool to streamline the payload development process for penetration testing. Furthermore, an examination of both the advantages and challenges that have emerged from this technological integration will be conducted.

4.9.1 User Interaction and Payload Generation

Users kick off the process of payload creation by submitting a text query that is connected to the requirements for their payload. Users are greeted warmly by the integration, and a user-friendly input interface is made available (Fig. 5).

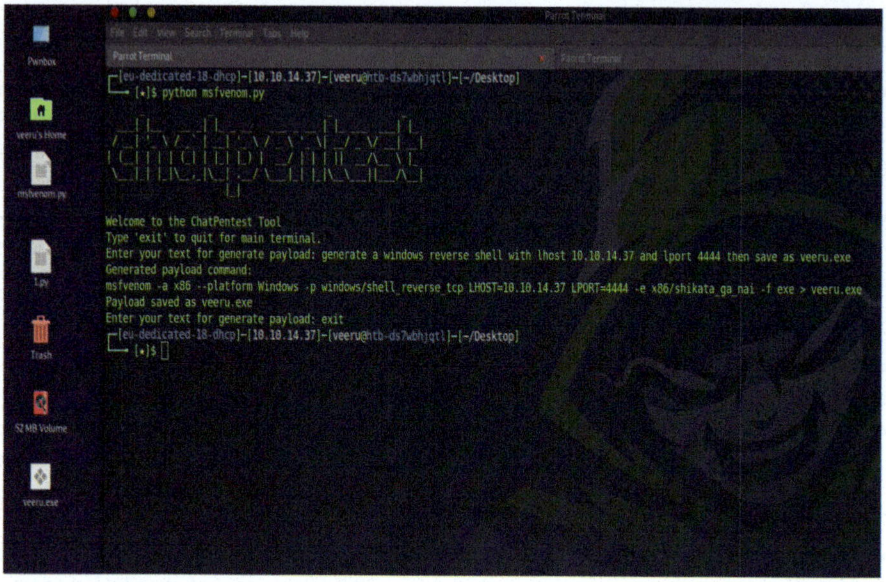

Fig. 5 ChatGPT integration to msfvenom

4.9.2 ChatGPT Parameters

The msfvenom command is assisted in producing its essential parameters with the help of ChatGPT, which is driven by OpenAI language model. Based on the user's input question, it will produce the necessary IP address, port number, and payload type for the user.

4.9.3 Parameter Presentation

The user is given the opportunity to evaluate the produced parameters, which include the IP address, port number, and payload type and output file name. Users are given the opportunity to validate the parameters here before moving on to the payload generating process.

4.9.4 Payload Generation

After receiving consent from the user, the msfvenom command is crafted with the help of the produced arguments, and it is then run with the help of the subprocess module in Python. The produced payload is stored in an output file based on user provided name (for example, payload.exe), which may then be analyzed in more detail.

4.10 Experimental Analysis of ChatGPT Integration to Msfvenom

4.10.1 Advantages

- The integration uses AI-generated parameters to automate the process of constructing msfvenom instructions, making use of them to streamline the process. Because of this automation, payload production accelerates up, which in turn reduces the amount of human labor needed by penetration testers.
- Users have reported seeing a more natural and straightforward interface when they engage with the system via the use of natural language inquiries. This strategy reduces the difficulty of entering the competition for persons who have varied degrees of experience in Metasploit and penetration testing.
- Users have reported seeing a more natural and straightforward interface when they engage with the system via the use of natural language inquiries. This strategy reduces the difficulty of entering the competition for persons who have varied degrees of experience in Metasploit and penetration testing.
- The system is able to cater to individualized payload needs in response to user inquiries. Because of this versatility, penetration testers are able to adjust the production of payloads to particular circumstances, ensuring that payloads fit with the testing goals they have set.

4.10.2 Challenges

- Users are encouraged to conduct an in-depth analysis of the produced parameters before moving on to the development of the payload, even though ChatGPT may aid in the process of producing payload parameters. A payload that is either ineffective or improper might be the result of incorrect specifications.
- The integration requires the production of payloads that might be used for malicious purposes. If the created payloads are utilised for activities that are not authorized, there is a possibility that they will be misused. It is very necessary to adhere to appropriate penetration testing practises and take ethical factors into account.
- The integration does not do any tests to determine whether or not the produced payloads are effective or appropriate for certain target systems. Users are encouraged to verify payloads against target environments to ensure that they are successful in accomplishing the desired goals.

4.10.3 Example-4 Experimental Results of Chatgpt Integration to Metasploit Framework

Users are now able to produce Metasploit commands via the use of natural language inquiries as a result of the integration of ChatGPT with Metasploit, which exhibits

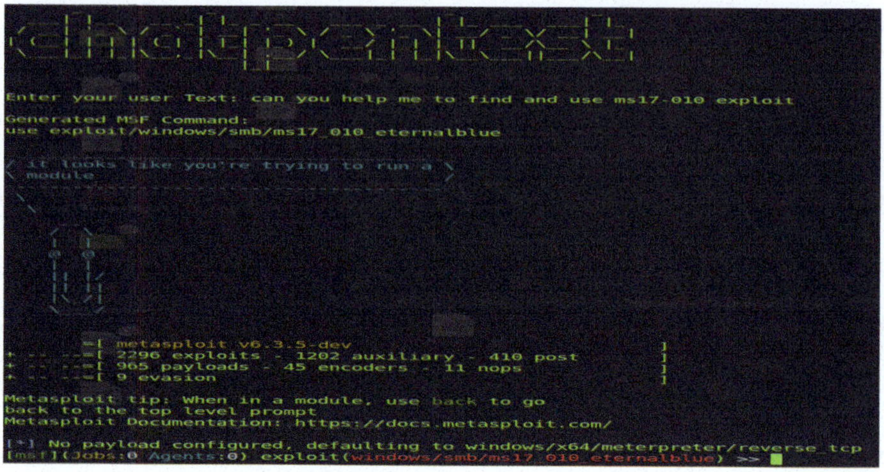

Fig. 6 ChatGPT integration to Metasploit

an intelligent and effective approach to penetration testing. The procedure and its outcomes are outlined below in the following steps:

4.10.4 User Interaction

Natural language interaction makes the ChatGPT and Metasploit penetration testing combination easy to use and approachable. The OpenAI language model takes user-entered queries describing desired penetration testing tasks and returns customized instructions for the hacking tool Metasploit. By automatically generating Metasploit commands in response to context, the gap between powerful penetration testing tools and regular users is closed. The feedback loop encourages users to work together with the AI-enhanced tool and provides chances for user education and ethical testing. The goal of the user-centric design of the integration is to make penetration testing more accessible to people of varying degrees of technical expertise (Fig. 6).

4.10.5 ChatGPT Interaction and Command Generation

The OpenAI Application Programming Interface (API) handles user inquiries and, depending on the information supplied, creates applicable Metasploit commands. Within the confines of the Metasploit framework, these commands have been crafted to cater to the goals and intentions of the user.

The user is provided with an opportunity to examine the produced Metasploit instructions. The integration guarantees that the instructions correspond with the user's purpose and appropriately represent the intended activities to be executed using Metasploit. This is achieved by ensuring that the commands are compatible with one another.

4.10.6 Metasploit Execution and Automated Actions

Python's subprocess module is used to carry out the execution of the produced Metasploit commands inside the framework that Metasploit provides. This makes it possible for users to engage with Metasploit directly and start the activities that have been defined. These actions include exploiting vulnerabilities, producing payloads, and managing sessions.

The connection makes it possible to automate a wide variety of tasks that may be carried out inside Metasploit, such as managing sessions, assessing vulnerabilities, exploiting vulnerabilities, and generating payloads. Users may take use of the connection to optimize and speed up the processes associated with their penetration testing.

4.10.7 Experimental Analysis of ChatGPT Integration to Metasploit Framework

In the context of penetration testing, the integration of ChatGPT with Metasploit carries with it a number of important benefits as well as challenges.

4.10.8 Advantages

- **Efficiency of automation:** Through the automation of the production of Metasploit commands, the integration speeds up the process of doing penetration testing. Users are able to rapidly design and execute complicated commands without needing to have an in-depth understanding of the command-line interface that Metasploit utilizes.
- **Accessibility and Usability:** Natural language interaction offers a straightforward and uncomplicated user interface that abstracts the complicated workings of the Metasploit framework. Because of its accessibility, Metasploit may be efficiently used by a larger audience, including security experts and hobbyists alike.
- **Personalization and Adaptability:** The capability of the system to produce individualized Metasploit commands in response to user inquiries enables personalization and adaptability. Users have the ability to customize the integration to accommodate a wide variety of penetration testing situations and goals.

4.10.9 Challenges

- ChatGPT creates Metasploit commands based on user questions; nevertheless, the knowledge of command context may be restricted due to this feature. Users should take care and verify automatically produced commands to ensure that they are in alignment with the desired aims and goals.
- Metasploit is a powerful tool that has the potential to be used unethically as well as for malicious objectives. This presents both a security and an ethical challenge. It is vital, in order to avoid abuse, to have appropriate user education, to take ethical issues into account, and to conform to legal and ethical requirements.
- Validation and Expertise in the integration could call for human validation and expertise, especially in situations that are very complicated. In order for users to successfully comprehend and make use of the produced commands, they need have a basic grasp of penetration testing and the Metasploit framework.

4.10.10 Conclusion

A new age of penetration testing has begun as a result of the merging of cutting-edge artificial intelligence technologies, such as OpenAI, with essential cybersecurity tools, such as WHOIS, msfvenom, Nmap, and Metasploit. This convergence has ushered in a new era. This synergistic merging of cutting-edge AI capabilities with time-tested security approaches has not only unlocked previously unimaginable levels of efficiency, but it has also spawned brand-new problems that make important contributions to the field of ethical hacking as a whole.

Throughout this research, the synergy between natural language processing and cybersecurity technologies has been thoroughly examined. This synergy facilitates user interaction with these tools in a conversational manner, thereby enhancing accessibility and usability. The implications of these connections are significant, offering both practical benefits and opportunities for intellectual exploration that could advance the field of ethical security assessments and penetration testing.

5 Evaluation and Comparison with Traditional Penetration Testing

5.1 Effectiveness and Vulnerability Discovery

Let us compare the Nmap command generation facilitated by ChatGPT, as demonstrated in Example 1, with the traditional Nmap scanning methods utilized in network scanning. By formulating a query to ChatGPT that requests, "Generate an Nmap command for discovering open ports on a target IP," a specialized Nmap command is swiftly provided in the form of: **nmap-p 1-65535 target_IP**. This command, which requires minimal effort to create, effectively identifies open ports on the target machine. In situations where a penetration tester may be uncertain about which command to employ for a vulnerability scan, ChatGPT or another large language model (LLM) can automatically generate the appropriate command. This automation enhances efficiency and assists testers in navigating the complexities of network scanning, allowing for a more streamlined approach to vulnerability assessment.

Traditional Nmap scanning, on the other hand, often includes the manual building of commands and the comprehensive examination of parameter settings. Traditional approaches are dependable; nevertheless, they may not attain the same level of efficiency and accuracy as ChatGPT fast recommendations, particularly when dealing with large-scale network evaluations.

5.2 Precision and False Positives

Consider a scenario in which ChatGPT is utilized to generate WHOIS queries, as illustrated in Example 2. In this instance, a domain name is provided instead of a URL, prompting ChatGPT to output a specific WHOIS command, such as "whois example.com." This generated command ensures accuracy and reduces the likelihood of errors, thereby minimizing the occurrence of false positives during the retrieval of WHOIS information.

Traditional WHOIS inquiries, on the other hand, may involve manual typing or copying and pasting, both of which may lead to transcription mistakes and probable false positives. The accuracy of the directives that are produced by ChatGPT provides an additional level of dependability that is difficult for older approaches to match.

5.3 Coverage and Scalability

The Metasploit framework, as demonstrated in Example 4, serves as a powerful tool for identifying vulnerabilities, exploits, and auxiliary modules. In this context, ChatGPT can be employed to generate a Metasploit (MSF) command to initiate an attack against a known vulnerable target. The suggested sequence of operations provided by ChatGPT may begin with the command exploit/windows/smb/ms17_010_eternalblue, followed by the necessary parameters.

There are instances when human testers may be uncertain about which exploit to employ for a specific vulnerability; in such cases, ChatGPT can assist in selecting the appropriate exploit, thereby enhancing the efficiency and effectiveness of the testing process.

In the conventional method, a penetration tester would manually search through Metasploit's large module library to locate an appropriate attack, enter the specifics of the target, and set up payloads. The automation provided by ChatGPT simplifies the process, which in turn makes it more scalable and efficient for testing numerous targets in a short amount of time.

5.4 *Time and Resource Efficiency*

Consider a real-world scenario, as illustrated in Example 3, in which exploitation payload instructions are constructed using ChatGPT. A query is inputted, requesting ChatGPT to "generate a payload command for a Windows reverse shell." In response, ChatGPT provides the following command:

msfvenom-p windows/shell_reverse_tcp LHOST=your_ip_address LPORT=your_port -f exe -o reverse_shell_payload.exe

When contrasted with the traditional approach of manually searching for and generating the payload, the use of automation significantly reduces the time and effort required for this task. This efficiency demonstrates the value of leveraging AI tools in the exploitation process.

Even though they are dependable, traditional techniques often need the manual selection and setting of payloads, which may be a time-consuming process. The prompt-based payload creation found in ChatGPT substantially speeds the process, making it an option that uses much less resources.

5.5 *Creativity and Unconventional Testing*

Imagine a scenario in which the Metasploit framework is employed to exploit a specific vulnerability in an Internet of Things (IoT) device. In this unique test, an attempt is made to perform an atypical exploitation. By querying ChatGPT for a Metasploit (MSF) command tailored to target the IoT device, a command is generated that effectively addresses the identified vulnerability.

The ability of the ChatGPT-generated command to successfully exploit the device's vulnerability underscores the innovative flexibility of the AI. The command is meticulously crafted, demonstrating a sophisticated understanding of the requirements necessary for exploiting the particular characteristics of the IoT device in question.

It may be difficult for traditional penetration testing to create such unorthodox ways since it relies on already discovered vulnerabilities. The capacity of ChatGPT to think creatively gives chances for uncovering vulnerabilities that may be missed by conventional approaches.

5.6 *Limitations and Challenges of Automation*

Within the context of Nmap and as illustrated in Example 1, consider an intricate network architecture that incorporates both firewalls and complex routing. When instructed to generate a comprehensive Nmap command for thorough network mapping, ChatGPT provides a command. However, it is important to note that ChatGPT lacks the contextual understanding of the specific complexities inherent in the network environment.

This limitation emphasizes the necessity for human expertise in interpreting the results and tailoring the Nmap command to effectively navigate the network's intricacies, ensuring that the generated command is appropriately adjusted to account for the presence of firewalls and complex routing configurations.

Traditional penetration testers are exceptional when it comes to analyzing the complexities of a network, gaining a knowledge of the routing configurations, and modifying scans appropriately. This example demonstrates how difficult it may be for ChatGPT to handle complicated situations, which emphasizes the need of human knowledge.

5.7 *Human Knowledge and Understanding of Context*

Let us examine a Metasploit scenario similar to that presented in Example 4. In this case, details of a specific vulnerability in an online application are provided to ChatGPT, along with a request for an MSF command to initiate an attack. While ChatGPT is capable of generating a command, it lacks a comprehensive understanding of the application's context and the various interaction sequences that may occur.

In contrast, a human penetration tester possesses domain expertise, which encompasses a thorough understanding of the application's architecture, behavior, and potential side effects. This contextual knowledge highlights the advantage of human insight, as it is crucial for the successful execution of exploitation and post-exploitation activities. Such expertise enables the tester to make informed decisions, adapt strategies in real-time, and effectively navigate the complexities associated with the application under assessment.

5.8 *Collaboration and Hybrid Techniques*

Think of a red team exercise that uses both conventional techniques and ChatGPT, as shown by the comparison between the several Metasploit commands in the Example-4. This would be an example of a hybrid approach. A human tester makes use of their knowledge and experience to choose the most appropriate exploit module, while ChatGPT lends a hand by automating ancillary activities such as payload development.

By collaborating, it becomes possible to leverage the effectiveness of ChatGPT while preserving human judgment and adaptability. This hybrid approach, which integrates AI-driven automation with human oversight, has the potential to enhance the overall efficiency of the penetration testing process.

5.9 Resource Requirements and Cost Analysis

First, let's have a look at the resources available. Traditional penetration testing requires the participation of a large group of knowledgeable testers, which may result in increased labor expenses. On the other hand, automation that is based on ChatGPT decreases the amount of human resources that are required, which has the potential to save expenses in the long term.

However, it is essential to keep in mind that the first training and integration of ChatGPT may result in financial obligations. It is essential to take into account both the initial investment and the resource savings that may be realized over the long run.

6 Future Directions and Conclusion

6.1 Future Work

Envision a web-based penetration testing chatbot that embodies the convergence of artificial intelligence and cybersecurity as we progress into the future. This chatbot represents the integration of various penetration testing tools, seamlessly combining functionalities from Nmap, WHOIS, Metasploit, and others. It is designed to be accessible to both novice users and seasoned professionals, thereby democratizing the tools and techniques of penetration testing.

Take the following as an Example: An e-commerce website's potential weaknesses will be evaluated by a security Analyst who is provided with a chatbot to assist in the process. The goals are communicated to the chatbot by the analyst via the use of natural language. In response, the chatbot conducts a symphony of AI-generated orders, including the orchestration of scans, the identification of open ports, the suggestion of exploitation methods, and even the simulation of a real-world breach scenario. This chatbot takes on the role of the virtual partner, blending the expertise of AI with the contextual knowledge of the analyst in a smooth manner.

The promise of the future extends much beyond the confines of existing limitations. Imagine groups of people interested in cybersecurity working together on different continents to use a chatbot in order to unravel the many levels of the penetration testing process together. An analyst in Tokyo starts the scan, and their colleague in New York makes adjustments to the settings depending on the input they get in real time. The chatbot acts as a bridge, overcoming the limits of physical location and helping to support the development of a worldwide community of penetration testers.

The road may not be without its difficulties, despite the fact that the perspective is encouraging. The need of resolving ethical problems, preventing unforeseen effects, ensuring the correctness and dependability of AI-generated instructions, and defending against unintended consequences remains unchanged. In addition, continuous research and improvement are required in order to perfect the chatbot's capacity to deal with a wide variety of circumstances, which range from standard vulnerability scans to newly discovered flaws.

The development of an AI-powered web-based chatbot for use in penetration testing mirrors the progression of AI more generally in the field of cybersecurity. It is a significant step towards democratizing the power of AI and making more difficult activities accessible to a larger audience, which is why it is such an important move. This chatbot symbolizes the future of penetration testing, which will be collaborative, dynamic, and based on ethical principles because of the synergistic dance that takes place between artificial intelligence and human knowledge that is sewn into the fabric of this chatbot.

6.2 Conclusion

A remarkable contrast has been produced in the field of cybersecurity as a result of the combination of cutting-edge AI technology with traditional penetration testing procedures. This is shown by the incorporation of ChatGPT into the landscape of penetration testing. This in-depth analysis has shed light on the unique qualities and subtleties of both ChatGPT-based techniques and conventional methodologies, therefore laying a solid groundwork for comparisons between the two.

The ChatGPT-based penetration testing technique, which is shown through the scripts that are included, has the potential to redefine the efficacy and efficiency of vulnerability finding. A tantalizing insight into a future when AI will become an important friend to penetration testers is provided by the speedy creation of instructions, inventive study of novel attack pathways, and simplified automation. This change in perspective has the potential to significantly improve both the scalability of penetration testing endeavors and the resource efficiency of such endeavors.

Traditional approaches, on the other hand, which are profoundly founded in human skill and contextual awareness, continue to be a powerful competitor. Human penetration testers have the ability to decipher the subtleties of complicated network infrastructures, adapt to changing settings, and apply ethical concerns in a way that may be absent from an AI-driven method. It is impossible to exaggerate the significance of human intuition and judgement, as well as the capacity to improvise in reaction to unforeseen circumstances.

An evolutionary leap that bridges the gap between tradition and innovation is represented by the use of artificial intelligence (AI) into penetration testing. One example of this is ChatGPT. The synergy between human knowledge and AI-driven automation will be the cornerstone of its efficacy. While the automation and quickness of AI have the ability to transform the landscape of penetration testing, the synergy between the two will be the cornerstone of its efficacy. A tantalizing glimpse into the future of penetration testing is provided by the envisioned web-based chatbot, which is positioned at the nexus of AI and human ingenuity. This future is one in which the sum of human and AI capabilities transcends individual prowess, ensuring the resilience and security of digital ecosystems worldwide.

References

1. Chu G (2021) Automation of penetration testing
2. Chu G, Lisitsa A (2018) Poster: agent-based (BDI) modeling for automation of penetration testing. Paper presented at the—2018 16th annual conference on privacy, security and trust (PST). pp 1–2. https://doi.org/10.1109/PST.2018.8514211
3. Li J, Meland PH, Notland JS, Storhaug A, Tysse JH (2023) Evaluating the impact of ChatGPT on exercises of a software security course. IEEE 4:1–6. https://doi.org/10.1109/esem56168.2023.10304857
4. openai. (a) openai Api references. openai. https://platform.openai.com/docs/api-reference
5. openai. (b) Openai documentation. openai. https://platform.openai.com/docs/introduction

Zero-Trust Blockchain-Based Digital Twin 6G AI-Native Conceptual Framework Against Cyber Attacks for e-Healthcare

Banirupa Chowdhury, Hamid Jahankhani, and Sangar Subramaniam

Abstract The increasing use of Internet of Medical Things (IoMT) devices in the e-Healthcare system is leading to a rise in cyberattacks. In the current cloud computing landscape, organizations are shifting from local infrastructure to cloud networks, which raises concerns about cloud security because sensitive data is stored there. Traditional cloud security solutions can suffer from centralized data vulnerabilities, rigid security policies, and inadequate access controls, leaving businesses vulnerable to new attacks. Attacks on the Dark Web have become more sophisticated, concentrating on cloud settings through a variety of techniques. While 6G is still in its early phases of development, the integration of its encrypted communications and decentralized computing will significantly improve digital healthcare systems. SplitFed Learning (SFL) can address data privacy and resource limitations in edge-based Internet of Things (IoT) systems, whereas Traditional approaches struggle with sequential data, which is frequently present in healthcare data streams. This paper offers a conceptual security framework that integrates AI-Driven Security Policy Management, SplitFed Learning, Blockchain technology, Digital Twin technology, and Zero Trust Network Access (ZTNA) principles to address these challenges.

Keywords 6G · e-health · Cybersecurity · Zero trust network access

---

B. Chowdhury
Prysmian Group, Milan, Italy

H. Jahankhani (✉)
Northumbria University London, London, UK
e-mail: Hamid.jahankhani@northumbria.ac.uk

S. Subramaniam
ISACA, Cheshire, UK

H. Jahankhani and B. Issac (eds.), *Cybersecurity and Human Capabilities Through Symbiotic Artificial Intelligence*, Advanced Sciences and Technologies for Security Applications, https://doi.org/10.1007/978-3-031-82031-1_23

1 Introduction

The word "eHealth" was first used in November 1999 during the 7th International Telemedicine Conference in London. An ICT-based system for medical analysis and diagnosis in the twenty-first century is what the WHO refers to as eHealth [1]. As internet connectivity grows, so does malware, which harms computers and networks. Cybercrime is rising rapidly, costing industries trillions [2]. US healthcare has the greatest data breach cost, $7.13 m, 84% greater than the global average. Healthcare cyberattacks costed $6 trillion worldwide by 2021, up from $3 trillion in 2015 [3]. IoT technology in healthcare makes use of smart items or devices to provide P2P, M2M, and P2M sharing and collection capabilities. Simplifying system vulnerability architecture and IoT device services and apps is a goal for researchers. Authorization of data access, intractability, anonymity and protection against identity theft attacks on security systems are important privacy and security features [4]. In the security of cloud computing and the internet of things, a number of factors are necessary to guarantee IoT-based cyber health security and privacy. IoT-based technology's ability to provide remote monitoring is a game-changer for the eHealth industry. Hospitals can remotely monitor patients and connected equipment with cloud-linked IoT health service models [5].

Medical device takeovers, data manipulation, exploitation, unauthorized hospital network access, and insufficient authentication are just a few of the ways that restricting patient medical and personal information can jeopardize healthcare security and privacy and impact potentially life-changing choices. Identity, communications, patient data, and IoMT integrity are all protected via mutual authentication. Among the numerous security techniques that can enhance user privacy is blockchain. The technology's decentralization, efficiency, tamper-proofness, traceability, and security make it perfect for Internet of Things applications. IoMT and blockchain enhance secure, real-time healthcare [6]. By combining digital twin technology and virtual representation with cloud architecture, new possibilities are made possible. Digital twin deployment from a distance and real-time data integration and analysis are made possible by cloud computing. Large computations and data management for complex analytics and predictive modelling are made possible by cloud resources [7]. Healthcare communication requires authentication that is secure, portable, and reliable. By storing user input locally and only distributing scattered learning data, federated learning (FL) secures networks. Edge computing and D2D paradigms could be advantageous for future 6G networks, but their dispersion could make them more vulnerable. Governments everywhere welcome AI platforms and services to enhance people's lives, even though research on 6G AI threat countermeasures is still in its early stages [8]. Decentralized Machine Learning Approach FL takes use of private information from businesses, clients, and hospitals. FL only uses a central server to aggregate local client weights and biases. The main application of FedAvg is aggregation. FL offers excellent privacy protection training, but its implementation isn't perfect. Because clients and the central aggregator are unaware of each other's data and habits, malicious users may interfere with FL training and system performance.

2 Literature Review

A budget friendly healthcare is always a priority to people when it comes to the question of accessing international assistance from healthcare professionals. Digital twins (DTs) can effectively act as a digital portrayal by providing a synthetic comprehension of physical substance or data flow that guarantees the equal rate convergence between the physical state and the virtual state in remote healthcare from the destination to the international hospitals.

The healthcare sector has seen a dramatic digital shift in the last few years as cutting-edge technologies like artificial intelligence (AI), blockchain, and zero-trust architecture (ZTA) have been implemented to solve changing security issues. ZTA provides a strong foundation for protecting sensitive medical data and managing access to healthcare resources because it places a strong emphasis on least privileged access and stringent authentication.

2.1 Zero Trust Architecture

Zero trust architecture addresses security needs in untrusted networks by following the "never trust and always verify" philosophy. Each access request is authenticated and monitored using dynamic access control security policy rules. Adaptive Security Architecture (ASA) from Gartner was updated into Zero Trust Network Access (ZTNA) in 2014. In 2014, Google created the BeyondCrop security model for network security. Statista forecast 90% company adoption of zero trust security in 2022, up from 60% in 2016. By 2027, the global Zero Trust market might be valued $60 billion. ZTA is becoming a more popular network security solution. ZTA is an extra security element that improves identity management and threat detection in healthcare networks. Other security features include user profiles, unique user IDs, and activity logs. In addition, the use of machine learning techniques for anomaly detection, like Quantile Regression Forest (QRF), facilitates the instantaneous observation of user actions and network traffic, allowing for the timely identification and resolution of security incidents. Healthcare organisations may strengthen data protection, guarantee regulatory compliance, and maintain patient trust in the face of new cyber dangers by utilising these cutting-edge security techniques and technology.

These capabilities improve healthcare network identity management, access control, and threat detection, reducing unauthorised access and malicious activities. Quantile Regression Forest (QRF) and other anomaly detection machine learning methods were used to increase healthcare system security by detecting aberrant user behaviour and security issues. User actions and network traffic are monitored in real time to discover and fix security issues, reducing cyber threats to patient data and healthcare operations. A centralised Security Operations Centre (SOC) using anomaly detection algorithms detects, investigates, and resolves security issues. However, this method couldn't solve the policy issues in an emergency mode.

In the context of healthcare 4.0, Almuseelem [9] suggests a safe authentication architecture for healthcare systems that makes use of lightweight HMAC (Hash-Based Message Authentication Code) and Zero-Trust Architecture (ZTA). The suggested framework minimises secret key length, lowers computing costs, and consumes less energy without sacrificing security as compared to current lightweight techniques. In Healthcare 4.0 systems, it overcomes resource limitation issues and dynamically updates HMAC keys to adapt to evolving threats. Data integrity and patient privacy are protected by the bidirectional D-HMAC authentication, which guarantees that only verified organisations exchange data. By granting access authorization based on identification privilege, security, scalability, and compliance are all improved. This framework concentrates on integrating the framework in real-time Healthcare 4.0 and 5.0 environments and expanding it to defend against broader threats in 5G healthcare scenarios but it doesn't mention the security adoption in robust 6G healthcare system. The study of Allouzi and Khan [10] highlights the threats to patient privacy and safety posed by wireless body area networks (WBANs) vulnerabilities and addresses the security difficulties faced by Internet of Medical Things (IoMT) devices.

Potential attack vectors, such as compromised clients, resource servers, credential service providers, or networks, are identified by the vulnerability analysis, which also suggests mitigating techniques. According to Al Taheri et al., [11] Agent-based Zero Trust Network Access (ZTNA) solutions change security paradigms. His proposed ZT Network Access protocol in patient monitors thoroughly compares these models to show agent-based ZTNA's pros and cons. Unlike perimeter-based security, ZTNA uses "never trust, always verify," with agents monitoring and transmitting logs without assuming trust. ZTNA uses micro segmentation, contextual access controls, and continuous authentication and verification, while legacy models rely on implicit trust. The benefits of agent-based ZTNA include improved security, adaptive and contextual security, fewer data breaches, and interoperability with modern IT infrastructure. However, sophisticated implementation, interoperability issues, and user experience concerns must be addressed including IoT and OT applications, AI-driven security, user-centric approaches, and access control and user experience improvements. These directions show how Zero Trust Security, guided by sent logs, adapts to new security threats and technologies. While in Wang et al. [12] Security and privacy concerns are addressed via a Trust-Based Authorization Access Control (TAAC) solution that uses blockchain and smart contracts to access medical data in permitted organisations.

2.2 Blockchain in Healthcare

Blockchain technology provides strong security and transparency in data sharing because of its decentralized, tamper-proof ledger [13]. It ensures correctness and openness by organizing data into immutable blocks that are validated by network nodes through consensus. Scalability, security protocols, and application support are critical factors for a successful implementation. Blockchain guarantees data availability in the face of interruptions, improves interoperability, and builds confidence through transparency. But issues like energy consumption and scalability need to be resolved [14]. Blockchain is a key factor in the development of data-sharing systems, including Dataspace 4.0 and 6G, due to its essential characteristics that pave the way for decentralized data-sharing platforms that are highly secure, efficient, and transparent [15]. Each node in a decentralized blockchain network maintains a copy of the ledger and takes part in transaction verification to uphold consensus. Due to the decentralized method, security and resilience are improved and the network is not controlled by a single body. In summary, the distributed processing of blockchain ensures security, transparency, and reliability in consensus processes among nodes, which supports the technology's promise for decentralized data-sharing across a range of applications.

A blockchain-based privacy-preserving approach [16] for IoMT healthcare data sharing is presented in this research. The system solves IoMT problems using advanced technologies like the Quad Merkle tree algorithm for data integrity, Zero-Knowledge Proof (ZKP) encryption via smart contracts for privacy, and the Multi Critic Deep Deterministic Policy Gradient (DDPG) algorithm with Markov Decision Process (MDP) for task offloading optimisation. Simulated evaluation reveals that the suggested model outperforms alternatives in efficiency, processing speed, data reduction, and storage.

This work offers an identity and access management solution for healthcare data exchange that is built on a private, secure blockchain [17]. Blockchain technology reduces fraud and illegal access by offering decentralised verification and authorization. Using Hyperledger Fabric, OAuth 2.0 guarantees robust authentication and trusted providers enhance security in a scalable system for managing access to sensitive data. Experiments demonstrated that as blockchain size increases, so do query and invocation times.

In order to improve data transmission security and real-time decision-making for the Internet of Medical Things (IoMT), Xiang et al. [18] suggests a trust-enhanced approach. Blockchain technology is used to monitor and verify data flow in a public manner. The trust management system is made up of blockchain, MSPs, CSPs, edge servers, and IoMT device requesters.

2.3 Digital Twin

An electronic version of a physical object, known as a digital twin (DT), allows information to be exchanged between the physical and digital counterparts. Design, predictive maintenance, risk assessment, and decision-making are just a few of the many tasks in the medical device sector that DT makes easier. This leads to improved safety, dependability, and cost effectiveness. The DT market is expanding rapidly and is expected to reach USD 16.44 billion by 2024 [19]. It was first introduced during the Industry 4.0 era. Despite the numerous benefits it offers, concerns about data ownership, privacy, and secrecy, as well as standardising data collecting and transmission, remain roadblocks. Despite its potential, significant investment in high-fidelity computer models and support from stakeholders are required for the full implementation of DT [20]. This approach to detect data poisoning attempts in distributed machine learning protects model parameter transmission privacy. The BL-FCL approach improves mobile distributed federated learning data privacy and security at minimal computational cost. The work also uses block-chain and distributed network learning architecture to create a secure DDS service model for distributed learning user privacy and security. This architecture does not improve DTN fault tolerance against internal attacks, particularly Byzantine attacks.

Digital Twin Application in Healthcare H4.0, the fourth healthcare revolution, integrates AI, big data, CPS, and the IoT to provide personalised and efficient treatments. Instead of in-person visits, H4.0 uses electronic records, telemedicine, and remote monitoring [21]. This shift offers real-time monitoring, accurate diagnosis, better treatment outcomes and better surgery. H4.0 relies on Digital Twin (DT), which allows real-time representations of physical systems to connect the virtual and physical worlds. Healthcare professionals can simulate treatments, make informed judgements, and develop medical procedures with DT.

2.4 AI Technology (SplitFed Learning) in Healthcare

The internet of medical things (IoMT) contains highly sensitive data and a single IoMT device is not enough to train sophisticated model. Federated learning models resolved most of these issues but due to resource-intensive nature with huge dataset it becomes costly and data privacy isn't maintained as it provides the raw dataset. In order to solve this problem Ni et al. [22] proposed a resource friendly SplitFed framework for healthcare where data privacy is maintained by intermediate shared features between servers and devices. However, it isn't fully distributed learning. To fill this gap Stephanie et al. [23] proposed a decentralized SplitFed learning approach which resolves the issues of non-independent and identically distributed data distribution in dynamic metaverse-driven healthcare system. It works by decentralizing server-side model leveraging the computational burden and offloading some client-side computational tasks to server. It incorporates with DT to enable real-time healthcare device status. However, this model requires further research in ensuring data privacy by implementing recent technologies such as homomorphic encryption.

2.5 Integration of Zero Trust Technology, SplitFed Learning, Digital Twin and Block-Chain Technology in 6G Healthcare System

The vision for the intelligent information society in 2030 is characterized by extensive digitization, intelligence-driven innovation, and a global reliance on data. This future is anticipated to be shaped by the transformative capabilities of 6G technology, which aims to establish seamless and limitless wireless connectivity. The comprehensive scope of 6G involves connecting all elements, ensuring pervasive wireless coverage, and integrating various functions like sensing, navigation, caching, communication, positioning, computing, control, radar, and imaging. This integration supports the implementation of diverse applications across different verticals. Furthermore, 6G is envisioned as an autonomous ecosystem with advanced intelligence and awareness, evolving from a focus solely on humans to a balanced consideration of both humans and machines. The communication and interaction with smart terminals in this era will be facilitated through various channels, including fingers, voice, eyes, and even brainwaves or neural signal.

Stephanie et al. [23] proposes Digital Twin-enabled Asynchronous SplitFed Learning (DT-ASFL) architecture to address Synchronous Federated Learning (FL) issues in e-healthcare systems. Synchronous FL promotes data privacy by communicating model parameters instead of raw data, however e-healthcare device capabilities can cause low resource utilisation and failure. DT-ASFL uses Digital Twin technology and asynchronous local model aggregation to improve resource utilisation, model performance, and convergence speed in e-healthcare systems. The architecture optimises model aggregation with a step-aware approach and edge device state clustering. Security Multi-party Computation (SMPC) improves training data reconstruction resistance. DT-ASFL architecture shares computing load between participants and a central server, and step-aware model aggregation to accelerate global model convergence. Advanced DT technology is used to optimise model aggregation using a clustering process based on predicted edge-cloud DT participant training times, enhancing efficiency.

By incorporating Federated Learning, AI-Driven Security Policy Management, Blockchain technology and Zero Trust Network Access principles, Kawalkar and Bhoyar [24] proposes an innovative approach to cloud security. The challenges associated with centralized data repositories, data privacy, and evolving cyber threats are all tackled within this comprehensive framework.

According to Shen et al. [25], big digital twin data (BDTD) is a blockchain-based system that combines blockchain with cloud computing to ensure the sharing of BDTD. To ensure BDTD confidentiality, technologies like cloud storage and proxy RE-encryption (PRE) are combined with hashing, timestamping, signatures, tokenization, and a suggested algorithm for determining the ideal sample rate for BDTD. We assess this approach using Hyperledger Fabric and the cloud. The sensitive healthcare industry was proposed to be solved by Manocha et al. [26] using CoT, IoT, FoT, DT and consortium blockchain technologies. CNN and the Gated Recurrent Unit (GRU) were utilized for stability. An algorithm for creating time-sensitive warnings was described, which is a hybrid approach for physical event recognition that is inspired by both CNN and BiGRU. To offer identity-based signatures, we employ a modified version of Reputation-based Byzantine fault tolerance (RBFT).

With the advent of quantum computing in 6G, post-quantum identity solutions are essential to secure digital identities. The need for cross-chain trust evaluations is highlighted for efficient security information sharing in a future metaverse with diverse blockchains. The article emphasizes the importance of exploring solutions for these challenges to advance the ZTA-6G framework.

3 Case Study Approach

The goal of this research is to propose a conceptual framework for securing patient data in healthcare system which is adopted in 6G communication system. This research leans more toward the deductive technique because it has a theoretical framework that may be used to test a hypothesis, conduct research, or validate as a workable framework. On the other hand, this research has no use of a combination of questionnaires and relevant literature to collect data and synthesize the most pertinent information. A critical analysis of current works will be utilized to find trends in the various study areas, which will subsequently lead to understand the gap and conceptualize a probable solution of the security breaches in IoT based healthcare system by utilising existing solutions.

The research will employ a case study approach. Descriptive analysis is employed to elucidate all data displays. Moreover, this research further elaborates on its findings. The Cyber Security framework will be established through the examination and understanding of the data. The anticipated research must follow applicable legal statutes and ethical considerations.

Case Study 1: How cyber-attacks impact users data privacy and healthcare system?

The iCare project (funded under POR FESR FSE 2014/2020 of the Calabria region, with participation of European Community Resources of FESR and FSE, of Italy and Calabria) included a qualitative evaluation of the effect based on the literature and interviews with medical specialists. Inspired by Suraci et al. [27] this approach was devised. The outcome of the survey identified the most common dangers, the most significant security needs, and the most susceptible assets in the eHealth area with the help of the literature study. Data acquired was enhanced by the information gathered from hospital doctors with different specialties. Especially useful were these interviews, which made it possible to do a cost-efficiency analysis that takes into account the security assurance of the most sensitive assets while using eHealth services. Following the approach often used in a qualitative risk study, the predictable impact can have a value of HIGH (range: 80%), MEDIUM (range: 50%), or LOW (range: 25%), depending on the nature of the research, most suitable for achieving the main aim of providing a general-purpose impact assessment. It's interesting that, given there isn't yet a infrastructure for the 6G technology, the suggested study just covers the techno-economic environment anticipated for 6G, not depending on any particular Radio Access Network (RAN) technology. The following four security requirement are examined: (i authentication, which means all stakeholders must agree on each other's credentials; (ii data protection, which means keeping data private and secure; (iii privacy, which means protecting identities; and (iv resilience, which means the service's most critical system must still work when things go wrong.

Based on the technologies in place, this calculates the effect of fixed attack conditions on the healthcare system for each service. This analysis focuses on the architecture described in Suraci et al. [27], which detects, processes, and stores patient health data using three levels. The lowest layer of the architecture is made up of sensors, which are utilized to detect data. Similar to SDN, controllers gather and process data and provide instructions to a group of sensors based on patient installation. Data storage is mainly the responsibility of data servers. The SDN scenario in Fig. 1 for the telemonitoring service demonstrates that whereas controllers and data servers are handled by the hospitals, sensors are maintained by outside vendors. Different stakeholders utilize sensors and controls for different things. We presume that they might be malevolent or targets of intrusions, which would be against the authentication protocol described in Suraci et al., [27]. The functional connection between a physical sensor and its controller may be abused by a hostile sensor, harming both the sensor and the stakeholder in charge of it. Risks are posed to sensors and their data by malicious controllers.

Case Study 2: Significant security breaches in healthcare and different types of threat levels of attacks

Data security and privacy are top priorities in healthcare. After EHRs took the role of paper patient records, data breaches have proliferated. Data breaches occur both internally and outside. Internal security breaches may be caused by inadequate infrastructure, software flaws, or illegal database access. Theft, ransomware, and hacking are examples of external breaches. Conferring to a data security study [28], data breaches are caused by human error (33.5%), hacking (14.8%), theft (16.3%), abuse (29.5%), and malware (10.8%).

With 113 million records compromised, 2015 was the worst year for healthcare including major data breaches at Anthem Inc., Excellus and Premera Blue Cross. Due to these flaws, 268,189,693 medical records were exposed. A data breach resulted in the theft of all patient data [29].

From the study of [30], "Cybersecurity in the Healthcare Industry," also named outdated operating systems and phishing as major threats. According to a recent

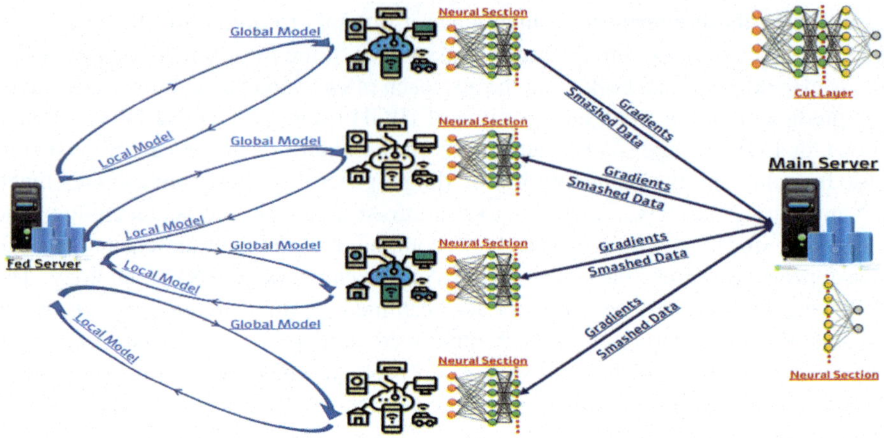

Fig. 1 SplitFed learning training (adopted from Hafi et al. [45])

Verizon survey, the majority of healthcare organizations use cloud storage and anticipate that, in five years, the majority of consumers would access data via mobile devices. The security of health data is more important than ever due to these changes and growing cyber threats. Cybercriminals are focusing more on hospital data since it has a higher black-market cost than credit card numbers and Social Security numbers.

3.1 Analysis on Cyber Incidents Based on Classical and Contemporary Security Challenges and Possible Solutions in 6G for IoMT

Due to cloud infrastructure's scalability and flexibility, businesses switched from local data centres to cloud at the start of the twenty-first century, which led to the growth of the IoMT ecosystem. Due to a lack of specified protocols, an emphasis on functionality over security, and inadequate security controls, a number of vulnerabilities have been exploited in major, well-known security events. Moreover, employing 6G technology in the healthcare sector creates security and privacy concerns due to the sensitive nature of health data. Furthermore, hostile attackers have the potential to compromise implantable medical devices (IMDs), including diabetic insulin pumps, pacemakers, and internal cardiac defibrillators, endangering human lives in addition to causing financial losses. A number of problems need to be fixed for healthcare based on 6G technology to be properly integrated. Notwithstanding these challenges, the healthcare sector stands to gain much from 6G, and it is anticipated that the technology would increase healthcare's reliance on AI and 6G connectivity.

3.2 *Adjustable Security Measures*

Adaptive security allows for solutions to be dynamically adjusted to address new threats and evolving network conditions. Adaptive security helps safeguard healthcare services and data by including techniques for identifying, assessing, reducing, and avoiding threats to privacy and security [31]. For the edge-cloud continuum, the existing security models which are event-driven, requirements-driven, game-based, and risk-based approaches need to be evaluated to detect new attack behaviour [32].

Because of the limited processing and memory capacity in IoT, handling massive volumes of data can be dangerous. Data breaches can occur from flaws in user interfaces, network services, sensors, and cloud servers, exposing private data and creating a risk when exchanging data. Due to large data quantities, inadequate policies, procedures, and authentications, IoMT also presents security vulnerabilities [33]. IoT vulnerabilities in sensors, cloud services, network services, and user interfaces may provide unwanted access, according to research. Internet of Things (IoT) devices can communicate data, however during interaction, software and hardware defects could result in harmful attacks [34].

3.3 *Sustainable Security Detection Techniques for Resource Efficient Healthcare*

One of the new features of IoMT-2030 is sustainability; security, privacy, and resilience have all been improved over 5G (IoMT-2020). 6G's ubiquitous connection will be made possible by improved mMTC and uRLLC from 5G, necessitating long-term security, privacy, and trust mechanisms [35, 36]. Conventional security techniques, such manual processes, can lessen risks and guarantee security, but they need more resources. AI-based Zero trust Architecture for healthcare needs to be considered in the present methods.

Medical device heterogeneity refers to standards and technology. 6G networks are expected to support a large number of linked devices and high-intensity data transport (Naseer et al. 2023) [37]. It is crucial to address the variety of IoMT devices to ensure the security of 6G healthcare services and data. Traditional security solutions are insufficient in the increasingly complex IoMT environment with changing technology and threats. The variety of future 6G networks may pose severe security concerns due to increased attack surface.

Distributed ledger technologies (DLT) like as blockchain, differential privacy, Digital Twin and federated learning offer future 6G networks the ability to be more secure and resilient, as suggested by the IMT-2030 recommendations. Blockchain technology has the potential to improve network efficiency by decreasing reliance on centralized servers. It can be used for distributed authentication and healthcare data security. Options for real-time threat identification and response are made possible by breakthroughs in artificial intelligence and machine learning. Smart contracts and protocols in blockchain technology improve security. Due to open access, public blockchain technology may compromise privacy and security. Lack of data tracking and management regulation can cause breaches. Blockchain processing speed and scalability are slower than traditional databases. Security concerns may arise when using distributed ledger systems and data synchronization.

3.4 Authentication Issues

Innovation in less secure situations has been hindered by the proliferation of medical IoT devices. New technology security governance, standards, norms, and regulations are unknown to patients, healthcare providers, and other stakeholders. Unnoticed security vulnerabilities might result from sharing a device without encryption, passwords, or a secure cloud. Customers' lack of familiarity with sophisticated technologies gives rise to privacy and security issues. Because FL uses wireless communication, users mistrust it. Concerns about data security are raised by users who are unfamiliar with wireless data management. Healthcare clients' identities and personal information are exposed by inadequate user ID identification [38]. Since context-awareness systems transfer personal data to third parties, which may be accessed, security and privacy must be given top priority. When end users share and access data, cloud vulnerabilities enable code injection attacks on the Internet of Things. When transferring sensor data to healthcare servers, hybrid cloud systems, whether public or private, may raise security concerns for users and physicians [39]. Hence, cloud data is susceptible to hacking, manipulation, and invasions of privacy.

3.5 Cloud Server Vulnerabilities

If the key used in a cryptography system is disclosed, the data cannot be effectively protected. There is a risk of data loss when health data is transferred from cloud servers to blockchain. Although cloud servers employ robust encryption and cryptography, the involvement of third parties can nevertheless introduce security vulnerabilities during data transmission [40]. Unauthorized interference poses possible dangers to users obtaining raw biological data from sensors. Authentication errors can cause delays in the encryption and decryption of data, which can result in denial-of-service (DOS) attacks. Cloud-based data is susceptible to harmful attacks, data tampering, privacy and security breaches, and hackers exploiting patient data. The database lacks security due to its limitations. Research has demonstrated that digital signatures are susceptible to forgery. Storing data on the public cloud has significant concerns regarding security and privacy. If there is a single failure, the involvement of a third party may lead to the disclosure of health data [41]. Storing personal health records (PHR) in unencrypted form in the cloud exposes them to potential security breaches by hostile actors. Attribute-based encryption (ABE) is used for access control and search encryption, but it has a security weakness that compromises data security. Handling data in big amounts can pose a danger to its security [41].

3.6 Limiting Access to Attacker

New security risks are emerging and becoming more sophisticated as the usage and popularity of IoMT devices continues to rise. Finding vulnerabilities in core technology and investigating potential alternatives helps strengthen the security of communication networks [42]. Adaptive security safeguards data in real-time, which is especially useful in IoMT contexts with limited resources. Reducing the geographic region of exposure to potential hostile agents can be achieved by utilizing effective and adaptive security solutions across the edge-to-cloud continuum. By providing Access control protocol the access can be customized.

4 Proposed Framework for Zero-Trust Blockchain-Based Digital Twin 6G AI-Native Healthcare

In this research a zero-trust, block-chain based digital framework is proposed by using existing solutions. Each part of the framework is explained using existing work done so far and connection them together to establish a better architecture. All the elements of the framework are critically analyzed and co-related with each other.

Model computational requirements and privacy during FL ML training at client side.

The main disadvantage of FL is that the whole architecture could not be supported by resource-constrained clients, as those are in the Internet of Things. It is usual for the ML models to be deep learning models. Furthermore, there exists a privacy problem from the model's point of view during training because the server and clients have total access to the local and global models. These concerns were addressed by the development of split learning (SL). Once the whole machine learning model is divided into numerous smaller network components and trained separately on a server, SL delivers the data to clients locally. Comparing running the entire network as in FL with training a subset of it at the client reduces processing load, which is crucial for machine learning computation on devices with limited resources A client cannot, however, access a server-side model and vice versa [43].

4.1 Integrating Federated Learning and Split Learning

The conventional SplitNN technique involves the use of a split layer, also known as a cut layer, to partition the neural network into two distinct sub-networks. During the transportation of training data, the client performs forward propagation by computing the output of its subnetwork up to the cut layer. The server receives the result of the cutting layer. The server calculates the outcome till the final network layer. The gradients are returned to the client after undergoing backpropagation from the last layer to the cut layer within the server's sub-network. The client has control over the residual backpropagation from the cut layer to the first layer of the network. This approach is repeated until the client has new training data. In a vanilla SplitNN design, the server does not have direct access to the clients' raw data or all of the model parameters. Only two pieces of data, namely the gradient from the server to the clients and the output of the cut layer from the clients to the server, are exchanged between the clients and the server [44].

SplitFed, also known as SFL, is a hybrid collaborative learning algorithm that combines SplitNN and FL, two distributed machine learning solutions, to provide an enhanced method that adheres to the learner-server paradigm. SplitNN and FL share the two benefits with one another. Each of the sub-models is trained in parallel, even if the model is divided into learner/client and server sides. In addition to the primary server that was part of the SplitNN design, there is a new part of the architecture called Fed Server. Figure 1, The working procedure is divided into multiple phases: The Fed Server sends each participating client the global learner-side model component to start the process. The forward propagation function is then carried out concurrently by each learner using their own local data up to the cut layer of learners, where they transfer the corrupted data to the cut layer of the main server. Under the assumption that the SplitNN concept is as described thus far, the server calculates the cost function, assumes control of the remaining forward propagation, and backpropagates to the server's cut layer. It's amazing that this server-side procedure may run simultaneously for several clients, as we will see later. Each client then carries out the back propagation on the learner's model segment. Some rounds are completed without the Fed Server in the cycle of forward and backward propagation between the students and the server. Subsequently, the learners send their modifications to the Fed Server, which compiles them and builds an international learner-side model [45].

4.2 How SplitFed Learning Can Help the Data Security Challenges in 6G

Connecting a huge number of intelligent devices (IoE) in IoMT is one of the main characteristics of 6G. Undoubtedly, this would result in a sharp increase in the number of users and new apps that place different performance demands: quite high data throughput, very high dependability, and very little latency. According the critical analysis done in Hafi et al. [45] Tables 1 and 2 shows how splitFed model helps to overcome the challenges in 6G over traditional solutions.

Centralised data sharing system limits data privacy and security concerns, interoperability issues and single points of failure. To address this problem Sachin et al. [24]

Table 1 How SFL can help to face the challenges in 6G

Challenges	Technology	Limitation	How SFL can help
Client model sharing	FL	Same replica of client model is sent to all client sides	Doesn't distribute the full model, The model is divided into two sections: one for the primary server and the other for the client. All clients have access to the same dedicate sub-model
Privacy	SplitNN	Centralized model training	The learning process is accelerated by using the raw data from end devices to train the client model
Privacy threats and Security threats	Deep learning Homomorphic encryption	False prediction with high confidence, Polluting the model's training data	The model is made more resilient and reliable against attacks by dividing it into sub-models and using multiple client side copies. The attacker cannot deduce the attributes and parameters of the model by attacking one sub-model and leaving the others unaffected
Efficiency	Zero trust technology	Limited performance with Huge model volume and, training the model with higher accuracy within short time	Combining with Zero Trust Architecture allows for more efficient management of big data sets in a short amount of time in a secure way
Interface-level security	ZTA	Less proficiency in dealing with Open API security threats, scalability, privacy	When a network has many users and multiple slices are given by multiple administrative domains, SplitFed can be deployed
Integration with DT recurring transfer of complex system data creates massive network traffic and data leaks	FL and traditional ML	Constrained network, bandwidth utilisation limitation, transfer of whole global model to industrial devices	Maintains a digital plane in sync with the physical system, Distributed parallel training of complex system over client and server side

Table 2 Security matrix of machine learning models (adopted from Hafi et al. [45])

ML model	Model privacy	Model aggregation	Training overhead	Computation response	Distributes computing	Access to raw data
Federated learning	L	H	L	H	H	L
Split learning	H	L	H	L	H	L
Split federated learning	H	H	M	L	H	L

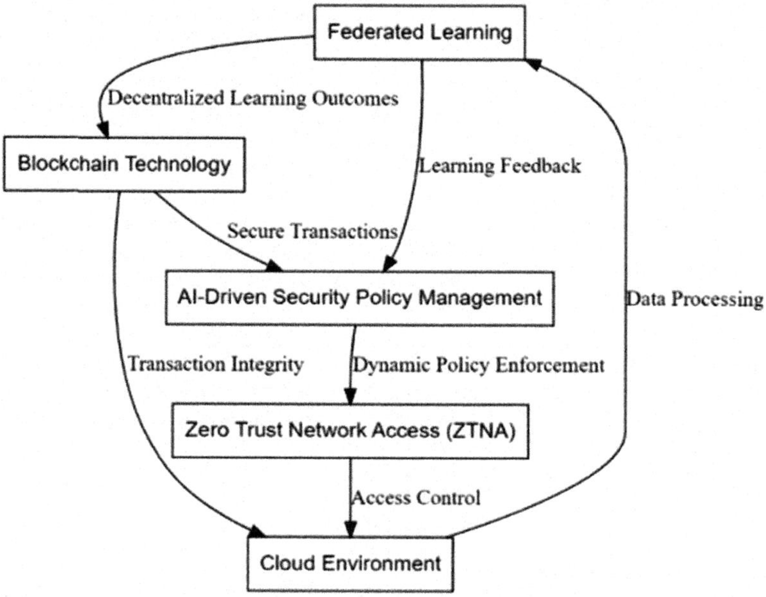

Fig. 2 Architecture for efficient cloud security model through federated learning

integrated Federated learning, AI-driven security policy management, Blockchain technology and zero trust network access (ZTA) principles (shown in Fig. 2).

4.3 *Blockchain, AI-Driven Policies, and ZTA Adopted from Sachin et al. [24]*

Split Learning, a recently created distributed machine learning paradigm, addresses the issue of privacy preservation by storing the data on the client devices. Split learning involves the division of tasks between a central server and client devices, each responsible for specific portions of the training process. After finishing the first computations, the client devices submit the obtained features to the server for further training of the model. On the other hand, federated learning involves training models on distributed devices like Edge nodes or mobile phones, rather than sending raw data to a central server. Only the alterations made to the model are consolidated on the server, while the model itself is trained individually on each device. Split learning has numerous benefits, such as enhanced privacy safeguards, reduced communication expenses, and improved scalability. It is utilised in various fields where limitations on resources and concerns about data privacy are important, such as edge computing, the Internet of Things, and healthcare. Split learning can incorporate privacy-preserving techniques, such as differential privacy, to offer strong guarantees of privacy. Secure aggregation approaches ensure the integrity of aggregated model updates. To summarise, Split Learning ensures the confidentiality of participants by providing a method of training models collectively without the need to directly exchange raw data. The integration of machine learning with an edge-cloud Digital Twin (DT) environment allows for remote monitoring to assure security using a real-time aggregation scheduling mechanism. Synchronous federated learning poses a potential threat to data privacy as the transmitted local model from an edge server may contain sensitive information that might be utilised to reassemble the training data. Attackers can effectively compromise the reconstruction of training data by leveraging the gradient of the local models. E-healthcare systems pose a significant threat to individuals' privacy due to the potential inclusion of sensitive health information in the training data.

By combining these above two models and DT enabled asynchronous SplitFed model by Stephanie et al. [46] an Asynchronous SplitFed Learning with Digital Twin, Blockchain, and Zero Trust (ASFL-DTBZ) is proposed in this work. Figure 3 depicts the framework's architecture, which is called as Asynchronous SplitFed Learning with Digital Twin, Blockchain, and Zero Trust (ASFL-DTBZ). Hospitals work together to build machine learning models in the ASFL-DTBZ healthcare system, but they do not share raw patient data. Private edge servers are owned by each hospital ($h \in H$) and are in charge of local model training and data collecting from Internet of Medical Things (IoMT) devices. Additionally, in order to give real-time updates on device condition and training progress, these edge servers communicate with Digital Twin servers (DT). Unlike traditional Federated Learning, (ASFL-DTBZ) provides a decentralized and asynchronous model-training mechanism. In (ASFL-DTBZ), edge servers train local models independently and asynchronously without awaiting input from other participants. Hospitals can train their models in this way, improving scalability, performance, and resource efficiency by doing it in parallel and asynchronously. As an added benefit, (ASFL-DTBZ) allows hospitals to contribute to the global model at their own pace by eliminating the synchronous aggregation bottleneck. While the deep model is trained on the cloud server, the shallow model is trained locally on the edge server of the hospital. By ensuring that sensitive data stays at the edge, this division improves privacy.

By building virtual versions of hospital settings, digital twin technology offers real-time insights on operational workflows, equipment usage, and patient health

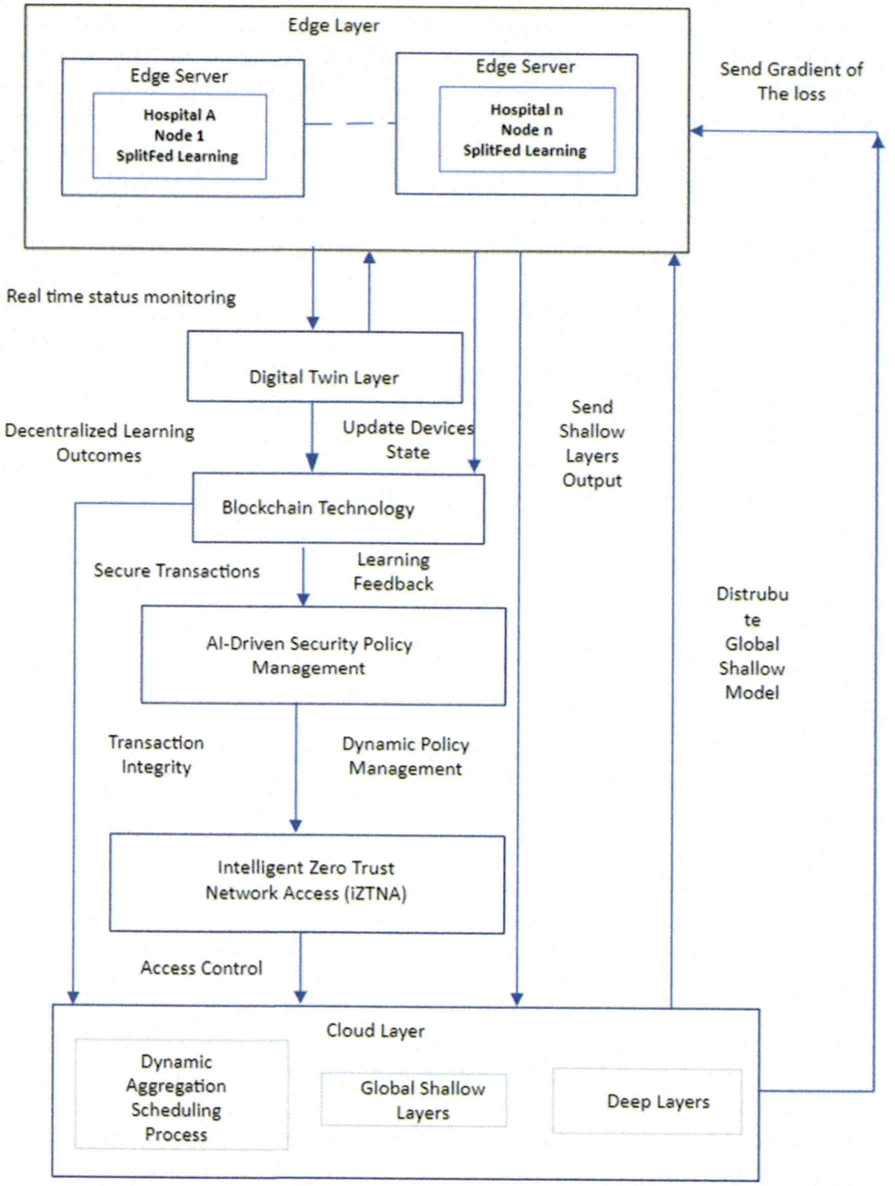

Fig. 3 Proposed conceptual framework

status. With the help of DT servers, edge servers and the cloud can communicate, allowing for the dynamic scheduling of model aggregation procedures in response to real-time data updates. A technique based on Secure Multiparty Computation (SMPC) is started when the prerequisites for aggregation are satisfied. The cloud server receives secret shares from the participants that are computed in parallel for their learned model parameters. These shares are combined to create global shallow layers, with the exception of the shallow model's initial hidden layer.

These data samples are stored in a blockchain which acts as an immutable, decentralized ledger to record hospital-to-hospital transactions and data exchanges. Transparency and confidence in data transactions are guaranteed by smart contracts, which automate access control and data-sharing agreements. The Digital Twins of each hospital produce transactions that are safely stored on the blockchain and pertain to data updates and model training. Within the healthcare network, blockchain technology improves data provenance, auditability, and integrity.

All the layers of the proposed model are used to ensure data security in order to protect from attacks. These layers are defined step by step with critical analysis focusing on data security and privacy.

4.4 Integration with Digital Twin Blockchain with SplitFed Model

The introduction of blockchain-enabled SplitFed learning has garnered significant interest from both academia and industry. It aims to address the issues associated with traditional FL, including data falsification, single point failure, and a lack of an incentive mechanism. Blockchain has been proven to provide security and privacy protection in Internet of Things situations. SplitFed is highly sophisticated because of the data privacy mechanism by creating client and server-side model, but it still requires a dependable central server to control the local devices and the model parameters that go with them. This could lead to single-point failure, man-in-the-middle attacks, etc. Using blockchain as the underlying architecture, a fully decentralized machine learning paradigm is accomplished to address this issue. The primary drawbacks of the existing blockchain-based DT solutions are their low efficiency, high communication cost, and lack of customized consensus methods. In order to jointly enable efficient and private SplitFed learning services for the DT empowered Edge-Cloud continuum, this paper is inspired by this and developed the Proof-of-Federalism (PoFe) custom consensus algorithm in this research together with additional state-of-the-art auxiliary mechanisms. This framework proposes a block-chained asynchronous SplitFed learning paradigm for the digital twin empowered cloud-based healthcare system to find the optimal trade-off between model accuracy, efficiency, and privacy issues of the EAaaS. In this paradigm, local learning tasks are taught by edge devices, while global aggregation is achieved by corresponding cloud-based

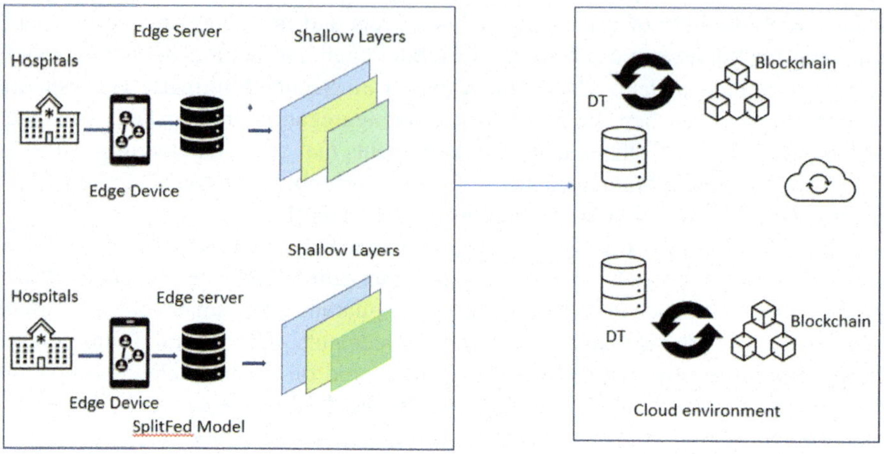

Fig. 4 Conceptual architecture of the proposed framework

digital twins using the blockchain's consensus mechanism. Both the local training and the global aggregate are executed asynchronously, Fig. 4.

The edge devices are trained locally and creates shallow layers which are stored on the Local database. The digital twins separately reach a consensus within the blockchain system and obtain the local model parameters. For every consensus round, the digital twined aggregator generates a candidate blockchain using the latest global aggregation model parameters. To improve efficiency, an adaptive asynchronous aggregation technique is applied, and the optimization is done considering several aspects like convergence efficiency, model accuracy, etc. In particular, a dynamic optimal blockchain generation rate can be attained. The global model parameters are then sent to the digital twins for cross-validation. The approved global parameters are used to rewrite the current ones up until convergence. It makes reasonable to consider local data as streaming data in the context of edge computing because data collection is ongoing. Consequently, the local model training is asynchronous and is contingent upon the block creation rate of each consensus round. This means that once the local training is finished, the edge device contacts its digital counterpart to confirm the consensus state. The digital twin evaluates the consensus process and then alerts the edge device to perform another round of local training or to immediately supply the local model parameters. This will help to alleviate the issues of idle edge devices and excessive edge-to-cloud communication overhead. Using the asynchronous global aggregate comes with a few residual local models that could be used in subsequent cycles.

4.5 How Blockchain Can Help to Ensure Privacy in Asynchronous Aggregation

The application of blockchain technology benefits in three ways. Blockchain, in the first place, makes digital twins reliable. With the decentralized learning system, problems like man-in-the-middle attacks and single-point failure are prevented. Second, blockchain technology can prevent other types of data manipulation as well as regional and global poisoning attacks. Thirdly, edge devices and digital twins with high processing power can be encouraged to engage by means of an incentive system provided by blockchain. Furthermore, blockchain is an excellent match for the asynchronous aggregating process. This consensus was reached by the digital twins, sparing the computing power of edge devices. Its modular design makes it easy to update and maintain in response to changing requirements. Every digital twin can take part in the Proof-of-Federalism (PoFe) consensus process by providing the best local model performance. Proof-of-Work (PoW) and PoFe are similar, however PoFe replaces the limitations in PoW by utilizing the computational resources utilized by FL's local training tasks. Since digital twins are hosted on the cloud and have access to ample storage, it directly stores the model parameters in the block to avoid increasing the complexity of the system. Since the digital twins are connected peer-to-peer via virtual network interface cards (VNICs), they also benefit from relatively low latency in their communication. In this case, the forking issues can be avoided. When a blockchain separates into two or more branches due to a delayed block propagation, this is known as a fork. With this method, block generation rate and propagation delay can be balanced. This whole integration of the model has following 8 stages Fig. 5.

4.6 Integration of AI Rule Based Blockchain

Integration of blockchain and AI creates a synergistic relationship that improves the shortcomings of each technology while maximizing the potential of both. Blockchain's secure and private features offer a solid foundation for AI algorithms by giving access to a multitude of high-quality data, which is essential for AI to function well. Blockchain speeds up AI performance through intelligent data tracking and analysis. It also streamlines manufacturing processes, optimizes energy usage in consensus algorithms, and delivers relevant information for decision-making tasks. On the other hand, AI increases blockchain's utility and efficiency by streamlining energy-intensive processes, spotting security threats, and expediting transaction times (Fig. 6). Machine learning can be applied by blockchain systems to enhance user privacy, prevent fraud, and verify data integrity all of which increase the scalability and reliability of the network. This integration improves the performance of both technologies and fosters a win–win partnership between blockchain and AI that strengthens system security, dependability, and efficiency, Fig. 6.

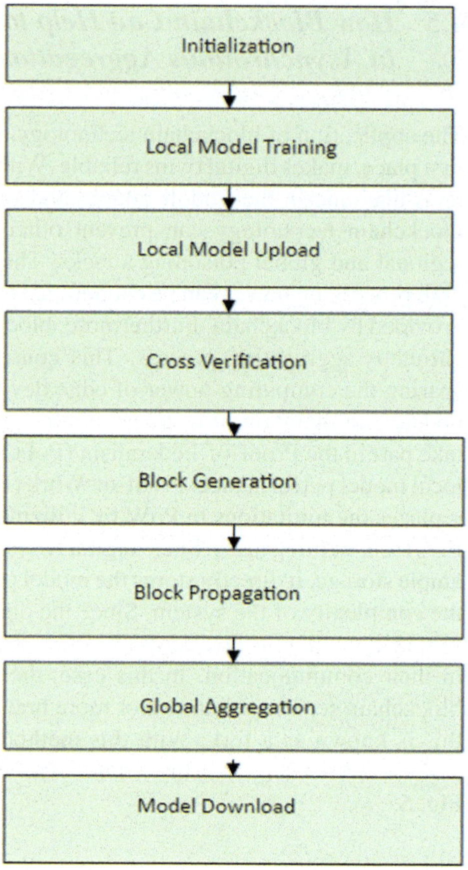

Fig. 5 Stages of blockchain aggregation

Rule-based artificial Intelligent systems that employ rule-based models to acquire artificial intelligence information are known as rule-based AI systems. Because these systems are deterministic, they require less data than other deep learning and machine learning systems. The intermediary between blockchain technology and multiple interconnected layers, such as Internet of Things sensors, clinics, and hospital layers, is an AI system based on rules. "A set of rules" and "a set of facts" are the two essential components that make up the rule-based AI system. These systemic layers aid in creating a fundamental comprehension of the facts and interpersonal rules. A three-layered, rule-based AI smart contract for distant sensor nodes are used in distant monitoring. First, by comparing contract address and observations with blockchain records, it validates IoT node registration and finds potentially malicious nodes. Second, it uses clever data processing to monitor sensor voltage fluctuations and send out alarms based on pre-established health conditions for measures like blood pressure, body temperature, and blood glucose. Finally, it finds faults such as NaN, empty, or zero values to diagnose sensor node difficulties. In order to improve trust and transparency in healthcare systems, the clinic layer also serves as a middleman between

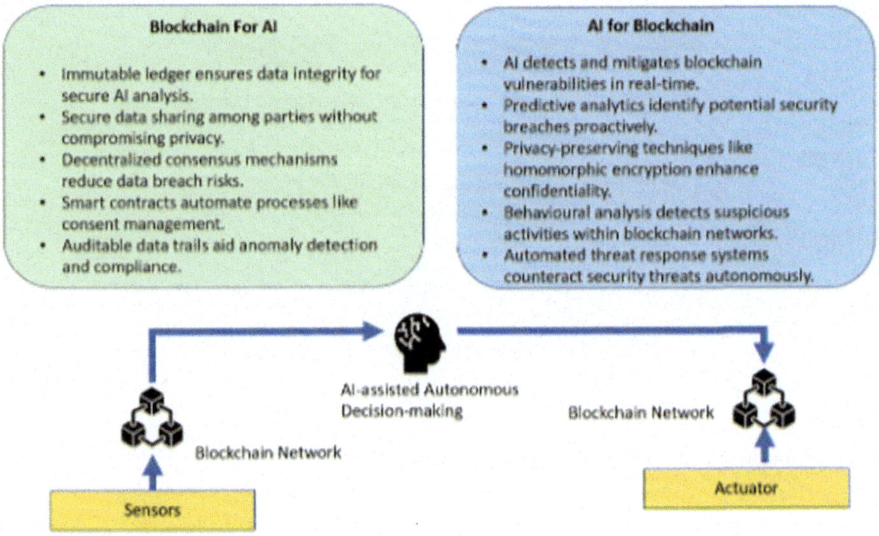

Fig. 6 AI-Assisted blockchain concept for healthcare

hospitals and IoT nodes, enabling patient information synchronization and registration on the blockchain network. Data verification is done by generating signature and validation of message.

4.7 AI Driven Security Policy Management Integrated with Zero Trust Architecture

This is a SplitFed-based iZTA approach consisting of two components PEP (policy enforcement point) and PDP (policy decision point), that ensures dynamic authentication and continuous diagnostics and mitigation (CDM). Figures 7 and 8 shows the policy engine and overall conceptual architecture of the ZTA framework. This model can be used for both centralized and decentralized SplitFed learning to ensure fast dynamic user authentication in a distributed scenario without cross-enterprise user ID exchange, hence preserving privacy. This solution addresses the privacy difficulties with conventional SplitFed identity management systems. The shallow layer training is done with a support vector machine (SVM)-based classifier for rapid authentication and least privilege-based access to corporate resources depending on topic characteristic.

The second component, shown in Fig. 7 adds threat intelligence and subject behaviours to the PDP/PEP system over time. The policy engine receives fine-grained access recommendations, including session time allocation and resource access level, for detection and mitigation. Internal and external asset databases, activity logs,

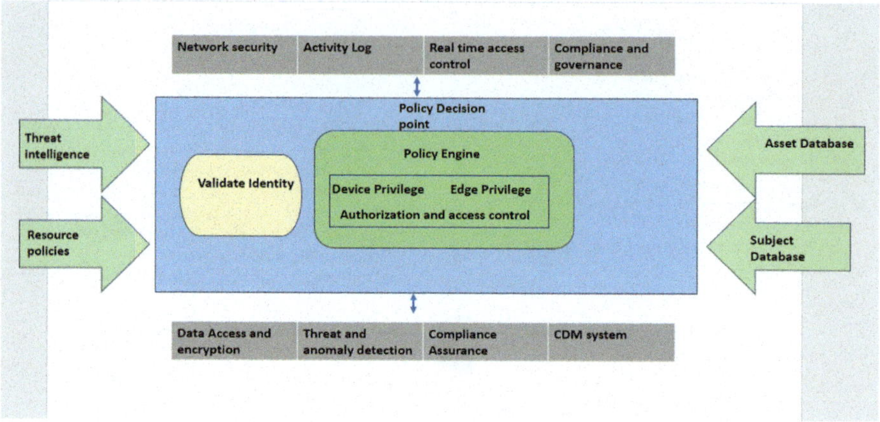

Fig. 7 Policy engine for iZTA in healthcare system

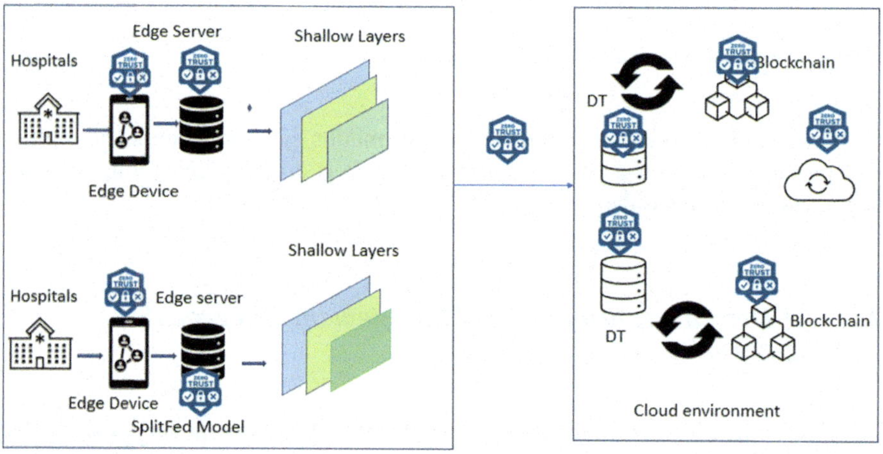

Fig. 8 Overall conceptual architecture of proposed ZTA

threat intelligence, and resource need policies inform these recommendations. For local model training, SVM-based classifiers are recommended to learn fine-grained access rules and natural language processing to transform policy documents into access rules.

Suggested ZTA relies resource access policies on subject database data. The PE recommender system further avoids ambiguity by ensuring the least privilege for each request using current data from activity logs and threat intelligence. Threat modelling successfully evaluates targeted attacks and mitigates ZTN risks. ZTN vulnerabilities in devices, services, and technologies create a large attack surface. Each layer-based abstraction component of the proposed system must comprehend ZTN threat modelling.

5 Conclusions and Future Work

This research presents a new conceptual IoMT cloud security framework that combines SplitFed Learning, AI-Driven Security Policy Management, Blockchain technology, Digital twin and Zero Trust Network Access (ZTNA) technologies. The hypothetical outcomes from this framework in several e-Healthcare cloud security scenarios exhibit its superior performance compared to current approaches, characterized by improvements in data privacy, accuracy, threat intelligence, efficiency and a substantial decrease in response delay. SplitFed Learning integration facilitates decentralized data processing, client model and server model separation, hence improving data privacy and mitigating the dangers of central data breaches in e-healthcare system. It also ensures the huge patient record handling in e-healthcare system. Blockchain technology offers a reliable and unchangeable record for cloud transactions, so strengthening the integrity of data. AI-powered security policy management, leveraging sophisticated algorithms, provides flexible and proactive security policy adjustment. By incorporating Zero Trust Network Access (ZTNA) principles, strict access control is implemented through continuous verification, which is essential for protecting cloud resources.

This study introduces a theoretical framework for safeguarding patient data in e-healthcare systems by employing privacy-preserving technologies like as blockchain, digital twin and zero trust architecture, in conjunction with SplitFed learning. Although significant progress has been made, this endeavour still faces limitations and obstacles that need to be addressed in future research. This framework is a fusion of state-of-the-art technologies and the most recent security solutions. It would have been an excellent accomplishment if the framework had been validated. As a result of time constraints, the work has been concluded as a conceptual framework. However, the next step will involve validating this framework.

References

1. Aazam M, Zeadally S, Harras KA (2018) Offloading in fog computing for IoT: review, enabling technologies, and research opportunities. Futur Gener Comput Syst 87:278–289
2. Ullah A, Aznaoui H, Sebai D, Abualigah L, Alam T, Chakir A (2024) Internet of Things and cloud convergence for eHealth systems: concepts, opportunities, and challenges. Wirel Pers Commun:1–51
3. Lallie HS, Shepherd LA, Nurse JR, Erola A, Epiphaniou G, Maple C, Bellekens X (2021) Cyber security in the age of COVID-19: a timeline and analysis of cyber-crime and cyber-attacks during the pandemic. Comput Secur 105:102248
4. Porkodi S, Kesavaraja D (2020) Integration of blockchain and internet of things. In: Handbook of research on blockchain technology. Academic Press, pp 61–94
5. Thilakarathne NN, Kagita MK, Gadekallu TR (2020) The role of the internet of things in health care: a systematic and comprehensive study. Available at SSRN 3690815.
6. Garg N, Wazid M, Das AK, Singh DP, Rodrigues JJ, Park Y (2020) BAKMP-IoMT: design of blockchain enabled authenticated key management protocol for internet of medical things deployment. IEEE access 8:95956–95977

7. Son S, Kwon D, Lee J, Yu S, Jho NS, Park Y (2022) On the design of a privacy-preserving communication scheme for cloud-based digital twin environments using blockchain. IEEE Access 10:75365–75375
8. Shoetan PO, Amoo OO, Okafor ES, Olorunfemi OL (2024) Synthesizing AI's impact on cybersecurity in telecommunications: a conceptual framework. Comp Sci IT Res J 5(3):594–605
9. Almuseelem W (2024) Continuous and mutual lightweight authentication for zero-trust architecture-based security framework in cloud-edge computing-based healthcare 4.0. J Theor Appl Inf Technol 102(1)
10. Allouzi MA, Khan J (2024) Enabling zero trust security in IoMT edge network. arXiv preprint arXiv:2402.10389
11. Al Taheri A, Rashed A, Al Marzooqi H, Al Hamadi H (2024) Synergizing security: designing a zero trust network access protocol in patient monitors (No. 12200)
12. Wang C, Wu W, Chen F, Shu H, Zhang J, Zhang Y, Wang T, Xie D, Zhao C (2024) A blockchain-based trustworthy access control scheme for medical data sharing. IET Inform Secur
13. Alsamhi SH, Myrzashova R, Hawbani A, Kumar S, Srivastava S, Zhao L, Wei X, Guizan M, Curry E (2024) Federated learning meets blockchain in decentralized data-sharing: healthcare use case. IEEE Internet of Things J
14. Radziwill N (2018) Blockchain revolution: how the technology behind bitcoin is changing money, business, and the world: 2016. Dan Tapscott and Alex Tapscott. Penguin Random House, New York, 348 p
15. Xu H, Klaine PV, Onireti O, Cao B, Imran M, Zhang L (2020) Blockchain-enabled resource management and sharing for 6G communications. Digit Commun Netw 6(3):261–269
16. Sutradhar S, Majumder S, Bose R, Mondal H, Bhattacharyya D (2024) A blockchain privacy-conserving framework for secure medical data transmission in the internet of medical things. Decis Anal J:100419
17. Sutradhar S, Karforma S, Bose R, Roy S, Djebali S, Bhattacharyya D (2024) Enhancing identity and access management using Hyperledger Fabric and OAuth 2.0: a block-chain-based approach for security and scalability for healthcare industry. Internet of Things Cyber-Phys Syst 4:49–67
18. Xiang X, Cao J, Fan W, Xiang S, Wang G (2024) Blockchain enabled dynamic trust management method for the internet of medical things. Decis Support Syst:114184
19. Business wire (2024) Global digital twin market size is projected to reach USD 16.44 billion by 2024, growing at a CAGR of almost 38%. https://www.businesswire.com/news/home/20201105005766/en/Global-Digital-Twin-Market-Size-is-Projected-to-Reach-USD-16.44-billion-by-2024-Growing-at-a-CAGR-of-Almost-38#:~:text=Global%20Digital%20Twin%20Market%20Size%20is%20Projected%20to,November%2006%2C%202020%2012%3A45%20AM%20Eastern%20Standard%20Time
20. Rowan NJ (2024) Digital technologies to unlock safe and sustainable opportunities for medical device and healthcare sectors with a focus on the combined use of digital twin and extended reality applications: a review. Sci Tot Environ:171672
21. Khan S, Alzaabi A, Ratnarajah T, Arslan T (2024) Novel statistical time series data augmentation and machine learning based classification of unobtrusive respiration data for respiration Digital Twin model. Comput Biol Med 168:107825
22. Ni W, Ao H, Tian H, Eldar YC, Niyato D (2024) FedSL: federated split learning for collaborative healthcare analytics on resource-constrained wearable IoMT devices. IEEE Internet of Things J
23. Stephanie V, Khalil I, Atiquzzaman M (2024) DSFL: a decentralized SplitFed learning approach for healthcare consumers in the Metaverse. IEEE Trans Cons Electron
24. Kawalkar SA, Bhoyar DB (2024) Design of an efficient cloud security model through federated learning, Blockchain, AI-driven policies, and zero trust frameworks. Int J Intell Syst Appl Eng 12(10s):378–388
25. Shen W, Hu T, Zhang C, Ma S (2021) Secure sharing of big digital twin data for smart manufacturing based on blockchain. J Manuf Syst 61:338–350

26. Manocha A, Afaq Y, Bhatia M (2023) Digital Twin-assisted Blockchain-inspired irregular event analysis for eldercare. Knowl-Based Syst 260:110138
27. Suraci C, Pizzi S, Molinaro A, Araniti G (2021) MEC and D2D as enabling technologies for a secure and lightweight 6G eHealth system. IEEE Internet Things J 9(13):11524–11532
28. Seh AH, Zarour M, Alenezi M, Sarkar AK, Agrawal A, Kumar R, Ahmad Khan R (2020) Healthcare data breaches: insights and implications. Healthcare 8(2):133
29. Puppala M, He T, Yu X, Chen S, Ogunti R, Wong ST (2016) Data security and privacy management in healthcare applications and clinical data warehouse environment. In: 2016 IEEE-EMBS international conference on biomedical and health informatics (BHI). IEEE, pp 5–8
30. Combs V (2020) The five biggest cybersecurity threats for the healthcare industry
31. Porambage P, Liyanage M (2023) Security and privacy vision in 6G: a comprehensive guide. John Wiley & Sons
32. Salehie M, Pasquale L, Omoronyia I, Ali R, Nuseibeh B (2012) Requirements-driven adaptive security: protecting variable assets at runtime. In: 2012 20th IEEE international requirements engineering conference (RE). IEEE, pp 111–120
33. Motwani A, Shukla PK, Pawar M (2022) Ubiquitous and smart healthcare monitoring frameworks based on machine learning: a comprehensive review. Artif Intell Med 134:102431
34. Kaur R, Shahrestani S, Ruan C (2024) Security and privacy of wearable wireless sensors in healthcare: a systematic review. Comp Netw Commun:24–48
35. Shen S, Yu C, Zhang K, Ni J, Ci S (2021) Adaptive and dynamic security in AI-empowered 6G: from an energy efficiency perspective. IEEE Commun Stand Mag 5(3):80–88
36. Halabi T, Bellaiche M, Fung BC (2022) Towards adaptive cybersecurity for green IoT. In: 2022 IEEE international conference on Internet of Things and intelligence systems (IoTaIS). IEEE, pp 64–69
37. Shanmugam B, Azam S (2023) Risk assessment of heterogeneous IoMT devices: a review. Technologies 11(1):31
38. Alsahafi YA, Gay V (2018) An overview of electronic personal health records. Health Policy Technol 7(4):427–432
39. Mahmoud AS, Mahmood NM (2021) A secure biomedical data sharing framework based on mCloud. Int J Nonlinear Anal Appl 12(2):1659–1671
40. Dwivedi AD, Srivastava G, Dhar S, Singh R (2019) A decentralized privacy-preserving healthcare blockchain for IoT. Sensors 19(2):326
41. Wang H, Ning J, Huang X, Wei G, Poh GS, Liu X (2019) Secure fine-grained encrypted keyword search for e-healthcare cloud. IEEE Trans Dependable Secure Comput 18(3):1307–1319
42. Ahmad I, Rodriguez F, Kumar T, Suomalainen J, Jagatheesaperumal SK, Walter S, Asghar MZ, Li G, Papakonstantinou N, Ylianttila M, Huusko J (2024) Communications security in Industry X: a survey. IEEE Open J Commun Soc
43. Thapa C, Arachchige PCM, Camtepe S, Sun L (2022) Splitfed: when federated learning meets split learning. Proc AAAI Conf Artif Intell 36(8):8485–8493
44. Abedi A, Khan SS (2024) Fedsl: federated split learning on distributed sequential data in recurrent neural networks. Multim Tools Appl 83(10):28891–28911
45. Hafi H, Brik B, Frangoudis PA, Ksentini A, Bagaa M (2024) Split federated learning for 6G enabled-networks: requirements, challenges and future directions. IEEE Access
46. Stephanie V, Khalil I, Atiquzzaman M (2023) Digital twin enabled asynchronous SplitFed learning in E-healthcare systems. IEEE J Sel Areas Commun

A Multimodal Approach to Biometric Authentication

Adrianna Kozierkiewicz, Adrian Graczyk, and Elias Pimenidis

Abstract A multimodal approach to biometric authentication leverages multiple physiological and behavioural characteristics to verify an individual's identity. Combining different modalities, such as facial recognition, fingerprint scanning, and voice recognition, can enhance the security and accuracy of authentication systems. Multimodal biometric systems are more resilient to spoofing and provide a higher assurance level than single-modal systems. They are increasingly adopted in various applications, from secure access control in corporate environments to personal device authentication, offering a seamless and robust user experience while safeguarding sensitive information. Our research focuses on developing an original method for multimodal biometric authentication, combining the recognition of keystroke dynamics and facial biometrics. In both cases, neural networks were used, and the necessary data was obtained from public databases. The VGG-Face2 database was used for facial recognition, while the "CMU Keystroke Dynamics Benchmark Dataset" was used for recognizing keystroke dynamics. These modalities were then fused at the match score level. Our developed method for the universal measures AUC and EER reached 0.999882 and 0.004515, respectively. The FAR and FRR measures were 0.004069 and 0.004412, respectively. These results mostly exceed previous achievements described in the literature. The developed solution is noninvasive, enabling its implementation in office environments as an additional layer of protection against unauthorized access to digital devices without requiring additional actions from end users.

Keywords Authentication · Biometric · Face reccognition · Keystroke dynamics

A. Kozierkiewicz (✉) · A. Graczyk
Faculty of Information and Communication Technology, Wroclaw University of Science and Technology, Wroclaw, Poland
e-mail: adrianna.kozierkiewicz@pwr.edu.pl

E. Pimenidis
School of Computing and Creative Technologies, University of the West of England, Bristol, UK
e-mail: elias.pimenidis@uwe.ac.uk

H. Jahankhani and B. Issac (eds.), *Cybersecurity and Human Capabilities Through Symbiotic Artificial Intelligence*, Advanced Sciences and Technologies for Security Applications, https://doi.org/10.1007/978-3-031-82031-1_24

1 Introduction

The ease with which information is transmitted today and the prevalence of storing sensitive data on personal digital devices make issues related to access security increasingly important. In the context of ensuring protection against unauthorized access to information, systems, or devices, three types of authentication methods for security purposes stand out [14]:

- **Something you have**—A physical object in the authorized person's possession, such as a key to a lock or an access card.
- **Something you know**—Information known only to the authorized person, such as a password or private key.
- **Something you are**—Unique characteristics that identify a person, such as fingerprints, are called biometrics.

Due to the increasing need for system security, while also considering the simplicity of authentication, biometric-based systems have gained the most popularity in recent years. Various biometric features can be applied in user verification and identification processes, ranging from fingerprints to iris and retina recognition, face recognition, and keystroke dynamics. A scent, hand geometry, DNA, ear shape, and vein patterns have also been studied, among other physical characteristics. Documented behavioural characteristics include gait, computer mouse movement dynamics, and eye movements during reading. All these features differ in effectiveness and ease of application, but the purpose of use most often dictates their selection [38]:

- **Mobile Phone Authorization**—Biometrics (mainly facial recognition and fingerprint scanning) provide an efficient and now widespread method of securing smartphones while being user-friendly.
- **Building Access Control**—Facial, iris, or fingerprint recognition allows for seamless entry to premises, enhancing security compared to traditional keys or access cards.
- **Home Assistants**—Home assistants, such as Google Home or Alexa, use voice recognition for user identification. This ensures security by allowing access only to authorized users, which is crucial when connecting the assistant to various Internet of Things devices at home.
- **Airports**—Biometrics are increasingly being implemented at airports, facilitating travellers from check-in to boarding while allowing authorities to monitor passenger flow.
- **Forensics**—Law enforcement and special agencies use biometrics to identify criminals in criminal investigations.
- **Banking**—Banks are adopting biometric solutions to enhance transaction security and simplify account management. Biometrics also helps combat identity theft and fraud.

In many situations, to ensure a higher level of security, multiple biometric features are used simultaneously. Multimodal biometric systems are those that combine two or more biometric modalities for the purpose of identification or authentication of individuals. Traditionally, their integration can occur at four levels: the sensor level (where the biometric data is captured), the feature extraction level (where unique features are extracted from the data), the match score level (where the similarity between the extracted features and the stored features is calculated), and the decision level (where the final decision about the identity of the individual is made).

A multimodal approach to biometric authentication provides relatively high recognition accuracy (usually above 90%) across various combinations of modalities. However, it is also important to note that many solutions rely on features that are captured in an invasive manner, meaning they require the direct involvement of the user, which can become cumbersome with regular repetitions. Invasive features may include those that require physical contact, such as fingerprint or iris recognition. Therefore, this study focused on a solution with the least invasiveness.

This paper aims to develop an original multimodal biometric authentication method. This goal is achieved through using neural networks for facial biometrics recognition and essential keystroke dynamics recognition, as well as combining these modalities at the match score level. The prepared solution is non-invasive, which would allow it to be implemented in office environments as an additional layer of protection against unauthorized access to digital devices, without requiring additional actions from end users.

The remainder of the article is organized as follows. Section 2 overviews related literature. The essential keystroke dynamics recognition methods, face recognition, and systems that combine these modalities are discussed. The paper's main contribution is provided in Sect. 3, where our developed method is presented in detail. Section 4 contains results of the experimental verifications that have been conducted. The final part, Sect. 5, summarises the paper and sheds some light on our upcoming research plans.

2 Related Works

Biometrics is a technique for automatically identifying individuals based on their unique characteristics. These characteristics can be divided into physical traits (fingerprints, facial features, voice patterns, iris or retina patterns, etc.) and behavioural traits, such as a distinctive gait or writing style. Due to the fact that we aim to propose a minimally invasive solution, this study focuses only on facial recognition and keystroke dynamics.

2.1 Keystroke Dynamics

In keystroke dynamics research, many classification methods have been employed over the past three decades. Keystroke dynamics recognition can be seen as a pattern recognition problem, and according to [35], most popular and widely used methods can be broadly divided into statistical approaches (61%), machine learning approaches (37%), and others (2%).

The most popular methods are also the simplest. They are based on directly using distance measures. For a new sample, distances in multidimensional space to user templates (created from analyzing training samples) are calculated based on which class membership is determined. Common measures used to calculate the distance score introduced in the literature include Euclidean distance [16, 30], Manhattan distance [4, 16], Bhattacharyya distance [12], Mahalanobis distance [15, 16], or even combinations of different measures. The article by Zhong et al. [41] presents an original combination of Manhattan and Mahalanobis distances. Statistical measures are also used due to their simplicity and ease of implementation.

Probabilistic methods assume each keystroke feature vector has a normal distribution [22]. They rely on determining the probability that a given keystroke profile belongs to a person registered in the database. Some techniques used include the Naive Bayes classifier [24], hidden Markov model [1], or Gaussian mixture model [11].

Machine learning is widely used in pattern recognition and is no different for keystroke dynamics. A decision tree is a pattern recognition technique that learns from examples and is mainly suitable for classification problems with a small number of output classes. Therefore, the primary tree-based learning method described in the literature is the random forest [33]. Another well-known classifier, described in the literature [20], is the support vector machine (SVM). Zack et al. [40] investigated the K-nearest neighbours classifier on long output texts. Evolutionary methods such as the genetic algorithm [26], ant colony algorithm [13], or particle swarm optimization [26] are techniques that have been studied to improve classification accuracy by finding optimal keystroke features for classification. A method that is gaining increasing popularity is deep neural networks, whose application in keystroke dynamics recognition is described in [3]. The operation of other network architectures such as recurrent [18, 36] and convolutional networks [18, 19] has also been investigated.

The research has shown that neural networks can achieve higher accuracy and performance compared to traditional statistical methods [5]. Notably, models like Bi-CNN and FFM-NN have demonstrated impressive accuracy rates of 96.8% and 94.7% respectively [2]. Moreover, recent studies highlighted that machine learning solutions, including the Random Forest model, can achieve accuracy levels as high as 99.9%. Additionally, deep learning techniques that combine CNN and GRU architectures have reached 99.31% accuracy, while the DBN model has shown strong performance with 98.01% accuracy. These advancements underscore the untapped potential of keystroke dynamics, demonstrating the remarkable effectiveness of neural networks and deep learning in significantly improving the precision and reliability of authentication systems.

2.2 Face Recognition

One of the first approaches to facial recognition was the "Eigenface" method, which generates a face vector by considering the standard features of face images stored in a database. Although more advanced techniques for facial feature extraction exist today, the concept of eigenfaces has not yet been abandoned [37]. Another method relies on Gabor wavelet transformations, used to extract facial features characterized to maintain variability due to differences in facial expression and lighting [25]. The hidden Markov model, a well-known stochastic model that includes a Markov chain with an unobserved or hidden sequence of states, is also frequently used in facial recognition methods [7]. Principal component analysis (PCA) has been explored, a technique used to reduce the number of variables or the dimensionality of features to improve the quality of facial recognition [17]. The support vector machine (SVM), based on the concept of decision space divided by constructing boundaries that separate samples of different class memberships, also finds application in this problem [8]. Deep convolutional networks like AlexNet [34], LeNet [39], GoogleNet [27], VGGNet [23] have proven to be highly effective in facial recognition (accuracy higher than 99%), surpassing the results achieved by holistic, geometric, and local-textural approaches.

2.3 Biometric Systems Based on Combining Face Recognition and Keystroke Dynamics

This section of the paper presents works focusing on multimodal biometric systems combining facial recognition and keystroke dynamics. There are only a few studies available in the literature on this topic.

In the paper [31], the K-nearest neighbors algorithm was used for keystroke dynamics recognition with a combination of Mahalanobis and Manhattan distance measures to normalize the dataset and remove outliers. Match score level fusion was proposed. The results were first normalized using the min-max method and then fused using sum, product, and weighted sum rules. The weighted sum rule achieved the best results, resulting in an EER of 1.76%.

In the article [10], Eigenface and PCA-based approaches were also used for facial recognition. For keystroke dynamics recognition hidden Markov models (HMM) were used. Fusion was performed at the confidence level according to the defined formula. The results of the experiments were presented using a false acceptance rate of 5.4% and a false rejection rate of 9.2%.

The goal of the authors of publication [9] was to create a biometric system that ensures acceptability and respects users' privacy. In the conducted analysis, many recognition and fusion methods at the confidence level were tested, but the most interesting approach that yielded the best results was using a genetic algorithm. Thus, the EER value equal to 2.22% was achieved.

In the paper [28], a continuous authentication system was presented consisting not only of keystroke dynamics and facial recognition but also of skin color. EigenFace and PCA algorithm were used for facial recognition. Skin color detection was based on the distribution of skin color in the YCbCr color space. Continuous keystroke dynamics recognition was based on a proprietary algorithm. Fusion was calculated at the match score level. On a private database, the system's performance achieved a false acceptance rate of 0% and a false rejection rate of 0.72%.

In summary, the literature lacks straightforward and cost-effective solutions to implement without the need for additional devices or sensors. What is missing are systems that may not demonstrate the highest security indicators but non-invasiveness in everyday operation.

3 Multimodal Method to Biometric Authentication

Our primary objective is to develop a non-invasive, multi-modal biometric system. To achieve this, we are integrating facial recognition and keystroke dynamics.

The process begins with the user logging into their computer using a password. During this login, both the camera image and keystroke data are recorded. These recordings are then sent to the respective facial recognition and keystroke dynamics modules for preliminary processing. The camera image is forwarded to the face detection module, while the keystroke dynamics module processes the raw keystroke data to generate a vector of keystroke timestamp markers.

Each module produces classification results independently. These results are then sent to a fusion module, which combines the facial recognition and keystroke dynamics data to generate a unified matching result. Based on this combined result and a predefined threshold, a decision is made regarding the user's authenticity. This overall process is illustrated in Fig. 1. The subsequent subsections provide a detailed explanation of the proposed approach.

3.1 Keystroke Dynamics Recognition Module

The keystroke dynamics recognition module gathers, processes, and analyses keystroke data to predict user identity based on typing patterns. Whenever a user presses or releases a key, the system logs the key name along with a timestamp of the

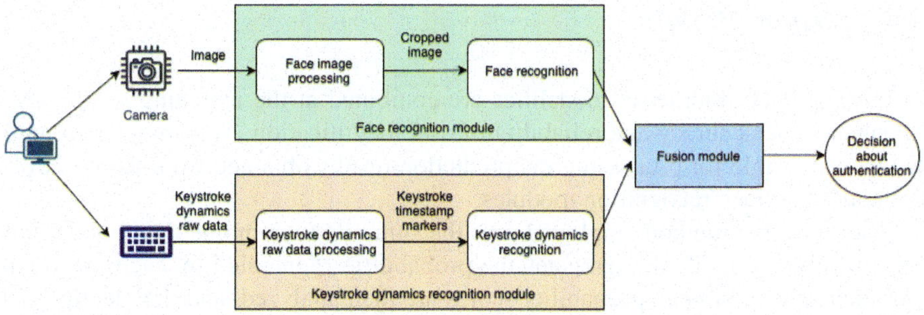

Fig. 1 General idea of the multimodal approach to biometric authentication

event. These timestamps are typically generated using an external reference clock to ensure high accuracy, forming a timestamp vector from the collected data.

The recognition process is a comprehensive one, employing a neural network of five dense layers and four batch normalization layers. The dense layers utilize the ReLU (Rectified Linear Unit) activation function, while the final layer employs a sigmoid activation function. During training, the Adam optimizer is used, and binary cross-entropy is chosen as the loss function. Once training is complete, the model predicts the probability of each user corresponding to a given password entry attempt.

3.2 Face Recognition Module

The face recognition module collects, processes, and analyses facial data to make user identity predictions. For face detection and classification in images, existing YOLOv8 models are utilized.

The process begins with capturing an image from the webcam, which is then processed to detect and crop the face, which is achieved using the YOLOv8n-face model, trained on the COCO dataset specifically for face detection tasks. This model is selected for its optimal performance and minimal resource usage, making it ideal for use on office devices without disrupting normal work activities.

Following preprocessing, the image is forwarded to a classifier based on the YOLOv8n-cls model, rigorously trained on the ImageNet dataset, to determine authentication. This model, known for its high accuracy, predicts the probability of each user being the one in the captured webcam image, ensuring reliable authentication.

3.3 Fusion Module

In our approach, biometric modalities are combined at the matching results level to enhance the accuracy and reliability of user identification. This integration module is responsible for processing the prediction results obtained from the keystroke dynamics and face recognition modules.

The ensemble method employed uses the sum rule to combine these modalities. Specifically, the method aggregates the probabilities generated by each individual classifier. The combined probability scores are then analyzed, and the identity with the highest cumulative score is selected from the database. If the declared user's identity matches the result produced by the fusion module, the identity is positively confirmed.

By cross-verifying the user's identity through both keystroke dynamics and facial recognition, the system significantly reduces the likelihood of false positives and false negatives, thereby enhancing security. This approach leverages the strengths of each biometric modality, providing a more robust and secure identification process.

4 Experimental Verification

The method described in Sect. 3 has been verified using two public benchmark datasets: VGGFace2 and CMU Keystroke Dynamics Benchmark Dataset.

The "CMU Keystroke Dynamics Benchmark Data-set" is a standard benchmark in many scientific studies on keystroke dynamics. It contains keystroke information from 51 users. All users typed the same password 400 times over eight sessions (50 repetitions per session). The breaks between sessions lasted at least one day to capture possible variations in participants' typing patterns depending on their psychological and physical state on a given day. The password ".tie5Roanl" was selected as an appropriate representation of a strong password of 10 characters in length.

VGGFace2 includes over 9000 individuals and more than 3.3 million face images. Most importantly, each individual in the database has an average of 362 face images. The distribution of faces across different identities varies, ranging from 87 to 843. However, this allowed for the easy extraction of 51 individuals with at least 400 images each from the database and matched to the available data from the "CMU Keystroke Dynamics Benchmark Data-set".

4.1 Keystroke Dynamics Recognition

First, we evaluated the effectiveness of the keystroke dynamics recognition module. The performance of the proposed neural networks was compared to other methods

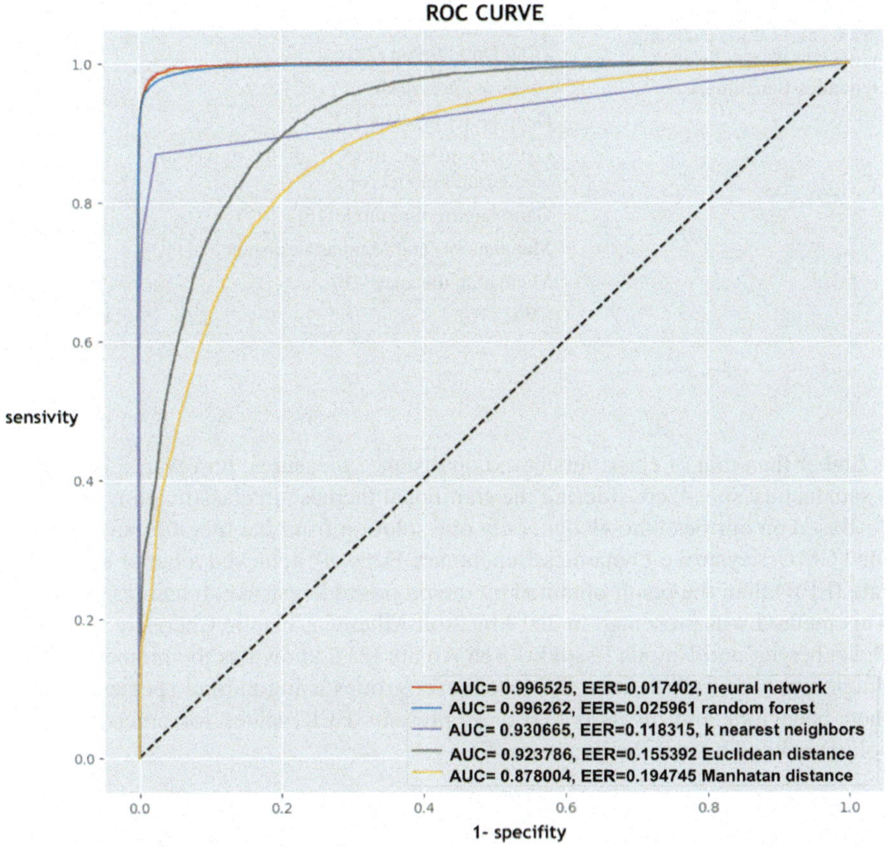

Fig. 2 Comparison of ROC curves of classifiers in the task of recognizing keystroke dynamics

used for this task. We determined the AUC and EER metrics to facilitate the comparison. These metrics are presented in Fig. 2, which also illustrates the performance of the individual classifiers based on the ROC curves.

While classifiers directly based on distance measures may have performed the worst in recognizing keystroke dynamics, their simplicity of implementation and speed of operation make them stand out. These unique features, despite their relatively low scores, are noteworthy in the context of this study.

The k-nearest neighbours and random forest algorithms, while achieving higher results than their predecessors for each metric, come with a significant trade-off. Their better adaptation to the task of recognizing keystroke dynamics in the studied database is evident, but this improvement is at the expense of a notable increase in algorithm execution time.

The classifier based on the proposed neural network performed the best, achieving an AUC of 0.9965 and an EER of 0.0174. The computational cost of such a solution

Table 1 EER metrics for various algorithms evaluated on the "CMU Keystroke Dynamics Benchmark Dataset"

Method	EER (%)
ZOO Doddington [21]	1.3
Proposed approach	1.7
Deep belief network [6]	3.5
Gaussian mixture model with the universal background model [6]	5.5
Gaussian mixture model [6]	8.4
Mahalanobis and Manhattan distance [41]	8.7
Manhattan distance [16]	9.6
SVM [16]	10.3

is higher than that of classifiers based on distance measures; however, it operates at a satisfactory speed, considering the significant increase in classification accuracy.

Based on our best knowledge, only one solution from the literature evaluated on the "CMU Keystroke Dynamics Benchmark Dataset" achieved a better equal error rate (EER) than the result obtained by the proposed keystroke dynamics algorithm. This method was presented in 2019 by Abir Mhennia, Estelle Cherriere, Christoph Rosenberger, and Najoua Essoukri Ben Amara [21], known as the biometric "Doddington zoo," which defines and labels user groups using animal species to reflect their behaviour and diversity. Table 1 presents EER values for other prominent solutions known from the literature.

4.2 Face Recognition

In the second stage, we verified the performance of the face recognition module. Pictures sourced directly from the VGGFace2 dataset were prepared accordingly. Faces were cropped to remove additional information contained in background patterns. Of utmost importance, the evaluation results are presented in a visually informative manner, using ROC curves and the calculated metrics of AUC and EER. These metrics were 0.99743 and 0.025, respectively, as shown in Fig. 3.

In the case of face recognition, since only a fragment of the entire VGGFace2 dataset was used for analysis, comparing the results of the classifiers with the solutions presented in the literature may be ambiguous and incomplete. Therefore, these comparisons have been omitted.

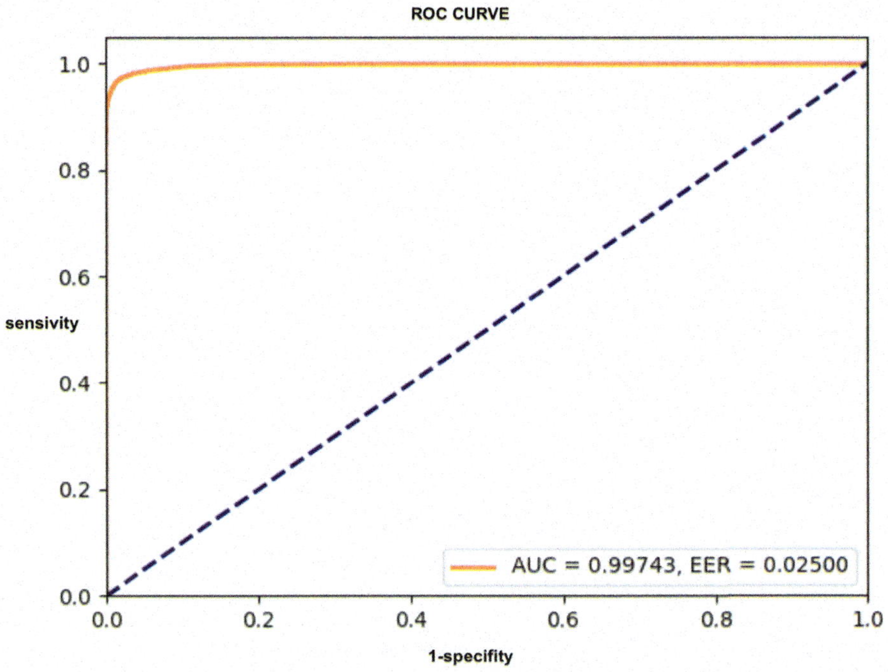

Fig. 3 ROC curve of YOLOv8-cls classifier in the task of face recognition

Table 2 Comparison of the results of single-modal modules with a multimodal solution

	AUC	ERR	FAR	FRR
Keystroke dynamics recognition	0.9965	0.0174	0.0168	0.0176
Face recognition	0.9974	0.0250	0.0203	0.0267
Multimodal approach	0.9999	0.0045	0.0041	0.0044

4.3 *Multimodal Approach*

The comparison of the developed multimodal solution with the results obtained separately by the keystroke dynamics module and the face recognition module is presented in Table 2 and Fig. 4. The specific numerical values of the predictions, reflecting the performance metrics of each module, are shown in Fig. 5 using confusion matrices.

Our research presents a promising multimodal approach that, as indicated by the results, surpasses solutions based solely on face recognition or keystroke dynamics. This approach holds great potential for further research and development in the field.

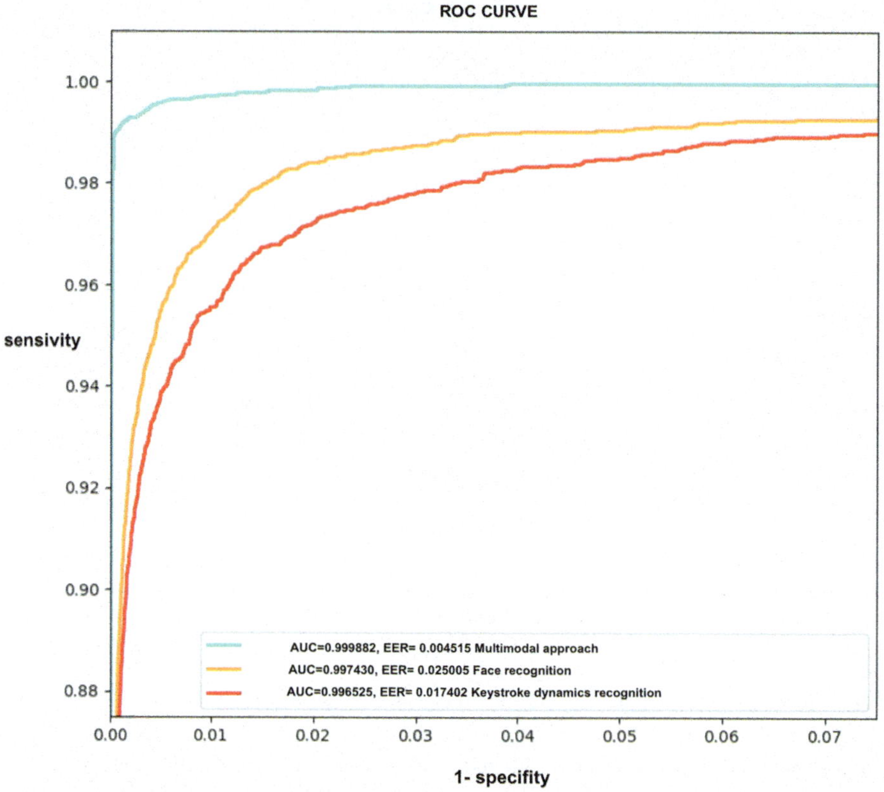

Fig. 4 Comparison of ROC curves of unimodal modules with the a multimodal solution

Our multimodal solution achieved better performance metrics in most cases compared to systems described in the literature, which also utilized the fusion of face biometrics and keystroke dynamics. The comparison of results is presented in Table 3. Although each study uses different data in their system (especially results obtained based on private databases are not very reliable), making direct comparison impossible, it is a relatively safe assumption that the use of neural networks and deep learning leads to better system effectiveness compared to traditional approaches.

Despite rigorous assumptions about non-invasiveness and acceptable speed of operation (achieved by avoiding the use of larger, more complex, and slower models), the developed solution provides competitive results compared to other biometric systems described in the literature, which use more invasive methods, such as fingerprint scanning or iris analysis.

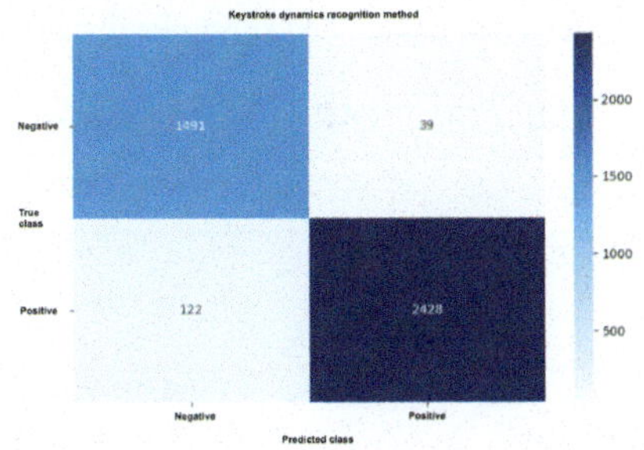

(a) Keystroke dynamics recognition method

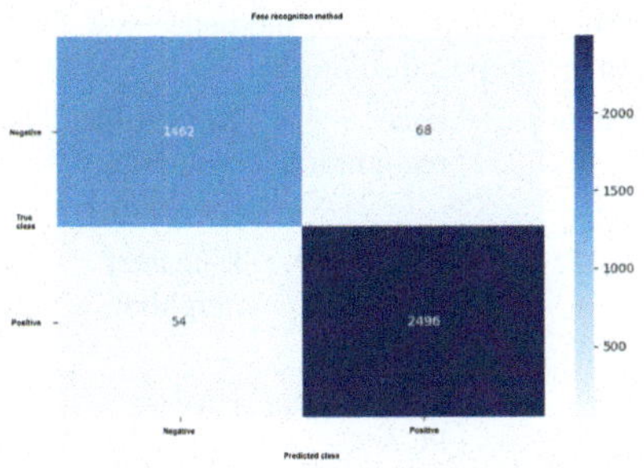

(b) Face recognition method

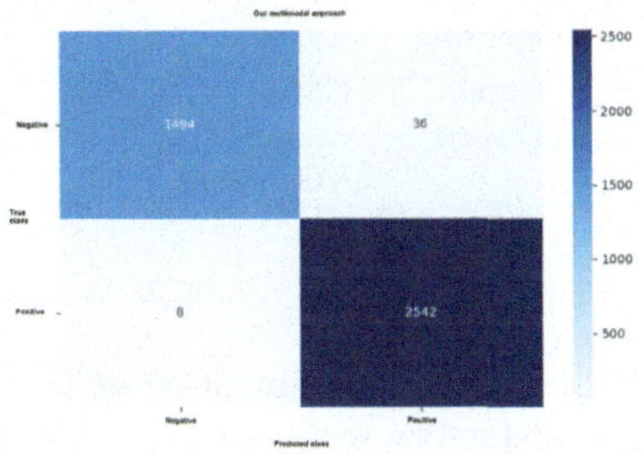

(c) Our multimodal approach

Fig. 5 Comparison of the confusion matrices of unimodal modules with a multimodal approach

5 Future Works and Summary

A review of existing biometric authentication solutions revealed a lack of non-invasive solutions while maintaining high effectiveness. Therefore, it was decided to propose a solution that stands out for its non-invasiveness, allowing it to be implemented in office environments as an additional protection against unauthorized access to digital devices, without requiring additional actions from end users. We proposed solution based on combination of keystroke dynamics and face recognition.

The "CMU Keystroke Dynamics Benchmark Dataset" was selected for the keystroke dynamics analysis. A single classifier based on the neural network achieved an AUC value of 0.9965 and an EER of 0.0174.

Data from the public VGGFace2 face database was used in the face recognition module. The YOLOv8n-cls model achieved results of 0.99743 for AUC and 0.025 for EER.

AUC and EER measures of 0.999882 and 0.004515, respectively, were obtained for the multimodal approach. FAR and FRR measures of 0.004069 and 0.004412, respectively, were also achieved. These results mostly exceed the current achievements in the literature.

Table 3 Comparison of biometric systems combining facial recognition and keystroke dynamics with the proposed solution

Source	Face recognition	Keystroke dynamics recognition	Dataset	ERR (%)	FAR (%)	FRR (%)
[32]	EigenFace, PCA	k-nearest neighbor	CMU dataset, ORL	1.76	–	–
[10]	EigenFace, PCA	HMM	Private database	–	5.4	9.2
[9]	Genetic algorithm	Genetic algorithm	GREYC, AR face	2.22	–	–
[29]	EigenFace, PCA	Author's algorithm	Private database	–	0	0.72
Proposed multimodal approach	Yolov8 model	Neural network	CMU dataset, VGGFace2	0.45	0.41	0.44

While our current focus on the combination of face recognition and keystroke dynamics has yielded satisfactory results, there are numerous avenues for further development. One particularly promising direction is the implementation of continuous authentication. In the potential office environments we've discussed, this would offer additional protection against unauthorized use in scenarios where a logged-out device is left unattended. This advancement would necessitate the creation of a proprietary training database and the execution of experiments based on test sessions with users.

Another potential development direction could be increasing the level of security. Possible approaches worth investigating could include using additional biometric features or adding pressure analysis to the keystroke dynamics recognition module.

References

1. Ali ML, Monaco JV, Tappert CC (2017) Hidden Markov models in keystroke dynamics. J Signal Process Syst 86:175–190
2. Arsh A, Kar N, Das S, Deb S (2024) Multiple approaches towards authentication using keystroke dynamics. Procedia Comput Sci 235:2609–2618
3. Bernardi ML, Cimitile M, Martinelli F, Mercaldo F (2019) Keystroke analysis for user identification using deep neural networks. In: International joint conference on neural networks (IJCNN), pp 1–8
4. Chandok R, Bhoir V, Chinnaswamy S (2022) Behavioural biometric authentication using keystroke features with machine learning. In: 2022 IEEE 19th India council international conference (INDICON), pp 1–6
5. Crawford H (2010) Keystroke dynamics: characteristics and opportunities. In: 2010 Eighth International conference on privacy, security and trust. IEEE, pp 205–212

6. Deng Y, Zhong Y (2013) Keystroke dynamics user authentication based on gaussian mixture model and deep belief nets. ISRN Signal Process
7. Dinkova P, Georgieva P, Manolova A, Milanova M (2016) Face recognition based on subject dependent hidden Markov models. In: 2016 IEEE International Black Sea conference on communications and networking (BlackSeaCom), pp 1–5
8. Ghazal M, Abdullah K (2020) Face recognition based on curvelets, invariant moments features and SVM, TELKOMNIKA Indonesian. J Electr Eng 18:733–739
9. Giot R, Hemery B, Rosenberger C (2010) Low cost and usable multimoal biometric system based on keystroke dynamics and 2d face recognition. In: 2010 20th International conference on pattern recognition. IEEE, pp 1128–1131
10. Gupta A, Khanna A, Jagetia A, Sharma D, Alekh S, Choudhary V (2015) Combining keystroke dynamics and face recognition for user verification. In: 2015 IEEE 18th International conference on computational science and engineering, pp 294–299
11. Hosseinzadeh D, Krishnan S (2008) Gaussian mixture modeling of keystroke patterns for biometric applications. IEEE Trans Syst Man Cybernet Part C Appl Rev 38:816–826
12. Janakiraman R, Sim T (2007) Keystroke dynamics in a general setting. In: Advances in biometrics: international conference (ICB), Seoul, Korea, August 27–29, 2007. Proceedings, Springer, pp 584–593
13. Karnan M, Akila M (2010) Personal authentication based on keystroke dynamics using soft computing techniques. In: 2010 Second International conference on communication software and networks. IEEE, pp 334–338
14. Khan HZU, Zahid H (2010) Comparative study of authentication techniques. Int J Video Image Process Netw Secur IJVIPNS 9–13
15. Killourhy K, Maxion R (2008) The effect of clock resolution on keystroke dynamics. In: Recent advances in intrusion detection: 11th International symposium (RAID), Cambridge, MA, USA, September 15–17, 2008. Proceedings 11. Springer, pp 331–350
16. Killourhy KS, Maxion RA (2009) Comparing anomaly detection algorithms for keystroke dynamics. In: 2009 IEEE/IFIP International conference on dependable systems and networks, pp 125–134
17. Kumar M (2021) Autofer: Pca and pso based automatic facial emotion recognition. Multimedia Tools Appl 80
18. Lu X, Shengfei Z, Shengwei Y (2019) Continuous authentication by free-text keystroke based on CNN plus RNN. Procedia Comput Sci 147:314–318
19. Maiorana E, Kalita H, Campisi P (2019) Deepkey: Keystroke dynamics and CNN for biometric recognition on mobile devices. In: 2019 8th European workshop on visual information processing (EUVIP), pp 181–186
20. Martono W, Ali H, Salami MJE (2007) Keystroke pressure-based typing biometrics authentication system using support vector machines. In: Computational science and its applications-ICCSA, International conference, Kuala Lumpur, Malaysia, August 26–29, 2007. Proceedings, Part II 7. Springer, pp 85–93
21. Mhenni A, Cherrier E, Rosenberger C, Essoukri Ben Amara N (2019) Analysis of Doddington zoo classification for user dependent template update: application to keystroke dynamics recognition. Future Gener Comput Syst 97:210–218
22. Monrose F, Rubin A (1997) Authentication via keystroke dynamics. In: Proceedings of the 4th ACM conference on computer and communications security, pp 48–56
23. Parkhi O, Vedaldi A, Zisserman A (2015) Deep face recognition. In: BMVC 2015-Proceedings of the British machine vision conference 2015. British Machine Vision Association
24. Pavaday N, Soyjaudah K (2007) Enhancing performance of Bayes classifier for the hardened password mechanism. In: AFRICON 2007. IEEE, pp 1–7
25. Rashid S, Abdullah A, Shihab M (2020) Face recognition system based on Gabor wavelets transform, principal component analysis and support vector machine. Int J Adv Sci Eng Inf Technol 10:959
26. Saini BS, Kaur N, Bhatia KS (2019) Performance comparison of particle swarm optimization and genetic algorithm for feature subset selection in keystroke dynamics. In: 2019 International conference on computational intelligence and knowledge economy ICCIKE, pp 684–689

27. Schroff F, Kalenichenko D, Philbin J (2015) Facenet: a unified embedding for face recognition and clustering. In: Proceedings of the IEEE conference on computer vision and pattern recognition, pp 815–823
28. Shen C, Zhang H, Yang Z, Guan X (2016) Modeling multimodal biometric modalities for continuous user authentication. In: 2016 IEEE International conference on systems, man, and cybernetics (SMC), pp 001894–001899
29. Simonyan K, Zisserman A (2014) Very deep convolutional networks for large-scale image recognition. arXiv:1409.1556
30. Singh S, Arya K (2011) Key classification: a new approach in free text keystroke authentication system. In: 2011 Third Pacific-Asia conference on circuits, communications and system (PACCS). IEEE, pp 1–5
31. Srivastava S, Sudhish PS (2016) Continuous multi-biometric user authentication fusion of face recognition and keystoke dynamics. In: 2016 IEEE region 10 humanitarian technology conference R10-HTC, pp 1–7
32. Stefan D, Yao D (2010) Keystroke-dynamics authentication against synthetic forgeries. In: 6th International conference on collaborative computing: networking, applications and worksharing (CollaborateCom 2010), pp 1–8
33. Syed Z, Banerjee S, Cheng Q, Cukic B (2011) Effects of user habituation in keystroke dynamics on password security policy. In: 2011 IEEE 13th International symposium on high-assurance systems engineering. IEEE, pp 352–359
34. Taigman Y, Yang M, Ranzato M, Wolf L (2014) Deepface: closing the gap to human-level performance in face verification. In: 2014 IEEE conference on computer vision and pattern recognition, pp 1701–1708
35. Teh PS, Teoh A, Yue S (2013) A survey of keystroke dynamics biometrics. Sci World J 2013:408280
36. Tse K-W, Hung K (2020) User behavioral biometrics identification on mobile platform using multimodal fusion of keystroke and swipe dynamics and recurrent neural network. In: 2020 IEEE 10th symposium on computer applications and industrial electronics (ISCAIE), pp 262–267
37. Wahyu Mulyono IU, Ignatius Moses Setiadi DR, Susanto A, Rachmawanto EH, Fahmi A, Muljono (2019) Performance analysis of face recognition using eigenface approach. In: 2019 International seminar on application for technology of information and communication (iSemantic), pp 1–5
38. Xiao Q (2007) Technology review—biometrics-technology, application, challenge, and computational intelligence solutions. IEEE Comput Intell Mag 2:5–25
39. Yin X, Yu X, Sohn K, Liu X, Chandraker M (2019) Feature transfer learning for face recognition with under-represented data. In: 2019 IEEE/CVF conference on computer vision and pattern Recognition (CVPR), pp 5697–5706
40. Zack R, Tappert C, Cha S-H (2010) Performance of a long-text-input keystroke biometric authentication system using an improved k-nearest-neighbor classification method. In:2010 Fourth IEEE International conference on biometrics: theory, applications and systems (BTAS), pp 1–6
41. Zhong Y, Deng Y, Jain A (2012) Keystroke dynamics for user authentication. In: 2012 IEEE computer society conference on computer vision and pattern recognition workshops, pp 117–123

The Role of EdTechs in the Commodification of Knowledge

Sukhvinder Hara

Abstract The role of Education Technology Startups (EdTechs) and neoliberalism in commodifying knowledge within higher education is examined, with a focus on the UK and India during the COVID-19 pandemic. The study explores how the pandemic accelerated EdTech adoption, transforming academic content into marketable commodities aligned with neoliberal ideals of privatisation and competition. While commodification is presented as a pathway to financial viability and enhanced accessibility, infrastructural and socioeconomic inequalities remain significant barriers. EdTechs' innovative business models have reshaped educational access, equity, and content quality, often exacerbating pre-existing disparities. Ethical concerns, including data privacy and the erosion of academic autonomy, are also discussed. Through a comparative analysis of the UK and India, the paper highlights initiatives addressing the digital divide and challenges related to digital poverty. While EdTechs offer potential benefits in broadening educational access, the study stresses the importance of balancing technological innovation with core educational values of equity and quality, ensuring that education remains a public good.

Keywords EdTech · Digital divide · Knowledge commodification

1 Introduction

This paper examines the impact of Education Technology Startups (EdTechs) and neoliberal ideology on the global commodification of knowledge, focusing on the UK and Indian higher education systems during the pandemic. It first examines the key drivers of knowledge commodification, the mechanisms of enablement, and the implications for the higher education sector and students. Supporters for the commodification of knowledge emphasise the benefits of such technologies providing alternate revenue streams to improve financial viability of Higher Educational Institutions

S. Hara (✉)
Middlesex University, London, UK
e-mail: s.hara@mdx.ac.uk

H. Jahankhani and B. Issac (eds.), *Cybersecurity and Human Capabilities Through Symbiotic Artificial Intelligence*, Advanced Sciences and Technologies for Security Applications, https://doi.org/10.1007/978-3-031-82031-1_25

(HEIs) and the increased accessibility they provide to learners. Nonetheless, there are barriers to education beyond the control of Higher Education Institutions (HEIs) and the education sector, requiring further discussion [1].

The recent global Covid pandemic event compelled HEIs globally to adopt educational technology solutions to deliver education, accelerating the commodification of knowledge. This convergence of economic necessity, technological innovation and digital pedagogical transformations [2], has once more opened up the debate on the commodification of knowledge and the impact of neoliberalism to push privatisation within higher education. Within this context, educational technologies have emerged as a significant force shaping higher education dynamics by enabling the commodification of academic content, resources, and experiences. This has led knowledge to become an applied mechanism for gaining skills and competencies [3, 4] in the "globalised economy" [5] for employment [6]. This transition shifts the focus of education in HEIs from the notion of doing public good to a marketable commodity that can be bought and sold, similar to any other commodity [7]. Thus, allowing HEIs to access more markets globally without needing physical infrastructures.

It is difficult for the higher education sector to refute the disruptive benefits of new educational technology solutions [8] and the lifeline these solutions provided to "fix" delivery and accelerate its adoption [9] in India and the UK. This was despite infrastructure failures leading to power cuts in India contributing as a significant factor in the inequality to access to online education and digital poverty in the UK. Along with EdTechs, communication and video conferencing companies became valuable stakeholders [10, 11] for the global education sector, who were not previously associated with delivering global education on a mass scale.

EdTechs not only provide technology solutions but combine these with innovative business models that monetise educational content [12, 13], which can drive knowledge commodification further in the post-pandemic era as new market-driven opportunities arise. However, it is still being determined how much innovation HEIs can incorporate without distancing themselves from their primary role of providing in-person education.

This transformation in digital pedagogy has drawbacks, as it could impact access, equity, and content quality in education due to the priority given to the neoliberal agenda. Potentially reducing the primary regulatory objectives of not-for-profit HEIs [14] and the "academic mission" [15].

In summary, this paper will focus on:

1. Impact of EdTech industry on higher education's push for knowledge commodification
2. The effect of neoliberalism on privatisation
3. The implications of commodification on access and equity.

2 Definition of EdTech

The Department for Education defines EdTech as *"The practice of using technology to support teaching and the effective day-to-day management of education institutions. It includes hardware (such as tablets, laptops, or other digital devices), and digital resources, software and services that help aid teaching, meet specific needs, and help the daily running of education institutions (such as management information systems, information sharing platforms and communication tools)"*, [16]. Although educational technologies are not a recent development and have long been integral to the delivery of education through synchronous and asynchronous learning modes, we have seen these develop to encompass different modes (on campus, online access) and through the development of mobile applications.

EdTech is a portmanteau for education and technology. EdTechs are corporate enterprises that harness technology to innovate in education [17]. This innovation combines existing and emerging technologies to create innovative learning solutions for the commercial market. These technologies include Virtual and Augmented Reality, Artificial Intelligence, Machine Learning, adaptive learner configurations and algorithms, conferencing technologies, podcasting, and gamification. This has developed into the hybrid learning model where traditional teaching methods are complemented with solutions from EdTechs for flexible, immersible [18] and personalised learning [19].

This transformation in educational reach has also led many academics to revisit pedagogical methodologies for digital delivery since the pandemic [20, 21]. It has also spurred EdTechs to develop more innovative learning solutions since the sector has made inroads into all levels of education, since the Covid pandemic.

A recent market report on "Pandemic Privatisation in Higher Education: Edtech and University Reform" [22] summarises the growth of EdTechs by key market segments: type, sector, deployment, and end user. Within each of these segments, there are EdTechs that focus on (hardware, software, content creation), specific educational age groups, deployment solution (cloud, on-prem and app) and the end user (business, individual consumer, HEI). Furthermore, the report introduces a new term in EdTech *edutainment*, where recent developments in solutions for entertainment and education are combined to create interactive content to maintain learner interest. Examples of these include online learning platforms (e.g., Udacity), adaptive learning software (e.g., Knewton platform) and educational applications (e.g., Quizlet).

The global investment into internet infrastructure, growth in mobile smart devices, and consumptive consumer behaviour have contributed to the success of EdTechs. With cloud infrastructure costs reducing, cloud storage increasing, and cloud deployment replacing on-premises solutions, it has contributed to the success of educational technology enterprises [23, 24]. The advancement of EdTech solutions has also enabled higher education to access new customers and compete in new markets. But it has also exposed the structural inequalities embedded with the education system that disproportionately affect marginalised communities, e.g., ethnic minorities and low earners. In the UK, the digital divide increased inequalities where low internet speed and access, rural locations, and device poverty were contributing factors [25, 26]. These issues were further complicated in India with electricity inequality, language obstacles and the lack of trained staff in India [27].

Considering these complexities, it becomes essential to examine the neoliberal agenda of higher education to foresee how privatisation has become driven by EdTech innovations whilst ensuring higher education remains true to its fundamental tenets. The market-driven approach has forced institutions to stay competitive despite offering similar services and educational experience (courses, state-of-the-art facilities, student experience) to differentiate themselves by capitalising on partnerships, knowledge, reputation and brand [28, 29].

3 Revisiting the Objective of Education and Market-Driven Education

The basic premise of education is that it should be accessible to everyone regardless of social status or economic wealth. It is considered a fundamental right by UNESCO under the Convention against Discrimination in Education. The UK accepted this on 14th March 1962 with subsequent ratification into UK law. In India, the Right to Education Act and The Right of Children to Free and Compulsory Education Act were enacted 1st of April 2010. Nevertheless, there are global variations in the accessibility of education concerning gender equality [30], particularly due to cultural and political decisions [31].

HEIs have played a pivotal role in society's progression by promoting the acquisition of knowledge, critical thinking and democratic principles [32]. In India and the UK, state funding of higher education ensures equitable access for all members of society. However, reduced state funding has impacted HEIs, requiring financially stressed HEIs to operate more as commercial organisations [33, 34].

There is universal agreement amongst academics over the positive impact of education on the development of an individual's human capital and the recognition that education is given in "developing economies like India" [35]. It is accepted that education can improve employability and earning potential over a lifetime. This contributes to the individual's progressive economic wealth and that of their nation's economic development, providing the foundations to compete internationally with other countries [36].

HEIs successfully make use of student completion statistics and student employability statistics [37]. By providing strategic in-house employability services to further develop student employment skills [37] and establishing links with industry partnerships [38], enables institutions to use these as marketing tools. Therefore, education for employability is not a new trend but rather an inevitable outcome of completing higher education, and this is an expectation from students [39]. At a national level it makes strategic sense for nations to invest in education to maximise the potential of their population for economic growth and development and for individuals to improve their financial situation and that of their communities. Furthermore, an educated population reduces their dependence on the state and their impact on public services, e.g., social services [40], healthcare [41], and the judiciary [42].

However, changing population demographics raises the debate about how institutions will respond to declining worldwide birth rates. This includes India, where a "steady decline in fertility is evident" [43]. In the UK, primary schools have closed or have been merged as the number of children have fallen [44], which is likely to impact universities in the future.

With an ageing global population, HEIs may need to market education to a different demographic, as the retirement age of nations is likely to increase. There is little doubt that this will not make it on future agendas for long-term strategy planning. In India and the UK, companies (e.g., HSBC, Goldman Sachs, Amazon) run Returnship programmes focused on getting people back into employment. In 2023, the UK government pledged £63.2 million for Returnships (skills boot camps, apprenticeships, sector-based work academy programmes) for adults over fifty years of age who are returning to work or changing careers [45]. Furthermore, the UK Lifelong Learning (Higher Education Fee Limits) Act 2023, enables learners to access student loans until the age of sixty for four years of post-18 education and will allow learners to pay proportionately for modules and short courses. While details are being finalised, the World Economic Forum identified "Upskilling: Training the workforce of the future" as a critical trend for EdTechs [46].

In the future, focus on employability is likely to increase further and access to learning rather than the distinction between education and training. HEIs will inevitably look to expand alternative services and develop partnerships with EdTechs and industry.

4 Neoliberalism

Neoliberal ideology, characterised by free-market principles advocating deregulation, privatisation, and state de-investment, has considerably influenced the higher education sector [47]. The ideology promotes competition between institutions, commodifying knowledge and shifting educational priorities from equity and accessibility towards profit-oriented metrics. According to Mintz [48], this has led to the "student-as-customer" trend. Furthermore, globally, the rise of for-profit educational institutions and the reduction of public funding for universities is an evident trend of neoliberalism [49], which was accelerated by the pandemic.

Under this ideology, higher education institutions adopt the characteristics of commercial organisations offering Education-as-a-Service (EaaS) for sale and evaluate organisational performance based on delivery, service, quality, and profitability, thus moving education away from the public good. Spruin [50] examined the strategy of Canterbury Christ Church University (CCCU) to stay competitive and flexible in the current higher education market. In which CCCU had identified new opportunities in online learning for STEM courses, retraining and continuing professional development (CPD) as sources of further revenue generation. This illustrates a shift in higher education institutions towards more profit-oriented objectives. Similarly, in India, there has been a marked increase in private HEI [51, 52] and the development of forward-looking curriculums that anticipate the impacts of future technologies on higher education [53].

In the pandemic, we experienced the robustness of many educational institutions that utilised EdTech solutions to pivot existing business models for economic survival and continuation of services. In that sense, we can see that the HE sector has already developed the neoliberal skillset and culture to survive [54]. This can be seen as moving away from the notion of public good favouring market-oriented principles of competition, consumer choice and operating efficiencies [55].

The ideology emphasises individual responsibility, limited government intervention, and the belief that market forces should drive decision-making across various sectors, including education. In higher education, neoliberal ideology has fuelled the expansion of privatisation, leading institutions to seek alternative funding sources (tuition fees, research grants, philanthropic donations, and corporate partnerships, changing the priority for education [15]. Governance structures are now geared towards leadership that directs commodity-led higher education. The emphasis on commodification, competition, consumerism, efficiency, and market-driven outcomes has transformed academia into a commercial education sector, despite HEIs being not for profit. HEIs are adopting market strategies in decision-making, from capitalising on brand and reputation to strategies on student enrolment and operating with lean workforces [38] rather than prioritising academic excellence.

Supporters of market dynamics believe competition acts as a catalyst for innovation and progress while enabling HEIs to identify and rectify inefficiencies through investments and leaner processes. Investments may include strategic faculty hires, developing careers, infrastructure projects, and technology investments to raise the profiles of HEIs in the sector [56].

With EdTech entities playing an increasingly prominent role in shaping educational policies and practices [57], the question of the aforementioned issues in equity, access, and quality of education are raised. Furthermore, EdTechs have become pivotal partners to higher education mainly for their demonstrable scalability and delivery of educational content, even though online digital content has been available through Massive Open Online Courses (MOOCs) since 2012 [58]. Partnerships have continued with EdTechs post-pandemic, as most institutions have opted for hybrid methods for the delivery of education. While online learning has expanded the reach of higher education institutions (HEIs), it has also revealed underlying structural inequalities, further widening the digital divide within communities that education was meant to reduce [1].

An essential facet of neoliberalism is privatisation, where market mechanisms for transferring goods and services are seen as more efficient than state intervention (policies and regulation). States are now unable and unwilling to fully fund higher education due to global circumstances, since the pandemic, with ongoing militarised conflicts, economic recessions, increasing social care costs due to ageing populations and funding of pensions. Together with an increased demand for international qualifications has resulted in the expansion of for-profit universities, alternative education providers, online education platforms [59] and the out-sourcing of services in higher education as in other business sectors [49].

Higher education institutions are now operating a market-driven education model [60] due to reduced state investment in public education by competing, selling knowledge as a commodity, offering consumers more choices and employing economic-driven operational efficiencies [61] but there are several concerns. Primarily, education has moved from its fundamental principle of education being for public good and societal development [48] to one of privilege where inequalities in quality, equity, access and social mobility exist despite the assertions that market-driven education fosters innovation and progress by providing individuals with opportunity [62]. The increasing penetration of EdTechs into the educational landscape further contributes to the privatisation and marketisation of education [16].

However, if market-driven policies benefit elite institutions that can purchase, partner and implement educational technologies, eroding the ethos of higher education for those HEIs unable to make these investments. This promotes a culture of corporatisation, where brand, ranking, reputation and purchasing power can be traded for the diversification of revenue streams over academic virtues. Furthermore, if privatisation is driven by neoliberal ideology and educational technology can improve an individual's capital development and autonomy. It raises crucial questions about the role of education in the context of society [63]. It removes from the consciousness the communal aspect of responsibility and democracy in a society as a collective [5]. Additionally, this may reduce democratic control and accountability as decision-making moves from public bodies to private companies focused on profit [64]. Neoliberal economic ideologies, which prioritise free-market principles, privatisation, and deregulation, have significantly shaped the commodification of higher education.

While the academic debate continues on privatisation and commodification, it needs to consider how user behaviour and on-demand content have become prevalent in all societies in accessing all types of information, including education [65]. This has become another factor to consider in the push to commodify knowledge.

5 Commodification of Knowledge

The commodification of knowledge results from the convergence of technological innovations [66], economic necessities, and digital pedagogical developments within higher education. It transforms education into a marketable commodity that can be bought and sold in global financial systems. Education content includes all subject matters and formats (e.g., presentations, videos, resources, books, articles, practical experiences, assessments) [67].

The process involves transforming educational content, resources, and experiences into marketable products and services that can be monetised through numerous business models. This is, in part, driven by the increased demand for flexible and accessible learning options [13], made possible by the expansion of global digital infrastructures, global internet programmes, access and ownership of devices, and lower cost of cloud storage [68, 69]. This enables HEIs to add further services to existing education delivery modes to appeal to diverse student demographics in local and international markets for revenue generation.

EdTechs employ several business models to monetise educational content and services. These reflect the diverse needs and preferences of the education systems users, institutions, and stakeholders in the education ecosystem. These business models often blend innovative pricing strategies, revenue streams, and partnership arrangements to generate sustainable revenue while delivering customer value [35]. Some of the standard business models in the EdTech industry include subscription-based models, freemium models, revenue-sharing arrangements.

Within this context, the role of EdTechs has emerged as a significant force shaping higher education dynamics via scalable products and services using business models "B2B, B2C, B2B-B2C, and C2C" [70] including online courses, learning platforms, user-adaptive learning systems to virtual reality simulations, catering to diverse learner needs, preferences, for all subjects. According to Bargavi and Shanmugam [70] the addition of AI, ML, and adaptive algorithms is the third wave in EdTech technologies focused on personalisation coinciding with the pandemic. These promise to revolutionise teaching and learning processes, improve educational outcomes, and expand access to knowledge on a global scale [22].

At the same time, they present challenges related to commercialisation, standardisation, customer data privacy, and the erosion of academic autonomy [71]. Moreover, commodification may exacerbate educational inequalities, reinforce knowledge production and dissemination, and perpetuate market-driven ideologies prioritising profit over traditional pedagogies [72]. In light of these complexities, it becomes imperative to critically explore EdTech start-ups' role in commodifying knowledge within higher education.

6 The Rise of EdTechs and Their Synergy with Neoliberalism

The EdTech industry has transformed education by offering innovative solutions and contributing to digital pedagogy, which learner behaviour, technology infrastructures and technology accessibility have supported. Its success and impact were demonstrated during the pandemic, transforming how knowledge was accessed, created, delivered, consumed, and assessed. In this period EdTechs were able to deliver on several services:

1. The delivery of digitised knowledge [64]
2. The purchase of digitised knowledge through platforms [57]
3. Content creation and curation [72]

4. Personalised and adaptive learning environments [1]
5. Retention of learners through analytics and interventions [22].

At the height of the pandemic, EdTechs demonstrated their scalability of solutions to meet the demands of various stakeholders (students, academics, and educational institutions). Their success has meant that they have been retained as key suppliers [22]. The education sector catalysed their growth, enabling EdTechs to pivot into being suppliers for different learning segments and end users [73, 74].

During this period, they also facilitated the expansion of for-profit educational models [22], blurring the boundaries between public and private education provision. We can surmise that states may select market-driven solutions to deregulate control and further support privatisation to supplement or replace traditional services. This has implications for education as institutions prioritise financial viability over the fundamental purpose of educational institutions.

As corporate entities, EdTechs align with the commercial objectives of neoliberal ideology in that they respond to market demands by developing innovative products and solutions that enable institutions to deliver learning and commodify knowledge. Furthermore, they offer numerous commercial models and revenue streams to the higher education sector [70]. Competition between EdTechs themselves for market dominance leads to more innovative development, partnership incentives and models for their customers. This monetisation of knowledge reinforces the perception that education is a consumer good, aligning it with neoliberal ideology and the privatisation agenda of higher education institutions.

EdTechs use several strategies in curating and creating knowledge for commodification, leveraging user-created content via Open Educational Resources (OER), and proprietary content development strategies [35]. EdTechs maintain control over the content they create, customise learning experiences, and align it with their unique value propositions and pedagogical approaches to compete with other EdTechs and HEIs [18]. This sector may itself become a serious competitor to the higher education sector [75].

EdTechs leverage learning algorithms and analytics to customise the user experience, which can be tailored to individual learning needs and abilities, leveraged by UK and Indian EdTech start-ups Potential.ly and CareerGuide.com [76]. This benefits learners who need to adjust options in learning environments for cognitive disabilities, ensuring cognitive maximisation leading to increased engagement [77]. Algorithms are also used to automate interventions when required and challenge higher learners [13], a contrast from higher education, where bureaucratic processes inform relevant departments within the institution to apply interventions. This enables a learner to take control of the learning environment, further reinforcing neoliberalism values of customer empowerment within a market-driven economy.

The fundamental repositioning of higher education towards market principles for institutional viability cannot suppress the concern for access to education, quality, and the broader societal role of universities over commercial interest.

7 Global Accessibility and Scalability of EdTech

EdTechs have emerged as transformative in reshaping the educational landscape through their accessible and scalable solutions. This enables learners to use learning technologies using existing infrastructures (internet, cloud-based access, smart devices). The internet by which these solutions are accessed has enabled these solutions to transcend geographical barriers [78] and socioeconomic constraints, as seen in the increase in the adoption of smart devices in India [70]. This has led to democratising access to education and enabling the underserved and marginalised to take part in education. It also provides access to those facing physical, financial and logistical barriers, increasing inclusion and diversity through its global reach [1, 13]. By providing learners with high-quality, educational content, personalised learning experience and instruction from domain experts.

Most learners access these solutions via online platforms, digital marketplaces, and application stores, for resources and tools developed by global EdTechs and content creators. This global marketplace is based on innovation, inclusion, diversity, and competition, which drives continuous improvement in the range and quality of educational offerings. It should be noted that unlike higher education institutions, many EdTech solutions focus on individual learning and offer periods of free subscriptions for users to trial the content and personalisation before fully committing to the solution [79]. This is unlike in higher education, where tuition fees must be paid to access credit-based academic content.

Unlike HEIs, EdTechs can also scale rapidly as they do not need physical locations or campuses [70]. This removes some barriers that higher education institutions face regarding timetabling, physical resources, and faculty. EdTech replicates the interaction between peers and academics in higher education institutions through their online platforms [80], reproducing the "cohort-based" experiences.

8 Investments Comparisons Between EdTechs and Higher Education

The global EdTech industry has had billions of dollars invested into it from venture capital, private equity investors, and corporate entities, which has had a considerable impact on the growth of this sector [81].

This is supported by recent EdTech success in India, with start-up Byju securing funding of $800 million in 2022 [82]. With global investors supporting this growth and confidence in the demand for digital learning, there is little doubt that investments have driven innovation, entrepreneurship, and market expansion. Investments have enabled EdTech start-ups to develop and scale innovative products and services utilising current and emerging technologies and engage with higher education stakeholders [62].

Whilst the EdTech sector is booming, the global higher education sector is under considerable pressure to deliver education without similar levels of investment. Global economic challenges (recession, pandemic, ageing populations, falling birthrates) for states have stayed the same. While HEIs face increased demand for international education, stagnant tuition fees have meant that state-funded education cannot deliver similar innovation or investment in higher education. Therefore, higher education institutions are increasingly expected to operate as corporate organisations to increase revenue and not be limited by state-set tuition fees. As corporate-run institutions, HEIs may need to consider education offerings to mature students, to economies such as India where the "working-age populace made up more than 68% of the nation's overall population as of 2021" [70], and to digital learners whilst reducing internal costs.

8.1 Implications of Reduced Funding

In contrast to investments in EdTech, there is a trend of decreasing public investment in education in the UK [83] and India [84] spearheaded by state shifts in economic priorities and agendas.

This includes several types of disinvestments, such as reductions in public funding for education, reduction of student financial aid, and the transfer of educational services to private companies. Multiple factors are involved in causing this trend, all pushing governments to reduce their spending on education. The pandemic stretched the fiscal strength of most global states, leading to the adoption of fiscal austerity measures to reduce debt and inflation [85].

Since then, reductions in public funding to reduce fiscal deficits have continued in most global economies. The trend of adopting neoliberal economic policies in higher education has limited public investment in education, deregulating the need for state intervention. With this reduction in investment, educational institutions must maintain quality standards and access to meet increasing student needs [86]. Governments are reluctant to raise taxes for individuals or corporations to cover additional costs of funding the educational sector, but also because of pressure to remain attractive to foreign investment [87].

Furthermore, ageing populations present fiscal challenges for governments in meeting healthcare and pension expenses. Consequently, governments prioritise social welfare programs over education, shifting resources away from schools and student support services. Additionally, heightened demand for healthcare and social services prompted governments to allocate funds away from education to address immediate healthcare needs and financial support packages to businesses (e.g., business loans and VAT deferrals) [88, 89].

Whilst access to EdTech solutions for individuals or HEIs is based on a financial transaction, the reduction in public funding undermines access to higher education. This, in turn, impacts an individual's success and social mobility and long-term impact on a nation's development, potentially setting a cycle for inequality [90]. In the UK the new Lifelong Learning Entitlement launching in 2025 provides funding to learners to access courses until the age of 60, it remains to be seen how many unregistered providers with Office for Students, will be able to provide courses.

9 Pedagogical Concerns

The commodification of higher education raises some significant pedagogical concerns relating to the quality of content, relevance, teaching and learning practices. There is a risk of commercially curated or created content becoming standardised, which risks creativity and limits intellectual inquiry in academia [91]. This may lead to a market where popular and on-trend topics are prioritised over other disciplines, limiting critical thinking skills [92]. Furthermore, the use of algorithms and adaptive learning technology raises concerns about the decline of human-centred pedagogy and the changing role of academics [93]. This leads to issues around academics' privacy, autonomy, and agency in developing and delivering content and may create "depersonalised and decontextualised learning environments" [94]. Critics argue it may incentivise "surface-level learning" and "rote memorisation" rather than deep understanding and critical engagement with course material [95].

Although retention and engagement metrics are used in higher education, EdTechs could increase their focus on completion and engagement metrics that may prioritise short-term outcomes over long-term educational objectives [13]. This shift risks undermining the development of intellectual skills [49].

9.1 Equitable Access

The commodification of knowledge in higher education may affect access and equity for the student demographic. While it has the potential to widen access via digital platforms and mobile applications, it presents new challenges. As previously discussed, it can intensify the disparities and unfairness that higher education institutions seek to eliminate.

During the pandemic, when digital delivery became the norm, it highlighted the inequalities in device poverty and internet connectivity across the different socioeconomic layers of society in the UK and India. This was particularly noticeable in rural areas, which affected access to education. Ofcom listed three aspects that increased digital exclusion during the pandemic and further "entrenched the digital divide". These were lack of access to the internet, the lack of digital skills and affordability for internet access [26].

Despite there being national initiatives in both countries to reduce these issues, e.g., "National Programme on Technology Enhanced Learning, Swayam Programme, DIKSHA" [70], and in the UK—Data Poverty All-Party Parliamentary Group, the

Table 1 Obstacles in India and the UK, adapted from [9, 26, 27]

Obstacles in India	Obstacles in UK
Electricity—power cut-offs and with reduced hours	Connection speed in rural areas
Smart devices are owned by less than 30% of the population	1 in 5 children in 2021 did not have suitable digital equipment/devices for learning
Internet penetration is low in rural areas	Contract affordability
Speed and connectivity issues	Financially excluded individuals
Insufficient funds to improve infrastructure	Low digital literacy
Medium of language	
Lack of skilled teachers in technology and resistance to change	
Examination oriented assessment	
Crowded EdTech space	

Digital Poverty Alliance, and the Good Things Foundation's Data Poverty Lab [96] (Table 1).

In Argentina this resulted in students accessing education via "non-edtech-tech", using WhatsApp [97]. It is, therefore, not difficult to surmise that non-affluent learners may not be able to access digital learning, and that affordability may become an issue to access premium content from subscription-based models.

Students from affluent backgrounds may benefit disproportionately from personalised learning algorithms, adaptive learning software, and other EdTech innovations, while students from disadvantaged backgrounds struggle to access essential educational resources. As a result, commodification can reinforce existing inequalities in educational attainment and socio-economic status. These issues further the gap between privileged and marginalised groups and support critics' view of neoliberal ideology [90].

9.2 *Quality and Accountability*

Higher education has robust quality assurance processes in the UK and India, overseen by the Office for Students (OfS) and University Grants Commission (UGC). Still, critics of privatisation contend that organisation efficiency processes may compromise the quality of education. Ball and Grimaldi argue that the emphasis on market-driven approaches often result in HEIs adopting business practices that prioritise cost-cutting and efficiency over educational outcomes which detract from the quality of education provided [98].

The proliferation of private higher education providers poses another significant challenge. Private providers often operate with greater autonomy than public institutions, which are subject to stringent regulatory oversight, leading to inconsistencies in educational standards. The lack of a centralised regulatory body overseeing multiple private institutions can result in varied quality of education across providers, although in India private institutions have less autonomy [99].

While privatisation in higher education aims to introduce efficiency and innovation, it also brings significant risks related to maintaining educational quality. The lack of comprehensive regulatory oversight further complicates the assurance of consistent educational standards across institutions. Therefore, this requires discussions between regulatory stakeholders, public and private institutions and EdTechs.

10 Risks from EdTechs

Educational technology has provided numerous benefits to all higher education stakeholders. However, there are some critical risks associated with EdTechs and their technologies. HEIs must conduct due diligence when incorporating EdTech solutions and or entering into partnerships with them to protect their brand and reputation, particularly when responding to a national emergency (e.g., pandemic).

The rapid emergence of new EdTech providers raises concerns about their long-term sustainability. Many start-ups in the education technology sector may need more stability or longevity to ensure continuous support and development of their products for their customers. This can lead to disruptions in the learning process if a company fails or pivots away from its initial mission [100], due to pressure from investors and or the market. This instability can undermine the trust and reliability essential for educational tools, as profit and growth may be the primary objectives of EdTechs. The market is also saturated with EdTechs, making it difficult for academic institutions to choose reliable partners, e.g., India has nearly 4500 EdTech start-ups [101].

EdTech solutions collect vast amounts of learner data, which can be utilised to personalise and improve educational interventions [102]. However, this raises ethical concerns about using student data and the potential for misuse, privacy violations and data security [103]. Furthermore, the algorithms used in these technologies can perpetuate existing biases, leading to unequal educational opportunities and outcomes [103]. Ensuring these systems are transparent, equitable, meet current legislation on data and security is crucial to HEIs' due diligence process to mitigate their liability in potential breaches.

11 Device Poverty and Digital Exclusion

Many of the issues relating to neoliberalism and the commodification of knowledge have been discussed. Analysing the two countries' responses makes it possible to detect concerns raised by academics and the benefits of digital education delivery, despite fears of EdTechs further commodifying education. Whilst the COVID pandemic was a global disaster, there may never be another opportunity to disrupt, test and validate the rapid mass delivery of digital education using business and specialised EdTech solutions. This period also highlighted the impact of digital poverty and exclusion in both countries.

In March 2020, the UK and India announced the closure of all education facilities, rapidly moving education online [104]. HEIs in both countries resorted to video communication solutions (e.g., Zoom, Google, Teams), already in institutions, to deliver online teaching to students [10]. The move also prompted academics to support their students by developing and updating teaching content (e.g., creating video lectures) for digital delivery that could be accessed on demand [10]. Furthermore, both countries introduced initiatives to help with technologies, services, and accessibility, e.g., "National Programme on Technology Enhanced Learning (NPTEL), Study Web for Active Young Expiring Minds (SWAYAM), e-Pathshala, DIKSHA portal, SWAYAM Prabha, National Repository of Open Educational" [70], and in the UK Get Help with Technology and Higher Education Covid-19 Operational Guide [105, 106]. An overview of the assistance provided is summarised in Table 2.

Despite the assistance provided by the states in both nations, inequity, inaccessibility and digital poverty/exclusion impacted learning [26, 102]. This was further impacted by national infrastructures and disparity in internet access, which are issues outside of the control of HEIs. Notably, wealth is directly correlated to access to education, perpetuating social disparities within social classes supporting the view held by academics against neoliberal ideology.

Furthermore, Kumar Das highlighted cultural issues where girls and vulnerable groups were coerced into "dropping" education in part because of digital delivery and due to device poverty [108]. This leads to another factor in which equity is measured in terms of "availability of electricity" and "internet connectivity" [109], again disproportionately affecting those with lower income and from rural settings. This is supported by the lack of suitable connectivity, electrical power and device poverty as the main obstacles to digital learning, despite the high use of information technology in education [27]. In India, where access is available, the language of instruction in EdTech solutions can also be a barrier [70]. At the same time, digital literacy was a barrier in the UK, along with those choosing to stay offline due to low incomes [26, 96].

12 Conclusion

In conclusion, EdTechs have accelerated the neoliberal agenda by speeding up the commodification of the knowledge process in higher education in both the UK and India by "hardening" neoliberal practices [110]. The pandemic accelerated this transformation towards market-driven principles, where education is seen as a commodity

Table 2 Overview of UK and Indian response to pandemic—adapted from [22, 66, 70, 82, 89, 105, 107, 108]

Impact area	India	UK
Access to higher education	Increased access through online platforms like SWAYAM, Coursera, and edX; universities rapidly moved to online delivery	Expanded access via platforms like FutureLearn and Open University; rapidly moving to online and blended learning models across universities
Digital divide	Initiatives by the state: PM eVIDYA and NDLI aimed to address the digital divide	State initiatives like "Get Help with Technology" and "Digital Skills Bootcamps" aimed to reduce digital divide
Quality of education	Platforms like SWAYAM and NPTEL provided structured services	Universities used business platforms (Zoom, Microsoft Teams, and Moodle) to maintain standards
Affordability	Online resources reduced costs for many students as many platforms provided free courses e.g., Coursera provided support	Financial aid and free resources through platforms like Open Learn and university support funds helped students manage costs
Educator training and development	Platforms provided courses for effective online teaching methods e.g., SWAYAM	Accelerated professional development through platforms like Advance HE and targeted online teaching support by universities
Curriculum standardisation	Aligned with UGC and AICTE guidelines for online courses; use of platforms like SWAYAM for standardised content	Alignment with QAA and OfS standards; universities adapted curricula for online delivery using platforms like Blackboard
EdTech adoption	Universities are rapidly adopting EdTech increasing the use of platforms like Zoom, Google Classroom, and Microsoft Teams for remote learning	There has been a significant uptake of EdTech; widespread use of Zoom, Microsoft Teams, and learning management systems like Moodle and Blackboard
Government policies	National Education Policy 2020 and UGC guidelines supported online learning; emergency measures for remote education	Government guidelines promoted remote learning; policies supported the continuation of education through online platforms
Innovative teaching methods	Adoption of AI, VR, and gamified learning increased; platforms like BYJU and Vedantu implemented innovative teaching methods	There has been increased use of VR, AR, and AI; platforms like Immersive VR Education and online web platforms provided innovative teaching solutions

[60]. EdTechs have played a pivotal role by providing scalable and innovative solutions that enable the delivery of educational content to the broader market of learners, with multiple revenue-generating business models [12, 22].

Academics have raised significant concerns regarding the commodification of education, highlighting issues for access, equity, and quality of content [57]. The argument is that commodification prioritises profit over educational objectives, transforming education into a commodity to be bought and sold in the academic marketplace [49]. This market-driven approach can undermine the intrinsic value of education as a public good, emphasising economic outcomes and the development of human capital for employability purposes rather than for intellectual curiosity [35].

It also raises ethical concerns over collecting and using student data for commercial purposes, which may be used to further product development and profit-driven objectives. This increases privacy, consent, and data ownership issues [103], potentially lead to stakeholder conflict.

Advocates defend commodification by arguing that technology enhances learning, increases access to education, and promotes innovation in teaching and assessment. EdTech start-ups leverage digital technologies, data analytics, and artificial intelligence to personalise learning, adapting to individual learner needs, and improve student engagement [1]. At the same time, they can create in-class cohort experiences for geographically dispersed learners [80], who may not be able to attend physically.

The commodification of knowledge offers financial benefits for HEIs and increased accessibility for learners. However, it also presents challenges, including exacerbating existing inequalities between social groups [90], compromising educational quality, and undermining the traditional role of education. The market-driven approach prioritises efficiency, profitability, and consumer choice, often at the expense of equity and academic integrity [98].

Higher education institutions have responded to these challenges by integrating EdTech innovations into their systems, forming strategic partnerships, and implementing policies to safeguard academic standards and ethical values [28, 29]. The global accessibility and scalability of EdTechs have democratised access to education [13], particularly in underserved regions. Still, they have highlighted the digital divide and issues of affordability in both the UK and India [9, 26, 27]. As the landscape of higher education continues to evolve, stakeholders need to navigate these complexities thoughtfully, ensuring that the benefits of technological advancements and market principles are balanced with the fundamentals of equity, access, and quality and that education does not leave behind the notion of public good [62].

The future of higher education will likely see a continued emphasis on employability and lifelong learning (Bhagwat, 2024), driven by neoliberalism and EdTech innovations. As this sector develops, discussions between stakeholders and EdTechs will need to be held to address the challenges and opportunities presented by the commodification of knowledge.

Disclosure of Interests The author has no competing interests to declare that are relevant to the content of this article.

References

1. Laufer M et al (2021) Digital higher education: a divider or bridge builder? Leadership perspectives on edtech in a COVID-19 reality. Int J Educ Technol High Educ 18(1). https://doi.org/10.1186/s41239-021-00287-6
2. Parra M, Marambio C, Ramírez J, Suárez D, Herrera H (2020) Educational convergence with digital technology: integrating a global society. In: HCI international 2020—late breaking posters, pp 303–310
3. Acton R (2023) Doing knowledge work differently: problem-based projects as encounters in coming-to-know. High Educ 86(1):225–242. https://doi.org/10.1007/s10734-022-00910-z
4. Winch C, Hugo W, Mcqueen R (2019) Book review teachers' know-how: a philosophical investigation. J Educ 77 [Online]. Available: https://journals.ukzn.ac.za/index.php/joe/article/view/1473/1605
5. Sakhiyya Z, Rata E (2019) From 'priceless' to 'priced': the value of knowledge in higher education. Glob Soc Educ 17(3):285–295. https://doi.org/10.1080/14767724.2019.1583089
6. Kitto K (2022) How can EdTech support graduate employability? ASCILITE Publ:e22184. https://doi.org/10.14742/apubs.2022.184
7. Williams G (2016) Higher education: public good or private commodity? London Rev Educ 14(1):131–142. https://doi.org/10.18546/LRE.14.1.12
8. Yadav K (2019) Disruptive innovative technologies in higher. Int J Adv Educ Res Int 4(2):43–54
9. Azad SM (2022) The rise of EdTech start-ups on India. Recent Trends Manag Soc Sci 2(2):24–32 [Online]. Available: https://www.researchgate.net/publication/358939954
10. Camilleri MA, Camilleri AC (2022) The acceptance of learning management systems and video conferencing technologies: lessons learned from COVID-19. Technol Knowl Learn 27(4):1311–1333. https://doi.org/10.1007/s10758-021-09561-y
11. Massner CK (2021) The use of videoconferencing in higher education. Intech 11(tourism):13 [Online]. Available: https://www.intechopen.com/books/advanced-biometric-technologies/liveness-detection-in-biometrics
12. Komljenovic J (2021) The rise of education rentiers: digital platforms, digital data and rents. https://doi.org/10.1080/17439884.2021.1891422
13. Komljenovic J, Hansen M, Sellar S, Birch K (2024) Edtech in higher education: empirical findings from the project 'Universities and unicorns: building digital assets in the higher education industry'. Center for Global Higher Education University of Oxford [Online]. Available: https://kclpure.kcl.ac.uk/portal/en/publications/edtech-in-higher-education-empirical-findings-from-the-project-un
14. Ofs, Securing student success: regulatory framework for higher education in England 24, no. November, 2022
15. Dianti Y (2017) Guidelines for development plans for HEIs. Angew Chemie Int Ed 6(11):951–952, 5–24 [Online]. Available: http://repo.iain-tulungagung.ac.id/5510/5/BAB2.pdf
16. Vicentini L, Day L, Gill V, Lillis J, Komers S (2022) Future opportunities for education technology in England. Department of Education, no. June [Online]. Available: https://eric.ed.gov/?id=ED624345%0A, https://files.eric.ed.gov/fulltext/ED624345.pdf
17. Khanna P, Sehgal R, Malviya M, Khanna A, Dubey AM (2022) EdTech start-ups: a mode of transformation in teaching and learning. In: Indian SMEs and start-ups, vol 13. World Scientific, pp 177–202
18. Huang R (2022) A guideline framework of hybrid education, learning, and assessment (HELA) version July 1, 2022
19. AlNajdi S (2014) Hybrid learning in higher education. In: Society for information technology & teacher education international conference, pp 214–220
20. Retention of Learners, Digital learning and digital institution in higher education. Educ Sci 13(1). https://doi.org/10.3390/educsci13010088

21. Otto S, Bertel LB, Lyngdorf NER, Markman AO, Andersen T, Ryberg T (2024) Emerging digital practices supporting student-centered learning environments in higher education: a review of literature and lessons learned from the Covid-19 pandemic. Educ Inf Technol 29(2):1673–1696. https://doi.org/10.1007/s10639-023-11789-3
22. Williamson B, Hogan A (2021) Pandemic privatisation in higher education. Edtech & University Reform, no. February
23. DaSilva CM, Trkman P, Desouza K, Lindič J (2013) Disruptive technologies: a business model perspective on cloud computing. Technol Anal Strateg Manag 25(10):1161–1173. https://doi.org/10.1080/09537325.2013.843661
24. Aydin H (2021) A study of cloud computing adoption in universities as a guideline to cloud migration. SAGE Open 11(3). https://doi.org/10.1177/21582440211030280
25. Townsend L, Sathiaseelan A, Fairhurst G, Wallace C (2013) Enhanced broadband access as a solution to the social and economic problems of the rural digital divide. Local Econ 28(6):580–595. https://doi.org/10.1177/0269094213496974
26. Ofcom (2022) Digital exclusion A review of Ofcom's research on digital exclusion among adults in the UK, March, pp 1–21 [Online]. Available: https://www.ofcom.org.uk/__data/assets/pdf_file/0022/234364/digital-exclusion-review-2022.pdf
27. Biswas N (2021) Role of information technology in education: obstacles in EdTech in India-a study. GRD J Eng 6(10):14–19 [Online]. Available: www.grdjournals.com
28. Ankrah S, AL-Tabbaa O (2015) Universities-industry collaboration: a systematic review. Scand J Manag 31(3):387–408. https://doi.org/10.1016/j.scaman.2015.02.003
29. Bakırlıoğlu Y, McMahon M (2021) Co-learning for sustainable design: the case of a circular design collaborative project in Ireland. J Clean Prod 279. https://doi.org/10.1016/j.jclepro.2020.123474
30. UNESCO (2020) A new generation: 25 years of efforts for gender equality in education
31. Saleem Mazhar M, Goraya NS (2020) Afghan women education: bottlenecks & future. A Res J South Asian Stud 35(1):47–72 [Online]. Available: https://www.unicef.org/afghanistan/education
32. Dias D, Soares D (2018) Civic learning outcomes: a step towards an inclusive higher education. Int J Incl Educ 22(4):360–374. https://doi.org/10.1080/13603116.2017.1365956
33. Johnson MT (2022) The knowledge exchange framework: understanding parameters and the capacity for transformative engagement. Stud High Educ 47(1):194–211. https://doi.org/10.1080/03075079.2020.1735333
34. Sattiraju V, Ligade VS, Muragundi P, Pandey R, Janodia MD (2023) National and higher education institutions (HEIs) IP policies: comparison of Indian HEIs' IP policies from a global perspective. J Knowl Econ 14(2):1979–2006. https://doi.org/10.1007/s13132-022-00915-0
35. Sharma H (2022) Mapping the global EdTech revolution during the pandemic: from 'determinism' to 'solutionism. In: Re-imagining Educational Futures in Developing Countries: Lessons from Global Health Crises, pp 119–137. https://doi.org/10.1007/978-3-030-88234-1_7
36. Chankseliani M, Qoraboyev I, Gimranova D (2021) Higher education contributing to local, national, and global development: new empirical and conceptual insights. High Educ 81(1):109–127. https://doi.org/10.1007/s10734-020-00565-8
37. Belchior-Rocha H, Casquilho-Martins I, Simões E (2022) Transversal competencies for employability: from higher education to the labour market. Educ Sci 12(4). https://doi.org/10.3390/educsci12040255
38. Durazzi N (2021) Opening universities' doors for business? Marketization, the search for differentiation and employability in England. J Soc Policy 50(2):386–405. https://doi.org/10.1017/S0047279420000276
39. Hewison J (2020) Student academic experience survey 2020—Jonathan Neves (Advance HE) and Rachel Hewitt (HEPI) [Online]. Available: https://policycommons.net/artifacts/2095951/student-academic-experience-survey-2020/

40. Baum S, Ma J, Payea K (2013) Education pays 2013: the benefits of higher education for individuals and society. Trends High Educ Ser:1–48 [Online]. Available: http://www.dev.skylinecollege.edu/sparkpoint/about/documents/education-pays-2013-full-report.pdf
41. Grossman M (2017) Determinants of health. In: An economic perspective. Columbia University Press, pp 42–110
42. Lochner L (2020) Education and crime. Econ Educ A Compr Overv:109–117. https://doi.org/10.1016/B978-0-12-815391-8.00009-4
43. Visaria L (2022) India's date with second demographic transition. China Popul Dev Stud 6(3):316–337. https://doi.org/10.1007/s42379-022-00117-w
44. Adams R (2023) Primary schools in England close, merge and shrink as pupil numbers fall. The Guardian
45. Department for Education (2023) Budget 2023: what are 'returnerships' and who are they for?
46. Bhagwat M (2024) These 5 key trends will shape the EdTech market upto 2030. World Economic Forum. https://www.weforum.org/agenda/2024/02/these-are-the-4-key-trends-that-will-shape-the-edtech-market-into-2030/. Accessed 8 May 2024
47. Thorsen DE (2009) The Neoliberal challenge: what is neoliberalism? pp 1–25
48. Mintz B (2021) Neoliberalism and the crisis in higher education: the cost of ideology. Am J Econ Sociol 80(1):79–112. https://doi.org/10.1111/ajes.12370
49. Younis AA-H, Hatim M (2021) Trends in higher education under neoliberalism: between traditional education and the culture of globalization. Educ Res Rev 16(2):16–26. https://doi.org/10.5897/err2020.4101
50. Spruin BE (2021) Creating a competitive advantage in the UK higher education sector through mirrored course delivery: a case example. October
51. Bray M (1995) Education and development in Asia. 15(2)
52. Saravanakumar AR, Padmini Devi KR (2020) Indian higher education: issues and opportunities. J Crit Rev 7(2):542–545. https://doi.org/10.31838/jcr.07.02.101
53. Aithal PS, Aithal S (2020) Promoting faculty and student-centered research and innovation based excellence model to Reimage universities. Int J Manag Technol Soc Sci:24–42. https://doi.org/10.47992/ijmts.2581.6012.0081
54. Omerzel DG, Biloslavo R, Trnavcevic A (2011) Knowledge management and organisational culture in higher education institutions. JEEMS 16(2):111–139. http://www.jstor.c
55. Marginson S, Yang L (2024) Has the public good of higher education been emptied out? The case of England. High Educ 88(1):297–319. https://doi.org/10.1007/s10734-023-01117-6
56. Dorhout PK (2016) Attracting and retaining competitive faculty—startups, core facilities, and investment strategies...Oh My!. Merrill Ser Res Mission Public Univ:105–112. https://doi.org/10.17161/merrill.2016.7778
57. Williamson B (2021) Making markets through digital platforms: Pearson, edu-business, and the (e)valuation of higher education. Crit Stud Educ 62(1):50–66. https://doi.org/10.1080/17508487.2020.1737556
58. Sultan N (2019) The disruption and global implications of massive open online courses (MOOCs) for higher education. In: Faghih N (ed) Globalization and development: economic and socio-cultural perspectives from emerging markets. Springer International Publishing, Cham, pp 201–215
59. Hunt SA, Boliver V (2023) The private higher education provider landscape in the UK. Stud High Educ 48(9):1346–1360. https://doi.org/10.1080/03075079.2023.2199317
60. Wingard J, Farrugia C (2018) Market-driven education 7. In: The 60-year curriculum, Routledge, pp 104–118
61. Ruiz-Mallén I, Heras M (2020) What sustainability? Higher education institutions' pathways to reach the Agenda 2030 goals. Sustainability 12(4). https://doi.org/10.3390/su12041290
62. Komljenovic J, Birch K (2023) Strengthening EdTech in higher education: policy recommendations and principles 1.0, no June, 2023
63. Smolentseva A (2023) The contributions of higher education to society: a conceptual approach. Stud High Educ 48(1):232–247. https://doi.org/10.1080/03075079.2022.2130230

64. Morris NP, Ivancheva M, Coop T, Mogliacci R, Swinnerton B (2020) Negotiating growth of online education in higher education. Int J Educ Technol High Educ 17(1). https://doi.org/10.1186/s41239-020-00227-w
65. Decuypere M, Grimaldi E, Landri P (2021) Introduction: critical studies of digital education platforms. Crit Stud Educ 62(1):1–16. https://doi.org/10.1080/17508487.2020.1866050
66. Bayne S, Gallagher M (2021) Near future teaching: practice, policy and digital education futures. Policy Futur Educ 19(5):607–625. https://doi.org/10.1177/14782103211026446
67. Fewella LN, Khodeir LM, Swidan AH (2021) Impact of integrated e-learning: traditional approach to teaching engineering perspective courses. Int J Eng Pedagog 11(2):82–101. https://doi.org/10.3991/IJEP.V11I2.17777
68. Wu J, Ping L, Ge X, Ya W, Fu J (2010) Cloud storage as the infrastructure of cloud computing. In: Proceedings—2010 international conference on intelligent computing and cognitive informatics, ICICCI 2010, pp 380–383. https://doi.org/10.1109/ICICCI.2010.119
69. Arpaci I (2016) Understanding and predicting students' intention to use mobile cloud storage services. Comput Human Behav 58:150–157. https://doi.org/10.1016/j.chb.2015.12.067
70. Bargavi R, Shanmugam K (2023) EdTech industry in India: revolution and challenges in the Indian market: teaching case study. J Inf Technol Teach Cases:1–10. https://doi.org/10.1177/20438869231189526
71. Komljenovic J, Williamson B, Williamson B (2024) Behind the platforms: safeguarding intellectual property rights and academic freedom in Higher Education [Online]. Available: https://www.ei-ie.org/en/item/28484:behind-the-platforms-safeguarding-intellectual-property-rights-and-academic-freedom-in-higher-education
72. Regele MD (2020) Pedagogy and profit? Efforts to develop and sell digital courseware products for higher education. Am Educ Res J 57(3):1125–1158. https://doi.org/10.3102/0002831219869234
73. Cowling MA, Crawford J, Vallis C, Middleton R, Sim KN (2022) The EdTech difference: digitalisation, digital pedagogy, and technology enhanced learning. J Univ Teach Learn Pract 19(2):1–13. https://doi.org/10.53761/1.19.2.1
74. Pinheiro R, Tømte CE, Barman L, Degn L, Geschwind L (2023) Digital transformations in Nordic higher education
75. Rof A, Bikfalvi A, Marques P, Rof A, Bikfalvi A, Marques P (2022) International forum of educational technology & society pandemic-accelerated digital transformation of a born digital higher education linked references are available on JSTOR for this article : Pandemic-accelerated Digital Transformation of a Born Digita 25(1)
76. Jäger T, Falk M, Lenz S (2021) Innovative business models for higher education: an exploratory analysis on education technology start-ups in selected countries IHF working paper 1/2021 innovative business models for higher education an exploratory analysis on education technology St
77. Standen PJ et al (2020) An evaluation of an adaptive learning system based on multimodal affect recognition for learners with intellectual disabilities. Br J Educ Technol 51(5):1748–1765. https://doi.org/10.1111/bjet.13010
78. City S et al (2024) Borderless learning environments. Int Multidiscip J Res Innov Sustain Excell 1(2):43–49
79. Satish Dingre S (2022) Understanding user journeys in EdTech startups: the role of analytics integration. J Mark Supply Chain Manag 1(1):1–3. https://doi.org/10.47363/jmscm/2022(1)123
80. Veletsianos G, Houlden S (2023) On the 'university of the future": a critical analysis of cohort-based course platform Maven. Learn Media Technol. https://doi.org/10.1080/17439884.2023.2164862
81. Williamson B, Komljenovic J (2023) Investing in imagined digital futures: the techno-financial 'futuring' of edtech investors in higher education. Crit Stud Educ 64(3):234–249. https://doi.org/10.1080/17508487.2022.2081587
82. Soulunii S (2022) BYJU'S raises $800M in funding led by founder Raveendran at $22B valuation

83. Drayton E, Farquharson C, Ogden K, Sibieta L, Tahir I, Waltmann B (2023) Institute for fiscal studies annual report on education spending in England
84. Tobenkin D (2024) India's higher education landscape. https://www.nafsa.org/ie-magazine/2022/4/12/indias-higher-education-landscape. Accessed 13 May 2024
85. Murgatroyd S (2019) Preparing for a different future-Learning and technology in an age of disruption. Paraguayan J Distance Educ 1(1):27–40 [Online]. Available: https://www.researchgate.net/publication/332410011_Preparing_for_a_Different_Future_-_Learning_in_an_Age_of_Disruption
86. QAA (2018) UKSCQA UK standing committee for quality assessment the revised UK quality code for higher education, no March, pp 1–5 [Online]. Available: www.ukscqa.org.uk
87. Öz-Yalaman G (2020) The effect of corporate tax on foreign direct investment. Erciyes Üniversitesi İktisadi ve İdari Bilim. Fakültesi Derg 55:221–244. https://doi.org/10.18070/erciyesiibd.597602
88. Moretti D, Leroy A (2020) Government financial management and reporting in times of crisis. OECD Policy Responses to Coronavirus, no. November, pp 1–14, 2020 [Online]. Available: https://www.oecd.org/coronavirus/policy-responses/government-financial-management-and-reporting-in-times-of-crisis-3f87c7d8/
89. Sibieta L, Cottell J (2020) Education Policy Responses across the uk to the pandemic. Educ. Policy Inst. 4(1):9–15
90. Balan A (2023) Neoliberalism, privatisation and marketisation: the implications for legal education in England and Wales. Cogent Educ 10(2). https://doi.org/10.1080/2331186X.2023.2284548
91. Shelton C, Schroeder S, Curcio R (2024) The teacher-to-teacher online marketplace of ideas: a new future in curriculum? Vol 1. INC
92. Fessel S (2006) The impact of academic freedom policies on critical thinking instruction. InSight A J Sch Teach 1:51–58. https://doi.org/10.46504/01200604fe
93. Chrisley R (2020) A human-centered approach to AI ethics: a perspective from cognitive science. In: The Oxford handbook of ethics of AI. Oxford University Press, 9 Jul, 2020. https://doi.org/10.1093/oxfordhb/9780190067397.013.29
94. Warren-Lee N, Smith L, Orchard J, Kelly L, James J, Coles A (2024) Bracing ourselves: embracing policy changes through a Long-Standing University–schools teacher education partnership in England. Educ Sci 14(2). https://doi.org/10.3390/educsci14020158
95. Li L, Ismail SM, Patra I, Lami D (2024) Not a passive learner but an active one: a focus on the efficacy of philosophy-based language instruction and its consequences on EFL learners' critical thinking, engagement, and academic achievement. BMC Psychol 12(1):1–19. https://doi.org/10.1186/s40359-024-01648-2
96. Faith B, Hernandez K, Beecher J (2022) Digital poverty sets up a vicious circle: those without digital connectivity lose out and become financially poorer. Digital Poverty in the UK, no. 202
97. Williamson B, Macgilchrist F, Potter J (2021) Covid-19 controversies and critical research in digital education. Learn Media Technol. https://doi.org/10.1080/17439884.2021.1922437
98. Ball SJ, Grimaldi E (2022) Neoliberal education and the neoliberal digital classroom. Learn Media Technol 47(2):288–302. https://doi.org/10.1080/17439884.2021.1963980
99. Raza R (2009) Examining autonomy and accountability in public and private tertiary institutions. Hum Dev Netw World Bank (November):5–24
100. Williamson B (2019) Williamson Big Data education, Chap 1 . Big Data Educ., no. September 2017, p 23 [Online]. Available: https://www.researchgate.net/publication/332496698_Williamson_Big_Data_Education_Chapter1_2017/link/5cb84db0299bf120976d3ffe/download
101. Sarma A (2024) Edtech in India: boom, bust, or bubble? https://www.orfonline.org/expert-speak/edtech-in-india-boom-bust-or-bubble. accessed 17 May 2024
102. Akintayo OT, Eden CA, Ayeni OO (2024) Evaluating the impact of educational technology on learning outcomes in the higher education sector: a systematic review. Int J Manag Entrep Res 6(5):1395–1422. https://doi.org/10.51594/ijmer.v6i5.1091

103. Regan PM, Jesse J (2019) Ethical challenges of EdTech, big data and personalized learning: twenty-first century student sorting and tracking. Ethics Inf Technol 21(3):167–179. https://doi.org/10.1007/s10676-018-9492-2
104. Spanoudaki M, Otermans P, Aditya D (2023) Individuals' inner wellbeing during the COVID-19 pandemic: a quantitative comparison of social connections and close relationships between the UK and India. World Acad Sci Eng Technol Int J Psychol Behav Sci 17(1):1–7
105. Department for Education (2022) Higher education: operational guidance, pp 1–22 [Online]. Available: https://assets.publishing.service.gov.uk/media/62f121ad8fa8f50335b35e50/_WITHDRAWN__20220223_HE_guidance.pdf
106. Department for Education (2022) Get help with technology for remote education—GOV.UK. https://www.gov.uk/government/collections/get-help-with-technology-for-remote-education. Accessed 19 May 2024
107. Selwyn N (2021) Education and technology, vol 2
108. Kumar Das P (2021) Impact of pandemic Covid-19 on higher education-Indian context. Trends J Sci Res 1(1):13–21. https://doi.org/10.31586/ujbm.2021.010102
109. Nag RK (2022) Is India ready to accept an EdTech-intensive system in post pandemic times? A strategic analysis of India's 'readiness' in terms of basic infrastructural support. Decision 49(2):253–261. https://doi.org/10.1007/s40622-022-00313-0
110. Watermeyer R et al (2021) 'Pandemia': a reckoning of UK universities' corporate response to COVID-19 and its academic fallout. Br J Sociol Educ 42(5–6):651–666. https://doi.org/10.1080/01425692.2021.1937058

Preserving User Autonomy in Smart Home Data Collection: Solutions and Ethical Responsibilities

Alexander Khalil Daou and Tasmina Islam

Abstract With the increasing ubiquity of Internet of Things (IoT) devices and the rapid rise of smart home environments, users are sharing more data than ever before. Though the adoption of smart home devices may bring many convenience benefits to the end user, the cost to personal autonomy should be considered. This paper discusses the means in which smart home user personal autonomy may best be preserved, and presents arguments for who should be ethically responsible for said preservation. Informed consent, user control, and legislation are identified as key factors integral to the preservation of user autonomy. The paper concludes that both the user and vendor are ethically responsible for preserving user autonomy from a consequentialism perspective, and that the government bears this responsibility from a deontology perspective.

Keywords Smart home · Privacy · IoT · Internet of Things · Information ethics · Autonomy

1 Introduction

The Internet of Things (IoT) has seen rapid growth over the past decade, and one major contributing factor is the large-scale adoption of smart homes. With the advent of ever-connected devices automating menial day-to-day domestic tasks, smart home environments are now more common than ever. From 2018 to 2023 alone, smart home usage has increased by 121% [1]. Alongside the benefits of such an ecosystem come a plethora of ethical concerns. Smart home devices collect sensory information in the user's most intimate environment, essentially digitising their life and transmitting it to the vendor's servers, where it may be processed further. Without sufficient

A. K. Daou (✉) · T. Islam
Department of Informatics, King's College London, London, UK
e-mail: alexanderdaou@outlook.com

T. Islam
e-mail: tasmina.islam@kcl.ac.uk

H. Jahankhani and B. Issac (eds.), *Cybersecurity and Human Capabilities Through Symbiotic Artificial Intelligence*, Advanced Sciences and Technologies for Security Applications, https://doi.org/10.1007/978-3-031-82031-1_26

awareness and available controls, smart home user personal autonomy is at great risk of subversion.

Whilst there is plenty of work researching smart home device security, there is an absence of research investigating the preservation of smart home user autonomy. This work uniquely contributes to the field by identifying some of the most effective means of preserving the personal autonomy of smart home users, and proposing arguments for who is ethically responsible for said preservation.

The remainder of the paper is structured as follows: Section 2 presents the literature review. Section 3 declares the research questions. Section 4 describes the methodology. Section 5 is the analysis. Section 6 offers the discussion. Section 7 concludes the paper.

2 Literature Review

2.1 The Sensitive Nature of Smart Home Data

In their paper detailing privacy and security challenges in smart homes, Bugeja et al. [2] describe how a smart home will naturally process a variety of sensitive information, such as personal photos, videos, or documents. The larger the variety of smart home devices connected to the home network, the larger the variety of personal information at risk. The paper references always-on smart home personal assistants (SPAs) such as Amazon Echo and Google Home, that are constantly listening through their built-in microphones. SPAs are amongst the most popular smart home devices due to their versatility and ease of use. They provide efficient automation of everyday tasks in a convenient manner. However, they also process some of the most sensitive information. Edu et al. [3] describe how the always-on nature of SPAs can result in the inadvertent leakage of sensitive, personal conversations. Their paper highlights an example in which an Amazon Echo recorded a couple's personal conversation and sent it to a registered contact, unbeknownst to the couple. In addition to constantly listening in on their users, SPAs are often integrated in the smart home environment to control other devices, exposing them to additional information. Apthorpe et al. [4] reinforce the concern of data leakage from smart home devices by demonstrating that sensitive data can be inferred even from encrypted IoT device traffic. In their paper, the authors detail an experiment in which they successfully illustrate how a passive network observer may obtain sensitive user information by examining the metadata and network patterns of encrypted smart home traffic. This was demonstrated against four commonly used smart home devices. To highlight one example, remote analysis of sleep monitor traffic allowed the authors to infer the sleep schedule of a user.

mind and by default [26]. This helps mitigate user oversight, and ensures less effort is required from the user in order to secure the device and minimise data exposure.

The final factor proposed is legislation. Though extrinsic to the smart home device and vendor, legislation can serve as a precursor to device design and vendor policies and procedures. This is exemplified in Sect. 5.3 with a brief analysis of the EU's GDPR and how its edicts translate to the preservation of smart home user autonomy. A government mandate is a guaranteed way to ensure smart home vendors comply with measures aimed at preserving user autonomy. As mentioned in Sect. 5.3, smart home vendors may be inclined to maximise data collection, and the end user is powerless to stop this without any legal backing. The only alternative would be to simply not use the device in question. A regulation such as GDPR would force vendors to embed autonomy-preserving measures into their product design and data handling, or risk incurring severe penalties. Such legislation is effective countrywide, or even across multiple countries in the case of GDPR, establishing widespread coverage that benefits large populations. This also means that smart home vendors from anywhere in the world who want to sell their products in a covered country must comply with these legal requirements, resulting in a global impact that encourages normalisation of such measures. A recent example of regional legislation that has prompted a global technology change is the EU's Directive 2022/2380 [29], which forced Apple Inc. to retire their proprietary Lightning computer bus in favour of USB-C. Though not all GDPR principles may be fully realised in a smart home environment [27], the regulation has demonstrated the pivotal role government bodies play in the protection of user autonomy and privacy rights, and in some cases also raised the awareness of end users directly [28].

6.2 RQ2: Who Is Ethically Responsible for Preserving the Personal Autonomy of Smart Home Users?

From a consequentialism perspective, it is argued that the ethical responsibility for personal autonomy preservation is shared between the user and the vendor, but with greater responsibility placed on the vendor. There will always be an effort required from the user to maintain their autonomy. The user must at least have the intention of minimising their data exposure, otherwise they are certain to undermine their own autonomy even without exploitative intervention by smart home device vendors. Users must endeavour to read privacy policies, examine the available privacy and data collection settings, and understand their rights as data subjects if there is any hope for true autonomy. A lack of motivation on the user's part will inevitably lead to a reduced capacity for self-determination, putting decision making into the hands of others. However, user motivation is often subverted by vendors who make little effort to support user understanding, and instead deter users by means of ineffectual privacy policies and a lack of privacy controls [12, 13, 19]. The lack of vendor transparency and the absence of privacy-preserving features oppresses users into privacy resignation. When users are left in the dark on the collection and usage of their data, with seemingly no means of obtaining clarity, they often give up altogether [7, 8]. Regardless of their motivation, there is little a user can feasibly do if the vendor makes no attempt to cater to their privacy. It is unreasonable to expect users to read and comprehend a highly legalistic 20-page privacy policy that can only be obtained by emailing the vendor, for example. Smart home vendors should take actions which yield the best outcome for their customer base, as this will impact their customer retention and brand reputation, the latter of which will ultimately affect their ability to acquire new customers. In essence, an outcome that is good for the user will also be good for the vendor. If a user discovers they can purchase an alternative smart home device that features greater privacy controls for the same price, they may consider moving to this alternative vendor. Brand reputation plays a significant role in the trust of the average smart home user [5, 7], and there is a strong stigma associated with companies that compromise the autonomy of their user base. Additionally, vendors risk facing litigation if there is any legislation in place requiring them to implement autonomy-preserving principles, further emphasising their accountability.

From a deontological standpoint, it is contended that the responsibility falls entirely on the government. Vendors are legally bound to legislation, so it could be argued that it is their responsibility by extension, but it is ultimately the government or other public body that is responsible for drafting and ratifying such legislation. Smart home vendors are generally corporate entities whose foremost duty is to their shareholders, whereas the principal duty of the government is to serve its country's citizens—it is these citizens that make up the smart home user base. One may argue it is also the duty of the user themself to preserve their personal autonomy, but this would be a duty they cannot feasibly fulfil. If a vendor decides to collect, retain, and distribute user data without informing the user and obtaining their consent, the user would be powerless to stop this and remain entirely unaware. With legislation in place, however, vendors can be prevented from operating in such a manner. Only the government has the legal power to ensure universal preservation of user autonomy. Privacy, which is intrinsic to personal autonomy, is protected under the Universal Declaration of Human Rights [30]. As noted by the International Service for Human Rights [31], though everyone has a responsibility to respect human rights, it is governments that have a duty to enforce them and ensure their fulfilment.

6.3 Limitations

Though the ethical concern of personal autonomy is examined and comprehensively discussed, the issue of anticipatory conformity is not addressed in this paper. By their very nature, smart home devices require user input to provide functionality. The more data an IoT device is fed, the more features it can offer the end user. For example, a user must share location data in order to receive web mapping functionality in return. Hence, it is inevitable that there will be some degree of anticipatory conformity, even without deliberate coercion. It is likely that many users will continue to "allow all" to avoid being disadvantaged by having less features available than what the device is capable of offering. This is a larger topic, however, and one which should be investigated for future work.

7 Conclusion

In this work, some of the key factors involved in the preservation of smart home user personal autonomy have been identified, and arguments have been made for who should be held responsible for such preservation from the perspectives of two disparate ethical theories.

To best preserve smart home user autonomy, it is essential that users are informed of what personal data is collected and how this data is used. This should be delivered in an intelligible fashion, with the smart home vendor making significant effort to relay this information and receive sincere, informed consent from the user. Users should have a means of controlling the degree of data collection, by way of opt-in/opt-out settings or a kill switch, for example. Vendors should be legally bound to preserve end user autonomy to the most feasible extent possible. This can be achieved by ensuring the ratification and enforcement of legislation that mandates smart home vendors to adhere to principles such as data minimisation and data protection by design.

Both smart home users and vendors are consequentially responsible for the preservation of user autonomy, but for different reasons. The user must have an intention to preserve their autonomy in the first place. If there is no effort made by the user to increase their awareness and minimise their data exposure, they will naturally undermine their own personal autonomy. However, the majority of users do indeed care about preserving their autonomy and protecting their privacy, but are often subverted by vendors who make little effort to support users in maintaining their agency. Instead, users are frequently coaxed into privacy resignation due to a lack of appropriate documentation and available privacy options. Though the user bears some responsibility from a consequentialism standpoint, greater obligation should be put on the vendor, who has the means of enhancing the feasibility of user autonomy preservation. Furthermore, vendors also have much to lose if they fail to improve their users' autonomy and privacy rights—reputational damage, customer retention, potential litigation.

Deontologically, governments are responsible. It is the duty of the government to serve the interests of its country's citizens, including their right to personal autonomy. This is not the responsibility of the vendor, who is beholden to its shareholders. Only the government has the authority to legally mandate and enforce autonomy-preserving features in smart home devices and vendor processes. The user cannot require this of the vendor, nor should the vendor necessarily feel duty bound to preserve user autonomy. The government is the protector of rights, and thus has the jurisdiction to ensure such rights are fulfilled.

References

1. Smart Home—Worldwide | Statista Market Forecast. https://www.statista.com/outlook/dmo/smarthome/worldwide#smart-homes. Last accessed 27 Mar 2024
2. Bugeja J, Jacobsson A, Davidsson P (2017) On privacy and security challenges in smart connected homes. In: 2016 European intelligence and security informatics conference (EISIC), pp 172–175
3. Edu JS, Such JM, Suarez-Tangil G (2020) Smart home personal assistants: a security and privacy review. ACM Comput Surv 53(6):1–36. Article 116
4. Apthorpe N, Reisman D, Feamster N (2016) A smart home is no castle: privacy vulnerabilities of encrypted IoT traffic. In: Workshop on data and algorithmic transparency 2016 (DAT'16)
5. Zheng S, Apthorpe N, Chetty M, Feamster N (2018) User perceptions of smart home IoT privacy. Proc ACM Human-Comp Interact (CSCW) 2(200):1– 20
6. Chalhoub G, Kraemer MJ, Nthala N, Flechais I (2021) It did not give me an option to decline: a longitudinal analysis of the user experience of security and privacy in smart home products. In: Proceedings of the 2021 CHI conference on human factors in computing systems (CHI'21), Article 555, pp 1–16
7. Zeng E, Mare S, Roesner F (2017) End user security and privacy concerns with smart homes. In: Proceedings of the thirteenth symposium on usable privacy and security (SOUPS 2017), pp 65–80
8. Haney JM, Furman SM, Acar Y (2020) Smart home security and privacy mitigations: consumer perceptions, practices, and challenges. In: HCI for cybersecurity, privacy and trust (HCII 2020), pp 393–411
9. Kuyucu MK, Bahtiyar S, İnce G (2019) Security and privacy in the smart home: a survey of issues and mitigation strategies. In: 2019 4th International conference on computer science and engineering (UBMK), pp 113–118
10. Marafa FM, Sa'ad S, Tukur A, Mohammed A (2021) A review on impact of Internet of Things (IoT) on individual privacy in smart home systems. In: 2021 2nd International conference on intelligent engineering and management (ICIEM), pp 127–131
11. Graef I, Petročnik T, Tombal T (2023) Conceptualizing autonomy in an era of collective data processing: from theory to practice. Digital Society 2(19):1–24
12. Biocco P, Keshavarz M, Hines P, Anwar M (2018) A study of privacy policies across smart home companies. In: An interactive workshop on the human aspects of smarthome security and privacy (WSSP 2018)
13. Manandhar S, Kafle K, Andow B, Singh K, Nadkarni A (2022) Smart home privacy policies demystified: a study of availability, content, and coverage. In: Proceedings of the 31st USENIX security symposium, pp 3521–3538
14. Zuboff S (2015) Big other: surveillance capitalism and the prospects of an information civilization. J Inf Technol 30(1):75–89
15. Bowen GA (2009) Document analysis as a qualitative research method. Qual Res J 9(2):27–40
16. Flick U (2018) An introduction to qualitative research, 6th ed. Sage
17. Kaptein M, Wempe J (2002) The balanced company: a theory of corporate integrity. Oxford University Press
18. O'Connor Y, Rowan W, Lynch L, Heavin C (2017) Privacy by design: informed consent and Internet of Things for smart health. Proc Comp Sci 113:653–658
19. Jensen C, Potts C (2004) Privacy policies as decision-making tools: an evaluation of online privacy notices. In: CHI '04: proceedings of the SIGCHI conference on human factors in computing systems, pp 471–478
20. McDonald AM, Cranor LF (2008) The cost of reading privacy policies. I/S: J Law Policy Inf Soc 4(3):543–568
21. Pollach I (2007) What's wrong with online privacy policies? Commun ACM 50(9):103–108
22. Bugeja J, Jacobsson A, Davidsson P (2020) Is your home becoming a spy? A data-centered analysis and classification of smart connected home systems. In: IoT'20: proceedings of the 10th international conference on the Internet of Things, Article 17, pp 1–8

23. Alexa, Echo devices, and your privacy—Amazon customer service. https://www.amazon.co.uk/gp/help/customer/display.html?nodeId=GVP69FUJ48X9DK8. Last accessed 21 Mar 2024
24. Privacy notice—Amazon customer service. https://www.amazon.co.uk/gp/help/customer/display.html?nodeId=201909010. Last accessed 21 Mar 2024
25. Pöhls HC, Rakotondravony N (2020) Dynamic consent: physical switches and feedback to adjust consent to IoT data collection. In: Distributed, ambient and pervasive interactions (HCII 2020), pp 322–335
26. European Parliament and the Council of the European Union (2016) Regulation (EU) 2016/679 of the European parliament and of the council of 27 April 2016 on the protection of natural persons with regard to the processing of personal data and on the free movement of such data, and repealing Directive 95/46/EC (General Data Protection Regulation)
27. Bastos D, Giubilo F, Shackleton M, El-Moussa F (2018) GDPR privacy implications for the Internet of Things. In: 4th Annual IoT security foundation conference
28. Dahl V, Österlin M (2020) Impact of GDPR on data sharing behavior of smart home users
29. European Parliament and the Council of the European Union (2022) Directive (EU) 2022/2380 of the European parliament and of the council of 23 November 2022 amending Directive 2014/53/EU on the harmonisation of the laws of the Member States relating to the making available on the market of radio equipment
30. Universal Declaration of Human Rights | United Nations. https://www.un.org/en/about-us/universal-declaration-of-human-rights. Last accessed 23 Mar 2024
31. Who protects human rights? | ISHR. https://ishr.ch/about-human-rights/whoprotects-human-rights/. Last accessed 23 Mar 2024

Building a Human-Centric SOC: A New Framework for Success

John Mwangi, Julie Wall, Umar Ismail, and Ameer Al-Nemrat

Abstract The Security Operation Centre is a hub where the Information Security Team monitors, detects, analyses, and prioritizes events from critical digital assets on an ongoing basis. The objective is to ensure that any malicious activities, indicators of attack are stopped and contained before having a major impact to an organization. Early detection is very important when trying to combat cyber threats. The Security Operation Centre is equipped with intelligent tools and skilled analysts that help detect such events. With a focus to constantly improve Security Operation Centre effectiveness, a thorough understanding of human factors and human errors that may lead to potential security breaches need to be investigated. Incorporating artificial intelligence and machine learning technologies has gone a long way to compensate for human error in the Security Operation Centre, through automation of routine tasks and incorporation within Security, Orchestration, Automation and Response. This has led to better rapid threat anomaly detection, incident response and a reduction of Security Analysts' cognitive load. That said, the existing literature suggests a lack of a systematic approach, for example in assessing Security Analysts' performance. There is a gap in the research regarding human factors and the limitations of human error within the Security Operation Centre, particularly given that it operates as a socio-technical environment where social interactions and technological systems are closely integrated. Effective collaboration, communication, and teamwork are essential in such a setting, and this research looks to further bridge that gap. Through a case study, current practices within the Security Operation Centre

J. Mwangi · J. Wall · U. Ismail (✉) · A. Al-Nemrat
University of East London, London, UK
e-mail: umismail@uel.ac.uk

J. Mwangi
e-mail: u1944405@uel.ac.uk; jkm.mwangi@aol.com

J. Wall
e-mail: julie.wall@uwl.ac.uk

A. Al-Nemrat
e-mail: ameer@uel.ac.uk

H. Jahankhani and B. Issac (eds.), *Cybersecurity and Human Capabilities Through Symbiotic Artificial Intelligence*, Advanced Sciences and Technologies for Security Applications, https://doi.org/10.1007/978-3-031-82031-1_27

will be explored from the personnel perspective. In addition, investigating transferable skills from other domains such as medical, aviation, and other sectors that manage complex environments under high stress are reviewed to determine if they offer valuable information. This paper utilizes Secure Tropos to produce the Security Operation Centre meta model. This novel approach forms the basis of a new proposed framework that looks to identify relationships and security requirements within the Security Operation Centre entity. Human centric design that accounts for human factors and human errors within the Security Operation Centre is crucial for maintaining a robust cybersecurity posture. By better understanding current practices within the Security Operation Centre, this research intends to contribute towards a more human centric approach.

Keywords Secure Tropos methodology · Human factor engineering · Security operation centre · Security requirements engineering

1 Introduction

The twenty-first century has witnessed the most technologically sophisticated threats that the world has seen such that cyber security incidents have asymmetrically evolved and threats to individuals, institutions and governments are high [1]. Maintaining Confidentiality, Integrity and Availability, the CIA triad, when it comes to business data is a high priority for the majority of organizations [1]. As such, Security Operation Centres (SOC) play a crucial role in safeguarding organizations by seeking to stay ahead of cyber threats [2]. They achieve this by continuously monitoring, detecting, and responding to cybersecurity threats. Armed with intelligent tools such as Security Information and Event Management (SIEM) systems, threat intelligence feeds, and intrusion detection/prevention systems along with skilled security analysts, they help contain/stop malicious activities. This prevents data breaches, protects sensitive information, minimizing financial loss, whilst maintaining the overall security posture of an organization. Having the latest tools and advances in technology is not the only consideration to ensure the SOC remains effective. There is a social dynamic requirement as the SOC ecosystem involves teamwork, which needs clear communication and collaboration [1].

Creating an effective SOC involves drawing insights from various related works in the field [3]. Organizations often adopt established frameworks such as the National Institute of Standards and Technology (NIST) Cybersecurity Framework or International Organization for Standardization (ISO) 27001 to structure their SOC operations effectively [4]. The People, Process, and Technology (PPT) Framework is also closely associated with the SOC from past research [5]. That said, this framework is not unique to the SOC and is used in various technology topics such as knowledge management [6] and customer relationship management [7], to list a few. SOC

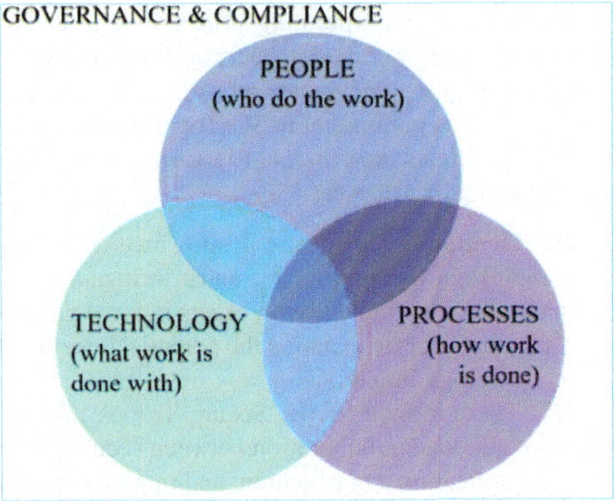

Fig. 1 Extended framework representing people, processes, technology, governance and compliance [8]

vendors within industry use this framework to summarize and structure their products and offerings. The existing PPT framework was expanded to integrate Governance and Compliance (PPTGC), elevating them from often subordinate processes to essential components within the SOC, reflecting their growing importance [8]. This framework has evolved (see Fig. 1).

Considering humans have a major role to play within the SOC, there is currently a gap to fully understand and resolve human factor challenges. The existing frameworks such as PPT, PPTGC do not account for human factors. Our research looks to address this gap by proposing a new framework which is anchored on the Secure Tropos Methodology. As a high-stakes, high pressure environment, SOC analysts' decisions could have critical operational impact to the organization. The volume of security events that a SOC analyst analyses in a day is high. Decisions need to be made quickly as to what is a false positive or negative. They also need to factor in how much time to dedicate to these events to reduce spending valuable time investigating events of little value. Consequently, getting this evaluation wrong could lead to a breach. In 2013, the retailer 'Target' was affected by a major cyber security breach where the SOC initially detected infiltration [9]. However, for an unknown reason, they decided not to investigate further which consequently allowed the threat actors to execute their objective [3]. The weight of getting decisions wrong whilst playing such a crucial role may at times leave SOC analysts feeling stuck in a 'thankless task' [10]. Due to the high pressure and at times lack of information that leads to the creation of a much clearer picture of events ongoing, erroneous decisions may lead SOC analysts to make errors [11]. Also, SOC environments can contain unwittingly vicious cycles that impact morale, causing burnout thus leading to low retention of SOC analysts [10].

This paper seeks to introduce a novel approach to establishing a Human-Centric SOC that aligns with the principles of the Secure Tropos methodology [12]. By leveraging insights from human factor engineering, which looks at how people use technology, and incorporating principles that consider human limitations, ability and expectations, our model incorporates user-centred design principles to minimize errors and improve incident response.

Efficiency is paramount in SOC operations, where timely detection and response to security threats are critical [13]. Considering this, this paper looks to investigate the following hypotheses:

> ***H1***: Integrating the Secure Tropos methodology into SOCs will lead to a more comprehensive identification and prioritization of security requirements, resulting in enhanced alignment with organizational goals and improved security posture. The intention is to examine the potential benefits of integrating the Secure Tropos methodology into SOCs.
> ***H2***: The adoption of the Secure Tropos methodology in SOCs will result in more effective collaboration between security stakeholders, leading to improved communication, coordination, and implementation of security requirements. The intention is to assess the potential impact of implementing the Secure Tropos methodology within SOCs.

In this paper, we utilise a modelling language which is part of ongoing research to create a framework for holistically modelling a secure SOC environment, which is anchored on security requirements but also considers the roles humans play within the SOC thus creating a more human centric model. Our proposed model emphasizes automation, orchestration, and integration of security tools to streamline workflows and optimize resource utilization.

The contributions of the paper are as follows:

> ***Human-Centric SOC Meta-Model Integration***—We propose a Human-Centric SOC Meta-Model that amalgamates concepts from security requirements engineering, human factor engineering, and the current SOC model. This meta-model serves as the foundational framework for designing and implementing security operations tailored to human behaviours and interactions within a SOC environment. By integrating human-centric considerations into the core of the SOC architecture, we aim to enhance overall security resilience by mitigating the impact of human errors and vulnerabilities.
> ***Enhanced Definitions for SOC Environments***—Our model provides comprehensive definitions for SOC concepts, relationships, and properties, with a focus on addressing the unique challenges of securing a SOC environment. These definitions encompass not only technical aspects but also human-related factors such as cognitive biases, decision-making processes, and behavioural patterns. By capturing these elements, our model enables a holistic understanding of the security landscape within SOC operations, facilitating more effective risk assessment and mitigation strategies.

By reducing manual tasks and automating routine processes, analysts can focus on high-priority threats and strategic security initiatives. In response, this paper presents a human-centric SOC model based on the Secure Tropos methodology, aiming to mitigate human errors, enhance effectiveness, and ensure privacy in multi-tenant environments. Multi-tenancy concept is explored in greater detail in the related works section. By utilizing the Secure Tropos methodology, this paper looks to extend the Secure Tropos concept within a SOC setting by identifying deeper interdependencies and relationships to better capture and analyse SOC security requirements.

The remainder of this paper is structured as follows: Sect. 2 describes the related work and Sect. 3 outlines the methodology. The proposed framework is explained in Sect. 4 and the results are presented and discussed in Sect. 5. Finally, we conclude the proposed approach and discuss future directions in Sect. 6.

2 Related Work

There exist two viewpoints when it comes to SOCs, industry and academic [14]. The academic viewpoint appreciates that the topic has a lot of drive from industry. There is also a lot of ambiguity within industry when referring to the SOC. That said it is important to note that there is agreement within the research community on the SOC capabilities but there is lack of consensus when it comes to what constitutes a SOC. There has been more work focusing on characteristics of the SOC without necessarily paying much attention to the overall picture. The lack of a commonly agreed holistic definition of the SOC and its composition can be a major challenge for both researchers and organizations [14]. There is a need for a commonly agreed terminology to advance the SOC research field further as lack of consensus hinders the development of efficient SOCs but also from a research point of view hinders further future innovation.

In [15], the SOC has been defined as the organizational unit at the heart of security operations. They argued that it is not a single entity or system but rather a complex structure that manages and enhances an organization's overall security posture whose core function is to detect, analyse and respond to cybersecurity incidents and threats whilst employing PPT. However, traditional SOC models often overlook human factors, leading to inefficiencies and human errors remaining a significant challenge in cybersecurity [2]. These errors may be down to fatigue [16], due to the volume of security events SOC analysts deal with daily, the by-pass of processes when investigating active incidents, misalignment in configuration, and the ever-changing Tactics, Techniques, and Procedures (TTPs) that threat actors use to deliver exploits, etc. Management of human risk factors will never be 100% effective considering you can only moderate human fallibility rather than eliminate it. During a review of the aviation [17] and medical domains, it was concluded that human rather than technical failures represented the greatest threat to complex and high-risk systems [18]. Considering the earlier definition of SOC as a complex structure [15], there is a case for comparison. The review carried out by Vielberth et al. [15] categorized errors into two groups: information-handling problems and violations (motivational problems). It also stated that different error types have different underlying mechanisms that occur in different parts of the organization and require different approaches for risk management. They also categorized mistakes as either rule-based or knowledge based. The study also looked at failures which can either be active or latent. In the medical domain, for example, active failures may be committed by people in direct contact with the patient. In contrast, latent failures, whose adverse effects may take longer to become evident, arise within the management and organizational sphere.

There is an argument that decisions made by senior management can create conditions in the workplace that subsequently promote individual errors and violations. Consequently, the likelihood of an unsafe act being committed is heavily dependent on the nature of the task and by the local workplace conditions, which in turn, are the product of "upstream" organizational factors [18]. Technology should amplify human capacity and capabilities to be creative and apply critical thinking to tasks and problems. Increasing automation helps decrease the number of mundane tasks [19, 20] by deploying tools. Automating specific tasks can also help increase operational efficiency. A counter argument by Reasons [18] states that automation and advancement in equipment is not a panacea for human factor problems as this merely relocates them. In contrast, training people to work effectively in teams incurs relatively low costs but has achieved significant enhancements of human performance within the aviation industry.

As mentioned earlier there is a lot of ambiguity when it comes to the commonly agreed holistic definition of the SOC and its composition [14]. In their research [15] sought to find the definition of the SOC, the design/architecture and its constituent parts. Their study broke the literature found into two distinct categories:

- General Aspects—This included SOC definitions, operational models and architecture.
- Building Blocks—Composition of the SOC.

From the SOC architecture perspective, three general architectural approaches applied to the SOC design were identified [15]:

- Centralized—Data is sent from different locations and subsidiaries to a central SOC for further analysis and processing [8, 21].
- Distributed—Resembles a single system operating across subsidiaries [13, 22]. From a user's perspective this may appear as if dealing with one entity. Such distributed systems enable all entities to retrieve, process, combine and provide security information and services to other entities [23, 24] allowing the workload and data to be spread evenly.
- Decentralized—Combines elements of both centralized and distributed design approaches [25]. A few SOCs possibly with limited capabilities report to a centralized SOC, either one or more.

From the security requirements capture perspective, it is important to map out the composition of the SOC and how each piece interacts, not forgetting the humans within the environment. Early SOC models, proposed over 15 years ago, were comprised of five modules [8, 21], which included an event generator, event collector, message database, analysis engine and reaction management software. This has moved on and there is the incorporation of digital forensics and proactive capabilities to prevent attacks in modern SOCs. For the SOC architecture, it has been suggested that this consists of a generation layer, acquisition layer, data manipulation layer and an output or presentation layer [25].

SOCs operation can be internal or external to a business [3, 14, 26, 27]. Five different operational models based on the size of organizations and authority have been proposed by Zimmerman [28]. These include Virtual, Large, National, Small and Tiered SOCs, which can be further clustered into four main categories; Dedicated, Virtual, Outsourced and Hybrid. Each model has advantages and disadvantages. Considerations should be made before choosing a model suitable to an organization. Factors such as company strategy, industry/sector in which the organization operates, size of organization, implementation costs, inhouse setup timescales versus outsourcing the whole service, regulations (based on sector), privacy, jurisdiction, availability requirements, management support considering the criticality of the business, integration and how well this service plugs into the existing infrastructure, and data loss concerns considering the vast amounts of data that converged in a central point for processing. It is also challenging to recruit and retain SOC experts, and all the above key points should help in determining the type of SOC an organization takes up.

Managing security, especially in outsourced multi-tenanted environments, presents unique privacy challenges [29]. Multi-tenancy, in the context of cloud services, refers to a software architecture where a single instance of the software application serves multiple customers, or "tenants." These tenants share the same underlying infrastructure and resources while remaining isolated from each other logically. In a multi-tenant environment, each tenant typically has its own set of customizable configurations, data, and user access controls, ensuring privacy and security [12]. This approach allows cloud service providers to achieve economies of scale by efficiently utilizing resources across multiple customers while providing cost-effective services [12]. Multi-tenancy is commonly used in cloud-based Software-as-a-Service (SaaS) applications, where it enables the provider to serve a large customer base efficiently while offering scalability and flexibility to individual tenants. It is important to ensure segregation between tenant instances from a security and privacy perspective [12].

Multi-tenancy is mostly relevant to Managed Security Service Providers (MSSP), who offer SOC services to multiple organizations under one tenancy. Tools used to monitor these different organizations in most cases are hosted in a tenant managed by the MSSP. Even though there are technical controls in place to ensure segregation of one organization environment from the next, there are privacy risks that would need to be addressed [12]. SOC analysts within an MSSP may have access to multiple environments as part of their job when it comes to the monitoring approach. Our proposed Meta-Model looks to highlight the need to capture security requirements whilst adopting privacy-by-design principles, ensuring that sensitive data is adequately protected, and access controls are strictly enforced. Encryption, anonymization, and data segregation techniques are employed to prevent unauthorized access and mitigate the risk of data breaches across tenant boundaries.

2.1 SOC Open Challenges

The number of cyber security related breaches is on the increase, and it is thought that the number of unreported cases could be higher than the reported ones [3]. The annual cost of a cyber-attack has also risen. The mean time to detect a cyber-breach is roughly 196 days as of 2018, with a further 69 days to contain the attack, which could be seen as ineffective detection and mitigation [15]. This also demonstrates how cyber breaches can go undetected for long periods and could be an indication of how complex it can be for an organization to fully understand assets that need protecting and what technology to implement that would result in effective controls. As part of the review, this paper, along with references [3, 7], and Crémilleux et al. [30], summarizes several challenges facing the SOC. Figure 2 also includes the proposed human factor category from our research. The challenges are outlined (see Fig. 2).

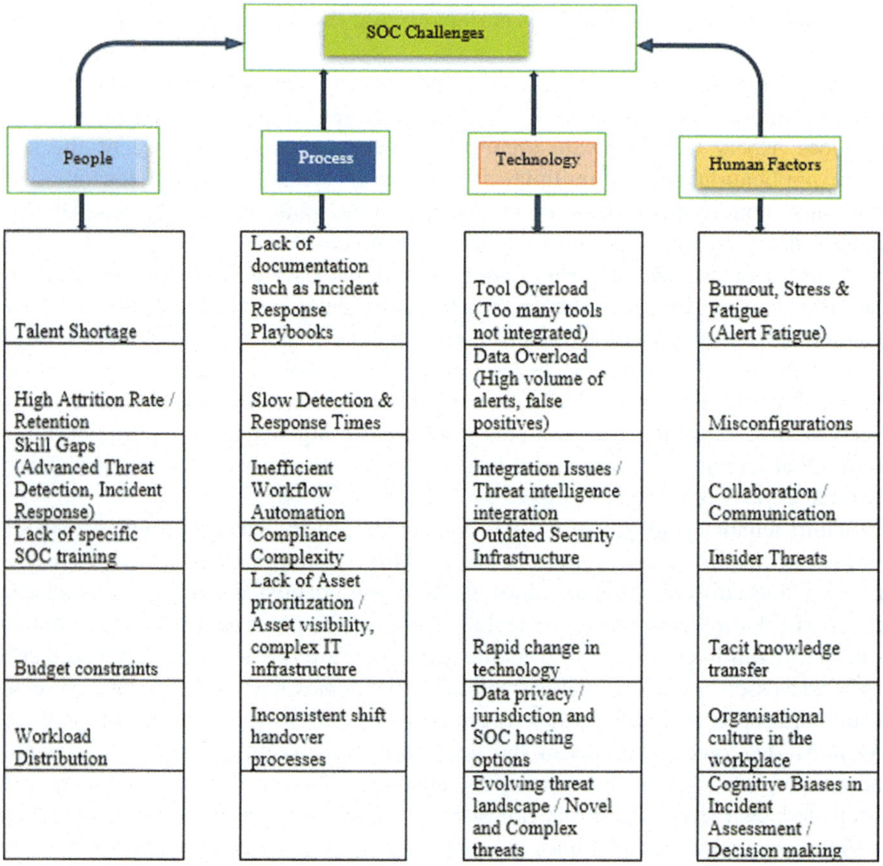

Fig. 2 Summary of SOC challenges from [3, 7, 30] and our contributions, with a focus on human factors as a critical addition to the PPT framework

Also, the PPTGC framework [8] frames open SOC challenges into the four categories of People, Process, Technology, Governance and Compliance. This is a more logical way to consolidate and categorize to better address and look to resolve issues.

People

Challenges include a lack of understanding of how a SOC analyst decides to investigate an incident and how much time to dedicate to it [31]. Many attempts to improve efficiencies within the SOC are not informed by a decision-making theory or even input from SOC analysts [32]. Improvements seem to be derived from organisational level methods through the intervention of governance structure, interconnectivity between business units, and the delegated authority of the SOC [8]. Even though the approaches are valid without a firm understanding of the cognitive and emotional nature of the SOC role, it is difficult to ensure that analysts are best equipped to make effective decisions. This also means there is little consideration for subjective factors such as fatigue and stress that affect SOC analysts.

To address the time limitation, cyber deception technologies such as honey pots are becoming more common [29]. They could be used as a tactic to buy more time so SOC analysts can reach decisions while under less pressure. The way this works is by simulating a real environment where the threat actor reveals their intentions in a safe environment. Cyber deception technology is increasingly becoming part of an active defence toolkit as it has the potential to increase uncertainty and fear in attackers such that they make mistakes, waste resources and leak information about their TTPs [29].

There is a need to understand the link between the decision making and behaviour of both threat actor and defender. Most cyber deception research tends to build from a computer science perspective where the scope is often truncated to misdirecting a threat actor on a network rather than impacting decision making and behaviour as suggested by Ashenden et al. [29]. A study by Sundaramurthy [10], which involved training several computer science students before embedding them in three different SOCs by applying an anthropological approach, found evidence that rather than the parallel process expected in normative decisions, many SOC analysts followed a "hunch", which in most cases are accurate despite the intuitive non-analytical nature of the approach. It may also be the case that due to the highly dynamic nature of the constant evolving threats, SOC analysts rely on skilled intuition. This type of decision-making process has the characteristics of being quick and heuristics are also often highly adaptive and effective [33]. These sorts of decisions are described as Type 1. Type 2 decisions are often 'rational', much slower and deliberate and on some occasions less effective in contrast to Type 1 [33]. Due to the high volume of alerts that need investigating and which are mostly false positive, it would be unrealistic to assume that SOC analysts would treat these alerts equally [5]. There is a tendency to fall back to Type 1 decisions to cope with the large volume of alerts.

The application of tacit knowledge within the SOC was investigated by Cho et al. [34]. This is often described as knowledge that is difficult to articulate or transfer. The application of tacit knowledge can be seen in other domains such as medicine and is beneficial in expediting problem-solving procedures during an emergency response. Within an IT setting, there is a lack of understanding when it comes to tacit knowledge [34]. During post-incident review within the SOC, and especially when reviewing the Root Cause Analysis (RCA), opportunities to identify areas where tacit knowledge have been applied may come to light. Also, the best way to integrate technologies like cyber detection into established SOCs is unclear, especially since analysts rely on tacit knowledge and have developed valuable habits over time to perform their roles effectively.

The role of a SOC analyst is cognitively demanding due to the myriads of activities they are responsible for [35]. Such activities require constant monitoring of security alerts and being in a state of perpetual vigilance. To better understand the quality of alerts produced by security tools, [35] carried out a series of interviews and an online survey with 20 SOC analysts. The findings indicated that most alarms were attributed to benign triggers or events that could be explained as being legitimate within an organisational context. The downside to this, from a SOC analysts' perspective, was high levels of stress. Each alert, false positive or not, involves some level of investigation. Thus, SOC analysts risk reprimand should they misattribute an alert as benign. It is important that the tools deployed within the SOC are tuned to reduce the number of false positive detections, which would then impact the SOC analysts' cognitive load.

As for training, this may take many forms, from vendor specific for all the various tools the SOC analyst interacts with daily, to learning whilst doing the job, such as shadowing a senior or more experienced colleague. There may be a need to develop a system that models triage actions of senior analysts to aid junior colleagues dealing with similar cases [36]. Playbooks provide an overview of actions and tasks based on the experience of senior SOC analysts [37]. When it comes to training, the creation of knowledge graphs/training matrix representing domain knowledge and gaps can help focus areas that need attention [38].

Processes

Within the SOC, processes are based on a framework, such as the Incident Response Lifecycle as the main goal is to prepare and respond to incidents [39–41]. The NIST Incident Response Lifecycle consists of four steps, "Preparation", "Detection and Analysis", "Containment, Eradication and Recovery" and "Post-Incident Activity". The view on processes based on the systematic literature review from [15] is that technical steps within the Incident Response Lifecycle are dealt with intensively, whilst those surrounding them are treated sporadically, highlighting an area for further research [15]. Their observation is based on the lack of SOC specific scientific publications especially relating to "Post-Incident Activity".

Preparation—Focuses on data collection but there is a lack of uniformity of steps composing this process [15]. As part of their investigation, [42] conducted interviews with 13 SOC professionals from 5 SOCs and concluded that SOC analysts relied on having the right toolsets to understand the data.

Detection and Analysis—A huge volume of data is collected within the SOC, and it can become overwhelming to even seasoned security experts. The focus here is to make sense of what has been collected by turning it into useful information [41]. This step also includes alert prioritization/triage, which serves two main purposes; ensuring that the most severe incidents are treated as a priority and looks at the distribution of incidents based on available resources.

Containment, Eradication and Recovery—The focus here is to stop harmful events in their tracks, then look to eradicate and recover. Several frameworks can be adopted during this stage of tackling incidents such as the Observe, Orient, Decide, Act (OODA) loop, which is an analytical framework for decision-making [43]. One could also adopt the Plan, Do, Check, Act loop [41].

Technology

Levels of data collection within the SOC vary from one organization to another depending on their scope, size, operational model, and architecture [15]. Prioritization of assets that need to act as sources of security events need to be determined prior to data collection. There is a fine balance to be struck when it comes to collecting too much or too little data to be presented to the SOC analyst. Collecting too much may impact operational efficiency, conversely collecting too little may mean malicious events are missed. Other elements that come into play include data retention, privacy, and regulations. Organizations need to determine the sensitivity of the data captured and look at ways to minimize the risk around it [15].

The integration of advanced technologies, like SIEM, SOAR (Security Orchestration, Automation, and Response), and threat intelligence platforms is crucial for enhancing SOC capabilities [44]. Critical data sources and assets must be configured to transmit security events to tools such as a SIEM [15]. These sources of data may include firewalls, Intrusion Protection Systems (IPS), Intrusion Detection Systems (IDS), antivirus software, identity and access management platforms, switches, routers, servers, or virtualized environments such as hypervisors [15]. It should be noted that people also play a major part as a data source. In the article by Vielberth et al. [45], they look at humans being employed as a sensor. In Human-as-a-Security-Sensor, [45] look at human abilities to detect anomalies, which at times are not detected by automated processes.

Governance and Compliance

Having SOC governance and compliance is key [15]. Governance is responsible for setting out the effective and efficient use of IT systems. This is achieved by providing strategic direction, which is reinforced by policies, standards, and procedures. Compliance focuses on the ability to abide by external regulations. The SOC lacks holistic standards or industry specific guidelines to help organizations to make decisions [46]. That said, the SOC can ensure that certain compliance requirements are met by following standards that focus on specific tasks [47, 48].

A SOC should have controls in place that are regularly audited to assess the current level of maturity, SOC capabilities and operational effectiveness. The Capability Maturity Matrix (CMM) is a framework for assessing the maturity of organizational process consisting of five stages: non-existent, initial, repeatable, defined process, reviewed and updated, and continuously optimized [27]. A recommendation by Schinagl et al. [49] would be to adopt this approach to assess how well SOC building blocks have been implemented.

Summary

The SOC continues to face several challenges. Key gaps identified in the review include difficulties in identifying SOC components, issues with security capture including the identification of relationships and interaction between SOC components, and challenges related to human error. Considering the SOC is a socio-technical environment where social interactions and technological systems are closely integrated, addressing human factor gaps may result in further creation of effective collaboration, communication, and teamwork, which are essential in such a setting. This paper aims to address security capture challenges using Secure Tropos and provide further insights into human error by conducting a case study.

3 Methodology

As seen in the related work section, different study methods have been used to conduct SOC research, such as semi-structured interviews, onsite visits, case studies, and ethnographic field work [3, 15, 30, 49]. This project utilized two phases. A case study in phase one focused on the role human factors and human error plays within the SOC. Phase two utilized Secure Tropos to extend the PPT framework to include human factors. This novel approach in phase two provides a means for both security experts and software/system designers to better visualize and integrate security requirements as part of the SOC design process. Other frameworks such as PPT, do not offer the same level of visualization achieved through the creation of a meta-model. Visualization through meta-models, which can be further decomposed to reveal detailed interactions between various entities, appears to be a more effective approach for creating deeper understanding of the SOC environment. The proposed framework focuses on ensuring that the human factors, are accounted for and central to the architecture's design, with security needs embedded into every layer of the SOC ecosystem.

The Secure Tropos methodology, as a concept, refers to a high-level abstraction used to represent recurring patterns or structures within the system being modelled [21]. Concepts help to simplify the modelling process by capturing common characteristics and relationships among elements in the system. They provide a way to modularize and organize the system's components, making it easier to analyse and reason about the system's behaviour and security requirements. Concepts can include various elements such as actors, agents, goals, dependencies, capabilities, trustworthiness, security requirements, threats, attacks, and countermeasures [12]. These concepts are used to model the system's architecture, interactions, and security aspects, enabling stakeholders to understand and address security concerns throughout the development process.

The Secure Tropos methodology is used in software development and requirements engineering and extends the Tropos Framework by incorporating security consideration into the early stages of software development. The Secure Tropos methodology focuses on developing secure software solutions by considering and incorporating security from the onset ensuring it is not an afterthought that gets bolted on at the end [50]. It provides the ability to define and identify high level goals and ways to further secure systems from external threats. The additional benefit for our proposed model is that it can be adopted to review existing SOC deployments as part of a risk-based continuous assessment lifecycle but also review how new tools deployed to the SOC would affect the overall posture. Secure Tropos is also used to identify relationships and security requirements within the SOC entity by generating a Meta-Model. Lucid Chart, which has a Unified Modelling Language (UML) module, is also used in the production of the Meta-Model proposed in this paper.

Secure Tropos, which can be used to identify security constraints, is an Actor (agent) oriented methodology that can be used to identify and model different actors and how they interact with the solution or system under development [50]. Actors can be systems, organisations, or humans. It delves deeper into understanding the dependencies on actors and the goals, they are looking to achieve. These goals can relate to functionality, security, or other aspects of the solution or system. Our approach takes on the "social perspective" by considering the involvement of humans and their organisational behaviours and the social phenomenon around them. It is an approach that focuses on security concerns and issues that might affect systems security. Security assets are resources used to protect against concerns. Through modelling, Secure Tropos also allows configuration of security policies and rules that govern how systems should behave to ensure security. Traceability and risk assessment are also included where developers can trace, for example, security goals and requirements back to the processes and actors they originated from. Identification of risk and prioritisation is handled via risk management. As security needs regular revaluation rather than a onetime concern, Secure Tropos supports this approach by being incremental and iterative during the development cycle.

For a qualitative survey, Quick and Hall [51] suggest a range of (4–50) participants due to the volume of data collected and the appropriateness of participants based on their knowledge of the research topic. Considering this, our study utilised a mix of 6 experts and SOC analysts. For the case study, we utilized a qualitative research design consisting of an online questionnaire sent out to 15 SOCs within the UK targeted at Subject Matter Experts (SMEs) from industry. Questions included a mixture of closed and open-ended questions. The questions aimed to elicit an understanding of current SOC practices and how human factors have affected SOC effectiveness. The average time to complete the questionnaire was 47 min (min: 27, max: 57). Ethical approval for this research was provided by the University of East London Ethics Committee. The participants represented various roles within a SOC, MSSPs providing services to both public and private organizations. We believe this provided a solid breadth of perspectives in our sample.

4 Proposed Extended PPT Framework

This section introduces the proposed mapping for a Human-Centric SOC in the context of the Secure Tropos methodology. Based on our research and on the literature review, the following Secure Tropos constructs, and terminology have been identified as essential. Table 1 outlines the language and constructs derived from Secure Tropos, such as actors, relationships, and goals. Table 2 outlines each agent or entity with specific goals to accomplish, detailing their characteristics and relationships with other agents. Table 3 details the dependencies between each agent and other agents, while Table 4 outlines various risks identified in a SOC environment and their corresponding mitigation strategies.

To effectively structure SOC operations, our proposed framework adopts an approach similar to NIST, with particular emphasis on considerations starting from the design phase, as highlighted in the introduction section when describing incident handling. Our proposed framework consists of the following stages/levels with an anchor to Secure Tropos Methodology.

Table 1 Secure Tropos terminology (language and constructs) for SOC

Category	Element	Description
Language	Goals	Objectives that agents aim to achieve
	Tasks	Specific activities that agents perform to achieve goals
	Resources	Assets and capabilities required by agents to accomplish tasks and achieve goals
	Agents	Entities (individuals, groups, or systems/software components) that have goals and perform tasks
	Actors	A subset of agents, typically referring to people or groups/ software components within the SOC
	Relationships	Interactions or dependencies between agents, tasks, and resources
	Constraints	Conditions and limitations that affect how tasks are performed, or goals can be achieved
	Risks	Potential events or conditions that could prevent achieving goals or performing tasks effectively
	Mitigations	Strategies or actions taken to reduce risks
Constructs	Goal models	Illustration of goals of SOC agents and the tasks required to achieve them
	Actor Mapping	Visual representations of SOC agents, their roles, and their relationships
	Relationship Mapping	Illustrations of how SOC agents depend on each other for resources, tasks, and goal achievement
	Risk models	Table outlining potential risks and mitigations

Table 2 Goal models—description of agents, their characteristics, goals, and relationships

Agent	Characteristics	Goals	Relationships
Security analysts	Skilled in threat detection and response, continuously trained, work under high pressure	Efficiently detect and respond to security incidents, minimize false positives, maintain situational awareness	Collaborate with IT staff, communicate with management, utilize tools and threat intelligence, and monitor end-user activities, share threat intelligence with external partners
IT staff	Technical expertise in infrastructure management, responsible for implementing security controls	Maintain and secure IT infrastructure, ensure seamless operation of security tools, support SOC operations	Support security analysts, manage hardware and software assets, interact with cloud providers, report infrastructure status to management, maintain tools
Management	Strategic oversight, responsible for resource allocation, ensure compliance with regulations	Oversee SOC effectiveness, align SOC activities with business goals, ensure regulatory compliance, allocates resources	Communicate with analysts and IT staff, make decisions based on SOC reports, interact with external auditors, and provide training to end-users
External partners	Provide specialized expertise, threat intelligence, and incident response support	Enhance SOC capabilities, offer additional resources and knowledge, assist in incident response	Share threat intelligence with SOC, assist in incident response, provide external audits, collaborate with management on compliance assessments
End-users	Employees whose activities are monitored for security, potential internal threat vector	Perform their job functions securely, adhere to security policies, increase security awareness, reports anomalies	Receive training and awareness programs from management, follow security guidelines set by SOC, report anomalies to security analysts
Threat actors	Individuals or groups with malicious intent, constantly evolving tactics	Exploit vulnerabilities, disrupt operations, steal data	Adversarial relationship with SOC, targeted by threat modelling and defence mechanisms, pose risks that SOC mitigates through continuous monitoring and response strategies

Conceptual—Similar to what we have covered in this research, this stage involves the proposing or describing the entire ecosystem of SOC and identifying what is in SOC. This also includes the capture of the composition of SOC operations ensuring SOC resources (Analysts, Managers etc.) understand SOC boundaries and any limitations that may be present.

Table 3 Dependency graph—identifying SOC dependencies

Agent	Dependency
Security analyst	Depend on IT Staff for infrastructure support and tools maintenance, depend on Management for strategic direction and resources
IT staff	Depend on Management for resource allocation, depend on Security Analysts for feedback on tool effectiveness
Management	Depend on Security Analysts for incident reports and metrics, depend on IT Staff for infrastructure status
End-users	Depend on Management for training and awareness programs, depend on Security Analysts for incident response

Table 4 Risk models—SOC risks and mitigations

Risk	Description	Mitigation
Human error	Mistakes made by SOC staff or end-users that could lead to security breaches	Regular training and awareness programs, automated checks, and balances
Burnout	High stress levels leading to decreased performance among security analysts	Implement stress management strategies, rotate shifts, provide mental health resources
Evolving threats	Constantly changing tactics by threat actors	Continuous threat intelligence updates, regular training, adaptive security measures
Resource shortage	Insufficient resources (staff, tools, budget) to maintain SOC operations	Strategic resource allocation by management, prioritize critical operations, leverage external partnerships

Logical—This stage involves the creation of the Meta model (see Fig. 3) identifying the relationships between People, Process, Technology and Human Factors interacting with each other.

Implementation—This stage looks to address technical elements that enhance the effectiveness of SOC operations. This could be the implementation of different technologies can be used to enhance the SOC operations. It also includes for instance the use of Artificial Intelligence and Machine Learning techniques in detection, monitoring and event analysis. Other considerations that facilitate the implementation of a human centric SOC such as gamification which ensures SOC Analysts remain engaged when investigating events are implemented here.

Validating this new framework may take several forms such as seeking expert opinion, conducting a SOC Maturity Assessment using similar frameworks such as NIST, reviewing Key Performance Indicators (KPI) and other Metrics. As part of future research, our model could go through further validation by extending it use to a wider audience.

Figure 3 represents the proposed extended PPT Framework. We used Lucid chart to produce the Meta-Model, which details the composition, interactions and checks to ensure CIA is maintained. The diagram also shows key considerations, which

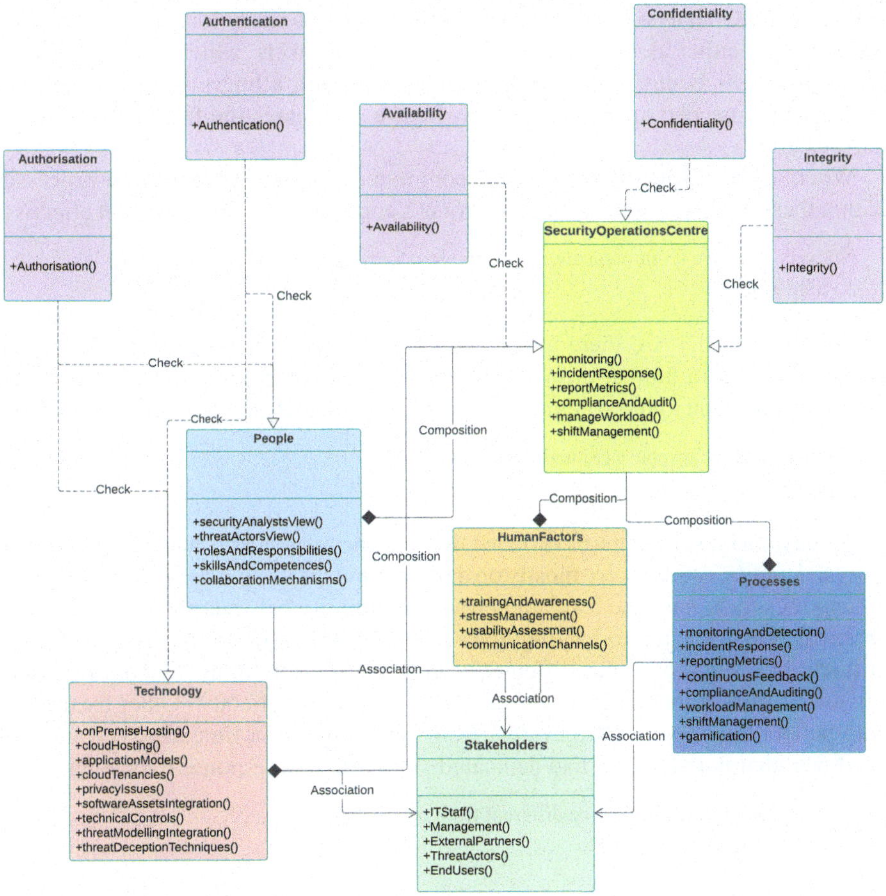

Fig. 3 Proposed extended framework representing people, processes, technology, and human factors

can further decompose into their own topics. For example, under "Technology", it is important to consider where hosting takes place. Is this in the cloud or on premise? What sort of risks are present with the options chosen and how does that impact privacy concerns. Other considerations, such as the location of where business data will reside/jurisdiction and regulations are also key.

5 Results and Discussion

R Studio's Text Mining module was used to qualitatively analyse the case study results. A sample of some of the questions posed during this study included "In your opinion, what are the primary causes of human errors in the SOC?". It was clear that some SOC analysts felt pressure was one of the contributors for human error. Pressure was either associated with a rush to close tickets, which also led to steps in set processes being skipped or having to deal with a huge volume of security events which needed analysing quickly. Figure 4 displays the feedback following this question.

When asked "How effective is the communication and coordination processes within the SOC in preventing human errors?", some analysts felt it was not effective.

> Not very effective (communication need improvement).
>
> —SOC Analyst

We also asked, "Are there any specific tasks or activities within the SOC that are more prone to human errors?". Event analysis and response was highlighted predominantly due to the volume of events that each analyst was dealing with daily.

> Event analysis & response (Situational Awareness)
>
> —SOC Analyst

A question on shift patterns revealed that half the respondents had an equal mix of days and nights. A third were mostly on day shift and the remainder had a broad range of activities to support the team and provide continuity between shifts in periods of absence. When it came to handing over from one shift to another, results revealed different SOCs approach this differently, with one SOC outlining high priority security events during the shift and the customers that were affected. Other used email with bullet points. It was also evident from the response for time allocated for shift handover, that not all SOCs had dedicated slots with one response indicating:

> There is no set time, this can be different from day to day.
>
> —SOC Analyst

Fig. 4 Question response displayed using word cloud

Based on the case study results, a SOC would benefit from having a protected slot for handover with a clear set of processes and procedures on how and what needs to be covered during the handover, ensuring the transfer of responsibility is clear from one shift to the next. This is also in contrast if compared to the medical and aviation industries. Results from the case study indicated that the SOCs were conducting training on tools they used. Also, SOC analysts were learning from past events. Where appropriate and available they were utilising automation to try and reduce human error.

After implementing the Secure Tropos methodology to address current challenges facing the SOC, our proposed framework could offer significant improvements across key areas such as the capture of security requirements during the design stage of the SOC but also conducting assessments on existing SOC deployments. The adoption of Secure Tropos creates a proactive approach to threat mitigation, enabling SOC teams to identify and address potential threats more effectively through dynamic threat modelling and automated security controls. This is in support of the hypotheses H1 and H2 as described in the introduction section. Moreover, Secure Tropos promotes visualisation, identification of clear goals, objectives and relationships, collaboration and communication among SOC stakeholders, leading to improved coordination and alignment of security initiatives with organizational goals. This enhanced collaboration enhances overall efficiency and effectiveness within the SOC. Additionally, the methodology facilitates the efficient capture and integration of security requirements into SOC operations, ensuring that security efforts are prioritized based on their impact on security posture and organizational objectives. Overall, the findings underscore the effectiveness of Secure Tropos in enhancing SOC capabilities and strengthening an organization's overall security posture.

6 Conclusion

The proposed Human-Centric SOC model offers a comprehensive framework for enhancing security within the SOC. Extending Secure Tropos concepts and creating a detailed view of the relationships between components provides a better understanding. By integrating human factor engineering, efficiency enhancements, and privacy considerations within the Secure Tropos methodology, organizations can establish resilient and effective security operations tailored to the needs of diverse stakeholders. This paper has also shown the importance of addressing the human element within the SOC as reliance on technology alone cannot resolve the existing issues. It is important to address social dynamic requirements within the SOC ecosystem as it involves teamwork, which also needs clear communication and collaboration between different groups.

This paper proposed to explore the development of a human-centric SOC that would integrate human factors and security requirement factor engineering while using the Secure Tropos methodology. This study faced minor limitations and challenges which included the complexity of integrating multidisciplinary concepts such as human factor engineering, limited prior research in human factors engineering within the SOC, and the need for long-term evaluation, which can be an area of future development. Despite these limitations, addressing these challenges could lead to significant improvements in SOC design and security outcomes, underscoring the importance of navigating these limitations effectively to ensure the validity and practicality of the research findings. Tailored to the needs of diverse stakeholders. Our model would also be useful in evaluating established SOCs, considering it is adopted to a risk-based continuous assessment approach. Without time limitation our study could benefit from wider worldwide engagement for further validation. This could be achieved in the future. The following areas have also been identified as potential topics for future research in the SOC context; tacit knowledge transfer, SOC specific training methods, targeted data collection and the utilisation of data protection techniques such as anonymization and pseudonymization.

References

1. Tetrick LE, Pfleeger SL, Horne WG (2024) GMU cybersecurity incident response team social maturity handbook updated 10.20.16, Scribd. Available at: https://www.scribd.com/document/527834317/GMU-Cybersecurity-Incident-Response-Team-Social-Maturity-Handbook-Updated-10-20-16. Accessed: 01 Sept 2024
2. Robinson N (2023) Human factors security engineering: the future of cybersecurity teams. EDPACS 67(5):1–17. https://doi.org/10.1080/07366981.2023.2211429
3. Kokulu FB, Soneji A, Bao T, Shoshitaishvili Y, Zhao Z, Doupé A et al (2019) Matched and mismatched SOCs. In: Proceedings of the ACM SIGSAC conference on computer and communications security, pp 1955–1970, Nov 2019
4. Miloslavskaya N (2017) Security intelligence centers for big data processing. In: Proceedings 5th international conference on future Internet of Things cloud workshops (FiCloudW), pp 7–13, Aug 2017
5. Feng C, Wu S, Liu N (2017). A user-centric machine learning framework for cyber security operations center. Paper presented at the 2017 IEEE international conference on intelligence and security informatics (ISI)
6. Bhatt GD (2001) Knowledge management in organizations: examining the interaction between technologies techniques and people. J Knowl Manage 5(1):68–75
7. Bourgeois J, Ganame A, Kotenko I, Ulanov A (2007) Software environment for simulation and evaluation of a security operation center. In: Information fusion and geographic information systems. Springer, Berlin, Germany, pp 111–127
8. Shahjee D, Ware N (2022) Integrated network and security operation center: a systematic analysis. IEEE Access 10:27881–27898. https://doi.org/10.1109/ACCESS.2022.3157738
9. Plachkinova M, Maurer C (2018) Security breach at target. J Inf Syst Educ 29:11–20
10. Sundaramurthy S (2017) An anthropological study of security operations centers to improve operational efficiency
11. Schinagl S, Schoon K, Paans R (2015) A framework for designing a security operations Centre (SOC). Paper presented at the 2015 48th Hawaii international conference on system sciences
12. Mouratidis H, Shei S, Delaney A (2019) A security requirements modelling language for cloud computing environments. Softw Syst Model 19(2):271–295. https://doi.org/10.1007/s10270-019-00747-8
13. Karim Ganame A, Bourgeois J, Bidou R, Spies F (2008) A global security architecture for intrusion detection on computer networks. Comput Secur 27(1):30–47

14. Chamkar SA, Maleh Y, Gherabi N (2022) The human factor capabilities in security operation center (SOC). In: Advances in information, communication and cybersecurity, pp 579–590. https://doi.org/10.1007/978-3-030-91738-8_53
15. Vielberth M, Bohm F, Pernul G, Fichtinger I (2021) Security operations center: a systematic study and open challenges [Online] Ieeexplore.ieee.org. Available at: https://ieeexplore.ieee.org/document/9296846. Accessed 2 Sept 2024
16. Reeves A, Delfabbro P, Calic D (2021) Encouraging employee engagement with cybersecurity: how to tackle cyber fatigue. SAGE Open 11:215824402110000. https://doi.org/10.1177/21582440211000049
17. Guest G, Bunce A, Johnson L (2006) How many interviews are enough? Field Methods 18(1):59–82. https://doi.org/10.1177/1525822x05279903
18. Reasons J (2015) Understanding adverse events: human factors. In: Proceedings 48th Hawaii international conference on system sciences, pp 2253–2262, Jan 2015
19. Islam C, Babar M, Nepal S (2019) Automated interpretation and integration of security tools using semantic knowledge. In: Advanced information systems engineering, vol 11483. Springer, Cham, Switzerland, pp 513–528
20. Kanemoto Y, Aoki K, Iwamura M, Miyoshi J, Kotani D, Takakura H et al (2019) Detecting successful attacks from IDS alerts based on emulation of remote shellcodes. In: Proceedings IEEE 43rd annual international computer software and applications conference (COMPSAC), pp 471–476, July 2019
21. Bidou R, Bourgeois J, Spies F (2004) Towards a global security architecture for intrusion detection and reaction management. In: Information security applications, vol 2908. Springer, Berlin, Germany, pp 111–123
22. Bourgeois J, Syed R (2009) Managing security of grid architecture with a grid security operation center. In: Proceedings of international conference on security and cryptography, pp 403–408
23. Syed RH, Pazardzievska J, Bourgeois J (2012) Fast attack detection using correlation and summarizing of security alerts in grid computing networks. J Supercomput 62(2):804–827
24. Syed RH, Syrame M, Bourgeois J (2013) Protecting grids from cross-domain attacks using security alert sharing mechanisms. Future Gener Comput Syst 29(2):536–547
25. Miloslavskaya N, Tolstoy A, Zapechnikov S (2016) Taxonomy for unsecure big data processing in security operations centers. In: Proceedings IEEE 4th international conference future Internet Things cloud workshops (FiCloudW), pp 154–159, Aug 2016
26. Forte D (2003) An inside look at security operation centres. Netw Secur 2003(5):11–12
27. Jacobs P, Arnab A, Irwin B (2013) Classification of security operation centers. In: Proceedings of information security South Africa, pp 1–7, Aug 2013. https://doi.org/10.1109/ISSA.2013.6641054
28. Zimmerman C (2014) Ten strategies of a world-class cybersecurity operations center
29. Ashenden D, Black R, Reid ID, Henderson S (2021) Design thinking for cyber deception. Paper presented at the 54th Hawaii international conference on system sciences, Hawaii
30. Crémilleux D, Bidan C, Majorczyk F, Prigent N (2019) Enhancing collaboration between security analysts in security operations centers. In: Risks and security of internet and systems, vol 11391. Springer, Cham, Switzerland, pp 136–142
31. Kokulu FB, Soneji A, Bao T, Shoshitaishvili Y, Zhao Z, Doupé A, Ahn G-J (2019) Matched and mismatched SOCs: a qualitative study on security operations center issues. Paper presented at the proceedings of the 2019 ACM SIGSAC conference on computer and communications security, London, United Kingdom
32. Dalal RS, Howard DJ, Bennett RJ, Posey C, Zaccaro SJ, Brummel BJ (2022) Organizational science and cybersecurity: abundant opportunities for research at the interface. J Bus Psychol 37:1–29. https://doi.org/10.1007/s10869-021-09732-9
33. Kahneman D, Klein G (2009) Conditions for intuitive expertise: a failure to disagree. Am Psychol 64:515–526. https://doi.org/10.1037/a0016755
34. Cho SY, Happa J, Creese S (2020) Capturing tacit knowledge in security operation centers. IEEE Access 8:42021–42041. https://doi.org/10.1109/access.2020.2976076

35. Alahmadi B, Axon L, Martinovic I (2022) 99% false positives: a qualitative study of SOC analysts' perspectives on security alarms. Paper presented at the 31st USENIX security symposium, Boston, MA, United States
36. Zhong C, Lin T, Liu P, Yen J, Chen K (2018) A cyber security data triage operation retrieval system. Comput Secur 76:12–31
37. Applebaum A, Johnson S, Limiero M, Smith M (2018) Playbook oriented cyber response. In: Proceedings of national cyber summit (NCS), pp 8–15, June 2018
38. Pingle A, Piplai A, Mittal S, Joshi A, Holt J, Zak R (2019) Relext: relation extraction using deep learning approaches for cybersecurity knowledge graph improvement. In: Proceedings IEEE/ACM international conference on advances in social network analysis and mining, pp 879–886
39. Chin-Ching Lin A, Wong H-K, Wu T-C (2005) Enhancing interoperability of security operation center to heterogeneous intrusion detection systems. In: Proceedings 39th annual international Carnahan conference on security technology, pp 216–221
40. Nabil M, Soukainat S, Lakbabi A, Ghizlane O (2017) SIEM selection criteria for an efficient contextual security. In: Proceedings international symposium on networks, computers and communications (ISNCC), pp 1–6, May 2017
41. Khyavi MH (2015) Isms role in the improvement of digital forensics related process in soc's. arXiv:2006.08255 [online] Available: https://arxiv.org/abs/2006.08255
42. Bridges R, Iannacone M, Goodall J, Beaver J (2018) How do information security workers use host data? A summary of interviews with security analysts. arXiv:1812.02867v1 [online] Available: https://arxiv.org/abs/1812.02867
43. Osinga F (2007) Science strategy and war: the strategic theory of John Boyd. Routledge, London, U.K.
44. Hámornik B, Krasznay C (2018) A team-level perspective of human factors in cyber security: security operations centers. In: Advances in human factors in cybersecurity, vol 593. Springer, Cham, Switzerland, pp 224–236
45. Vielberth M, Menges F, Pernul G (2019) Human-as-a-security-sensor for harvesting threat intelligence. Cybersecurity 2(1):35
46. Goodall JR, Ragan ED, Steed CA, Reed JW, Richardson GD, Huffer KMT et al (2019) Situ: identifying and explaining suspicious behavior in networks. IEEE Trans Vis Comput Graphics 25(1):204–214
47. Onwubiko C (2015) Cyber security operations centre: security monitoring for protecting business and supporting cyber defense strategy. In: Proceedings international conference cyber situational awareness data analytics assessment (CyberSA), pp 1–10, June 2015
48. Miloslavskaya N (2017) SOC-and SIC-based information security monitoring. In: Recent advances in information systems and technologies, vol 570. Springer, Cham, Switzerland, pp 364–374
49. Schinagl S, Schoon K, Paans R (2015) A framework for designing a security operations centre (SOC). In: Proceedings 8th Hawaii international conference on system sciences, pp 2253–2262, Jan 2015
50. Salnitri M et al (2019) Modelling the interplay of security, privacy and trust in sociotechnical systems: a computer-aided design approach. Softw Syst Model 19(2):467–491. https://doi.org/10.1007/s10270-019-00744-x
51. Quick J, Hall S (2015) Part two: qualitative research. J Perioperative Pract 25(7–8):129–133. https://doi.org/10.1177/1750458915025007
52. Wiegmann DA, Shappell SA (2017) Human error perspectives. In: A human error approach to aviation accident analysis, pp 20–44. https://doi.org/10.4324/9781315263878-2

Comparative Analysis of Heart Rate (HR) and Heart Rate Variability (HRV) Signals for User Authentication Using a Machine Learning Approach

Eniola Sharon Ajewole

Abstract This study uses Heart Rate (HR) and Heart Rate Variability (HRV) as biometric modalities for user authentication. Traditional knowledge-based and possession-based authentication methods suffer from significant limitations, such as vulnerability to security breaches and the inconvenience of physical devices. Biometrics, such as HR and HRV, provide a unique and potentially more secure alternative. By collecting HR and HRV data from 16 participants using a wearable device, this study evaluates the feasibility of using these physiological signals for continuous and transparent authentication. The data was processed using feature extraction techniques, including continuous wavelet transformation and statistical analysis, to identify critical features for classification. A Siamese Neural Network (SNN) is employed to classify the data, achieving significant accuracy improvements when HR data is integrated with HRV. The results demonstrate that HR-based biometric authentication provides higher accuracy than HRV alone, with odds ratios confirming the system's ability to correctly authenticate legitimate users while minimising errors when HR and HRV data is merged.

Keywords Biometric authentication · Heart rate (HR) · Heart rate variability (HRV) · Wearable devices · Continuous authentication · Siamese neural network (SNN) · Machine learning · Feature extraction · Continuous wavelet transformation (CWT) · Signal processing · Physiological signals · Transparent authentication · Classification accuracy · Data augmentation · Smart health devices

1 Introduction

In an increasingly digitalised world, authentication has become a cornerstone of security, ensuring access to systems, data, and services is restricted to authorised individuals. With the proliferation of online services, electronic transactions, and

E. S. Ajewole (✉)
Sheffield Hallam University, Sheffield, UK
e-mail: b8009528@hallam.shu.ac.uk

H. Jahankhani and B. Issac (eds.), *Cybersecurity and Human Capabilities Through Symbiotic Artificial Intelligence*, Advanced Sciences and Technologies for Security Applications, https://doi.org/10.1007/978-3-031-82031-1_28

interconnected devices, ensuring that individuals are accurately and securely identified is paramount to protecting sensitive information and maintaining trust in digital systems [1]. Knowledge-based authentication, such as passwords, is prone to security breaches due to weak passwords, phishing attacks, and social engineering [2]. Users often need help remembering complex passwords, leading to poor password practices, such as reusing passwords across multiple platforms [3]. This makes systems highly vulnerable to attacks. Possession-based methods, like smart cards and tokens, offer higher security but can be lost, stolen, or cloned [4]. Carrying physical devices can also be inconvenient for users, particularly in environments requiring quick and seamless authentication [5].

Biometric authentication, which relies on unique physical or behavioural traits, is increasingly popular due to its potential for high accuracy and user convenience [6]. However, biometric systems have challenges. Fingerprints can be lifted from surfaces, facial recognition can be spoofed using high-resolution photos or videos, and iris scans can be compromised under certain conditions [7]. Additionally, biometric data cannot be easily changed once compromised, unlike passwords or tokens. These limitations highlight the need for more secure and reliable authentication methods that can overcome the challenges posed by current systems. To address these concerns, the use of biometrics like Heart Rate (HR) and Heart Rate Variability (HRV) for authentication presents a promising approach, offering several advantages over traditional biometric methods [3].

HR and HRV are pivotal physiological markers that have long been used to assess cardiovascular health, stress levels, and overall autonomic nervous system (ANS) balance [3]. HR refers to the number of heartbeats per minute and provides a direct measure of cardiac activity [8]. At the same time, HRV quantifies the variation in time intervals between consecutive heartbeats, offering insights into the dynamic interplay between the sympathetic and parasympathetic branches of the ANS [9]. Recently, the potential of HR and HRV as biometric modalities for human identification and authentication has gained significant attention [9].

This study is designed to simulate real-world conditions where users may be in varying physical and emotional states, thus providing a robust evaluation of HR and HRV as biometric modalities. Data is collected continuously over 16 days; this approach is critical for assessing the stability and reliability of HR and HRV under different conditions, which is essential for developing a reliable biometric system.

The remainder of this paper is structured as follows: Sect. 2 reviews the relevant literature on biometric authentication systems, highlighting recent advancements in using bio-signals such as HR and HRV. Section 3 details the data collection, processing, and analysis methodology, including feature extraction and classification techniques. Section 4 presents the results, including model performance and accuracy metrics. Section 5 discusses the implications of these findings. Finally, Sect. 6 concludes the paper with a summary of key insights, suggestions for future research, and potential practical applications.

2 Literature Survey

Bio-signal-based biometric systems offer several advantages, including enhanced user experience due to their convenience and speed and the uniqueness of the data, as each individual has a distinct set of biometrics. Unlike other behavioural biometric systems, such as keystroke, gait, signature, and voice recognition, the proposed system in this research operates independently of user activity, making it more versatile and less intrusive.

In the realm of cognitive biometrics for authentication, recent advancements have focused on utilising multiple bio-signals for enhanced security [10]. For example, a method [11] was developed to use electroencephalogram (EEG), electrocardiogram (ECG), and electrodermal activity (EDA) signals for user authentication. This approach highlights the capability of wearable devices to continuously collect and analyse multiple bio-signals, including ECG, HRV, and EDA, providing a robust and seamless authentication process [11]. These developments demonstrate the potential of integrating various physiological signals within a single device to improve authentication accuracy and user convenience.

Pinto et al. [12] highlighted the critical need for continuous user authentication in smart health monitoring devices, emphasising the shift from reactive to proactive healthcare models enabled by such technologies. Pinto et al. [12] proposed an innovative method using photoplethysmography (PPG) via a Galvanic Coupling (GC) channel, achieving a high accuracy rate of 98.66%, thereby demonstrating the potential for significantly enhancing the reliability and security of smart health devices. Research on using photoplethysmography (PPG) for authentication has shown promising results [13], with Discrete Cosine Transform (DCT) employed for feature extraction. The study [13] tested three supervised learning algorithms, achieving accuracy rates of 93% with Decision Tree (DT), 98% with K-Nearest Neighbors (KNN), and 99% with Random Forest (RF). PPG signals were normalised using maximum-minimum normalisation and spline interpolation to enhance the results further.

An ECG-based authentication system employing incremental learning was proposed to address the challenge of authenticating users across different emotional or physical states [14]. This system was tested by collecting data from 11 participants over 10-min sessions across six days. Results showed that after five days of incremental learning, the system effectively reduced the false acceptance rate from 6.49 to 4.39% and increased the true acceptance rate from 61.32% to 87.61%. Another study [15] developed an ECG system to authenticate users based on nonlinear normalisation across various conditions and activities. This system achieved an accuracy of 99.05% in resting states and 88.14% in non-resting states. These studies [14] and [15] collectively demonstrate the potential of using bio signals for continuous authentication across different physical and emotional states. However, further research involving more participants over extended periods in uncontrolled environments is needed to produce more robust results.

Continuous and transparent authentication systems offer significant convenience for users and reduce the risk of identity theft by eliminating the need for repeated manual authentication. To achieve this, a trustworthy, continuous, and contactless authentication system based on heartbeat biometrics was developed [16]. This system utilises a cardiac scan, which measures the unique automatic heart deformation caused by cardiac muscle self-excitement, leading to a high Balanced Accuracy (BAC) of 98.61% and an Equal Error Rate (EER) of 4.42%. These results [16] demonstrate the feasibility and effectiveness of using heartbeat signals for continuous and transparent authentication. Similarly, another system [17] was proposed that leverages chest vibrations from a person's heartbeat, known as a seismocardiogram, for mobile phone user authentication. This system allows users to authenticate by placing their mobile phone on their chest for a few seconds. With an accuracy of 96.49% achieved using discrete wavelength feature extraction and a Support Vector Machine (SVM) classifier, this method presents a novel approach that is difficult to replicate. However, as the trend in authentication systems moves towards continuous and implicit authentication, the focus is increasingly on methods that require no additional user actions for authentication, making such systems even more user-friendly and secure.

The exploration of biometric authentication using HR and HRV data offers promising advancements over traditional physical and behavioural biometrics, which have inherent limitations such as susceptibility to forgery and the need for active user participation. Bio-signal-based systems, which leverage the unique physiological characteristics of individuals, provide a more secure and user-friendly alternative. Recent studies have demonstrated the feasibility of using multiple bio-signals—such as ECG, PPG, and EEG—in various combinations to enhance authentication accuracy and robustness. The proposed study on using HR and HRV data from 16 participants over several days for authentication and validation builds on these findings, aiming further to explore the potential of continuous and transparent biometric systems. By collecting and analysing HR and HRV data under different conditions and over extended periods, this study seeks to address current gaps in research, particularly regarding the variability of physiological signals in various emotional and physical states.

The results from previous research underscore the viability of using bio-signals for continuous authentication, with various systems achieving high accuracy rates in controlled and semi-controlled environments. However, the proposed study's focus on HR and HRV data could provide more insight into their advantages and limitations, especially in dynamic and uncontrolled settings.

2.1 Research Contribution

This research integrates HR and Heart Rate Variability HRV as biometric modalities for enhancing user authentication, providing a continuous and non-intrusive mechanism. The uniqueness of this work is summarised as follows:

- Evaluating HR and HRV as biometric signals introduces a novel approach to user authentication, ensuring enhanced security without requiring active user participation.
- This work proposes a methodology for combining HR and HRV data, potentially improving data consistency and accuracy across diverse operating conditions. Analysing and representing the dataset through advanced feature extraction techniques makes the system more resilient to variations and less susceptible to external influences, such as environmental noise or device inconsistencies. This helps ensure the authentication process remains robust, even in dynamic conditions.
- Data augmentation simulates real-world variability by introducing synthetic data that mimics noise and other external factors. This approach enhances the generalisation capabilities of machine learning models, making the authentication system more resistant to external noise and improving its overall performance in recognising legitimate users.

3 Proposed Model

This model outlines a structured approach for collecting, processing, and analysing HR and HRV data using wearable devices to develop a robust authentication framework. The proposed model (Fig. 1) includes advanced signal processing techniques for extracting relevant features, machine learning algorithms for classification, and validation procedures to assess system performance. The aim is to create an authentication system that balances security and usability, leveraging the uniqueness of physiological signals to deliver reliable and seamless user verification.

The process begins with data collection. If successful, the process moves to data segmentation; otherwise, it restarts. After data segmentation, the data undergoes augmentation to enhance the dataset. The augmented data is then transformed using a continuous wavelet method to extract detailed time–frequency information, followed by feature extraction. The most relevant features are selected and classified using a Siamese Neural Network (SNN). Finally, validation is performed by analysing the results, and the process repeats.

3.1 Data Collection

Data is collected using a Fitbit Sense device, a health-focused smartwatch known for its advanced biometric sensors that measure HR and HRV [18]. The device utilises a multi-path optical heart rate sensor and an electrodermal activity (EDA) sensor, which work together to enhance data accuracy and reliability [19]. The HR sensor continuously tracks beats per minute (BPM). At the same time, HRV is calculated by analysing the root mean square of successive differences between normal heartbeats

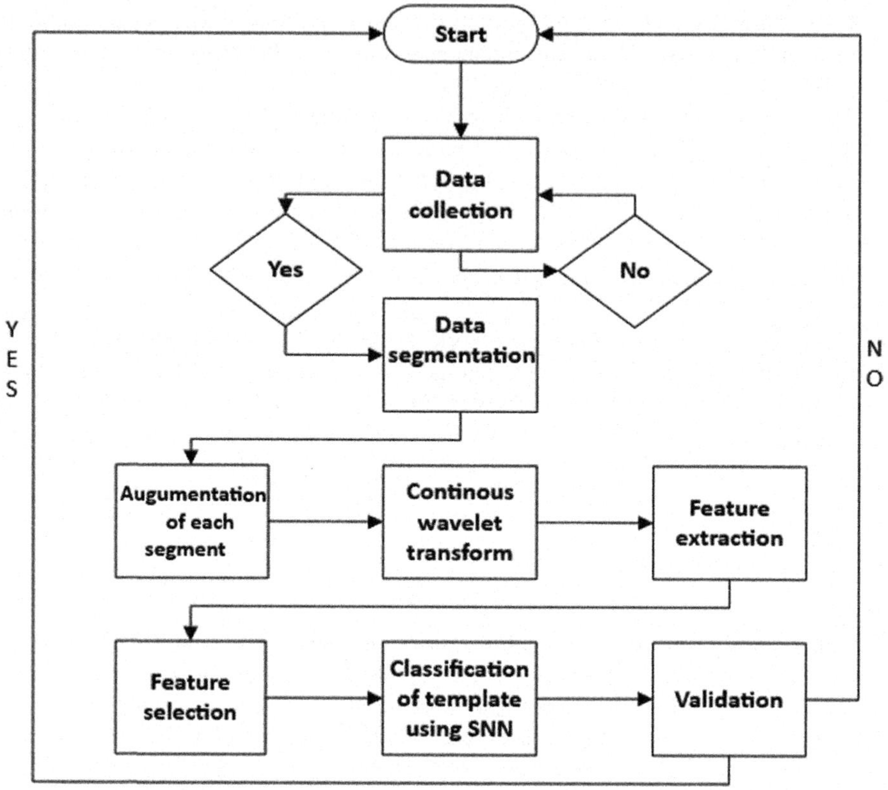

Fig. 1 Process flow for the proposed model

(RMSSD) during sleep when readings are more stable [20]. Compared to competitors like the Apple Watch, Fitbit Sense excels in ease of use and user engagement, making it accessible to a broader audience due to its affordability and specialised focus on everyday health metrics [21, 22]. Its unique Readiness Score feature provides a comprehensive wellness assessment by integrating various health indicators [23]. Participants are advised to put on the smartwatch day and night.

HRV data is collected only during sleep when a participant is settled mainly at night, while HR data is collected during the day and at night. To ensure a balance in the data extracted for both the HR and HRV data signals, the data extracted and processed was between 9 pm and 6 am to match the time HRV data is recorded for most participants. Data was collected from 16 participants over 16 days. HR data was recorded consistently for all participants across the full 16 days. However, HRV data was only available for 10 participants over ten days. This discrepancy in data collection occurred because HRV data was not recorded on certain days when HR data was, potentially due to factors like participants' sleep stability, which might have affected the availability of HRV data on those days.

In this study, data was segmented based on the sampling frequency. HR data was recorded every 5 s, while HRV data was sampled at one data point every 5 min. This segmentation strategy allowed for more accurate and detailed analysis of the temporal patterns in HRV, facilitating a more precise examination of biometric signals over time. The sampling rate for both HR and HRV is calculated using Eq. (1):

$$\frac{1}{collection\ rate\ in\ seconds} \tag{1}$$

The sampling rate for HR = 0.2 hz.

The sampling rate for HRV = 0.0033 hz.

The time-based segmentation approach divides HR data into segments containing 10 data points each. To ensure continuity and capture overlapping trends, each segment overlaps by 5 data points, resulting in 1 min of data per segment based on a 0.2 Hz sampling rate. For HRV data, each segment consists of 6 data points, with an overlap of 3 data points between segments. This setup results in each segment representing 30 min of data, assuming a 0.0033 Hz sampling rate. These frequencies were chosen to accommodate the natural physiological characteristics of the two metrics. HR, being more dynamic and rapidly changing, requires a higher sampling rate (0.2 Hz) to capture the fast variations in heartbeats. In contrast, HRV, which measures variations between heartbeats and is inherently more stable, can be sampled at a lower rate (0.0033 Hz) without losing meaningful information.

3.2 *Data Augmentation*

In the context of HR and HRV data, augmentation introduces variations that mimic real-world noise and variability, which helps models become more resilient to inconsistencies in physiological data, such as noise and artefacts [24, 25]. One specific technique, data jittering, involves adding minor random variations to the timing or magnitude of data points, enhancing the robustness of models trained on such data by simulating real-world fluctuations [25]. The author [26] explained that in the context of time series classification, which HR and HRV fall under, data jittering can be helpful in several ways, such as enhancing model generalisation, improving robustness to noise, capturing intra-individual variability, and preventing overfitting. This function applies to jitter, a data augmentation technique. Gaussian noise (with a standard deviation of 0.01) is first added to the data using a jitter function. The standard deviation (std) of 0.01 in the data jittering process was chosen because it strikes a balance between introducing sufficient noise to generalise the model without distorting the underlying physiological patterns of the HR and HRV signals.

3.3 Signal Processing

Wavelet Transformation is widely used in biosignal analysis, including HR and HRV data [27]. CWT is chosen for this research due to its suitability for HR and HRV, which are non-stationary signals. CWT provides a continuous representation of a signal in both time and frequency domains, making it ideal for analysing non-stationary signals with transient events and oscillatory patterns. The Complex Morlet Transform (CMT), a variant of CWT, is used. The Morlet wavelet, a complex-valued waveform related to the Gaussian function, is effective for analysing oscillatory signals [26].

CMT w can be defined in Eq. (2) as the product of a complex sine wave and a Gaussian window:

$$w = e^{2i\pi ft} e^{\frac{-t^2}{2\sigma^2}} \tag{2}$$

where I is the imaginary operator using Eq. (3):

$$(i = \sqrt{\{-1\}}) \tag{3}$$

f is frequency in Hz, and t is time in seconds.
σ is the width of the Gaussian, which is defined by using Eq. (4):

$$\sigma = \frac{n}{2^{\pi} f} \tag{4}$$

CMT retains phase information, offering insights into signal dynamics, and provides a time–frequency representation that captures both time-domain and frequency-domain characteristics. This multi-resolution analysis is crucial for identifying physiological phenomena at different temporal and spectral resolutions [27]. CMT also enhances feature extraction by identifying relevant time–frequency components associated with specific physiological processes in HR and HRV data [28]. By reducing the high-dimensional feature space created by CMT, feature extraction retains essential information while discarding irrelevant details, improving computational efficiency and simplifying subsequent analysis tasks [29].

3.4 Feature Extraction and Selection

Feature extraction is essential in processing HR and HRV data after applying the CMT, as it simplifies analysis by reducing dimensionality, identifying key biomarkers, and enhancing interpretability. Eight features are extracted for this study from HR and HRV data, categorised into time-domain and frequency-domain features. These extracted features serve as input variables for machine learning algorithms, enabling accurate pattern recognition and classification of physiological states. Time-domain features include the mean, which reflects the average value of the signal; variance, which indicates the spread or variability, skewness describing the asymmetry of the signal distribution; root mean square (RMS) representing the signal's magnitude or energy; and kurtosis, which indicates the peakiness or flatness of the signal distribution.

Frequency-domain features encompass the dominant frequency, the frequency with the highest spectral power, the spectral centroid as the weighted average frequency of the signal spectrum, and spectral spread, representing the spectral power's dispersion around the spectral centroid. These features capture both statistical and spectral properties of HR and HRV signals, contributing to comprehensive analysis and interpretation of physiological data.

Given that HRV and HR data can include numerous features from both time-domain and frequency-domain analyses, feature selection helps reduce the dataset's dimensionality. This reduction enhances computational efficiency, prevents overfitting, and improves the interpretability of the results. By focusing on the most relevant features, feature selection minimises the risk of overfitting, which occurs when a model learns to recognise noise rather than the underlying data relationships, leading to poor generalisation of new data [30]. The SelectKBest method from scikit-learn is used in this research to select the top 5 features based on ANOVA F-values. This method separates features (X) from the target variable (Y), scales them using MinMaxScaler, and then applies ANOVA to assess the significance of each feature. The k features with the highest F-values, indicating the most significant differences across target variable classes, are selected, ensuring uniformity across all users and data files.

ANOVA F-value can be expressed using Eq. *(5):*

$$\mathrm{Fi} = \frac{between - group\ variance}{within - group\ variance} = \frac{explained\ variance}{residual\ variance} \tag{5}$$

Between-group variance is calculated using Eqs. (6) and (7):

$$\mathrm{SSBi} = \sum_{j=1}^{k} nj\left(y^{-j}. - y^{-}..\right)^2 \tag{6}$$

$$\mathrm{MSBi} = \frac{SSEi}{N - k} \tag{7}$$

where:

- SSBiSSBi is the sum of squares between groups for feature XiXi,
- y¯j·y¯ j· is the mean of the target variable yy for group jj,
- y¯··y¯ ·· is the overall mean of the target variable,
- njnj is the number of samples in group jj,
- kk is the number of groups (or classes),

- MSBiMSBi is the mean squares between groups.

Within-group variance is calculated using Eqs. (8) and (9):

$$\mathrm{SSEi} = \sum_{j=1}^{k} \cdot \sum_{l=1}^{nj} (y_{jl} - y_{j.})^2 \tag{8}$$

$$\mathrm{MSEi} = \frac{SSEi}{N - k} \tag{9}$$

where:

- SSEiSSEi is the sum of squares of residuals (within groups) for feature XiXi,
- yjlyjl is the ll-th observation in group jj,
- y¯j·y¯j· is the mean of the target variable yy for group jj,
- NN is the total number of observations,
- MSEiMSEi is the mean squares of residuals.

The ANOVA F-value Fi is then computed using Eq. (10) as the ratio of the mean squares between groups (MSBiMSBi) to the mean squares of residuals (MSEiMSEi):

$$\mathrm{Fi} = \frac{MSBi}{MSEi} \tag{10}$$

Features with higher F-values are considered more significant and are selected for inclusion in the model or further analysis. The SelectKBest method retained the top 5 features with the highest ANOVA F-values, ensuring that only the most significant features were kept for classification. This choice of "k = 5" was driven by the need to balance model complexity with performance. By focusing on the most relevant features, the model reduces the risk of overfitting while still capturing critical information, improving computational efficiency and ensuring robust performance.

3.5 Classification

Siamese Neural Network (SNN), a variant of the Neural Network, is used for classification. SNN is a supervised learning algorithm that requires labelled data during training. It learns a similarity function by analysing pairs of input instances with corresponding labels (e.g., similarity or dissimilarity).

In a Siamese Neural Network (SNN), two identical network branches (Fig. 2) have input layers, each processing one of the input instances to be compared. These branches share the same layers, known as the Siamese tower, which typically include convolutional, pooling, and possibly fully connected layers. These shared layers

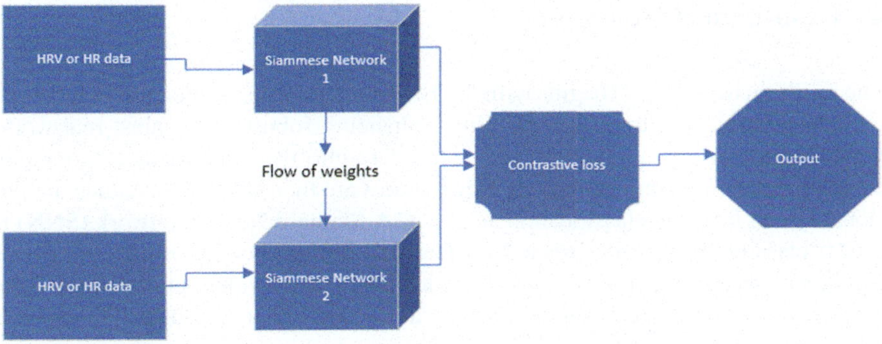

Fig. 2 SNN architecture

encode the inputs into fixed-length vector representations. The similarity or dissimilarity between these vectors is measured by calculating the Euclidean distance or cosine similarity.

The network is trained using a contrastive loss function to differentiate between pairs of similar and dissimilar samples. The load_and_process_data(user_folders) function scales and aggregates data from feature-selected and feature-extracted templates while create_balanced_pairs(user_data) assigns positive labels to pairs from the same user and negative labels to pairs from different users, ensuring balanced training data. The model consists of three fully connected layers with 128, 64, and 32 neurons, using the ReLU activation function to prevent vanishing gradients. A batch size of 32 and a learning rate of 0.0001 ensure efficient and stable training, with early stopping after five epochs without improvement to prevent overfitting. Data is split 80/20 for training and validation, with pairs created from the same or different users for the Siamese Neural Network (SNN). After training, the model is evaluated on a test set using metrics such as accuracy, precision, recall, F1 score, and individual user-level metrics. Euclidean distance measures vector similarity, and contrastive loss determines pair similarity, ensuring model performance with early stopping based on validation loss.

- **Accuracy**: The proportion of correctly classified instances among all instances.
- **F1 Score**: The harmonic mean of precision and recall, balancing them.
- **Precision**: The proportion of true positive predictions among all positive predictions.
- **Recall**: The proportion of true positive predictions among all actual positives.
- **Confusion Matrix**: A table showing the counts of true positive (TP), true negative (TN), false positive (FP), and false negative (FN) predictions.

4 Result and Discussion

The result section (Fig. 3) begins with the HR and HRV data segmentation, followed by data jittering to introduce variability, enhancing robustness against real-world noise. Next, CWT is applied to convert the data into the time–frequency domain, enabling detailed feature extraction. Key features are then selected to reduce dimensionality and improve classification accuracy. A Siamese Neural Network (SNN) is used to classify the feature-extracted and feature-selected data. The results from these classifications are compared to identify which template provides better performance. Finally, individual user accuracy metrics are generated to evaluate the system's effectiveness in authenticating users.

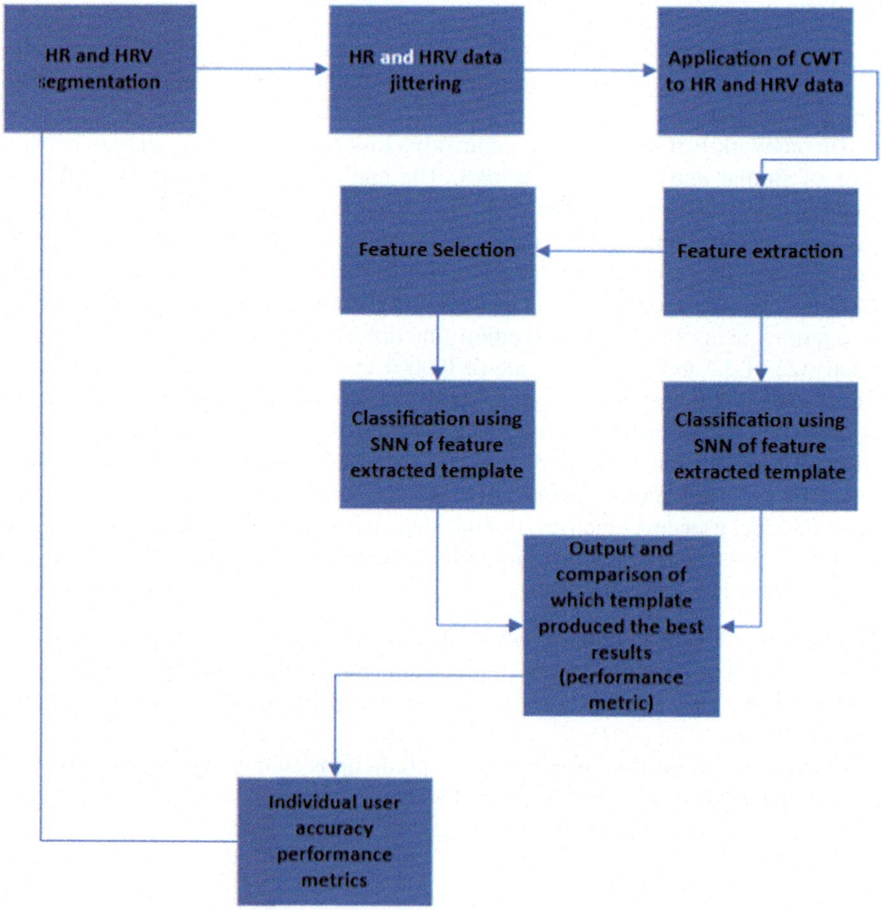

Fig. 3 HR and HRV result process flow

The classification process is divided into the following core steps:

- The initial phase is to use the HR and HRV feature-extracted data template from all users for all days to get a general idea of the system's metric outcomes.
- The next step is to use the HR and HRV features selected data from all users for all days to get the following metric result.
- Individual user metrics are analysed using HR and HRV feature-selected data templates separately to achieve results for each user.
- HR and HRV data are merged, and results are achieved using feature-extracted data.
- HR and HRV data are merged, and results are achieved using HR and HRV feature-selected data templates.
- Individual user HR and HRV merged feature selected data template is classified to achieve individual user metrics.

4.1 *Trained Output Using HRV and HR Feature Extracted Data Template*

The model trained using the HRV feature extracted data template shows relatively low accuracy (Fig. 4), ranging from 49 to 50% across different epochs. The F1 score also fluctuates between 49 and 55%, indicating inconsistency in the model's ability to balance precision and recall. Precision is stable at around 50%, but recall is notably lower, especially in the 15-epoch result (36.03%). This suggests that the model struggles to correctly identify true positive cases, leading to a high number of false negatives (FN).

The HR feature extracted data template shows much better performance in Fig. 5 when compared to the HRV output in Fig. 4, with accuracy ranging from 62 to 64%

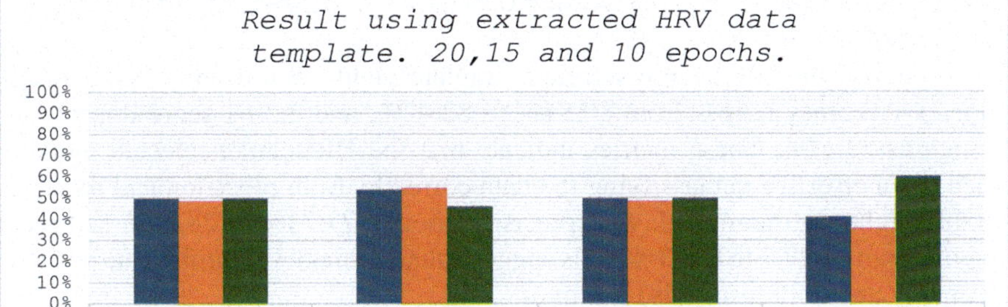

	Accuracy	F1 score	Precision	Recall
20	50%	54%	50%	41%
15	49%	55%	49%	36%
10	50%	46%	50%	60%

Fig. 4 Result using extracted HRV data template

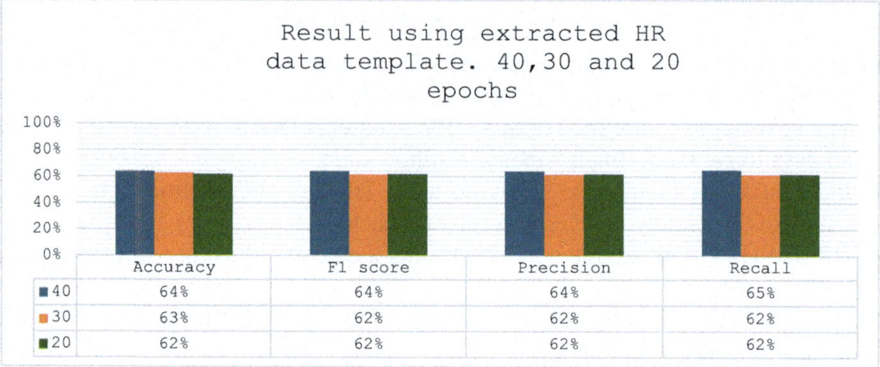

Fig. 5 Result using extracted HR data template

and F1 scores consistently above 61%. This suggests that the HR features are more effective in distinguishing between different classes. Both precision and recall are higher than the HRV model, with precision around 63–64% and recall around 61–65%. This indicates a better balance in the model's ability to correctly identify true positives while minimising false positives (FP).

4.2 Trained Output Using HRV and HR Feature Selected Data Template

The accuracy for the HRV feature selected data template fluctuates around 50%, with the highest being 51.00% (Fig. 6). The F1 scores similarly hover around 50%, indicating that the model struggles to distinguish between different classes effectively. The precision and recall metrics could be more consistent, with recall varying between 44.21 and 54.33%, suggesting difficulties in consistently identifying true positive cases.

In contrast, the HR feature-selected template yields significantly better results (Fig. 7). Accuracy ranges from 80.54% to 82.02%, and the F1 score improves to as high as 83.17%. These metrics indicate that the HR feature-selected model is much more effective in classifying the data correctly. Both precision and recall are notably higher in this model, with precision around 77.73% to 78.20% and recall ranging from 87.27% to 88.81%. This suggests that the model identifies true positives accurately and minimises false positives.

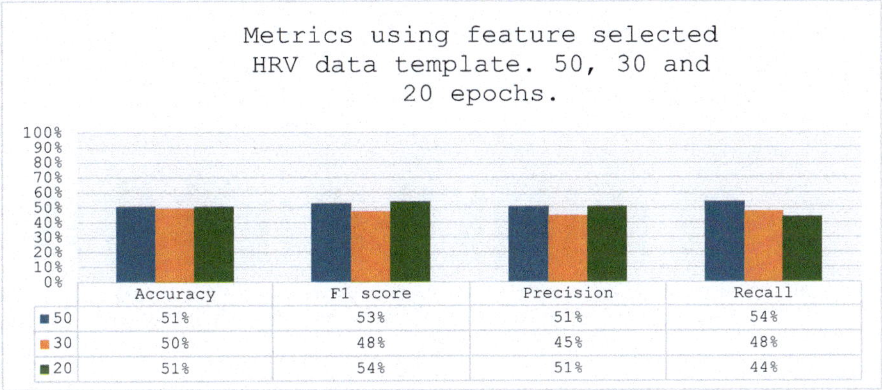

Fig. 6 Metrics using feature-selected HRV data template

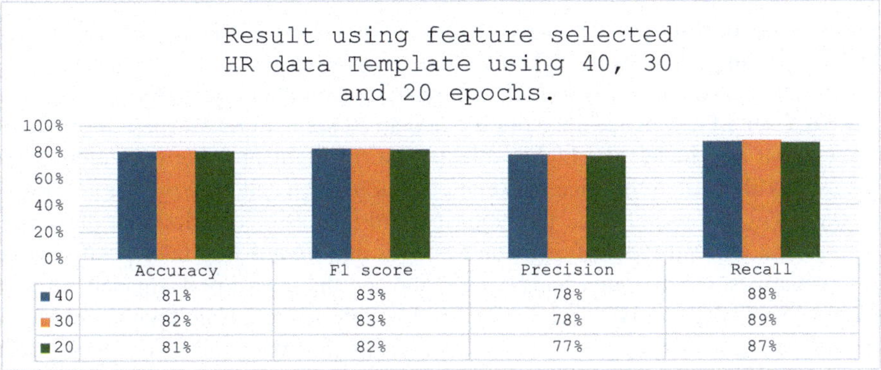

Fig. 7 Result using feature selected HR data template

4.3 Trained Output Using Individual User HRV and HR Feature Selected Data Template

In this section, individual user accuracy results are processed using the create_pairs_with_user_ids function. This function generates positive pairs (data from the same user) and negative pairs (data from different users) for training the Siamese Neural Network (SNN), with user IDs tracked to evaluate performance per user.

The HRV-based model shows a relatively poor performance across all users (Fig. 8). Accuracy hovers around 45% to 54%, with F1 scores reflecting a similar trend. User 5 achieved the highest accuracy of 54%, and user 3 had the highest F1 score of 60%, indicating slightly better performance in balancing precision and recall. The precision and recall metrics vary significantly, with user 3 again showing the highest recall at 66%. This indicates that while user 10 was better at identifying true

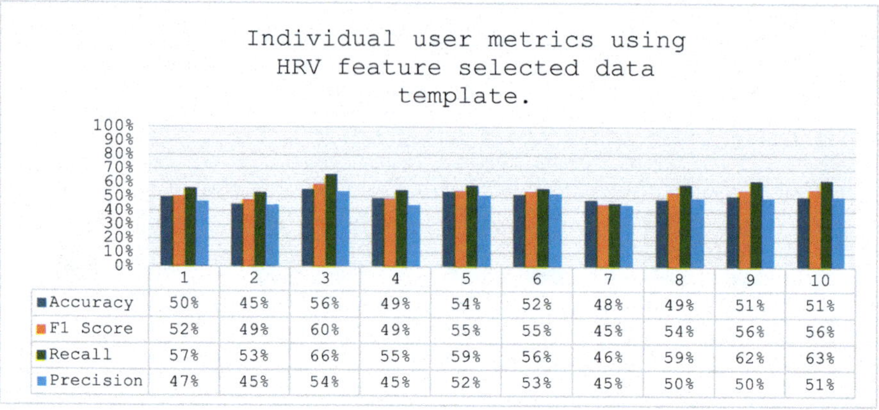

Fig. 8 Individual user metrics using the HRV feature selected data template

positives, the overall model struggled with distinguishing between users accurately, leading to many false positives (FP) and false negatives (FN).

The HR feature-selected model demonstrates significantly better performance (Fig. 9). User 14 has an accuracy of 89% and an F1 score of 89%. These high scores suggest that the model was highly effective at distinguishing between classes for this user, with a strong balance between precision and recall. The precision for user 14 was 88%, and user 15 had the highest recall of 90%, indicating that the model was good at correctly identifying true positives and minimising false positives and false negatives. On the other hand, user 6 showed the lowest performance, with an accuracy of 77% and an F1 score of 79%. Although still relatively high compared to the HRV model, this suggests that even within the HR data, there are variations in how well the model performs depending on the user's data.

4.4 Trained Output Using HRV and HR Merged Feature Extracted Data Template

The accuracy of the model ranges between 66 and 67% across different epochs, with the highest F1 score being 69% at 60 epochs (Fig. 10). These metrics indicate that the merged HR and HRV feature extracted data template leads to a moderate improvement in performance compared to using HRV data alone. The model shows a precision ranging from 58 to 65% and recall values from 69 to 74%. The recall rates, particularly at 60 epochs, suggest that the model is better at identifying true positives, but the relatively lower precision indicates challenges in minimising false positives.

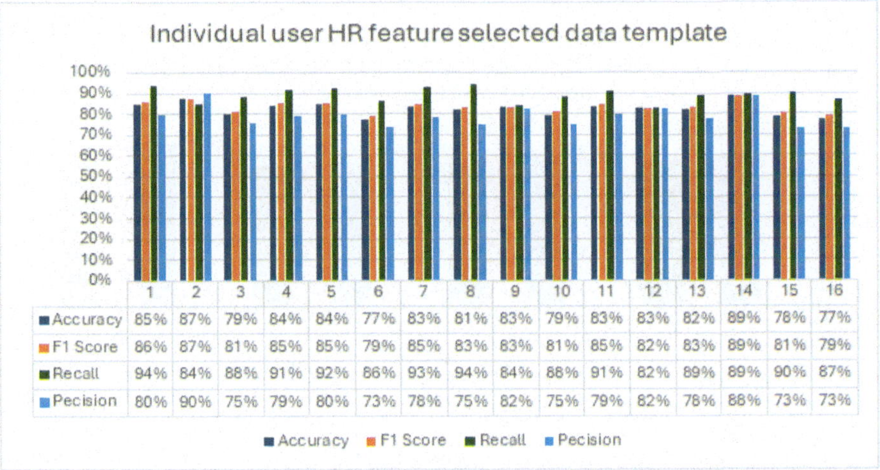

	1	2	3	4	5	6	7	8	9	10	11	12	13	14	15	16
Accuracy	85%	87%	79%	84%	84%	77%	83%	81%	83%	79%	83%	83%	82%	89%	78%	77%
F1 Score	86%	87%	81%	85%	85%	79%	85%	83%	83%	81%	85%	82%	83%	89%	81%	79%
Recall	94%	84%	88%	91%	92%	86%	93%	94%	84%	88%	91%	82%	89%	89%	90%	87%
Pecision	80%	90%	75%	79%	80%	73%	78%	75%	82%	75%	79%	82%	78%	88%	73%	73%

Fig. 9 Individual user HR feature selected data template

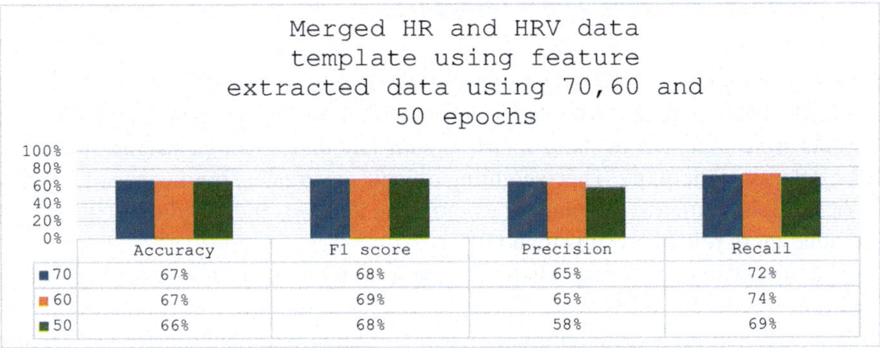

	Accuracy	F1 score	Precision	Recall
70	67%	68%	65%	72%
60	67%	69%	65%	74%
50	66%	68%	58%	69%

Fig. 10 Merged HR and HRV data templates using feature-extracted data

4.5 *Trained Output Using HRV and HR Merged Feature Selected Data Template*

The accuracy of the model ranges from 82 to 84%, with the F1 score ranging from 83 to 85% (Fig. 11). These metrics indicate that the merged HR and HRV data, when feature selection is applied, leads to a substantial enhancement in model performance compared to using either HRV alone or even the basic combined HR and HRV model without feature selection. Precision ranges from 79 to 80%, and recall values are particularly high, from 89 to 91%. This balance suggests the model is highly effective at identifying true positives while keeping false positives relatively low. The high recall is especially noteworthy, as it indicates that the model is less likely to miss true positive instances, making it more reliable in real-world applications.

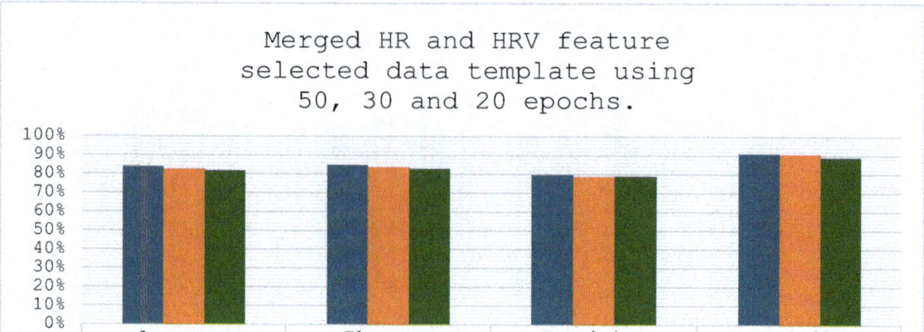

	Accuracy	F1 score	Precision	Recall
50	84%	85%	80%	91%
30	83%	84%	79%	91%
20	82%	83%	79%	89%

Fig. 11 Merged HR and HRV feature selected data template

4.6 *Trained Output Using Individual HRV and HR Merged Feature Selected Data Template*

Two separate Siamese Neural Network (SNN) models are developed to train on HR and HRV data templates. To create a combined metric that accounts for HRV data, weights were assigned to these metrics, with HR data given a weight of 0.8 and HRV data a weight of 0.2. This weighting reflects the greater significance of HR data in the combined metric calculation, as previous analyses indicated that HR metrics provided more reliable results than HRV.

The merged accuracy across users ranges from 69 to 84%, with User 6 achieving the highest accuracy of 84% and F1 score of 84% (Fig. 12). This indicates that the merged HR and HRV data template, when weighted and feature-selected, significantly enhances the model's ability to classify instances for most users correctly. Conversely, User 2 has the lowest accuracy (69%) and F1 score (71%), suggesting variability in the model's effectiveness across different users. The recall rates are generally high, with most users scoring above 70%, indicating that the model effectively identifies true positives. User 6 also excels in precision, achieving 87%, highlighting the model's strong performance in minimising false positives. On the other hand, users with lower precision, such as User 7 (64%), show a greater likelihood of false positives, indicating that while the model is generally reliable, there are variations in its effectiveness depending on the user's data.

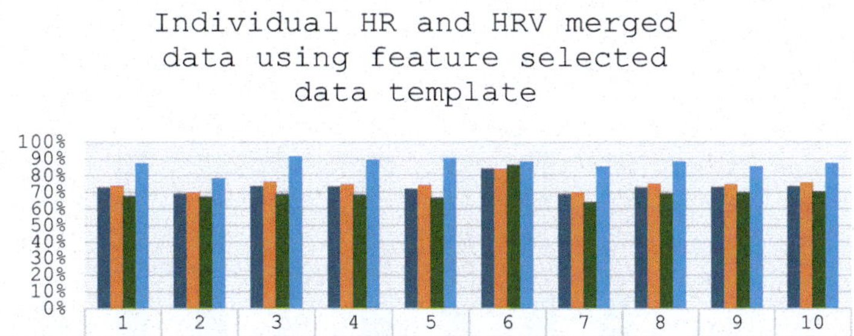

	1	2	3	4	5	6	7	8	9	10
Merged_Accuracy	73%	69%	74%	74%	72%	84%	69%	73%	73%	74%
Merged_F1_Score	75%	71%	77%	75%	75%	84%	71%	76%	75%	76%
Merged_Precision	68%	68%	69%	69%	67%	87%	64%	69%	70%	70%
Merged_Recall	88%	79%	92%	90%	91%	89%	86%	89%	86%	88%

Fig. 12 Individual merged HR and HRV merged metrics using feature-extracted data

5 Evaluation

The evaluation of various models trained using the fusion of HR and HRV reveals significant insights into the effectiveness of using these biometric signals for user identification and data validation. It is essential to compute threshold values for key performance metrics to establish an average benchmark for using HR and HRV data to identify and validate users. In machine learning and authentication systems, odds ratios (OR) measure a system's ability to correctly identify a legitimate user rather than an imposter. The OR is calculated using the TN, TP, FN, and FP values (Figs. 13 and 14).

Figures 15 and 16 display the TN versus FN percentage across the 10 users, highlighting the proportion of correctly identified negative cases (imposters) relative to the total negative predictions (both true and false negatives). TN versus FN percentage is calculated using Eq. 11

$$\frac{TN}{TN + FN} \times 100 \tag{11}$$

The chart (Fig. 16) shows that the system performs consistently well for most users, with percentages ranging between 84 and 89% for users User 1, User 3, and User 8. This indicates that the system is generally reliable in rejecting imposters while keeping false rejections of legitimate users relatively low. The system's best performance is observed for User 6, with a TN versus FN percentage of 90%, suggesting high reliability in rejecting imposters for this user. On the other hand, User 2 shows the lowest rate at 77%, indicating a higher occurrence of false negatives, meaning the system incorrectly rejects legitimate users more often for this user. The radar chart (Fig. 17) further complements this analysis by comparing the absolute values of TN

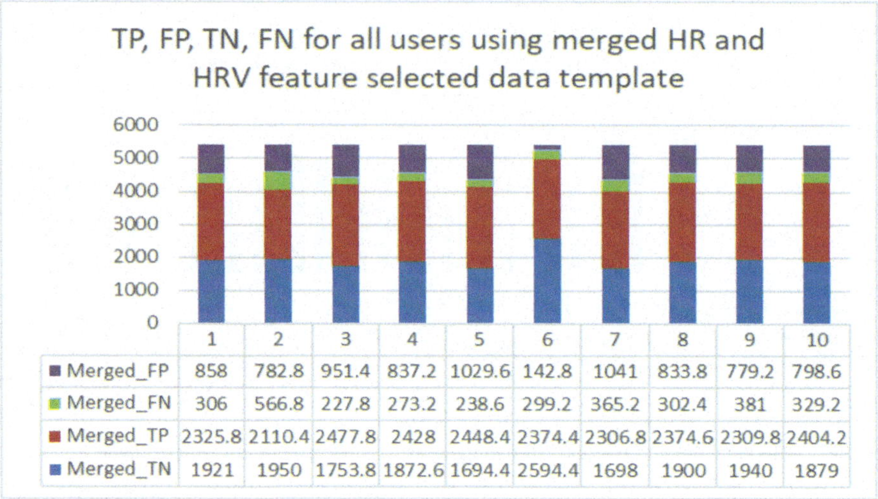

Fig. 13 TP, TN, FP FN values when HR and HRV data are merged using feature-selected template data for all users

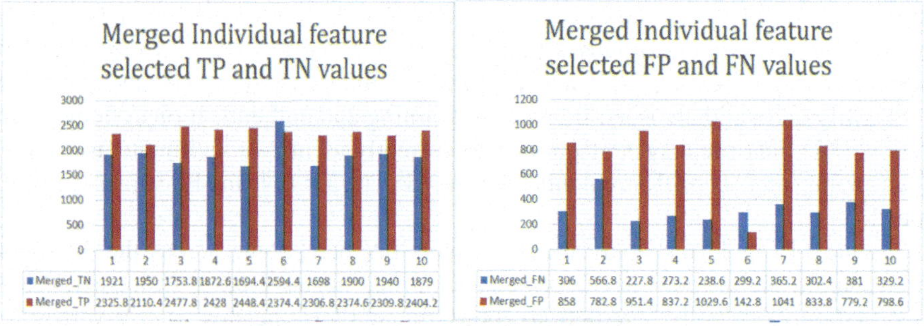

Fig. 14 TP and TN, FP and FN plot

and FN across all users. The True Negatives, represented by the blue line, consistently surpasses the False Negatives, represented by the red line, for every user. This clearly shows the system's ability to correctly identify imposters while maintaining a relatively low false negative rate. The dominance of True Negatives, especially for User 6, reinforces the system's high accuracy in identifying imposters.

Figures 18 and 19 provide valuable insights into the ability of the proposed system to authenticate users while minimising the number of false positives correctly. The results indicate that the system generally performs well across most users, with some variability in its effectiveness. TP versus FP percentage is calculated using Eq. (12):

$$\frac{TP}{TP + FP} \times 100 \tag{12}$$

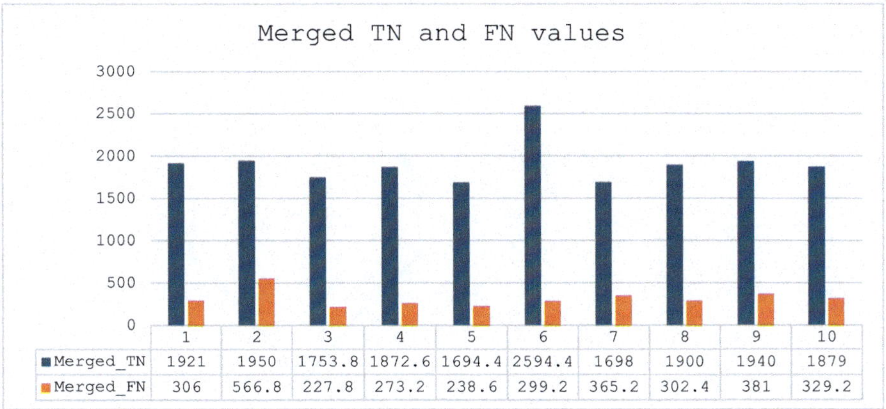

Fig. 15 Merged individual feature selected TN and FN values

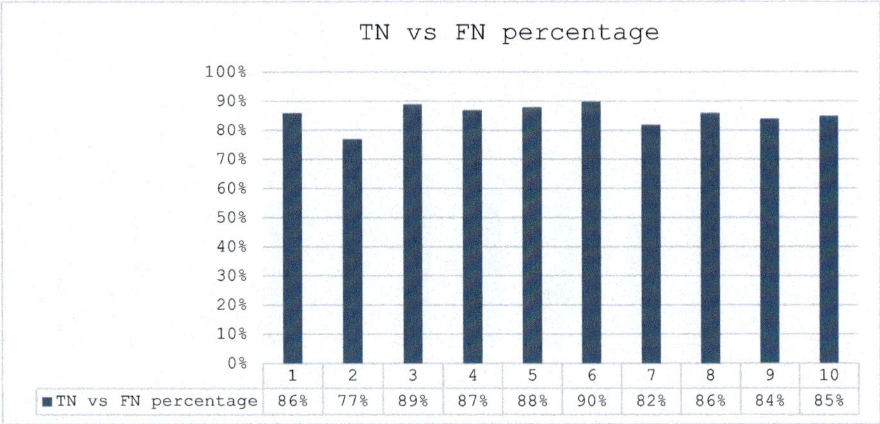

Fig. 16 Merged TN versus FN

For most users, the True Positive (TP) rate relative to False Positives (FP) is consistently high, ranging from 69 to 94%, indicating that the system is reliable, with few incorrect authentications of imposters. This high score shows that the system functions near-optimal for specific individuals. The radar chart (Fig. 20) shows that the true Positives consistently surpass False Positives for all users, emphasising the system's overall capability in correctly identifying legitimate users.

A high odds ratio indicates that the system is reliable [31]. While specific thresholds for odds ratios in machine learning may not be universally agreed upon, an OR above 10 is typically considered a very high level of reliability for binary classification systems, especially in areas like fraud detection or biometric authentication, where the goal is to favour legitimate users over imposters significantly [32]. This

Fig. 17 Merged TN versus FN

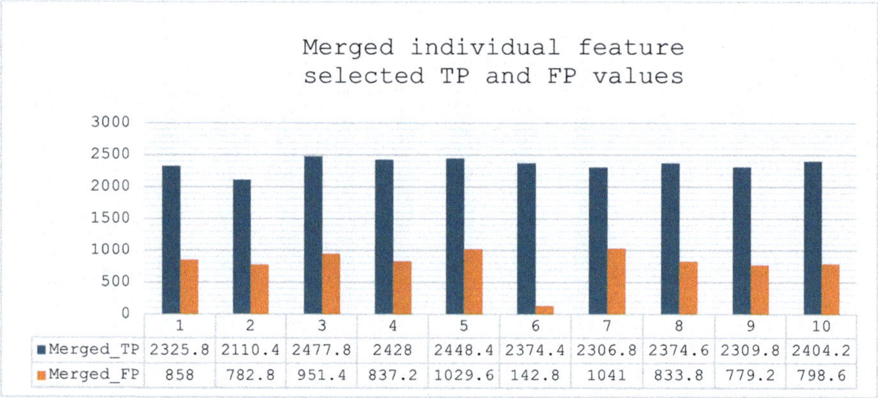

Fig. 18 Merged individual feature-selected TP and TN values

study calculates the odds ratio to assess the system's ability to distinguish between legitimate and unauthorised users.

The odds ratio is calculated using Eq. (13):

$$\text{OR} = \frac{\left(\frac{\textit{True Positives (TP)}}{\textit{False Negatives (FN)}}\right)}{\left(\frac{\textit{False Positives (FP)}}{\textit{True Negatives (TN)}}\right)} \tag{13}$$

The odds ratio values demonstrate the overall robustness of the authentication system, as shown in Fig. 21. Most cases show a strong ability to authenticate legitimate users while minimising incorrect imposter authentications. The system

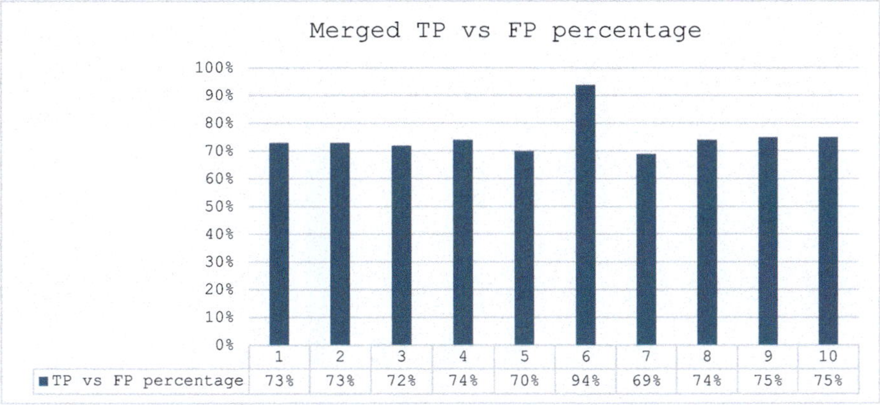

Fig. 19 Merged TP versus FP percentage

Fig. 20 Merged TP versus FP

performs exceptionally well for certain users, as seen with User 6, which has an odds ratio of 144.18. This means the system is 144 times more likely to correctly authenticate the user than to mistakenly authenticate an imposter, highlighting the system's efficiency in maintaining security for high-performing cases.

For other users, such as User 1, User 3, and User 4, the odds ratios fall within a solid range of 17 to 20, indicating that the system is still highly reliable. In these cases, the system is about 17 to 20 times more likely to authenticate legitimate users than to allow unauthorised access correctly. This reflects consistent and dependable performance, ensuring that the system balances security and accessibility effectively. The system maintains a positive outcome even for users with lower odds ratios, such as User 2, with an OR of 9.28. An odds ratio of 9 still indicates that the system is

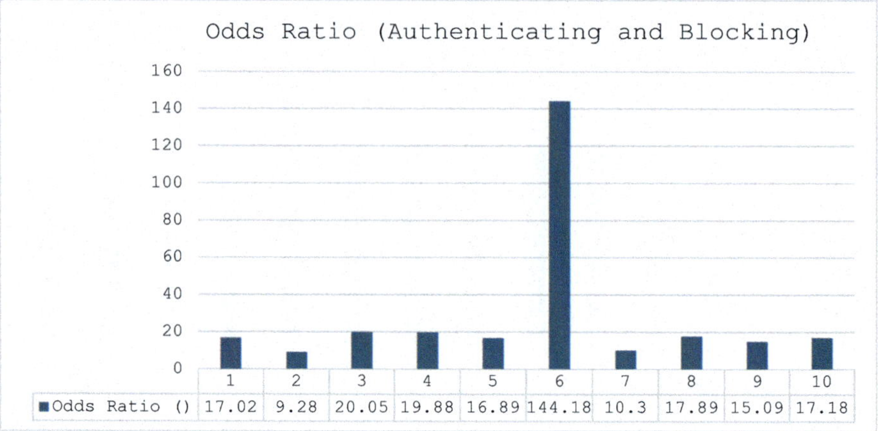

Fig. 21 Odds ratio

nine times more likely to correctly authenticate the users than to make an error with an imposter.

The study [33] explored HRV time-domain features for biometric authentication, reporting moderate performance in controlled environments with an accuracy of around 85%. In contrast, this research integrates HR and HRV data with machine learning techniques like SNN, yielding superior results, such as an odds ratio of 144.18 for User 6, indicating more robust user-authentication performance.

Overall, the odds ratio analysis illustrates that the system operates with high accuracy and security, successfully authenticating legitimate users far more frequently than allowing imposters. The system's performance is highly effective across the user base, with opportunities for enhancement in specific cases to increase the odds ratios.

6 Conclusion

The results of this study highlight the potential of using HR and HRV as biometric signals for user authentication. HR data proved to be a more reliable metric, achieving higher accuracy and F1 scores than HRV. When the HR and HRV data were combined, and feature selection was applied, the model's performance was further enhanced, resulting in balanced precision and recall. The odds ratio analysis demonstrated that the system performed exceptionally well for most users, with certain users achieving odds ratios as high as 144.18, indicating strong reliability in identifying legitimate users. One challenge encountered was the different sampling rates of HR and HRV, which complicated merging their outputs. To overcome this, two separate Siamese Neural Network (SNN) models are developed for HR and HRV, with class weights applied to merge their outputs. This approach ensured that the system could leverage the strengths of both signals, improving the overall classification accuracy while maintaining robust performance across diverse users.

Exploring the practical applications of the developed machine learning models in user authentication scenarios could provide valuable insights into their potential for real-world deployment. Using HR and HRV data, the developed machine learning model can be applied in practical user authentication scenarios, such as continuous, non-intrusive authentication for wearable devices like fitness trackers and smartwatches. These devices can continuously collect physiological signals, allowing real-time identity verification without user intervention. This approach is ideal for securing access to sensitive healthcare, finance, or mobile application systems. By analysing subtle variations in heart patterns, the model can identify users based on their unique biometric signatures. It offers a seamless and secure authentication method while reducing the risks associated with passwords or physical tokens.

Future research could focus on expanding the sample size and testing the system in more diverse and uncontrolled environments to assess its robustness further. Exploring other physiological signals with HR and HRV could also lead to more accurate and secure authentication systems. Incorporating incremental learning algorithms may also enhance the system's ability to adapt to changing physical or emotional states of users, ensuring reliable performance in dynamic conditions. Finally, efforts should be made to reduce the number of false positives for certain users to enhance overall system security, including further feature extraction and selection techniques. This is suggested as it is realised in this research that feature selection significantly improved the results achieved.

References

1. Ezugwu A, Ukwandu E, Ugwu C, Ezema M, Olebara C, Ndunagu J, Ofusori L, Ome U (2023) Password-based authentication and the experiences of end users. Scientific African 21:e01743. https://doi.org/10.1016/j.sciaf.2023.e01743
2. Billman GE, Huikuri HV, Sacha J, Trimmel K (2015) An introduction to heart rate variability: methodological considerations and clinical applications. Front Physiol 6:55. https://doi.org/10.3389/fphys.2015.00055
3. Hinojosa-Laborde C, Rickards CA, Ryan KL, Convertino VA (2011a) Heart rate variability during simulated Hemorrhage with lower body negative pressure in high and low tolerant subjects. Front Physiol 2. https://doi.org/10.3389/fphys.2011.00085
4. Feng Q, He D, Zeadally S, Khan MK, Kumar N (2019) A survey on privacy protection in blockchain system. J Netw Comput Appl 126:45–58. https://doi.org/10.1016/j.jnca.2018.10.020
5. Nanda A, Jeong JJ, Shah SWA, Nosouhi M, Doss R (2023) Examining usable security features and user perceptions of physical authentication devices. Comput Secur 103664 https://doi.org/10.1016/j.cose.2023.103664
6. Agrawal V, Hazratifard M, Elmiligi H, Gebali F (2023) Electrocardiogram (ECG)-based user authentication using deep learning algorithms. Diagnostics 13(3):439. https://doi.org/10.3390/diagnostics13030439
7. Shamshirband S, Fathi M, Chronopolous A, Monteiri A, Palumbo F, Pescape A (2020) Computational intelligence intrusion detection techniques in mobile cloud computing environments:

review, taxonomy, and open research issues. J. Inf. Secur. Appl. 55:102582. https://doi.org/10.1016/j.jisa.2020.102582
8. Chatzikokolakis K, Johnson A (2020) Editors' introduction. Proc Privacy Enhanc Technol 2020(1):1–4. https://doi.org/10.2478/popets-2020-0001
9. Lombardi F, Stein PK (2011) Origin of heart rate variability and turbulence: an appraisal of autonomic modulation of cardiovascular function. Front Physiol 2. https://doi.org/10.3389/fphys.2011.00095
10. Nakayama F, Lenz P, Banou S, Nogueira M, Santos A, Chowdhury KR (2019) A continuous user authentication system based on galvanic coupling communication for s-health. Wirel Commun Mob Comput 2019:1–11. https://doi.org/10.1155/2019/9361017
11. Khatri KK (2020) Research paradigm: a philosophy of educational research. Int Jo English Literat Soc Sci 5(5):1435–1440. https://doi.org/10.22161/ijels.55.15
12. Pinto L, Tavares T, Rosário D, Cerqueira E, Santos A, Nogueira M (2020) Double authentication model based on PPG and ECG signals. In: 2020 International Wireless Communications and Mobile Computing (IWCMC). https://doi.org/10.1109/iwcmc48107.2020.9148521
13. Kim J, Yang G, Kim J, Lee S, Kim KK, Park C (2021) Efficiently updating ECG-based biometric authentication based on incremental learning. Sensors 21(5):1568. https://doi.org/10.3390/s21051568
14. Hwang HB, Kwon H, Chung B, Lee J, Kim IY (2021) ECG authentication based on non-linear normalization under various physiological conditions. Sensors 21(21):6966. https://doi.org/10.3390/s21216966
15. Lin F, Song C, Zhuang Y, Xu W, Li C, Ren K (2017) Cardiac Scan. In: Proceedings of the 23rd annual international conference on mobile computing and networking. https://doi.org/10.1145/3117811.3117839
16. Wang L, Huang K, Wang W, Tian C, Sun K, Xie L, Gu Q (2018) Unlock with your heart: heartbeat-based authentication on commercial mobile phones. Proc ACM Interact Mob Wearable Ubiquitous Technol 2(140). https://doi.org/10.1145/3264950
17. Revett K, Deravi F, Sirlantzis K (2010) Biosignals for user authentication—towards cognitive biometrics? In: Proceedings of the 2010 international conference on emerging security technologies. https://doi.org/10.1109/est.2010.32
18. Shaffer F, Ginsberg JP (2017) An overview of heart rate variability metrics and norms. Front Publ Health 5(258). https://doi.org/10.3389/fpubh.2017.00258
19. Huhn S, Axt M, Gunga H-C, Maggioni MA, Munga S, Obor D, Sié A, Boudo V, Bunker A, Sauerborn R, Bärnighausen T, Barteit S (2022) The impact of wearable technologies in health research: scoping review. JMIR Mhealth Uhealth 10(1):e34384. https://doi.org/10.2196/34384
20. Kummer M (2023) What is a good heart rate variability (HRV)? Michael Kummer. https://michaelkummer.com/health/what-is-a-good-hrv/
21. Stoumpos AI, Kitsios F, Talias MA (2023) Digital transformation in healthcare: technology acceptance and its applications. Int J Environ Res Public Health 20(4). https://doi.org/10.3390/ijerph20043407
22. Ferguson T, Olds T, Curtis R, Blake H, Crozier AJ, Dankiw K, Dumuid D, Kasai D, O'Connor E, Virgara R, Maher C (2022) Effectiveness of wearable activity trackers to increase physical activity and improve health: a systematic review of systematic reviews and meta-analyses. The Lancet Digital Health 4(8):e615–e626. https://doi.org/10.1016/s2589-7500(22)00111-x
23. Hajj-Boutros G, Landry-Duval M-A, Comtois AS, Gouspillou G, Karelis AD (2021) Wrist-worn devices for the measurement of heart rate and energy expenditure: a validation study for the apple watch 6, polar vantage V and fitbit sense. Eur J Sport Sci 23(2):1–36. https://doi.org/10.1080/17461391.2021.2023656
24. Cohen MX (2019) A better way to define and describe Morlet wavelets for time-frequency analysis. Neuroimage 199:81–86. https://doi.org/10.1016/j.neuroimage.2019.05.048
25. Iwana BK, Uchida S (2021) An empirical survey of data augmentation for time series classification with neural networks. PLoS ONE 16(7):e0254841. https://doi.org/10.1371/journal.pone.0254841

26. Ileşan RR, Cordoş C-G, Mihăilă L-I, Fleşar R, Popescu A-S, Perju-Dumbravă L, Faragó P (2022) Proof of concept in artificial-intelligence-based wearable gait monitoring for Parkinson's disease management optimization. Biosensors 12(4):189. https://doi.org/10.3390/bios12040189
27. Siepmann M, Weidner K, Petrowski K, Siepmann T (2022) Heart rate variability: a measure of cardiovascular health and possible therapeutic target in dysautonomic mental and neurological disorders. Appl Psychophysiol Biofeedback 47(4):273–287. https://doi.org/10.1007/s10484-022-09572-0
28. Rouf N, Malik MB, Arif T, Sharma S, Singh S, Aich S, Kim H-C (2021a) Stock market prediction using machine learning techniques: a decade survey on methodologies, recent developments, and future directions. Electronics, 10(21):2717. mdpi. https://doi.org/10.3390/electronics10212717
29. Hegde J, Rokseth B (2020) Applications of machine learning methods for engineering risk assessment—a review. Saf Sci 122(1):104492. https://doi.org/10.1016/j.ssci.2019.09.015
30. Can YS, Chalabianloo N, Ekiz D, Ersoy C (2019) Continuous stress detection using wearable sensors in real life: algorithmic programming contest case study. Sensors 19(8):1849. https://doi.org/10.3390/s19081849
31. Tharwat A (2018) Classification assessment methods. Appl Comput Inf 17(1). https://doi.org/10.1016/j.aci.2018.08.003
32. Agresti A (2014) Categorical data analysis. Hoboken Wiley
33. Akhter N, Gite HR, Rabbani G, Kale KV (2015) Heart rate variability for biometric authentication using time-domain features. Commun Comput Inf Sci 168–175. https://doi.org/10.1007/978-3-319-22915-7_16
34. Lee S-W (2019) Wearable bio-signal(PPG)-based personal authentication method using random forest and period setting considering the feature of PPG signals. J Comput 14(4):283–294. https://doi.org/10.17706/jcp.14.4.283-294
35. Nagpal A, Singh V (2018) A feature selection algorithm based on qualitative mutual information for cancer microarray data. Procedia Comput Sci 132:244–252. https://doi.org/10.1016/j.procs.2018.05.195
36. Bouchlaghem Y, Akhiat Y, Amjad S (2022) Feature selection: a review and comparative study. E3S Web Conf 351:01046. https://doi.org/10.1051/e3sconf/202235101046
37. Credico AD, Perpetuini D, Izzicupo P, Gaggi G, Cardone D, Filippini C, Merla A, Ghinassi B, Baldassarre AD (2022) Estimation of heart rate variability parameters by machine learning approaches applied to facial infrared thermal imaging. Front Cardiovasc Med 9. https://doi.org/10.3389/fcvm.2022.893374

Theoretical Insights into Leveraging Machine Learning for Dynamic Cyber Risk Assessment and Enhancing Cyber Situational Awareness in SMEs

Mansour Almalki, Liqaa Nawaf, and Fiona Carroll

Abstract Small and mid-sized enterprises (SMEs) play a role in economies, but they are increasingly challenged to protect their digital assets. Unlike companies that have specialised cybersecurity teams and resources, SMEs often lack the knowledge and infrastructure needed to defend themselves against cyber threats. This vulnerability highlights the importance of taking an approach to cybersecurity which allows SMEs to stay vigilant about risks and the changing threat landscape. This theoretical paper investigates how AI (Artificial Intelligence) can enhance cyber situational awareness (CSA) and enable dynamic cyber risk assessment (DCRA) in SMEs. It establishes the criticality of CSA for effective cybersecurity measures, discusses existing CSA approaches and their limitations in the SME context, and explores the concept of cyber risk assessment and its challenges for SMEs, highlighting the need for dynamic approaches. This critical insight highlights the importance of adaptability in the cyber environment and enlightens the audience on its significance. In detail, the paper presents a literature review that delves into the role of AI in cybersecurity, highlighting its advantages for resource-constrained SMEs. It explores how AI can be harnessed to enhance DCRA and improve CSA. In conclusion, the paper summarises the essential findings and highlights the transformative potential of AI in empowering SMEs to take a proactive stance in managing their cybersecurity posture.

Keywords Cyber situational awareness · Dynamic cyber risk assessment · Machine learning · Small and medium-sized enterprises

M. Almalki (✉) · L. Nawaf · F. Carroll
Cardiff Metropolitan University, Cardiff, Wales, UK
e-mail: st20252905@outlook.cardiffmet.ac.uk

L. Nawaf
e-mail: lllnawaf@cardiffmet.ac.uk

F. Carroll
e-mail: fcarroll@cardiffmet.ac.uk

H. Jahankhani and B. Issac (eds.), *Cybersecurity and Human Capabilities Through Symbiotic Artificial Intelligence*, Advanced Sciences and Technologies for Security Applications, https://doi.org/10.1007/978-3-031-82031-1_29

1 Introduction

The lack of CSA among computer users, technology developers, system managers, and leadership in organisations can lead to cyber disasters and financial loss. For that reason, the stakeholders' knowledge about security threats, vulnerabilities, and risks should be enhanced [1].

According to Bada and Nurse [2], small and medium-sized enterprises (SMEs) face greater exposure to cyber threats and vulnerabilities from giant enterprises. For that reason, SMEs need to frequently assess cyber risks to enhance their cybersecurity state and resilience. Carías et al. [3] introduced the concept of cyber resilience as a comprehensive approach to cybersecurity for SMEs, which would enable cybersecurity specialists to anticipate, detect, resist and recover from cyber incidents.

SMEs must create resilience capabilities to mitigate risks and grow with minimal disruption. Furthermore, cyber resilience self-assessment tools help SMEs assess their cyber resilience capabilities and enhance their readiness to log on to the digital world. Such tools can help SMEs identify and mitigate cyber risks, thus contributing to their overall cybersecurity preparedness.

In this paper, we examine leveraging AI to enhance DCRA and CSA in SMEs. We will give an overview of the research background, including the research aim and objectives. As shown in. Fig. 1, we will focus on the three research areas: CSA, DCRA, and AI. In the end, we will concentrate on drawing conclusions based on our findings.

The main aim is to explore how a predictive AI can enhance DCRA practices, thus contributing to a deeper understanding and improvement of CSA for SMEs. The specific areas of discussion are the relationship between CSA and DCRA and then how we can improve both CSA and DCRA by using AI.

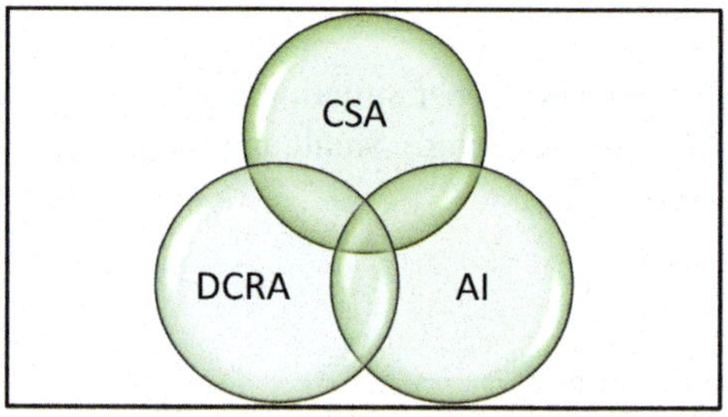

Fig. 1 The paper areas

2 Background

Prioritising CSA is crucial, particularly for decision-makers inside organisations; individuals are the weakest link in cybersecurity [4] Click or tap here to enter text.. Correspondingly, non-ITs, such as top management or board members, are usually not highly technically educated in cybersecurity and may need more awareness. Therefore, CSA should be considered more in order to improve cybersecurity and privacy [5].

One domain that can contribute to the improvement of CSA is artificial intelligence (AI). These days, AI is applied in many aspects of our lives, making it a vital component in various businesses and industries. With the rising importance of information and communications technology (ICT) in business, many SMEs need it in their operations, resulting in an increased demand for AI-based cybersecurity measures. As a result, recent technologies supporting cybersecurity have heavily integrated AI into their roles and technologies [6].

In contrast, risk management is a vital cybersecurity field that is crucial in the current business environment. With ever-evolving cyber-attacks, organisations, especially those with critical infrastructure, must understand and assess the risks for a good implementation of appropriate cybersecurity controls for business continuity and service delivery [7]. Moreover, various situations, such as rising international tensions, rapid digital advancements, and a surge in cybersecurity incidents which threaten cyberspace for business, have led to a growing need for cyber risk assessment [8].

On the other hand, many studies have confirmed a deficiency in the risk assessment aspect. According to Akinrolabu et al. [9] the authors conducted a thorough literature review to identify possible models for cloud risk assessment. They evaluated three existing models using specific criteria and compared them to their proposed model. Despite the numerous approaches developed to tackle cloud risks, the research revealed a significant gap in the study of risk assessment models. The authors suggest that future researchers could utilise this study as a foundation for developing models in the realm of risk assessment. There is a clear gap in knowledge regarding the automation of the risk assessment process, which could improve CSA and support decision-making. This paper will further investigate the implementation of AI to address this gap and the issues mentioned above.

3 Cyber Situational Awareness

To understand CSA, we should first know the meaning of situational awareness, which means being aware of the environmental elements, including their time, location, and significance. Situational awareness consists of three levels: perception of environmental elements, understanding of the present situation, and prediction of future status to have decision-making [10].

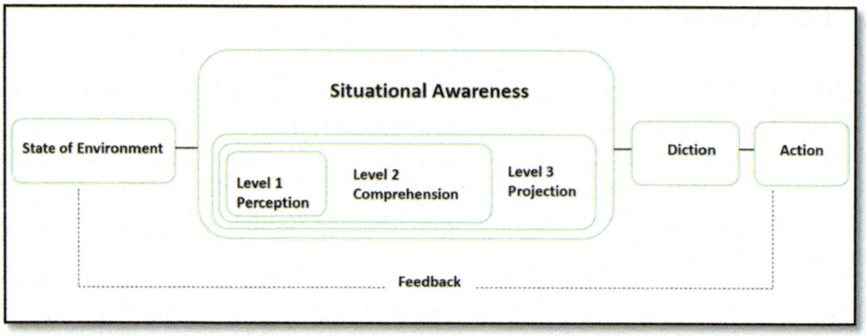

Fig. 2 Three levels of situational awareness [12]

On the other hand, Chandra et al. [11] demonstrate that situational awareness has both technical and cognitive characteristics. The technical aspects include gathering, organising, analysing, and merging data. Meanwhile, the cognitive aspect pertains to an individual's mental awareness and ability to comprehend the technical implications of a situation, enabling them to arrive at sound conclusions and make appropriate decisions.

As shown in Fig. 2, Jiang et al. [12] demonstrate the three situational awareness levels such as:

- Level 1 involves perceiving the cyber environment by identifying or detecting its conditions.
- At Level 2, specialists can understand the meaning of the current situation by gathering important information from perception to achieve their objectives.
- Level 3 requires projecting the near future to make informed decisions, which is achieved by extrapolating information from an understanding of the cyber environment to determine the impact of the current status on future conditions.

According to Jiang et al. [12] the importance of CSA is to apply what has been learned to practice. This ongoing process requires adjustments in subsequent iterations to enhance usability and sustainability. Cybersecurity is not just about technology; it also includes the people who interact with technology and are responsible for properly implementing and operating it. Similarly, Chaudhary et al. [13] showed that many studies have identified people's behaviours and actions as responsible for most cybersecurity incidents. Human behaviour and actions often cause most cybersecurity incidents; raising the end-user CSA is paramount as the first step in handling human factors.

Based on the studies above, cybersecurity is a multifaceted and ever-evolving field. Despite implementing advanced technology and sophisticated processes, these studies have revealed that individual users often represent the weakest point in the cybersecurity triad. The weakness is mainly due to their susceptibility and lack of awareness of potential risks in digital interactions. Consequently, it is imperative to enhance the organisation's CSA by providing cybersecurity training and education to staff and fostering a cybersecurity awareness culture. Employees can be empowered to identify and mitigate potential cyber threats. Thus, the risks of security breaches and data loss can be minimised by implementing security measures.

3.1 Cyber Situational Awareness Tools

There are many CSA tools used for identifying, understanding, and anticipating incoming cyber threats [14]. A study by [15] found that CSA plays a crucial role in improving cybersecurity practices. CSA tools scope a wide range of capabilities, including network-wide CSA (NwCSA) for network security administrators [16]. In contrast, cyber ranges can be used to enhance CSA and knowledge, mainly in the context of IoT security [17].

Additionally, visualisation tools for cybersecurity have rapidly matured, offering various techniques for information visualisation to address cybersecurity challenges, particularly in network traffic analysis [18].

Furthermore, the usage of honeypot and Security Information and Event Management (SIEM) tools has been offered for cybersecurity education models, highlighting the importance of technical skills in addition to increasing awareness of human weaknesses and social engineering [19, 20]. On the other hand, the use of cyber defence exercises (CDX) has been underlined as a significant tool for improving protection awareness in cyberspace, testing organisational resistance and response capabilities, and providing practical training for cybersecurity experts [21].

In the context of cyber-physical systems, CSA tools such as the adversary view security evaluation (ADVISE), the cybersecurity modelling language (CySeMol), and (CyberSAGE) have been developed to focus on probability analysis of attackers and automatic security assessment of cyber-physical systems [22, 23]. Additionally, the development of a cognitive and concurrent cyber kill chain model has emphasised the significance of tools addressing anonymity, deception, and uncertainty as significant components of cyber threats [24].

3.2 Cyber Situational Awareness in SMEs

According to Tsiodra et al. [25], SMEs face a challenging situation in cybersecurity, representing around 90% of companies globally and more than 50% of employees, according to the World Bank. Having sensitive and personally identifiable information, SMEs are becoming an easy target for exploitation by threat actors.

SMEs are businesses with less than 250 staff members [26] who are familiar with businesses and are crucial to economic development. In the UK, SMEs represent (99.9%) of the business population, with most (99.3%) having fewer than 50 employees [27]. SMEs are increasingly vulnerable to cyberattacks and must prioritise cybersecurity to protect their businesses. Cybersecurity threats are ever-increasing, and SMEs are desirable targets for cybercriminals. Therefore, SMES must have a robust cybersecurity strategy [28].

A recent study by Renaud and Ophoff [29] shows that SMEs are considered to be more susceptible to cyberattacks, mainly due to their limited resources and lack of situational awareness. They presented the current status of CSA in SMEs and suggested a model created based on Endsley's theory. In detail, they used data from an online survey of 361 SMEs in the UK, and partial least squares modelling was implemented to validate the suggested model.

Despite only focusing on one country, the research provides valuable insight into SMEs' vulnerabilities in the UK. Interestingly, the research results show that having a better understanding of potential cyber threats and adequate resources significantly influences SMEs' ability to implement cybersecurity measures and protect their cyber environment. Also, despite the limitations of resources, SMEs can effectively handle the most significant challenge of cybersecurity threats by enhancing their situational awareness and adopting appropriate measures.

In another study, Ncubukezi, Mwans, and Rocaries [30] conducted a study in South Africa to analyse the cybersecurity situation of SMEs. The study results were reliable and comprehensive because of the method of implementing the sampling when the researchers carefully selected 30 SMEs from different business sectors, ensuring that various perspectives and experiences were represented.

The findings show that cybersecurity practices in SMEs differ significantly across different business sectors. A key factor contributing to these differences is the absence of regulations, standards, procedures or guidelines aimed at fostering robust cybersecurity measures within SMEs. Consequently, SMEs often exhibit weak cybersecurity that can lead to significant economic and reputational damages. This study also identified shortcomings in current security measures and their implementation. Furthermore, it was noted that employees of SMEs have limited knowledge when it comes to managing cyber-attacks, thereby posing significant threats.

Importantly, all the studies discussed above have recommended several ways to maintain good cybersecurity in SMEs. Including, SMEs should develop detailed rules, regulations, and guidelines to promote good cybersecurity practices among employees. Additionally, they should invest in comprehensive cybersecurity training programs to equip employees with the essential knowledge and skills to identify and counter cyber threats. Furthermore, SMEs should continually monitor and update their security measures to counter the evolving nature of cyber threats, and that desire has not happened without enhancing the CSA for SMEs.

To summarise, there are different types of CSA tools that can be effective when dealing with cyber threats. Some types can give a better understanding of the network and help visualise it, while others are more specialised and can be used for cyber-physical systems. These tools are essential to increase security awareness, improve cybersecurity practices, and stay updated with the ever-changing cyber threat landscape.

4 Cyber Risk, Cyber Risk Assessment and Its Challenges

There are different definitions of cyber risk, but the most comprehensive definition is provided by Eling and Schnell [31]. They define cyber risk as "risk emerging from the use of IT that compromises the confidentiality, availability or integrity of data or services" [31, p. 483]. Cyber risks are not limited to specific firms and can affect organisations of all sizes and models, as per [32]. The exposure to cyber risks depends on a company's vulnerability, which means that the existing weaknesses in IT security increase the probability of a cyber incident, according to Marotta [33].

Figure 3 shows cybersecurity risk management and includes seven elements: establishing context, identifying the risk, analysing the risk, evaluating the risk, risk treatment, communication and consultation, and monitoring and review. In comparison, CRA includes only three elements: identifying, analysing and evaluating the risk, as shown in Fig. 4.

Risk management is described as "coordinated activities to direct and control an organisation with regard to risk [34]". It involves three main steps: (1) establishing the context, (2) assessing the risk, and (3) treating the risk. These steps should be carried out continuously and integrated into regular communication and consultation, monitoring and review, and documentation and reporting.

Managing cyber risks involves identifying, assessing, and prioritising potential risks, which is followed by the coordinated and economical application of resources to decrease, monitor, and control the probability and impact of any adverse

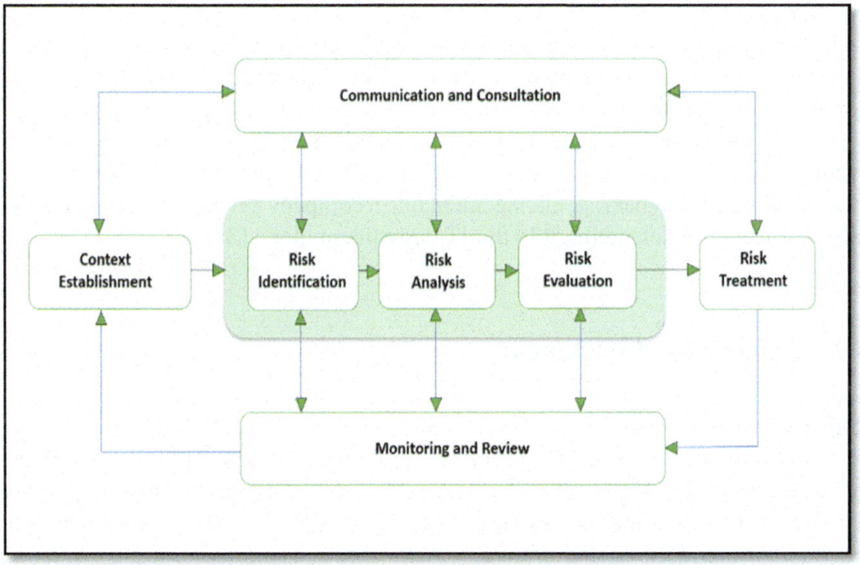

Fig. 3 Cybersecurity risk management process of ISO 27005 [34]

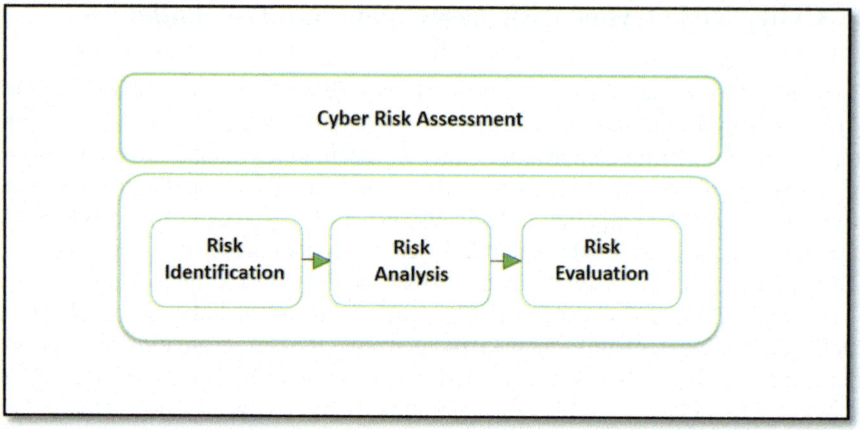

Fig. 4 Cybersecurity risk assessment process of ISO 27005 [34]

outcomes. Developing a cyber risk management framework that improves on existing frameworks and quantitative approaches to risk assessment is crucial [35].

Businesses face a rapidly changing threat landscape with evolving sophisticated cyber-attacks, which makes cyber risk management crucial to their safety and success [7]. The implementation of cyber risk management depends significantly on the level of cyber risk consciousness and general risk awareness [36]. Moreover, the relationship between cybersecurity culture and cyber risk management needs further investigation [36].

SMEs often operate within limited IT budgets, making them more susceptible to cyber threats; as highlighted by Renaud and Ophoff [29] these risks are not confined to IT failures but can be categorised into different types, such as people's actions, systems and technology failures, failed internal processes, and external events. Studies indicate that most incidents are caused by people [13]. Therefore, it is crucial to integrate cyber risk management into a company's overall risk management process instead of delegating it to the IT department alone [31].

4.1 *Cyber Risk Assessment*

Risk assessment is defined as the "overall process of risk identification, risk analysis, and risk evaluation" [37]. According to Craigen et al. [38] the importance of cybersecurity risk assessment is that it assists in detecting and mitigating potential risks to cyber infrastructure, as shown in Fig. 4. CRA is done to evaluate cyber risks and vulnerabilities within cyber environments, using tools, frameworks, and methodologies to protect critical assets from cyber threats. Moreover, the cybersecurity risk assessment should include the usage of cybersecurity maturity assessment frameworks [39], vulnerability scanning of critical cyber systems [40], cybersecurity metrics [41], multi-criteria decision-making frameworks [42], and dynamic simulation approaches [43], to estimate the level of risk within cyberinfrastructure and enhance the security state. These approaches allow for dealing with cyber threats and enhance security in different environmental contexts, such as SME environments.

4.2 Challenges of Cyber Risk Assessment in SMEs

Cybersecurity risk assessment in SMEs is challenging because of the limitation of resources, expertise, and the rapidly developed threats [44, 45]. The growing cyber risks for SMEs because of the growing cybercrime necessitate further investment in cybersecurity solutions [46]. As well as the lack of resources and suitable in-house expertise in SMEs further complicates the adoption of cybersecurity solutions and the awareness of cyber risk [47, 48].

Chandna and Tiwari [49] show that surviving online risk can be highly difficult for new entrepreneurial ventures that lack the resources and capabilities of founded companies. The lack of protection processes and awareness of the value of cybersecurity compounds the challenges SMEs face in dealing with cybersecurity issues [47]. Similarly, the lack of resources and relevant in-house expertise in SMEs further problematises adopting cybersecurity solutions and the awareness of cyber risks [45, 47]. Neri et al. [50] showed that SMEs are recently beginning to start the way for organisational cyber readiness and take many more proactive steps to handle cyber risks.

To handle these risks, several studies have emphasised the need for a more holistic approach to cybersecurity, like cyber resilience, which could allow SMEs to anticipate, detect, resist and recover from cyber incidents [3, 51]. Besides, developing customised cybersecurity risk assessment approaches to handling the specific requirements of SMEs is essential [44].

5 Dynamic Cyber Risk Assessment (DCRA)

In the digital age, cybersecurity is essential, and DCRA can play a crucial role in safeguarding individuals, organisations, and nations against cyber threats. Rea-Guaman et al. [52] designed a comprehensive framework for identifying and assessing cybersecurity risks. The framework was validated through a case study, and it provides a structured approach to identify threats, assets, vulnerabilities, and impacts.

Moreover, the framework also presents a risk assessment matrix to prioritise risks and drive informed decisions. Also, in their work, Zare, Zare and Azadi [53] focused on analysing information technology (IT) and computer systems' vulnerabilities. The investigation offered a way to identify, classify, and prioritise vulnerabilities based on likelihood and impact. The investigation likewise delivered guidance to mitigate the identified vulnerabilities and improve cybersecurity.

Likewise, Cremer et al. [54] emphasised the demand for better data availability in cyber risk management. Their research underscored collecting, analysing, and sharing data about cyber risks, incidents, and vulnerabilities to enhance CSA and decision-making. Similarly, their research presented a data governance framework for confidentiality, integrity, and availability.

From this research, it seems evident that integrating proactive functionality into DCRA frameworks can significantly enhance an organisation's security posture by enabling proactive threat preparedness. Cyber threat intelligence (CTI) knowledge has been identified as a crucial tool for proactive cyber risk mitigation [55]. However, according to Cheimonidis and Rantos [56], most current models have significant limitations, as they lack proactive capabilities due to the limited integration of information related to CTI. DCRA models mainly focus on real-time data related to ongoing activities within an organisation's environment, usually sourced from systems like IDS/IPS.

DCRA models mainly concentrate on monitoring ongoing activities, lacking historical context, limiting their ability to derive insights from past incidents [56]. However, integrating relevant CTI data can proactively reassess the existing risk level. Distributed data fusion is pivotal to comprehensively understanding a given environment or system.

Another issue is the need to enhance the quality of input data assimilated, leading to more consistent and accurate outcomes. According to Liu et al. [57], using data fusion techniques has demonstrated effectiveness in various applications. By employing these techniques, DCRA models can enhance the quality of input data assimilated, leading to more consistent and accurate outcomes. In addition to data fusion, trust evaluation techniques can be added to improve the accuracy and precision of fused results. Guo [58] demonstrated that the Lightweight Privacy-Preserving Trust Evaluation (LPPTE) scheme could improve DCRA models to amalgamate disparate, inconsistent, uncertain, or redundant data.

Additionally, this threat intelligence extraction and fusion framework can assist DCRA models in correlating and unifying data. Furthermore, the LPPTE framework can aid in extracting and merging threat intelligence, thereby helping the DCRA models to correlate and integrate data. By utilising these techniques, DCRA models can improve the quality of data assimilation, leading to more reliable and precise results.

Moreover, incorporating trust evaluation techniques can improve the accuracy of data fusion. Trust evaluation techniques are even more crucial when deploying DCRA models in a zero-trust environment (ZTA). In such a scenario, DCRA models are responsible for processing large amounts of data, which will be used as input for real-time risk-based decision-making. Due to the various range of data sources, including untrusted and potentially malicious ones, the use of trust evaluation techniques by DCRA models can enhance their reliability [57].

Guo [59] suggests that trust evaluation methods, specifically TFL-DT, effectively identify malicious users and should be incorporated into DCRA models. Furthermore, trust evaluation models can help determine the source of information malicious users provide and assess its accuracy, thus improving the precision and reliability of input data used by DCRA models [60].

In overview, these studies highlight the significance of solid frameworks and data availability in improving DCRA. By embracing these best practices, cybersecurity specialists can mitigate cyber risks effectively and efficiently and protect enterprises from threats.

6 Artificial Intelligence

In the world of cybersecurity, we are noticing more cybersecurity specialists turning to AI techs like Machine learning (ML) and Deep Learning (DL) to tackle cyber threats. AI is making a big difference in tasks like spotting intruders, finding malware, filtering out spam, managing vulnerabilities, predicting risks of breaches, responding to incidents, assessing risks, and preventing cyber threats. That is according to recent studies [61–63].

Additionally, Dilek et al. [64] show how AI approaches for combating cybercrimes have been acknowledged as a powerful mechanism for detecting and preventing attacks. Furthermore, the usage of AI in cybersecurity has been associated with new opportunities and greater efficiencies, particularly in AI skills related to cognitive intelligence or IQ Ariffin and Maskat [65].

Furthermore, Jawaid [62] highlights that techniques like AI, natural language processing (NLP), behavioural analytics, and DL have effectively enhanced cyber defence by safeguarding against threats like malware infections, phishing attempts, and insider attacks. According to Daniel and Andreas [63]. AI integration in cybersecurity extends further than defensive measures; it can likewise strengthen cybersecurity solutions against risks.

The reviews of AI and cybersecurity solutions emphasise the importance of various categories of AI, such as AI, DL, decision trees, k-nearest neighbour, support vector machines (SVM), and artificial neural networks, for cybersecurity applications [66]. The concept of explainable AI as a security measure for cyber threat hunting has been proposed to underscore the importance of transparent and interpretable AI models in enhancing cybersecurity [67].

The studies above highlight the various ways in which AI can be applied to cybersecurity, ranging from using AI to tackle cybersecurity challenges to using it in specific cases in SMEs. The studies also demonstrate how AI can improve cyber defence mechanisms, stressing its relevance and importance. By analysing these studies, we can understand how AI can strengthen cybersecurity measures through practical applications, specific techniques, and transparent and interpretable models.

Daniel and Andreas [63] carried out a study to explore how AI is utilised to enhance cybersecurity for SMEs. The research delves into the security features driven by AI that can significantly improve the security posture of SMEs. While leveraging AI-based security solutions offers benefits, the paper also addresses obstacles and challenges in their implementation. The study concludes that there is a growing need for user-efficient, AI-driven cybersecurity solutions tailored to SMEs. Radanliev et al. [68] introduced an adaptive AI system designed for predictive cyber risk analytics in challenging environments, hinting at AI's potential future applications in DCRA.

6.1 AI to Enhance DCRA

AI has been a professional tool for improving DCRA. A study conducted by Radanliev et al. [68] emphasised the potential of AI in DCRA, highlighting its role in improving organisational resilience and understanding cyber risk. Furthermore, the integration of AI in cyber risk impact assessments has been identified as a crucial step in comprehending economic impact assessment for cyber risk. Giudici and Raffinetti [69] also underscored the increasing prevalence of AI approaches in cyber risk management, particularly in cases where traditional statistical models cannot be made due to insufficient data.

Likewise, the application of AI in DCRA has been connected to enhancing resilience and an adequate understanding of cyber risk [70]. Additionally, AI has been shown to make complementarities between AI and cybersecurity, focusing on trust in AI techniques and enhancing the categorisation of risks for quantifying the cascading effects of cyber risk [71].

6.2 Advantages of Using AI in SMEs

Utilising AI in SMEs presents opportunities and benefits. AI can enhance decision-making processes by offering insights and predictions derived from data analysis, which is especially advantageous for SMEs without resources for complex optimisation models [72]. Moreover, implementing AI techniques can give SMEs an edge by enhancing performance, productivity and efficiency [73]. Research has shown that SMEs can increase their competitiveness through learning and imitating strategies simultaneously, highlighting the role of learning orientation in creating a competitive advantage [74]. Integrating AI in SME operations can foster innovation capability learning and overall performance improvement [75].

AI can assist SMEs with tasks such as predicting demand, foreseeing risks, and making informed decisions [76]. Integrating AI into SME operations can result in decision-making, increased competitiveness, enhanced innovation capacity, and streamlined operational efficiency. By using AI technologies in SMEs, we can gain an advantage in the marketplace and make way for long-term growth and prosperity.

AI shows benefits in evaluating risks. A key advantage is that AI models can get the connections among risk elements, resulting in forecasts of results [77]. These models can learn complex patterns from data, allowing them to make predictions about unobserved data, which is crucial in risk assessment [78]. Likewise, AIAF models have been shown to surpass traditional risk factor-based models in predicting future risks in different domains [79].

Moreover, AI significantly reduces costs associated with risk assessment processes by eliminating the need for a group of experts to conduct the assessment [80]. Additionally, AI models can extract unexpected risk factors from complex data [81]. Overall, the advantages of employing AI in risk assessment include enhanced accuracy through capturing complex interactions, outperforming traditional models, reducing costs by automating processes, and extracting novel risk factors from data. These benefits establish AI as a valuable tool in various fields for enhancing risk assessment processes and CSA.

6.3 Using AI in CSA

AI plays a crucial role in enhancing cyber situational awareness. By utilising those techniques, cyber specialists can effectively classify and combat malware [82]. AI-based frameworks can detect network entities, assess vulnerabilities, and allocate them to appropriate network segments in real-time, thereby improving cybersecurity [83]. Additionally, AI aids in cyber team communication and metacognitive cyber situational awareness, leading to better judgments and enhanced communication [69].

Moreover, AI enables the development of comprehensive cyber-physical-social awareness platforms, enhancing overall situational awareness by processing multiple data sources for better threat detection [84]. The utilisation of AI in conjunction with frequency-based representations of alerts allows for improved anomaly detection and threat identification [85]. Furthermore, AI plays a crucial role in cyberattack detection in complex systems like power grids [86].

AI algorithms, such as random forest and neural networks, have achieved popularity in cybersecurity for their effectiveness in identifying and mitigating cyber threats [87]. These algorithms can detect patterns in data, aiding in the identification of potential vulnerabilities and threats [62]. AI has been widely applied in cybersecurity for tasks like intrusion detection, malware detection, and user authentication [88].

7 Discussions

The cybersecurity gap among SMEs has made them prime targets for hackers due to their limited resources. This paper highlights and discusses the vital role of AI in enhancing CSA and DCRA for SMEs in tackling their current cybersecurity challenges. In detail, SMEs often face resource limitations, resulting in a cybersecurity personnel shortage, making it challenging to consistently monitor potential threats. In addressing this, CSA offers SMEs a perspective of their cyber landscape, enabling them to prioritise threats and allocate resources efficiently. In times of a cyberattack, SMEs require responses and quick decision-making abilities. CSA helps to make well-informed decisions by understanding the current status of systems and networks. Through defence strategies, like continuous monitoring of their environment, SMEs can proactively identify vulnerabilities before they become exploitable, allowing them to take preventive actions effectively.

Moreover, DCRA helps SMEs keep pace with these changes by continuously evaluating their risk exposure. Resource optimisation involves understanding the most critical assets and potential threats so that SMEs can focus their limited resources on mitigating the highest risks. DCRA helps SMEs demonstrate compliance by providing a structured approach to risk management. By implementing CSA and DCRA, SMEs can significantly improve their cybersecurity posture, mitigate risks, and protect their critical assets. Finally, the authors highlight the importance of AI to create promising opportunities for enhancing both CSA and DCRA within SMEs.

7.1 Initial Recommendations

By harnessing these AI techniques, SMEs can significantly enhance their ability to proactively manage cyber risks and strengthen their overall security posture. Below are some initial recommendations:

- Machine Learning to Enhance CSA:

Anomaly detection using algorithms like Autoencoders or Isolation Forest can identify unusual network traffic patterns or system behaviour indicative of potential threats. On the other hand, behaviour analysis using reinforcement learning can model normal user behaviour and detect deviations signalling compromise. Likewise, Natural Language Processing (NLP) can be used to analyse threat intelligence reports and extract actionable insights.

- Machine Learning to Enhance DCRA:

Regression models can be used to predict the probability and impact of specific cyber risks based on various factors. Decision trees or random forests can classify risks into different categories (e.g., high, medium, low). Also, AI models can analyse system logs and network data to identify potential vulnerabilities. Image recognition techniques can be applied to analyse system configurations for misconfiguration. Additionally, reinforcement learning can be used to optimise resource allocation for risk mitigation based on the dynamic threat landscape.

7.2 *Additional Considerations for Using AI*

There are some additional considerations for using AI, such as the critical nature of cybersecurity, which makes it crucial to use machine learning models that can explain their decisions. The quality and quantity of data are essential for training effective machine learning models. SMEs might face challenges in data collection and preparation. Machine learning models require continuous training and updates to adapt to evolving threats.

8 Conclusion

Integrating the AI model into CRA and DCRA has the potential to advance the efficiency and accuracy of threat detection and response. As this paper shows, AI-driven tools can quickly spot difficult-to-notice patterns, enabling quick detection and response to potential security incidents, which enhances the effectiveness of risk assessment methods. Indeed, SMEs often face challenges in maintaining cybersecurity defences due to resources and expertise in the field. However, AI can offer a cost solution for monitoring enterprise networks and pinpointing vulnerabilities, empowering SMEs to safeguard their data and business assets better. By utilising AI-powered security tools, SMEs can gain insights into network activities, swiftly identify and address threats, and enhance their cybersecurity readiness. Leveraging AI in cybersecurity operations helps SMEs tackle resource limitations by automating security tasks, reducing intervention, and allowing their IT resources to focus on strategic initiatives more effectively. In conclusion, the authors of this paper believe that AI-driven tools present an approach for network monitoring, vulnerability identification, and attack prevention, and this cutting-edge technology equips SMEs with the means to establish an environment that shields their businesses from cyber threats with severe consequences. Using AI-driven tools, SMEs can proactively resolve possible security breaches, preventing them from growing into significant issues; this helps enhance the CSA, maintain the security of their networks and ensure business operations.

References

1. Wang K, Guo X, Yang D (2022) Research on the effectiveness of cyber security awareness in ICS risk assessment frameworks. Electronics (Switzerland) 11. https://doi.org/10.3390/ELECTRONICS11101659
2. Bada M, Nurse CJR (2019) Developing cybersecurity education and awareness programmes for small- and medium-sized enterprises (SMEs). Inf Comput Secur https://doi.org/10.1108/ics-07-2018-0080

3. Carías JF, Borges MRS, Labaka L et al (2020) Systematic approach to cyber resilience operationalization in SMEs. IEEE Access 8:174200–174221. https://doi.org/10.1109/ACCESS.2020.3026063
4. Alotaibi F (2019) Evaluation and enhancement of public cyber security awareness
5. Franke U, Brynielsson J (2014) Cyber situational awareness—a systematic review of the literature. Comput Secur 46:18–31
6. Ansari MF, Dash B, Sharma P, Yathiraju N (2022) The impact and limitations of artificial intelligence in cybersecurity: a literature review. IJARCCE 11: https://doi.org/10.17148/ijarcce.2022.11912
7. Kure HI, Islam S, Mouratidis H (2022) An integrated cyber security risk management framework and risk predication for the critical infrastructure protection. Neural Comput Appl 34:15241–15271. https://doi.org/10.1007/s00521-022-06959-2
8. Timmers P (2019) Ethics of AI and cybersecurity when sovereignty is at stake. Minds Mach (Dordr) 29:635–645
9. Akinrolabu O, Nurse JRC, Martin A, New S (2019) Cyber risk assessment in cloud provider environments: current models and future needs. Comput Secur 87
10. Endsley M, Endsley MR (1995) Toward a theory of situation awareness in dynamic systems. Human Factors J 37(1):32–64. Human Factors: J Human Factors Ergonom Soc 37:32–64. https://doi.org/10.1518/001872095779049543
11. Chandra NA, Ramli K, Ratna AAP, Gunawan TS (2022) Information security risk assessment using situational awareness frameworks and application tools. Risks 10. https://doi.org/10.3390/risks10080165
12. Jiang L, Jayatilaka A, Nasim M et al (2022) Systematic literature review on cyber situational awareness visualizations. IEEE Access 10:57525–57554
13. Chaudhary S, Gkioulos V, Katsikas S (2022) Developing metrics to assess the effectiveness of cybersecurity awareness program. J Cybersecur 8
14. Jirsík T, Čeleda P (2020) Cyber situation awareness via IP flow monitoring. https://doi.org/10.1109/noms47738.2020.9110327
15. Khamzina B, Roza N, Zhussupbekova G et al (2022) Determination of cyber security issues and awareness training for university students. Int J Emerg Technol Learn 17:177–190
16. Jirsik T, Celeda P (2018) Toward real-time network-wide cyber situational awareness. In: NOMS 2018 - 2018 IEEE/IFIP network operations and management symposium, pp 1–7
17. Balto KE, Yamin MM, Shalaginov A, Katt B (2023) Hybrid IoT cyber range. Sensors 23. https://doi.org/10.3390/s23063071
18. Goodall JR (2011) An evaluation of visual and textual network analysis tools. Inf Vis 10:145–157. https://doi.org/10.1057/ivs.2011.2
19. Brahma M, Kumar Kalita H (2023) An ontology of cyber security automation. In: Data science and intelligent computing techniques. Soft computing research society, pp 135–147
20. Ariffin MAM, Darus MY, Haron H et al (2022) Deployment of honeypot and SIEM tools for cyber security education model in UITM. Int J Emerg Technol Learn 17:149–172. https://doi.org/10.3991/ijet.v17i20.32901
21. Seker E, Ozbenli HH (2018) The concept of cyber defence exercises (CDX): planning, execution, evaluation. In: 2018 International conference on cyber security and protection of digital services (Cyber Security), pp 1–9
22. Dinh Vu Q, Tan R, Yau DKY (2016) On applying fault detectors against false data injection attacks in cyber-physical control systems. In: IEEE INFOCOM 2016—The 35th annual IEEE international conference on computer communications, pp 1–9
23. He S, Li Z, Wang J, Xiong NN (2021) Intelligent detection for key performance indicators in industrial-based cyber-physical systems. IEEE Trans Industr Inform 17:5799–5809. https://doi.org/10.1109/TII.2020.3036168
24. Kim H, Kwon H, Kim KK (2019) Modified cyber kill chain model for multimedia service environments. Multimed Tools Appl 78:3153–3170. https://doi.org/10.1007/s11042-018-5897-5

25. Tsiodra M, Panda S, Chronopoulos M, Panaousis E (2023) Cyber risk assessment and optimisation: a small business case study. IEEE Access 11:44467–44481. https://doi.org/10.1109/ACCESS.2023.3272670
26. Hutton G (2022) Business statistics. briefing paper number 06152
27. Business population estimates 2020 - GOV.UK. https://www.gov.uk/government/statistics/business-population-estimates-2020. Accessed 19 Nov 2023
28. Lewis S (2020) Cyber attack warning for SMEs as risks heighten amid Covid-19 pandemic
29. Renaud K, Ophoff J (2021) A cyber situational awareness model to predict the implementation of cyber security controls and precautions by SMEs. Org Cybersecurity J: Practice, Process People 1:24–46. https://doi.org/10.1108/ocj-03-2021-0004
30. Ncubukezi T, Mwansa L, Rocaries F (2020) A Review of the current cyber hygiene in small and medium-sized businesses. In: 2020 15th international conference for internet technology and secured transactions, ICITST 2020. https://doi.org/10.23919/ICITST51030.2020.9351339
31. Eling M, Schnell W (2016) What do we know about cyber risk and cyber risk insurance? J Risk Fin 17:474–491. https://doi.org/10.1108/JRF-09-2016-0122
32. Lloyd G (2020) The business benefits of cyber security for SMEs. Comput Fraud and Secur 2020:14–17. https://doi.org/10.1016/S1361-3723(20)30019-1
33. Marotta A, Martinelli F, Nanni S et al (2017) Cyber-insurance survey. Comput Sci Rev 24:35–61. https://doi.org/10.1016/j.cosrev.2017.01.001
34. ISO/IEC 27005:2022. In: Information security, cybersecurity and privacy protection—Guidance on managing information security risks. https://www.iso.org/standard/80585.html. Accessed 5 Jan 2024
35. Lee I (2020) Internet of things (IoT) cybersecurity: literature review and IoT cyber risk management. Future Internet. https://doi.org/10.3390/fi12090157
36. Gatzert N, Schubert M (2022) Cyber risk management in the US banking and insurance industry: a textual and empirical analysis of determinants and value. J Risk Ins 89:725–763. https://doi.org/10.1111/JORI.12381
37. Ross R, Winstead M, McEvilley M (2018) Systems security engineering: considerations for a multidisciplinary approach in the engineering of trustworthy secure systems
38. Craigen D, Diakun-Thibault N, Purse R (2014) Technology innovation management review defining cybersecurity
39. Yusuff Marican MN, Razak SA, Selamat A, Othman SH (2023) Cyber security maturity assessment framework for technology startups: a systematic literature review. Ieee Access. https://doi.org/10.1109/access.2022.3229766
40. Sviličić B, Kamahara J, Rooks M, Yano Y (2019) Maritime cyber risk management: an experimental ship assessment. J Navig. https://doi.org/10.1017/s0373463318001157
41. Sherif E, Yevseyeva I, Basto-Fernandes V, Cook A (2023) The SMART approach to develop good cyber security metrics. https://doi.org/10.21203/rs.3.rs-3125926/v1
42. Tusher HM, Munim ZH, Notteboom T et al (2022) Cyber security risk assessment in autonomous shipping. Maritime Econ Log. https://doi.org/10.1057/s41278-022-00214-0
43. Armenia S, Angelini M, Nonino F et al (2021) A dynamic simulation approach to support the evaluation of cyber risks and security investments in SMEs. Decis Support Syst. https://doi.org/10.1016/j.dss.2021.113580
44. van Haastrecht M, Sarhan I, Shojaifar A et al (2021) A threat-based cybersecurity risk assessment approach addressing SME needs. https://doi.org/10.1145/3465481.3469199
45. Trantidou T, Bravos G, Valoggia P et al (2022) SENTINEL—approachable. Tailor-Made Cybersecurity and Data Protection for Small Enterprises. https://doi.org/10.1109/csr54599.2022.9850297
46. Hasani T, O'Reilly N, Dehghantanha A et al (2023) Evaluating the adoption of cybersecurity and its influence on organizational performance. Sn Bus Econ. https://doi.org/10.1007/s43546-023-00477-6
47. Shojaifar A, Järvinen H (2021) Classifying SMEs for approaching cybersecurity competence and awareness. https://doi.org/10.1145/3465481.3469200

48. Moneva A, Leukfeldt R (2023) Insider threats among dutch SMEs: nature and extent of incidents, and cyber security measures. J Criminol. https://doi.org/10.1177/26338076231161842
49. Chandna V, Tiwari P (2021) Cybersecurity and the new firm: surviving online threats. J Bus Strateg. https://doi.org/10.1108/jbs-08-2021-0146
50. Neri M, Niccolini F, Pugliese R (2022) Assessing SMEs' cybersecurity organizational readiness: findings from an italian survey. Online J Appl Knowl Manage. https://doi.org/10.36965/ojakm.2022.10(2)1-22
51. Carias JF, Arrizabalaga S, Labaka L, Hernantes J (2021) Cyber resilience self-assessment tool (CR-SAT) for SMEs. IEEE Access. https://doi.org/10.1109/access.2021.3085530
52. Rea-Guaman AM, Mejía J, San Feliu T, Calvo-Manzano JA (2020) AVARCIBER: a framework for assessing cybersecurity risks. Cluster Comput 23:1827–1843. https://doi.org/10.1007/s10586-019-03034-9
53. Zare H, Zare MJ, Azadi M (2018) Cybersecurity vulnerabilities assessment (A systematic review approach). Adv Intell Syst Comput 738:61–68. https://doi.org/10.1007/978-3-319-77028-4_10/COVER
54. Cremer F, Sheehan B, Fortmann M et al (2022) Cyber risk and cybersecurity: a systematic review of data availability. Geneva Papers on Risk and Insurance: Issues and Practice 47:698–736. https://doi.org/10.1057/s41288-022-00266-6
55. Naumov S, Kabanov I (2016) Dynamic framework for assessing cyber security risks in a changing environment, pp 1–4
56. Cheimonidis P, Rantos K (2023) Dynamic risk assessment in cybersecurity: a systematic literature review. future internet 15
57. Liu Z, Ma J, Weng J, et al (2021) LPPTE: A lightweight privacy-preserving trust evaluation scheme for facilitating distributed data fusion in cooperative vehicular safety applications. Information Fusion 73:144–156. https://doi.org/10.1016/j.inffus.2021.03.003
58. Guo Y, Liu Z, Huang C, et al (2023) A framework for threat intelligence extraction and fusion. Comput Secur 132:103371. https://doi.org/10.1016/j.cose.2023.103371
59. Guo J, Liu Z, Tian S et al (2023) TFL-DT: a trust evaluation scheme for federated learning in digital twin for mobile networks. IEEE J Sel Areas Commun 41:3548–3560. https://doi.org/10.1109/JSAC.2023.3310094
60. Schaberreiter T, Kupfersberger V, Rantos K, et al (2019) A Quantitative evaluation of trust in the quality of cyber threat intelligence sources. In: Proceedings of the 14th international conference on availability, reliability and security. association for computing machinery, New York, NY, USA
61. Zhang Z, Al HH, Damiani E et al (2022) Explainable artificial intelligence applications in cyber security: state-of-the-art in research. Ieee Access. https://doi.org/10.1109/access.2022.3204051
62. Jawaid SA (2023) Artificial intelligence with respect to cyber security. Jaai. https://doi.org/10.18178/jaai.2023.1.2.96-102
63. Daniel K, Andreas J (2022) Evaluation of AI-based Use cases for enhancing the cyber security defense of small and medium-sized companies (SMEs). Electronic Imaging
64. Dilek S, Çakır H, Aydın M (2015) Applications of artificial intelligence techniques to combating cyber crimes: a review. Int J Artif Intell Appl 6. https://doi.org/10.5121/ijaia.2015.6102
65. Mohd Ariffin NH, Maskat R (2021) A proposal of ethical competence model for cyber security organization. Indonesian J Electr Eng Comput Sci. https://doi.org/10.11591/ijeecs.v24.i3.pp1711-1717
66. Hebbar A, Kumar SA (2022) Artificial intelligence in cyber security
67. Kumar P, Wazid M, Singh DP et al (2023) Explainable artificial intelligence envisioned security mechanism for cyber threat hunting. Sec Privacy
68. Radanliev P, Roure D De, Walton R et al (2020) Artificial intelligence and machine learning in dynamic cyber risk analytics at the Edge. SN Appl Sci
69. Giudici P, Raffinetti E (2021) Explainable AI methods in cyber risk management. Qual Reliab Eng Int

70. Ghillani D (2022) Deep learning and artificial intelligence framework to improve the cyber security
71. Radanliev P, Roure D De, Maple C, Ani U (2022) Super-forecasting the 'technological singularity' risks from artificial intelligence. Evolving Syst
72. Park KJ (2021) Determining the tiers of a supply chain using machine learning algorithms. Symmetry (Basel). https://doi.org/10.3390/sym13101934
73. Baritto M, Billal M, Nasim S et al (2020) Supporting tool for the transition of existing small and medium enterprises towards industry 4.0. https://doi.org/10.1109/icdabi51230.2020.9325704
74. Efendi S, Sugiono E, Guritno E et al (2020) Building innovation and competitiveness for low technology Manufacturing SMEs through imitating capability and learning: the case of Indonesia. Cogent Soc Sci. https://doi.org/10.1080/23311886.2020.1803515
75. Naser V, Rezaei S, Wan Ismail WK (2017) Examining learning strategies, creativity, and innovation at SMEs using fuzzy set qualitative comparative analysis and PLS path modeling. J Bus Res. https://doi.org/10.1016/j.jbusres.2016.08.016
76. Carbonneau RA, Laframboise K, Vahidov R (2008) Application of machine learning techniques for supply chain demand forecasting. Eur J Oper Res. https://doi.org/10.1016/j.ejor.2006.12.004
77. Weng S, Reps J, Kai J et al (2017) Can machine-learning improve cardiovascular risk prediction using routine clinical data? PLoS ONE. https://doi.org/10.1371/journal.pone.0174944
78. Murdoch WJ, Singh CD, Kumbier K et al (2019) Definitions, methods, and applications in interpretable machine learning. Proc Natl Acad Sci. https://doi.org/10.1073/pnas.1900654116
79. Gao Y, Li S, Jin Y et al (2022) An assessment of the predictive performance of current machine learning-based breast cancer risk prediction models: systematic review. JMIR Public Health Surveill. https://doi.org/10.2196/35750
80. Krechowicz M, Krechowicz A (2021) Risk Assessment in energy infrastructure installations by horizontal directional drilling using machine learning. Energies (Basel). https://doi.org/10.3390/en14020289
81. Villamor E, Monserrat C, Del RL et al (2020) Prediction of osteoporotic hip fracture in postmenopausal women through patient-specific fe analyses and machine learning. Comput Methods Programs Biomed. https://doi.org/10.1016/j.cmpb.2020.105484
82. Milošević N, Dehghantanha A, Choo KR (2017) Machine learning aided android malware classification. Comput Electr Eng. https://doi.org/10.1016/j.compeleceng.2017.02.013
83. Nikoloudakis Y, Kefaloukos I, Stylianos K et al (2021) Towards a machine learning based situational awareness framework for cybersecurity: an SDN implementation. Sensors. https://doi.org/10.3390/s21144939
84. Mirza IB, Georgakopoulos D, Yavari A (2023) Cyber-physical-social awareness platform for comprehensive situation awareness. Sensors. https://doi.org/10.3390/s23020822
85. Park H, Choi Y-J (2022) Frequency-based representation of massive alerts and combination of indicators by heterogeneous intrusion detection systems for anomaly detection. Sensors. https://doi.org/10.3390/s22124417
86. Sahu A, Mao Z, Wlazlo P et al (2021) Multi-source multi-domain data fusion for cyberattack detection in power systems. IEEE Access. https://doi.org/10.1109/access.2021.3106873
87. Oyetoro A, Mart J, Amah U (2023) Using machine learning techniques random forest and neural network to detect cyber attacks. https://doi.org/10.14293/pr2199.000059.v1
88. Dasgupta D, Akhtar Z, Sen S (2020) Machine learning in cybersecurity: a comprehensive survey. J Defense Modeling Simul Appl Methodol Technol. https://doi.org/10.1177/1548512920951275

Safe Cyber Watch: Cyber Bully Alert and Parental Advisory System

Rojal Varghese and Muhammad Ali

Abstract Cyberbullying is a major concern in the digital age, especially for teenagers. The increased accessibility and usage of social media platforms have widened the issue, with incidents of cyberbullying in English experiencing a substantial rise over the past decade. This system provides a solution involves advanced AI to detect inappropriate content and involves parents in real-time monitoring. This approach embeds strong deep learning model to recognize content like hate speech, aggression, insults, toxicity, etc. It will create a safe online environment for children by detecting such contents. The architecture of this application is divided into four modules: Child, Parent, ML Model and Admin. This allows teens to share content from multiple sources that ML model will analyses for harmful content. The Parent module provides transparency and security by allowing parents to monitor their children's online activities and predictions made by the ML model. An alert mechanism notifies parents via email about cyberbullying instances. The Journal/Diary function offers a platform for teens to express their feelings. The admin module performs user management and ensures the integrity of the ML model, which is constructed with TensorFlow and Keras with an LSTM architecture, classifying cyberbullying into six classes: race, gender, ethnicity, sexuality, religion. After training on a diverse dataset, the model achieved an accuracy of approximately 94% on the training set. Project demonstrates, the potential of combining AI and web technologies to foster safer online spaces. By offering real-time alerts, engaging parents in monitoring, it embodies a holistic approach to tackling cyberbullying.

Keywords Cyberbullying · Machine learning · Social media sentiment analysis · LSTM · Recurrent neural network

---

R. Varghese · M. Ali (✉)
Anglia Ruskin University, Cambridge, UK
e-mail: muhammad.ali@aru.ac.uk

H. Jahankhani and B. Issac (eds.), *Cybersecurity and Human Capabilities Through Symbiotic Artificial Intelligence*, Advanced Sciences and Technologies for Security Applications, https://doi.org/10.1007/978-3-031-82031-1_30

1 Introduction

Cyberbullying has become a global concern that primarily affects children and teenagers because of social media's extensive reach. The "Safe Cyber Watch: Cyber Bully Alert & Parental Advisory System" combines deep learning algorithms with state-of-the-art technology to address this pressing problem. A complex machine learning model, created with TensorFlow and Keras, can recognize different types of cyberbullying, such as aggressive and hateful comments, which is essential to its operation. The system, which is divided into four components (Admin, ML Model, Child, and Parent), encourages proactive intervention. Parental participation is recommended as it enables the models. System performance and integrity are guaranteed by administrative control. All things considered, the "Safe Cyber Watch" project represents a major step forward in combating cyberbullying, highlighting the potential of technology to make children's digital environments safer.

1.1 Statement of the Problem

There are serious problems associated with this digital absorption, most notably cyberbullying. Cyberbullying has serious repercussions that can range from psychological anguish, low self-esteem, and difficulties in school to more serious ones like suicidal thoughts and self-harm.

1.2 Purpose of the Study

The purpose of the "Safe Cyber Watch: Cyber Bully Alert & Parental Advisory System" is to tackle the growing issue of cyberbullying, especially with kids and teenagers. The project aims to establish secure online environment that promotes healthy digital interactions by web technologies and sophisticated machine learning algorithms. The project aims to encourage proactive digital safety measures by collaboration among children, parents, and administrators.

1.3 Scope of the Study

The "Safe Cyber Watch" study's scope includes creating and putting into place a thorough system for identifying and stopping cyberbullying. Using cutting-edge machine learning techniques and contemporary web technologies for front-end and back-end development are essential elements. The system main modules Child, Parent, ML Model, and Admin enables a team-based approach to digital safety. Youngsters can directly enter text from social media, and the ML Model validate that text of cyberbullying. Parents receive alerts at the same moment, allowing for real-time monitoring and assistance.

2 Literature Review

2.1 Overview of Cyberbullying Detection

A cyberbullying detector's primary function is to recognize and flag potentially harmful content that might be construed as bullying. This is accomplished by creating algorithms that can analyze text determine whether the content contains bullying-related elements. The deep learning models, especially those built on neural networks have shown to be successful in processing and comprehending enormous volumes of unstructured text data that are "Safe cyber watch: Cyber bully alert and parent advisory system" produced on social media. For example, a study showed how task-specific word embeddings can be used to improve detection performance on multiple social media platforms while accommodating the distinct slang and language styles of each platform [1]. The difficulty increases in the context of non-English languages because there is a dearth of data and a greater complexity of linguistic features. To address these issues, studies have created deep learning models for the Bangla language [3]. These models make use of architectures like Deep Neural Networks and Bi-LSTM to meet the unique needs of the language.

2.2 Pros and cons of Cyberbullying Detection

Pros:

Evolving online behaviors: Deep learning models are well known for their capacity to analyses and learn from sizable datasets. This allows them to identify minute details and patterns in data that may point to cyberbullying [1].

Platform Adaptability: These models are able to adjust to various linguistic styles, slang terms, and communication conventions that are present on various social media platforms because they can be trained on a variety of data sets spanning these platforms [1].

Explainability and Transparency: With the introduction of explainability in deep learning models due to recent advancements, these models are now able to justify their predictions.

Cons:

Data and Privacy Concerns: Training deep learning models requires access to large amounts of data, which can raise significant privacy concerns.

Contextual and Cultural Challenges: While deep learning models are effective at learning from data, they often struggle with understanding context and cultural nuances.

Bias and Fairness: There is an inherent risk of bias in AI models, which can propagate or even amplify existing prejudices found in the training data. For instance, biases in the dataset, such as overrepresentation or underrepresentation of certain languages or demographics, can affect the model's fairness and accuracy [2, 3]

2.3 Deep Learning Models for Cyberbullying Detection

Deep learning models have become a cornerstone in the development of cyberbullying detection systems due to their ability to process and learn from large datasets, identifying complex patterns in textual data that are indicative of bullying behavior.

Convolutional Neural Networks (CNNs): Convolutional Neural Networks (CNNs), typically known for their success in image processing, have also been effectively applied to text analysis tasks, including cyberbullying detection. These models can capture spatial hierarchies in data, allowing them to identify important textual patterns such as word groupings and phrase structures that may signify bullying. By applying convolutional layers to word embeddings, CNNs can efficiently process and classify large volumes of text data on social media platforms [1].

Recurrent Neural Networks (RNNs) and LSTM Models: RNNs and their advanced variant, Long Short-Term Memory (LSTM) networks, are particularly adept at handling sequences, making them ideal for text-based applications where context and the order of words play critical roles. LSTMs have been extensively used in cyberbullying detection to understand the context within user messages and comments [4].

Hybrid Models: Hybrid models combine the strengths of multiple neural network architectures to enhance detection capabilities. For example, integrating CNNs with LSTMs allows systems to exploit both local features of text data and long-range dependencies within the data. This combination is particularly powerful in analyzing social media posts and comments, where both the immediate context and the broader conversational thread may contain cues vital for accurate cyberbullying detection [5].

2.4 Need for New Solutions in Cyberbullying Detection

While existing literature and methodologies have made significant strides in detecting cyberbullying, several compelling reasons necessitate the development of new solutions in this area:

Evolving Online Behaviors: As digital communication evolves, so too do the methods and forms of cyberbullying. Existing models often struggle to keep pace with rapid changes in online slang, coded language, and emerging social media platforms. New solutions are required to adapt dynamically to these changes, ensuring that detection methods remain effective and relevant [1].

Handling Data Scarcity and Quality: Effective cyberbullying detection relies on high-quality, annotated datasets, which are scarce for certain languages and demographics. Innovative approaches to semi-supervised learning, synthetic data generation, and active learning could help mitigate these data issues, enabling robust model training even in the face of limited data.

Improving Real-Time Detection: Real-time intervention is critical in preventing the escalation of cyberbullying incidents. However, current systems often suffer from latency issues or require extensive computational resources. Developing lightweight, efficient models that can operate in real-time, possibly on edge devices, is essential for timely and effective cyberbullying intervention.

3 Methodology

3.1 Overview

The development of the "Safe Cyber Watch: Cyber Bully Alert & Parental Advisory System" project utilized a methodical approach to tackle the intricate problems associated with identifying and preventing cyberbullying. The system's requirements were carefully gathered through stakeholder engagement, starting with an extensive design phase that made sure the system met the needs of kids, parents, and teachers.

Ensuring user-friendly interfaces and robust data management, the development process included both frontend and backend components. The front-end design prioritized making an intuitive user interface that children and their parents could both use. Because of their adaptability and support for responsive design, HTML, CSS, Bootstrap, and JavaScript were selected. The backend was designed to manage data, facilitate user interactions, and incorporate the machine learning model. It was powered by Python and Django.

The deep learning model was developed and trained primarily with the help of TensorFlow and Keras, with additional support for data manipulation, analysis, and evaluation coming from libraries like Numpy, Pandas, Matplotlib, Seaborn, and Scikit-learn. Regular expressions and NLTK were essential for text preprocessing, which increased the model's precision and effectiveness.

3.2 Data Collection

To detect cyberbullying, we used a Kaggle dataset that was designed to identify insults in social media conversations. Enumerating the many offensive words and behaviors that are frequently used in online conversation. The dataset, which includes annotations indicating the presence of insults, makes it easier to develop a strong classifier that can pick up on second nuances in language. The dataset provides a rich source of training examples for our machine learning model, even though there are occasionally missing timestamps.

Sample data

S. No.	Comment/tweet	Label
1	Good looking	0
2	Bull shit!	1
3	So skinny	1
4	Free gift for you	0

example: "**Good looking**": Labelled **0** (not bullying).
"**Bull shit!**": Labelled **1** (indicates offensive language, likely bullying).

3.3 Data Cleaning and Preprocessing

Data cleaning is the first and most crucial step that must be taken in every procedure. It involves transforming unprocessed data into the format needed to successfully train the model. For example, raw data such as "You look so ugly, and fat change the style" is pre-processed to become "look ugly fat change style." All unnecessary terms and special characters, such as @, (), []? /; etc., that are not needed for model training, are removed during this stage.

Data Preprocessing includes operations such as stop word removal to eliminate common words with minimal meaning and tokenization, which divides text into smaller chunks.

Tokenization: Text is tokenized by separating it into groups of words, numbers, and punctuation. Furthermore, insignificant components are used in place of non-sensitive ones. Thorough testing guarantees the tokenization method's security while upholding confidentiality requirements.

*stop words***:** Frequently occurring words in a language that are frequently disregarded in text analyses because of their low semantic significance are known as stop words. We extracted stop words from the dataset using our approach, including common terms like "a," "the," "for," and "from." The documents that have been processed are prepared for additional processing after removal.

Lemmatization is the process of reducing words to their most basic or dictionary form while maintaining their semantic content. By converting inflectional forms back to their base forms, lemmatization contributes to the quality of text analysis results while preserving the integrity of vocabulary in our methodology.

Stemming: This process involves stripping words of their affixes to return them to their root or base form, thereby reducing the number of different word types (Fig. 1).

3.4 Application of Long Short-Term Memory (LSTM)

LSTM is a popular Recurrent Neural Network (RNN) model used with speech, video, text, and time series datasets. It is particularly effective for sequential datasets as it can handle a single data point at a time. Based on an extension of the Simple RNN model, it was designed to identify long-term dependencies in text data. Unlike Simple RNN, it can remember previous value points. An RNN architecture consists of three layers: the input layer, hidden layer, and output layer.

3.5 Model Selection: RNNs and LSTM Networks

RNNs have certain setbacks because of the excessive number of network layers; they are a class of artificial neural networks that have connections between nodes. The LSTM network is an answer to this challenge because of its chain structure, which is like that of multiple neural network modules with RNN. It was a difficult task, so we had to look at a recent study. The LSTM architecture is shown in Fig. 2 shows the LSTM architecture, which includes a forget gate used inside the LSTM model in addition to the first and second gates, which are the input and output gates, respectively. These gates have allowed us to control which information is accepted and rejected throughout the network.

Forget gate

The forget gate extracts unnecessary information from the cell state by multiplying a filter. This step is crucial in optimizing the LSTM network's output by discarding less important or irrelevant data.

Input gate

The input gate adds information to the cell state. This is done using a sigmoid function, which controls the values that are allowed into the cell state.

Output gate

The output gate is responsible for selecting critical cell state information. After applying a tan function to the cell state to scale values into a manageable range, a vector is produced.

3.6 AWS EC2 Windows Server

The requirement for a dependable and expandable hosting environment led to the choice to utilize an AWS EC2 t2.micro instance running Windows Server. The Django application and the machine learning model are supported by this environment, which

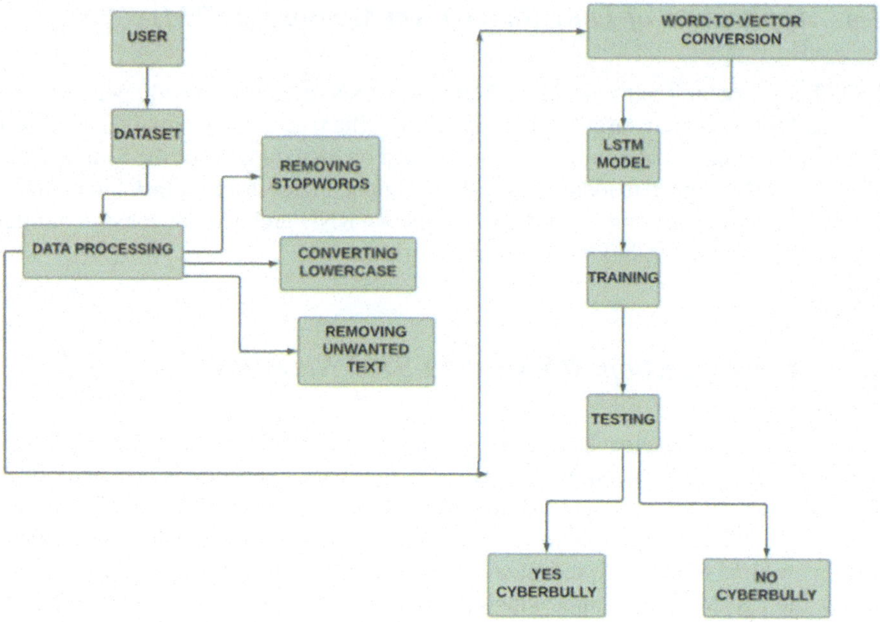

Fig. 1 Data collection and preprocessing

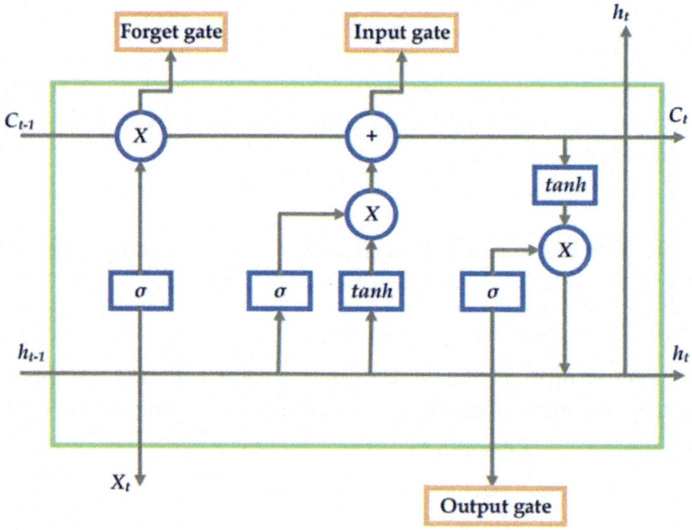

Fig. 2 Long short-term memory architecture

can scale to meet changing user loads as the project expands. The "Safe Cyber Watch: Cyber Bully Alert & Parental Advisory System" uses specialized RNNs known as LSTMs to detect cyberbullying in online conversations. Because of their memory gates, which retain context and filter out extraneous information, LSTMs are particularly good at identifying linguistic subtleties and improve their understanding of online language. LSTMs are used in this study to provide more detailed analyses than keyword detection by analyzing text inputs from youngsters or social media. By using feedback to continuously improve accuracy, they advance attempts to make online settings safer. This integration demonstrates how AI is being used progressively to shield people from cyberbullying. Python and Django: The backend of the system was developed using Django and Python to handle user interactions, data flow, and the integration of the machine learning model. Django's strong security features and scalability make it perfect for managing sensitive user data and expanding with the system's user base.

3.7 *Tools*

JavaScript, HTML, CSS, and Bootstrap: Bootstrap ensured responsive design, CSS improved styling, and HTML structured content for best viewing. JavaScript improved the user experience by introducing dynamic elements and interactivity.

TensorFlow and Keras: TensorFlow expedited model construction and training by effectively computing tensors, while Keras made intricate neural network implementations simpler.

Pandas and Numpy: Numpy handled numerical calculations, whereas Pandas handled data organization.

Seaborn and Matplotlib: These programmers generated perceptive graphics that aided in the analysis of data.

With the help of Scikit-learn, a full range of machine learning tools, cyberbullying detection feature selection, validation, and model evaluation were made easier.

3.8 *Installing on a Windows Server on AWS EC2*

AWS EC2 Windows Server "Safe Cyber Watch" deployment required several important tasks. At first, security settings were set up to permit HTTP communication on port 80 to strike a compromise between stringent security standards and accessibility. Installing Django and its dependencies, setting up a virtual environment, and installing Python created the Windows Server environment. For quick testing and deployment, port 80 of the Django development server was utilized, guaranteeing smooth application access. To guarantee correct serving, care was taken while managing static and media files within Django settings. Security was of the utmost importance, with access managed by AWS security groups and regular upgrades to prevent vulnerabilities. AWS EC2's t2.micro instance was used to accomplish scalability while preserving performance and efficiency.

4 Design and Implementation

4.1 Development of Fundamental Modules

The architectural design, which placed a strong emphasis on a user-focused platform, was realized during the implementation phase.

Setting up of Child Modules

The main user interface for younger users is the Child Module, designed to be both engaging and easy to use. The user interface (UI), created with JavaScript and Bootstrap, features a vibrant color scheme and intuitive icons to ensure cross-platform responsiveness. AJAX calls connect the Django backend with the frontend, allowing seamless text submission and analysis by the machine learning model. Children receive immediate reassurance and are encouraged to report instances of cyberbullying.

Setting up Parent Module

Parents can monitor alerts, review submissions, and leave feedback using the Parent Module, which functions as a dashboard. The dashboard, built using HTML5 and CSS3, provides clarity and simplicity. When potentially harmful content is detected, Django's email feature notifies users. A feedback system, implemented using Django forms, allows parents to confirm or challenge the system's assessments, enhancing the model's accuracy over time.

Setting up Administration Modules

System administrators can oversee user accounts, keep an eye on activity, and update the Machine Learning Model with the help of the Admin Module. AWS CloudWatch integration enables real-time system performance monitoring, while Django's integrated admin features make user management easier.

Architectural Components

There are two main layers in the architecture of the cyberbullying detection system: the Application Layer and the Machine Learning Layer.

Machine Learning Layer

Because of its superior sequence data handling capabilities, an LSTM (Long Short-Term Memory) model is chosen. This layer is in charge of validating output to guarantee accuracy and efficiency, optimising parameters, and training the machine learning model with historical data.

Application Layer

The web application, created using Django, provides functionalities such as data submission, login, user registration, administrative tasks, and results display. RESTful APIs facilitate real-time data handling, enabling smooth data processing and integration. Robust security protocols ensure data privacy, user authentication, and authorization, while maintaining compliance with ethical and legal requirements.

Data Flow

Prior to being processed by the LSTM model, data is first gathered via APIs or direct user input, cleaned, and formatted. The web interface displays the model's predictions, which are kept in a database.

Scalability and flexibility

The web server and processing units, for example, can scale independently of one another as the system's load increases thanks to its horizontal scalability design. Because of its modular architecture, which facilitates the easy integration of new technologies and updates, system capabilities are continuously improved.

The algorithm used affects how well the system can distinguish cyberbullying incidents from textual data. In this process, a number of machine learning models are assessed, the best model is chosen based on performance metrics, and the model is then improved to maximize its efficacy.

Algorithm Selection and Development

Selection Criteria for Algorithms

Performance: The main factors in choosing an algorithm are its performance, specifically its recall, accuracy, precision, and F1-score. The chosen algorithm must reduce false positives and false negatives while effectively differentiating texts that constitute cyberbullying from those that do not.

Scalability: As data volumes and complexity increase, the algorithm must continue to function well.

Adaptability: The algorithm ought to be able to adjust to novel patterns and trends, considering the dynamic nature of language and cyberbullying strategies. This flexibility can be attained by simple retraining or ongoing education.

Selected Algorithm: Long Short-Term Memory (LSTM).

After testing a number of algorithms, such as neural networks, SVM, and Naive Bayes, LSTM was selected due to its prowess in managing sequential data and its ability to retain long-term dependencies. When it comes to text processing tasks, LSTMs work especially well when sentence semantics and context are important factors.

4.2 Train -Test Split

Data Segregation:

Training Set: This set comprises 70–80% of the data and is used to train the model by adjusting weights in response to prediction errors.

Validation Set: It consists of 10–20% of the data and is used periodically to evaluate and fine-tune the model's hyperparameters during training, helping to avoid overfitting.

Testing Set: Comprising 10–15% of the data, this set is used solely to assess the model's performance after training and validation.

Random sampling: Data points are randomly selected to reduce unintentional bias, ensuring that the training, validation, and testing sets accurately represent the full dataset.

5 Result

The Safe Cyber Watch system combines modern machine learning with an intuitive user interface. After being trained on a variety of datasets, the LSTM model achieves a 94% training accuracy and a 73% test accuracy.

The model's performance is summarized in a confusion matrix for six categories. 'Not_cyberbullying' and 'ethnicity' show areas for improvement, while 'gender' and 'age' exhibit promising results. The classification report reveals that the system has an overall accuracy of 73%, with most categories demonstrating high precision and recall. Children can enter text through the interface, parents can review alerts, and administrators can effectively manage the system.

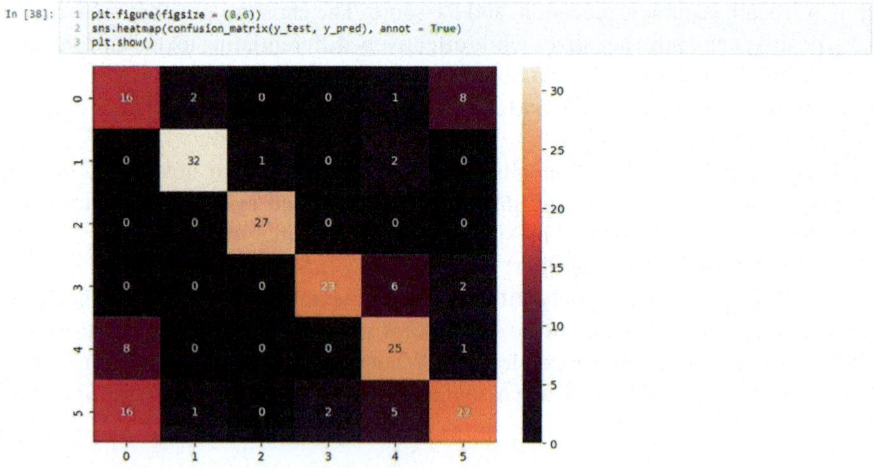

ers and structure. If the data format is not correct, it can be corrected through the pre-processing stage, where manual feature extraction can be done.

4.1 The Case for Network Flows

If normal traffic packet capture datasets were to remove the application layer (payload) data, signature-based detection methods are then rendered unsuitable, as the content-matching rules have no content to operate on. Although, this would address the privacy concerns but would only leave the packet headers for analysis. Using such a method, without any application data to compare known signatures, makes it very difficult to identify a bot's network communication. Hence, the flow-based features are becoming increasingly popular in traffic analysis research, as they are inherently exempt from privacy concerns by not including raw payload data, but instead extract various features from each network flow [33]. Moreover, research by Haddadi et al. [25] shows that flow-based detection systems outperform packet-orientated rule-based detection systems. Packets belonging to the same communication connection between a source and destination are aggregated together with this method, which is then processed into a representation of the whole connection. This takes the form of a 'network flow'. Therefore, this can provide a method of representing bot communication behaviours despite the bot's employing evasion techniques like packet fragmentation and encryption.

The process of obtaining flow data is generally performed through the aggregation of raw packets into flows, then extracting uni or bi-directional features from the flow data, which is exported to a collection client for analysis. For example, this can be conducted using a flow-exporting enabled network device, such as a Cisco router exporting net flows, which are then collected by a network administrator's computer (the flow collector).

The resulting flow output can take the form of a comma separated Value (.csv) text file, which is a common file format and is supported in most data analysis software and ML libraries. Additionally, as no raw payload data is stored, such as in a .pcap file, the file sizes of network captures, in a flow format are considerably smaller and take up a fraction of the storage space. This also can help to decrease ML training time due to the smaller dataset size [21].

4.2 Selected Datasets

Considering the factors mentioned in the previous subsection, the datasets that are most suitable are used in this work as they are collected within a physical environment (i.e. not generated through a simulation tool). They are realistic as they contain real bot traffic and are used by other researchers. Also note that multiple datasets are fundamental to sufficient evaluation of a detection system [65]. The following

Table 1 Dataset class distributions—CTU-13

CTU-13 scenarios	Bots(s)	Normal flows	Botnet flows	Total flows	Botnet (%)
1	Neris	2783675	40961	2824636	1.45
2	Neris	1787181	20941	1808122	1.16
3	Rbot	4683816	26822	4710638	0.57
4	Rbot	1118496	2580	1121076	0.23
5	Virut	128931	901	129832	0.69
6	Menti	554289	4630	558919	0.83
7	Sogou	114014	63	114077	0.06
8	Murlo	2948103	6127	2954230	0.21
9	Neris	1902521	184987	2087508	8.86
10	Rbot	1203439	106352	1309791	8.12
11	Rbot	99087	8164	107251	7.16
12	NSIS.ay	323303	2168	325471	0.67
13	Virut	1885146	40003	1925149	2.08

selected datasets contain a range of bot samples exhibiting their network-based communication and operational characteristics. Moreover, the resulting class balances are more representative of real scenarios, wherein the proportion of bot traffic is negligible.

Once all datasets have been pre-processed and labelled, the resulting class distributions can then be counted and analysed. This is to ensure that the datasets are consistent with the requirements to ensure that the data is representative of real world scenarios, with a realistic class balance (depending upon included traffic) and will aid in future analysis of evaluation and experimentation results that utilise these datasets. The label classes are in the form of binary categorical strings ('Normal' and 'Botnet'), which are suitable for binary classification.

(1) *CTU-13*: The CTU-13 dataset was captured in 2011 at the Czech Technical University (CTU) and includes 13 scenarios, in both .pcap (botnet capture only) and labelled bi-directional network flow formats (including all background, normal and botnet traffic) [20]. It includes 7 different bot samples each exhibiting unique operational behaviour. The scenarios cover botnet usage areas including DDoS, spam, click-fraud (CF) and port scanning (PS), each of which utilises different communication protocols and architectures, such as IRC, HTTP and P2P. The total size of the 13 extended feature set network flow datasets equals 4.58GB. Table 1 shows the class distribution of CTU dataset.

(2) *ISOT 2010*: The ISOT 2010 dataset is a merged product of malicious and normal traffic datasets [56]. The malicious traffic, from the French chapter of the Honeynet Project, comprises of activity from the Storm, Waledac and Zeus botnets. The normal background traffic datasets are the product of merged datasets from Ericsson

Table 5 Sample script for extended network flow feature extraction for ISCX Botnet 2014 dataset

Line no.	Linux Command
1	#!/bin/bash
2	echo "Starting ISCX pcap Feature Extraction into Extended Network Flow format …"
3	sudo argus -r ISCX_Botnet-Training.pcap -w ISCX_Training.argus
4	ra -r ISCX_Training.argus -c ',' -n -s saddr daddr proto sport dport state stos dtos swin dwin shops dhops stime ltime sttl dttl tcprtt synack ackdat spkts dpkts sbytes dbytes sappbytes dappbytes dur pkts bytes appbytes rate srate drate label > Netflow_Extended/ISCX_Training.binetflow
5	echo "Feature extraction complete!"

5.3 Dataset Preparation and Pre-processing

To facilitate ML in Python 3.7, using the sklearn library [48], the datasets must first be loaded from the .csv file into the program's Random-access memory (RAM), into a format suitable and supported by sklearn. This is achieved via the Pandas library with its DataFrame data structure object [45], chosen for the wide range of object attributes and data manipulation functions available. Dataset loading is conducted via pandas.read_csv [46], which returns the data as a DataFrame object.

Additionally, the CTU-13's labels contain verbose descriptions of the exact network activity, such as flow=From-Botnet-V50-4-TCP-WEB-Established-SSL and flow=From-Normal-V50-Stribrek, which are not suitable for categorical binary classification due to the range of descriptions. Therefore, during the pre-processing iterations, the 'Label' feature vector fields for CTU-13 were normalised to only contain 'Normal' and 'Botnet' categorical binary string labels. In the case of the ISOT 2010 and ISCX Botnet 2014, which are not directly labelled due to their packet capture file format, labels must be added to identify the network flow traffic, depending upon the identified infected machines' IP addresses, detailed in each dataset's documentation.

6 Feature Selection/Reduction

Through a process of feature selection and reduction, the dataset dimensionality can be reduced, thus reducing ML classifier fitting times, whilst retaining the more important features, without inducing a significant negative impact on classifier performance [16, 30]. As a side effect of such a process being carried out, feature importance metrics can be computed and subsequently help to provide an avenue down and the various behavioural characteristics can be analysed. This will help to gain an understanding of how different bot behaviours can be identified via the specific subsets through network flow features, providing the most variance in the classification process.

6.1 Manual Feature Exclusion

Features vectors from the dataset that are not essential were removed from the training data, such as the network flow metadata, dates, connection start and end time, along with the source and destination IP addresses. This is a cardinal part of data preprocessing for the design of the model's capabilities, as feature vectors that are manually excluded ensure that the classifiers are trained, evaluated and operated in a desirable way. The aim is to develop a detection system that is completely independent of the port numbers and protocols that are being utilised by bots in the network flow-based communications data [36]. Such considerations must be applied in the context of the range of botnet detection evasion methods that are being increasingly utilised and therefore must be taken into account, such as the dynamic port fluxing methods [7]. This can also allow to train a model to allow generalisation to only the behavioural feature vectors [25].

When these considerations were applied and the related features were excluded, the original extended dataset feature set was reduced to 24 instead of 33 (shown with a checkmark in Table 4), which only contains integer and float data types. This exclusion takes place within the dataset splitting function, via defining the feature labels that are to be maintained, as a result of the manual feature exclusion process, in an array and then utilising the 'pandas.DataFrame.loc' attribute [47], allowing the resulting 'X' feature vector DataFrame to be returned containing only the relevant features.

6.2 Computing Feature Importance

Each feature in the feature set exhibits a 'weight' in the classification of data and thus weaker or irrelevant features can be removed to reduce the dimensionality of the dataset. This allows training time to significantly reduce whilst keeping the classifier performance to acceptable values [58]. One of the most prominent methods is Random Forest (RF) classifier that proves to be effective and will be utilised in this work [11]. Further exploration of feature selection methods with a comparative evaluation was carried out by [50], whereby tree-based feature selection via a RF provides the highest F1-score of 0.99/1.0 with 10-fold cross-validation on ISOT dataset.

In this work, the feature ranking was performed using RF in sklearn, fitting the dataset results in a model object that contains a feature_importances_ attribute array. In this methodology, only the top 15 features of each dataset were retained for training and testing of each classifier's performance.

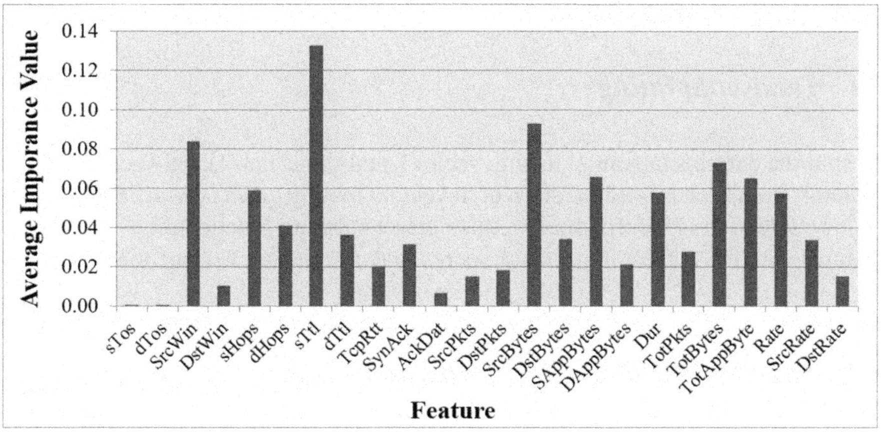

Fig. 3 Computed feature importance values; averaged across all datasets

6.3 Feature Importance Analysis

Once the feature selection methodology was applied to each dataset, the feature importance values were collected, averaged and presented in Fig. 3. It is clear from this analysis that sTos, dTos, ackDat and DstWin result in the lowest average feature importance values.Therefore, we can conclude that the general botnet network communication behaviour in these particular weak features are close to the normal network traffic behaviour and thus cannot be used to differentiate the traffic well. On the other hand, botnet behaviour can be more sufficiently characterised by the high feature importance values present in the sTtl, SrcBytes, SrcWin and TotBytes features.

This analysis proves that the feature selection should not be done on an averaged value basis, and instead should be conducted on a per-dataset basis. This is because different bots exhibit different operational behaviours, network communications and types of attacks. When conducting the investigation, in CTU-13's scenario 2 (Spam and Port Scanning behaviours using HTTP communication), SrcWin proved to be the most important feature with a value of 0.19, whereas, in CTU-13's scenario 10 (UDP DDoS attacks with an IRC communication protocol), SrcWin is one of the weakest features with a value of 0.004.

Computing feature importance can help in the dataset and bot characteristic analysis and thus can also help remove features with minimal importance in the classification process. As a result, the training and classifier evaluation times can be reduced with only a small classifier performance deterioration.

7 Product Methodology

7.1 Dataset Splitting

To split the datasets into an X feature vector DataFrame and y label vector array for training, a function is used to return both vectors for any given network flow dataset. Feature selection is incorporated in this dataset splitting functionality, returning 15 features with the highest importance socres in the X feature vector for each dataset.

7.2 Stratified 10-Fold Cross Validation

Instead of the normal 70:30 training-testing single fitting, which can result in an inaccurate evaluation of actual classifier performance and is especially weak when utilised in class imbalanced datasets [50], this methodology utilised stratified k-fold cross-validation for an accurate classifier performance.

This method operates by splitting the dataset into k folds, whereby $k - 1$ folds are used for training and the remaining fold is used for testing; this is repeated k times such that each fold is used for testing at least once and for training $k - 1$ times. Additionally, it is stratified in the splitting behaviour which means that the class balances are preserved in each fold of dataset. In this work, k is set to 10 to allow a more thorough evaluation of the classifier performance on the class imbalanced datasets.

As the sklearn cross_validate function includes a multi-threading argument n_jobs, each fold can be allocated with a CPU core to allow the 10 folds for a single dataset to run in parallel.

7.3 Evaluation Metrics

Due to the class imbalances present in the datasets, this must be taken into account when choosing the metrics which are to be used in the evaluation of binary classification performance of each classification method. Due to these class imbalances, accuracy alone will not be a suitable metric due to the positive class samples accounting for the minority of the total samples. The accuracy metric takes into account the True Negative (TN) results which causes the TN predictions to have a large weight on the accuracy metric. Accuracy can be calculated to compare the metrics as in Eq. (2):

$$\mathtt{Accuracy} = \frac{(\mathtt{TP} + \mathtt{TN})}{(\mathtt{TP} + \mathtt{TN} + \mathtt{FP} + \mathtt{FN})} \tag{2}$$

Table 10 Random Forest

Dataset		Fit time (s)	Precision	Recall	F1 score	Accuracy
CTU-13	1	34.01	0.9818	0.9698	0.9759	0.9977
	2	27.59	0.9915	0.9854	0.9884	0.9981
	3	80.07	0.9991	0.9991	0.9991	0.9996
	4	13.82	0.9863	0.9833	0.9848	0.9983
	5	0.87	0.9895	0.9496	0.9692	0.9901
	6	7.21	0.9949	0.9980	0.9964	0.9993
	7	0.68	0.9113	0.7987	0.8513	0.9799
	8	75.71	0.9725	0.9729	0.9727	0.9993
	9	20.96	0.9720	0.9800	0.9760	0.9907
	10	21.66	0.9996	0.9988	0.9992	0.9996
	11	0.77	0.9996	0.9985	0.9991	0.9997
	12	4.31	0.9641	0.9096	0.9361	0.9889
	13	33.58	0.9839	0.9836	0.9838	0.9980
ISOT		31.11	0.9929	0.9846	0.9887	0.9953
ISCX		4.822	0.9912	0.9930	0.9921	0.9913

Table 11 Random Forest—parameter selection

Parameter	Search space
N estimators	(low = 10, high = 200)
Criterion	['gini', 'entropy']
Min samples split	(low = 2, high = 5)
Min samples leaf	(low = 1, high = 4)
Min weight fraction leaf	(low = 0.0, high = 0.1)
Class weight	['balanced', none]

Artificial Neural Networks. But by using sequences of classifier instances to reduce the error rate in the next iteration, the number of classifier instances to be used in the training can be defined in the n_estimators hyperparameter, which defaults to 50 trees. Standard Adaboost algorithm can be denoted by Eq. (9) [34]:

$$\mathrm{H}_{\mathtt{Ada}}(I) = \mathtt{sign}\left(\sum_{t=1}^{T} \alpha_t h_t(I)\right) \tag{9}$$

where I is the datapoints/labels within the dataset, α_t is the weight of the classifier, h_t is the classifier and T is the number of cycles.

Table 12 shows the metrics output for Adaboost, with highest F1-score of 0.9991 for CTU-13-scenario 11, along with fit-time as 11.22 sec, precision as 0.9996, recall as 0.9985 and accuracy as 0.9991.

Table 12 Adaboost

Dataset		Fit time (s)	Precision	Recall	F1 score	Accuracy
CTU-13	1	396.06	0.9593	0.9142	0.9362	0.9565
	2	303.21	0.9906	0.9683	0.9793	0.9889
	3	781.93	0.9969	0.9972	0.9971	0.9977
	4	141.87	0.9645	0.9071	0.9349	0.9477
	5	11.55	0.9757	0.9373	0.9561	0.9767
	6	94.31	0.9918	0.9951	0.9935	0.9964
	7	10.06	0.9415	0.7511	0.8356	0.9618
	8	748.56	0.9652	0.9707	0.9679	0.9944
	9	229.98	0.9442	0.9377	0.9409	0.9551
	10	301.55	0.9991	0.9981	0.9986	0.9990
	11	11.22	0.9996	0.9985	0.9991	0.9991
	12	37.52	0.9551	0.7254	0.8245	0.8709
	13	253.61	0.9684	0.9470	0.9576	0.9714
ISOT		196.59	0.9693	0.9511	0.9601	0.9665
ISCX		42.46	0.9802	0.9813	0.9808	0.9800

Table 13 Adaboost—parameter selection

Parameter	Search space
N estimators	(low = 5, high = 100)
Learning rate	(low = 0.1, high = 0.1)
Algorithm	('SAMME', 'SAMME.R')
Random state	(low = 1, high = 50)

Table 13 shows the hyperparameters to be considered to optimise the performance of Adaboost. The default parameters used are [50, 1.0, 'SAMME.R', None].

8.5 *Linear Support Vector Machines*

Support Vector Machines (SVMs) are a method of classifying data by fitting a hyperplane to the feature vectors, creating a split in the data where it best determines the data labels can be effectively split. As a result, they are useful for datasets that have large dimensionality. A linear kernel is applied to the SVM classifier, allowing faster operation due to a simpler function to fit the data [68].

Equation (10) shows the primal problem by utilising the hinge loss:

$$\max_{\mathbf{w},\mathbf{b}} \frac{1}{2} w^T w + C \sum_{i=1} \max(0, y_i(w^T \phi(x_i) + b)) \tag{10}$$

Table 14 Linear SVC

Dataset		Fit time (s)	Precision	Recall	F1 score	Accuracy
CTU-13	1	1715.69	0.5831	0.6131	0.5977	0.6106
	2	626.35	0.5458	0.6945	0.6113	0.6173
	3	1514.78	0.4971	0.4956	0.4963	0.4965
	4	931.57	0.5284	0.6992	0.6019	0.6101
	5	35.88	0.5019	0.5552	0.5286	0.5400
	6	407.23	0.6498	0.9131	0.7593	0.7615
	7	23.19	0.4997	0.4954	0.4975	0.5726
	8	886.48	0.5308	0.7854	0.6335	0.6508
	9	625.14	0.6884	0.7862	0.7341	0.7452
	10	781.95	0.9363	0.9756	0.9555	0.9559
	11	28.82	0.9277	0.9678	0.9473	0.9479
	12	107.18	0.5036	0.5192	0.5113	0.5401
	13	556.60	0.5845	0.6752	0.6266	0.6356
ISOT		568.31	0.6283	0.6634	0.6454	0.6497
ISCX		111.19	0.7135	0.6634	0.6876	0.6870

Table 15 Linear SVC—Parameter selection

Parameter	Search space
Loss	['hinge', 'squared hinge']
Tol (tolerance)	(low = 1e-5, high = 0.1)
C (Regularisation parameter)	(low = 1, high = 5)

where x is the datapoint, ϕ is the identify function and C is the penalty term.

Table 14 shows the metrics output for Linear SVC, with highest F1-score of 0.9555 for CTU-13-scenario-10; along with fit-time as 781.95 sec, precision as 0.9363, recall as 0.9756 and accuracy as 0.9559.

Table 15 shows the hyperparameters to be considered to optimise the performance of Linear SVC. The default parameters used are ['squared_hinge', 1e-4, 1.0].

8.6 K Nearest Neighbors

The k-Nearest-Neighbours (KNN) classifier is an algorithm that calculates the distance of a data sample point from the nearest k number of data sample neighbours in the dataset's trained feature space. The nearest k labels to the new unseen data are then determined to be the best match for the new data to be labelled [41]. Sklearn with its default value for metric as 'minkowski' and power parameter as '2', is similar to a normal euclidean metric, taking a data point (x) and the predicted point (y) into

Table 16 K Neighbors

Dataset		Fit time (s)	Precision	Recall	F1 score	Accuracy
CTU-13	1	24039.29	0.9575	0.9391	0.9482	0.9687
	2	15635.49	0.9868	0.9728	0.9798	0.9894
	3	52461.87	0.9950	0.9985	0.9967	0.9972
	4	6526.29	0.9759	0.9350	0.9550	0.9680
	5	12.51	0.9454	0.9297	0.9375	0.9577
	6	2444.85	0.9892	0.9941	0.9917	0.9947
	7	7.12	0.6799	0.5618	0.6152	0.7081
	8	28655.06	0.9633	0.9570	0.9601	0.9864
	9	7335.46	0.9457	0.9421	0.9439	0.9582
	10	8345.15	0.9989	0.9979	0.9984	0.9988
	11	8.38	0.9976	0.9982	0.9979	0.9985
	12	43.16	0.9070	0.7776	0.8373	0.8845
	13	8643.61	0.9718	0.9571	0.9644	0.9783
ISOT		1355.26	0.9599	0.9763	0.9680	0.9744
ISCX		42.70	0.9752	0.9808	0.9780	0.9772

account with n number of dimensions, Euclidean metric can be denoted by Eq. (11) [14]:

$$f(x, y) = \sqrt{\sum_{i=1}^{n}(x_i - y_i)^2} \tag{11}$$

If the power parameter is set to '1' with 'minkowski' as default, the algorithm takes Manhattan metric denoted by Eq. (12) [14]:

$$f(x, y) = \sum_{i=1}^{n} | x_i - y_i | \tag{12}$$

For arbitrary value of power parameter, the algorithm uses minkowski and it is denoted by Eq. (13) [14]:

$$f(x, y) = (\sum_{i=1}^{n}(| x_i - y_i |)^q)^{\frac{1}{q}} \tag{13}$$

Table 16 shows the metrics output for KNN, with highest F1-score of 0.9984 for CTU-13-scenario 10, along with fit-time as 8345.15 s, precision as 0.9989, recall as 0.9979 and accuracy as 0.9988.

Table 17 shows the hyperparameters to be considered to optimise the performance of K-Neighbors. The default parameters used are [5, 'uniform', 'auto', 30, 2].

Table 17 K-Neighbors—parameter selection

Parameter	Search space
N neighbors	(low = 1, high = 10)
Weights	['uniform', 'distance']
Algorithm	['ball tree', 'kd tree']
Leaf size	(low = 1, high = 50)
P (power)	(low = 1, high = 5)

8.7 Artificial Neural Networks

Artificial Neural Networks (ANNs) utilise the concept of biologically-inspired mathematical neurons arranged in groups of units, called layers. Neurons also called perceptrons can be arranged in fully connected layers, where each perceptron has a set of input edges that are connected to a set of output edges produced by the previous layer. Similarly, a perceptron's output edges connect as an input to each perceptron in the next layer. Layers can be constructed in Keras library, using the sequential model to allow various layers (like 'dense' and 'dropout' layers) into a network.

The learning is facilitated by passing input training data to the classifier, whereby each feature vector is assigned an input neuron. Therefore, the input dimension must match feature vector length, which in the case of the extended network flow feature set, is 24. The neural network then passes this data through the layers of fully connected neurons in the network over one or more epochs (iterations), calculating the error of the predictions against a loss metric at each batch, which is then used in the backpropagation to update the weights of the neurons in the network to reduce the cumulative loss.

The feature selection is not performed in NN due to its ability to autonomously discern feature importance during training via backpropagation [51]. For the ANNs, the 'default hyperparameters' do not exist unlike previously discussed classifiers, hence, two network configurations are utilised with reasonably performing hyperparameters following some initial grid-search exploration with the KerasClassifier wrappers. Two ANNs are utilised, one with a single hidden layer of 24 units and one with two hidden layers, with 24 units in the first layer and 40 units in the second layer.

(1) *Neural Network 1*: Table 18 shows the metrics output for ANN#1, with highest F1-score of 0.9176 for CTU-13-scenario-13; along with fit-time as 102.98 sec, precision as 0.9228, recall as 0.9124 and accuracy as 0.9308. When all the scenarios are considered, the performance is poor.

Table 19 shows the hyperparameters to be considered to optimise the performance of ANN#1. The default parameters used are [5, 100, 100, 'relu', 'sigmoid', 'binary_crossentropy', 'adam'].

Table 18 Neural network 1

Dataset		Fit time (s)	Precision	Recall	F1 score	Accuracy
CTU-13	1	159.02	0.6761	0.6189	0.6462	0.6602
	2	105.75	0.6902	0.6295	0.6585	0.6649
	3	265.01	0.7429	0.9038	0.8155	0.8159
	4	69.69	0.6349	0.5935	0.6135	0.6218
	5	14.56	0.5341	0.5964	0.5636	0.5757
	6	53.57	0.9010	0.8580	0.8790	0.8816
	7	18.66	0.6497	0.6497	0.6497	0.7478
	8	175.85	0.9125	0.8048	0.8553	0.8787
	9	109.35	0.7252	0.7401	0.7326	0.7437
	10	98.38	0.6680	0.7150	0.6907	0.6910
	11	19.97	0.6463	0.6898	0.6673	0.6677
	12	30.59	0.5724	0.5779	0.5752	0.6076
	13	102.98	0.9228	0.9124	0.9176	0.9308
ISOT		70.52	0.4613	0.4975	0.4787	0.4819
ISCX		17.17	0.6605	0.7106	0.6846	0.6841

Table 19 NN1—parameter selection

Parameter	Search space
Epochs	(low = 2, high = 10)
Batch size	(low = 10, high = 1000)
Hidden layers	(low = 10, high = 1000)
Layers activation	['softmax', 'selu', 'softplus', 'relu', 'tanh', 'sigmoid', 'linear']
Output activation	['softmax', 'selu', 'softplus', 'relu', 'tanh', 'sigmoid', 'linear']
Loss function	['mean_squared_error', 'mean_absolute_error', 'squared_hinge', 'categorical_crossentropy', 'binary_crossentropy']
Optimise function	['SGD', 'RMSprop', 'Adagrad', 'Adadelta', 'Adam', 'Adamax', 'Nadam']

(2) *Neural Network 2*: Table 20 shows the metrics output for ANN#2, with highest F1-score of 0.7998 for CTU-13-scenario-7; along with fit-time as 27.30 s, precision as 0.7997, recall as 0.8 and accuracy as 0.9206. When all the scenarios are considered, the performance is poor.

Table 21 shows the hyperparameters to be considered to optimise the performance of ANN#2. The default parameters used are [5, 100, 100, 'relu', 0.25, 500, 18, 'relu', 0.25, 'sigmoid', 'binary_crossentropy', 'adam'].

Table 20 Neural network 2

Dataset		Fit time (s)	Precision	Recall	F1 score	Accuracy
CTU-13	1	598.39	0.6715	0.5434	0.6007	0.6133
	2	368.91	0.4942	0.4999	0.4970	0.5018
	3	908.09	0.4971	0.5	0.4985	0.4988
	4	247.62	0.6401	0.5951	0.6168	0.6252
	5	29.89	0.6564	0.5605	0.6047	0.6177
	6	141.03	0.9354	0.6087	0.7375	0.7397
	7	27.30	0.7997	0.8	0.7998	0.9206
	8	527.50	0.8579	0.6159	0.7170	0.7366
	9	321.23	0.7494	0.6193	0.6782	0.6884
	10	307.27	0.6998	0.7415	0.7201	0.7205
	11	31.61	0.6551	0.6933	0.6737	0.6741
	12	69.99	0.5566	0.5535	0.5551	0.5864
	13	280.27	0.5791	0.5016	0.5376	0.5453
ISOT		212.00	0.4613	0.4991	0.4795	0.4827
ISCX		73.01	0.6394	0.6642	0.6516	0.6511

Table 21 NN2—parameter selection

Parameter	Search space
Epochs	(low = 2, high = 10)
Batch size	(low = 10, high = 1000)
Hidden layers	(low = 10, high = 1000)
Layers activation	['softmax', 'selu', 'softplus', 'relu', 'tanh', 'sigmoid', 'linear']
Dropout 1	(low = 0.1, high = 0.5)
Hidden layers 2	(low = 10, high = 1000)
Input dim 2	(low = 2, high = 24)
Layers activation 2	['softmax', 'selu', 'softplus', 'relu', 'tanh', 'sigmoid', 'linear']
Dropout 2	(low = 0.1, high = 0.5)
Output activation	['softmax', 'selu', 'softplus', 'relu', 'tanh', 'sigmoid', 'linear']
Loss function	['mean_squared_error', 'mean_absolute_error', 'squared_hinge', 'categorical_crossentropy', 'binary_crossentropy']
Optimise function	['SGD', 'RMSprop', 'Adagrad', 'Adadelta', 'Adam', 'Adamax', 'Nadam']

9 Optimization

9.1 Random Search

In a random search methodology, stochastic values are chosen for each hyperparameter type from a set for a given classifier. This method is extremely inefficient and may not result in an optimal hyperparameter set in a reasonable amount of computational time for guaranteed improvements [57].

9.2 Grid Search

In this method of hyperparamets exploration, possible hyperparameter values are arranged in a grid, whereby all possible combinations of hyperparameter sets are constructed and then exhaustively traversed to evaluate each set. Although being slightly better in terms of computational efficiency than the random search, certain instances give low F1 score and take more time [57]. However, the grid search can explore a more effective combination of hyperparameters outlying from the default values. However, this method is limited to the grid initially defined. Moreover, the resolution of the hyperparameter values is limited to initial values, resulting in a hyperparameter set that may not be near-optimal due to the resolution limitations. Additionally, the efficiency of the grid search method exponentially decreases with the increase in the number of hyperparameters. Large data sets take more time which deteriorates the grid search time efficiency. Table 22 shows the grid search parameter combinations for all the classifiers.

It is worth noticing that the fitting was a single iteration performed (against the largest dataset which is 1 GB in size) with the default hyperparameters for the particular classifier. As changing hyperparameters fundamentally alters the algorithmic operations which can lead to varied fitting times. For example, the RF classifier

Table 22 Classifier grid search combinations and times

Classifier	Grid search combinations	Largest dataset fit time (with default hyperparameters) (s)	Total grid search time (h)
Decision Tree	1408	55.6	21.75
Random Forest	3456	80.1	76.90
Ada Boost	840	781.9	182.44
Linear SVC	200	1514.8	84.16
N-Neighbors	2200	52461.9	32060.05
Neural Network #1	3025260	265	222692.75
Neural Network #2	88942644000	908.1	22435781949

utilises an n_estimators integer value hyperparameter, which is used to control the number of separate trees utilised in the creation of the forest. As a result, increasing the forest size will also increase processing time required for the classifier to fit the data.

9.3 Genetic Algorithm

This research proposes utilising a more efficient method of hyperparameter optimisation. The natural biological evolution tends to favour the fitter individuals in a population, in that they survive longer and breed wherein the genes in the offspring are crossed over from the parents allowing mutation in the genetic makeup. This natural selection process allows the desirable fittest characteristics to be propagated through the future population generations. In the case of hyperparameter optimisation, this research has applied Genetic Algorithm (GA) as a method for converging towards more optimal hyperparameter sets to increase the training efficiency [32].

The GA algorithm has proven to be effective in the research done in intrusion detection and malware detection. The authors in the paper [16], experimented to detect malware on an Android platform using GA. The algorithm provides promising results in reducing feature space. The paper also states the capability of GA to work with reduced feature dimensional along with reduction in training time. The paper [40] worked with optimised RF with GA and found it to be more effective compared to a standalone classifier for botnet detection based on DNS queries, using TI-2016 DNS dataset. In the paper [54] on adaptive anomaly-based intrusion detection system using GA and profiling, the authors were able to achieve a high detection rate and low false positive rate, when used with CICIDS-2017 dataset. These work and others justifies the use GA in our research, as a potential optimisation approach.

(1) *Gene Pools*: Before a GA can operate, a search space is defined for hyperparameter maximum and minimum numerical values along with categorical strings. Therefore, as each hyperparameter takes the form of a 'gene' in the GA, this search space can be thought of as the 'gene pool', defining the states or range of states in which they can exist. This gene pool is not defined as a grid and thus is not constrained to the resolution and intervals present in a grid, in terms of integer and float data-typed hyperparameters.

Although more hyperparameters are available for each of the classifiers in Scikit-Learn library, suitable consideration was taken to ensure that a range of relevant hyperparameters were chosen and that the constraint values of each (depending on data type) are such that the resultant pool is able to encompass a highly dimensional gene pool search space. The gene pool for the classifiers is shown in Table 23.

(2) *Operation*: Each individual 'chromosome' in a population consists of a hyperparameter set, with each gene containing the relevant values in the range defined for the hyperparameter in the gene pool. As a result, each chromosome exhibits different lengths and hyperparameter data types [32].

Table 23 The Gene pools defined for each classifier

Classifier	Gene (Parameter)	Search space
Decision Tree	Criterion	['gini', 'entropy']
	Splitter	['best', 'random']
	Min_samples_split	(low = 2, high = 5)
	Min_samples_leaf	(low = 2, high = 4)
	Min_weight_fraction_leaf	(low = 0.0, high = 0.1)
	Class_weight	['balanced', 'None']
Random Forest	N_estimators	(low = 10, high = 200)
	Criterion	['gini', 'entropy']
	Min_samples_split	(low = 2, high = 5)
	Min_samples_leaf	(low = 2, high = 4)
	Min_weight_fraction_leaf	(low = 0.0, high = 0.1)
	Class_weight	['balanced', 'None']
Ada Boost	N_estimators	(low = 5, high = 100)
	Learning_rate	(low = 0.1, high = 1.0)
	Algorithm	['SAMME', 'SAMME.R']
	Random_state	(low = 1, high = 50)
Linear SVC	Loss	['hinge', 'squared$_h$inge']
	Tol (Tolerance)	(low = 1e-5, high = 0.1)
	C (Regularisation Parameter)	(low = 1, high = 5)
K-Neighbors	N_neighbors	(low = 1, high = 10)
	Weights	['uniform', 'distance']
	Algorithm	['bal_tree', 'kd_tree']
	Leaf_size	(low = 1, high = 50)
	P (Power)	(low = 1, high = 5)
Neural Network 1 (24-x1)	Epochs	(low = 2, high = 10)
	Batch_size	(low = 10, high = 1000)
	Hidden_layers	(low = 2, high = 1000)
	Layers_activation	['softmax', 'selu', 'softplus', 'relu', 'tanh' 'sigmoid', 'linear']
	Output_activation	['softmax', 'selu', 'softplus', 'relu', 'tanh' 'sigmoid', 'linear']
	Loss_function	['mean_squared_error', 'mean_absolute_error', 'squared_hinge', 'categorical_crossentropy', 'binary_crossentropy']
	Optimiser_function	['SGD', 'RMSprop', 'Adagrad', 'Adadelta', 'Adam', 'Adamax', 'Nadam']
Neural Network 2 (24-x1-d1-x2-d2-1)	Epochs	(low = 2, high = 10)
	Batch_size	(low = 10, high = 1000)
	Hidden_layers	(low = 2, high = 1000)
	Layers_activation	['softmax', 'selu', 'softplus', 'relu', 'tanh' 'sigmoid', 'linear']
	Dropout_1	(low = 0.1, high = 0.5)
	Hidden_layers_2	(low = 10, high = 1000)
	Input_dim_2	(low = 2, high = 24)
	Layers_activation_2	['softmax', 'selu', 'softplus', 'relu', 'tanh' 'sigmoid', 'linear']
	Dropout_2	(low = 0.1, high = 0.5)
	Output_activation	['softmax', 'selu', 'softplus', 'relu', 'tanh' 'sigmoid', 'linear']
	Loss_function	['mean_squared_error', 'mean_absolute_error', 'squared_hinge', 'categorical_crossentropy', 'binary_crossentropy']
	Optimiser_function	['SGD', 'RMSprop', 'Adagrad', 'Adadelta', 'Adam', 'Adamax', 'Nadam']

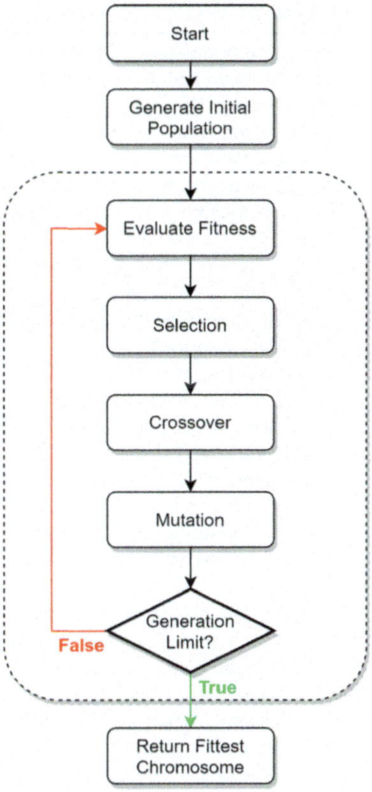

Fig. 4 The operation of genetic algorithm

In this implementation, a heuristic approach is taken to the initial population generation. This is where the initial population is injected with a known-performance chromosome of default hyperparameter values. This is advantageous to a random initialisation as it gives a place to start if the rest of the population is not initially fit.

In each generation, each member of the population undergoes an evaluation via a fitness function. In this case, the stratified 10-fold cross-validation methodology is utilised to maintain classifier F1 score performance evaluation consistent with the previous experiment. Therefore, the maximisation of this F1 score is desired, and thus in the selection process, the two fittest are chosen as the parents for the next generation. The parents' genes are then crossed over using one central point so that the offspring have the genes of each parent. The population is then filled to the population limit with the offspring and other randomly injected chromosomes. The mutation takes place in a random gene of each chromosome in the population to alter its value, helping to avoid arriving at a local optimum and increase the chances of convergence towards fitter chromosomes.

Figure 4 shows the operation of GA through a flow chart. Algorithm 1 presents a pythonic-style pseudo-representation of the GA class implementation.

Algorithm 1: Genetic Algorithm for hyper-parameter tuning

```
population = initialise population(populationSize);
chromosomeLength = len(population[0]);
fittestOfGen = [];
while generation < generationLimit do
    fitnessValues = [];
    //Evaluate the fitness F1 scores for chromosome in population do
        metrics = CrossValidate(classifier, dataset, cv=k);
        f1Score = calcF1Score(metrics);
        fitnessValues.append(f1Score);
    end
    //Selection of the fittest chromosomes sorted = sort(fitnessValues);
    parent1 = sorted[0];
    parent2 = sorted[1];
    fittestOfGen.append(parent1);
    //Crossover of genes from the fittest parent offspring1 = [];
    offspring2 = [];
    for gene in chromosomeLength do
        if gene < (chromosomeLength/2) then
            offspring1.append(parent1[gene];
            offspring2.append(parent2[gene];
        else
            offspring1.append(parent2[gene];
            offspring2.append(parent1[gene];
        end
    end
    //Generate a new population from the offspring population = [];
    for index in populationSize do
        if index < populationSize/4 then
            chromosome = offspring1.copy();
        else if index < populationSize/2 then
            chromosome - offspring2.copy();
        else
            chromosome = generateRandomChromosome();
        end
        population.append(chromosome);
    end
    //Mutate a random gene in each chromosome
    for each chromosome in population do
        mutate(chromosome);
    end
end
bestChromosome = sort(fittestOfGen);
return bestChromosome[0];
```

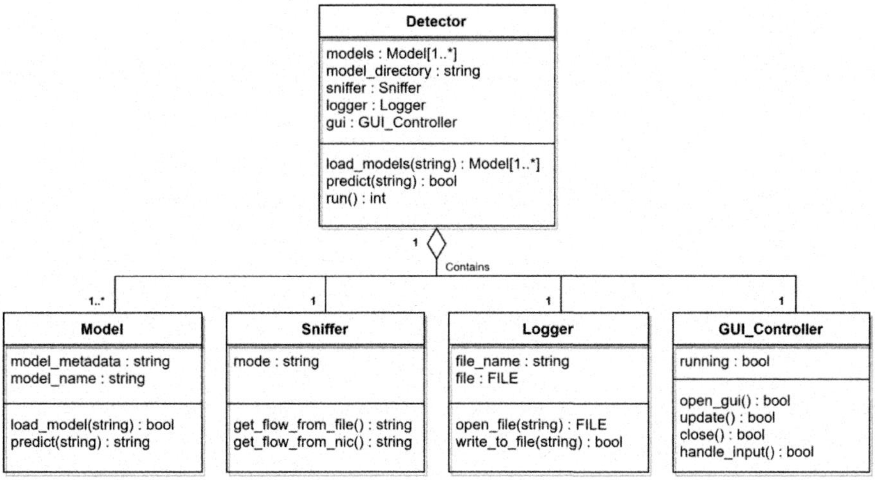

Fig. 7 Class association diagram

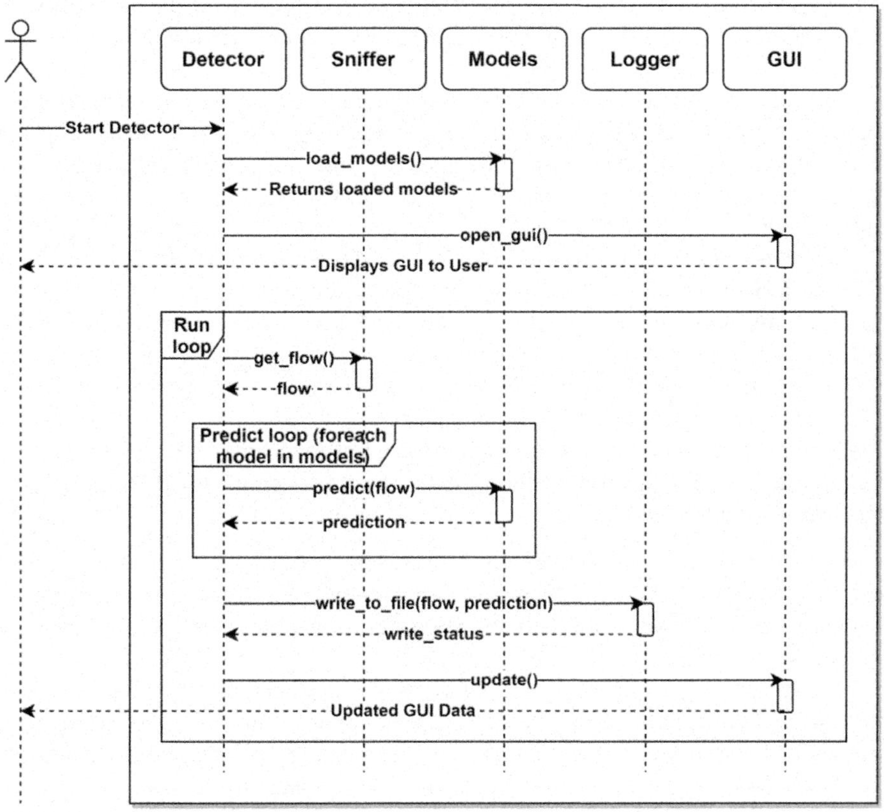

Fig. 8 System interaction diagram

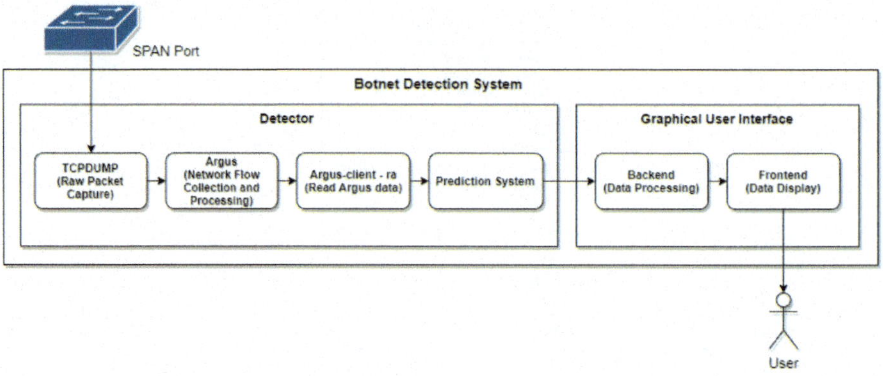

Fig. 9 The Botnet detection system's high-level architectural design

(1) *Operation*: This was originally implemented via a bash script with tcpdump, Argus and the 'ra' client running, with the stdout of each being piped into the stdin of the next program. The final section of the pipe was the stdin of the main detector Python process. This method worked, via a function to read stdin within the sniffer module.

However, the sniffer class should be able to run these processes by itself, without having to rely on the input pipeline being correct in bash script. Therefore, the Python 'subprocess' module was implemented to increase control and integrate the data sniffing and processing components required for successful injection into the sniffer module.

(2) *Adaptation to Support Multiple Formats of Data Input*: To allow support for multiple formats of data input, i.e. in the case of .pcap or pre-processed network flow files, to achieve requirements F-D-2 and F-D-3 as in Table 24, respectively, the sniffer was designed to utilise an argument, which is initially set to false for normal network sniffing operation, whereby .pcap or network flow files can be specified in the detector via a command-line argument (-r).

If a file is specified, the sniffer is configured at load time as to the mode it is required to run in. This means altering the subprocess commands that are running depending upon the mode.

(3) *Subprocess Management*: Some issues were encountered during the development and initial testing whereby the subprocesses would become 'orphans' i.e. they would hang, and keep running perpetually in the background even when the main detector process was terminated.

This was discovered to be a result of having to include the *shell=true* argument in the *Popen* function, as this is required to enable the successful calling and subsequent execution of commands, similar to how they are executed and behave on a command-line shell. This works via the forking of a child process and first executing an sh shell process, which then subsequently calls the command provided as an argument to *Popen*.

Table 24 Product requirements (functional) specification

ID	Type	Priority	Requirement
F-R-1	Functional	Required	The detection engine must be able to process raw network data, received from the machine's NIC from a SPAN port on a switch, into network flow data with the required features present
F-R-2	Functional	Required	The detection engine must load (de-serialise) the serialised machine learning model objects from the files
F-R-3	Functional	Required	The detection engine must be able to predict the classification of each network flow using all of the loaded models
F-R-4	Functional	Required	The detection engine must generate an alert and information for any network flows that have been labelled as 'botnet' by one or more of the models that gave the predictions
F-R-5	Functional	Required	The detection engine must log generated alerts to file
F-R-6	Functional	Required	The detection engine must log labelled network flow data to file
F-R-7	Functional	Required	The detection engine must have the ability to send the labelled flow data to the Graphical User Interface (GUI)
F-R-8	Functional	Required	The GUI must be able to receive data from the detection engine
F-R-9	Functional	Required	The GUI must be able to display alerts
F-R-10	Functional	Required	The GUI must be able to display network flow data
F-D-1	Functional	Desired	The detection engine may have the ability to run standalone (i.e. without the GUI)
F-D-2	Functional	Desired	The detection engine may have the ability to read and process inputted packet capture '.pcap' files (instead of reading from the network)
F-D-3	Functional	Desired	The detection engine may have the ability to read and process inputted network flow files (such as the '.binetflow' or '.csv' dataset files)
F-D-4	Functional	Desired	The detection engine may have the ability to add more bot models to the system (similar to how new rules can be added to the Snort Intrusion Detection System) and dynamically load any new models at load time
F-D-5	Functional	Desired	The GUI may have the ability to display the detection results as a real time network flow activity chart

The issue subsequently arises during the exiting of the main detection engine python process, whereby the sniffer object's destructor _del_ is called. This destructor originally contained the code below, calling terminate on the subprocess objects, resulting in each of the subprocess' process ID to be sent a SIGTERM signal.

```
def_del_(self):
# The Sniffer object's destructor
# Kills the subprocesses via their parent shell PIDs
if(self.read_from_file == False): if self.tcpdump is not None:
# Kill tcpdump
os.killpg(self.tcpdump.pid, signal.SIGTERM)
if self.argus is not None:
# Kill argus
os.killpg(self.argus.pid, signal.SIGTERM)
if self.ra is not None:
# Kill ra
os.killpg(self.ra.pid, signal.SIGTERM)
print("Deleting sniffer object and any relevant running subprocesses.")
```

However, this PID belongs to 'sh' process that is initially forked, which gets properly terminated as expected, but the child process that it forks does not, resulting in the orphaned processes. This is an undesirable characteristic for a final product and caused memory issues during testing, with many orphaned processes maintaining persistence in the background.

To solve this, the orphaned subprocesses must also somehow be sent a SIGTERM signal, and the solution must be properly implemented as to not involve a 'hacky' fix, which could be calling *os.kill(subprocess.pid + 1)*. The subprocess module's *Popen* function was found to support a *preexec_fn* argument, whereby we can specify a system call in the child 'sh' process, after the fork from the parent, but before 'sh' is executed.

This can be used to call *os.setsid*, which creates a new session and process group that results in the calling process to become the process group leader. In this case, the calling process is the sh shell which forks a child process of one of the components, such as the *tcpdump* or *argus* subprocesses, which subsequently inherit the process group ID (PGID) of the calling parent 'sh' process.

As a result, the sh calling parent process and the callee share the same PGID, and thus, all processes within the group can be killed using *os.pgkill*, passing in the PID (as the calling shell process is the process group leader, PID == PGID) of the parent subprocess and SIGTERM. This functionality is subsequently implemented in the Sniffer's *_del_* object destructor.

11.2 Scalable Model Sets

Scalable model sets are implemented to meet F-D-5 as in Table 24. This allows new models to be added to the directory and their metadata to be entered into the metadata JSON storage file and they will subsequently be dynamically loaded at runtime—and this is similar functionality to the scalable ruleset design in Snort.

The functionality is implemented in the detector, whereby it finds the models listed in the metadata file and then attempts to find the same name model file in the models directory. Any valid models found are then de-serialised into separate model objects, allowing the detector to maintain an array of successful model objects for its prediction operations.

11.3 Graphical User Interface (GUI)

Electron was chosen over methods such as Tkinter for the GUI development due to its rising popularity for enabling the creation of dynamic desktop-based web-style applications, with a NodeJs server backend and a JavaScript-powered frontend. This allowed a dynamic rendering of CSS-styled HTML pages and interaction with the backend via an exposed send and receive to the frontend sandboxed JavaScript renderer process.

This method enables the power and familiarity of web development to allow desktop applications to be more usable and functional without the overhead of another GUI library. Additionally, this method gives the ability to allow the GUI to be hosted and on another machine on the network for monitoring purposes, as it is disconnected from the running detection engine process, as it runs over a client-server websocket interaction.

To allow only one page being in memory and being rendered at a time, an API call is invoked at each page load within the renderer process, which requests all the data stored in a data handling class created in the backend to track all received flows and alerts. When this request is received in the main backend process, the event handler calls Data_Controller.get_all_data() which returns all the data stored in the storage arrays and subsequently sends the data back to the frontend. As these are in the backend, they are isolated from the frontend and are not lost when new pages are loaded.

The renderer code for each page then handles the array of received data, as opposed to a single item as was the case in the beginning, and thus iterate through each index and render the specific data correctly. This helps to maintain a perceived persistence and continuity of data between the specific pages available.

11.4 WebSockets

Since the user interface is running as a separate process from the detection engine, the network flow data and alerts must somehow be sent from the detector to the GUI. Initially, Inter Process Communications (IPC's) were considered, but the power, ease and flexibility of a websocket implementation over localhost allowed a quick and reliable method of transferring data using the WebSocket client and server libraries.

The GUI's backend, therefore, hosts a websocket server to listen for incoming connections, and then includes an event handler to handle each message received. Additionally, this method allows for future flexibility, as the GUI could be running on another machine on the network and so the detector on the network edge can connect to the GUI running elsewhere in the network, via its IP address. This is a common use case for dashboard-style monitoring systems, wherein the dashboard is connected over the network to the detector for remote monitoring.

11.5 Product (Botnet Detection Software)

The main features of the software are as follows. The network flow classification predictions are done via serialised bio-optimised RF classifier models. The model is scalable and the sniffer supports multiple input types like raw live network sniffing, raw packet capture files (.pcap) and pre-processed (extended feature set) network flow text files (.binetflow). The extensible desktop-based dashboard GUI is made using the Electron framework, with file logging for alerts, labelled network flows and desktop notification alerts.

12 Detection Testing

12.1 Unit Testing

As per the V-Model development process, first the unit testing was carried out via the use of the unit test module in Python.

A mixture of black and white box testing is utilised. The black box testing intends to ensure that the functions do not fail at a high level, whereas the white box testing ensures that as many statement execution branches in each function are executed and tested for their impacts, whether being assertions of return statement values or against expected variable values after the execution of a specific function.

To help ensure that the black and white box unit tests achieve a large coverage, a report was generated after the script runs. This was to ensure and validate that the unit tests included relevant actions which enable the majority of code statements in all branches to be executed, where reasonably was feasible. Statements that are not directly triggered during the creation of white-box tests generally required excessive modification of the running environment, such as removing other files or processes running on the system.

12.2 Integration Testing

The integration testing intends to test high-level functionality, with direct relation to the analysis and architecture design phase, wherein the high-level architecture is defined. These tests can test what white box unit testing do not. The tests were conducted using test data utilising the main run.sh script to enable testing of the various modes available, ensuring that the expected behaviours are present in the actual functionality of both the detector and the GUI.

12.3 Acceptance Testing

Acceptance testing directly maps onto the original requirements, and therefore the requirements are the test cases. These acceptance tests are generally what helps to ensure that integration was successful in meeting the required and possibly the desired requirements. In this case, all required and desired specifications were met, with some areas for improvement identified even in this stage for future versions of the software.

13 Results

Table 25 shows the summary of the results for all classifiers without optimisation using 10-fold cross-validation. From the results, it can be observed that the highest averaged F1 score and accuracy across all of the datasets is achieved by the RF classifier, with an F1-score of 0.9742 and accuracy of 0.9964, closely followed by the DT classifier with an F1-score of 0.9618 and accuracy of 0.9837. The highest average fitting times are present in the kNN classifier which also presents the fourth-highest F1 score.

The poor performance exhibited by the Linear SVM classifier may be due to the non-linearly disparate distribution of the botnet samples in the datasets, making it more difficult for the classifier to gain a higher precision.

In their default configuration, the two neural network models do not perform well, both performing under an F1-score of 0.70, which is surprising but can be due to the nature of the networks used. Optimisation of the various hyperparameters will be explored in the next section.

Table 25 Classifier performance without optimisation—averaged across all datasets

Classifier	Fit time(s)	Precision	Recall	F1 score	Accuracy
Gaussian Naive Bayes	3.4279	0.5176	0.6780	0.5781	0.5912
Decision Tree	11.8037	0.9597	0.9638	0.9618	0.98372
Random Forest	23.8155	0.9820	0.9670	0.9742	0.9964
Ada Boost	237.3703	0.9734	0.9320	0.9508	0.9725
Linear SVC	594.6946	0.6213	0.7003	0.6556	0.6705
K Neighbors	10370.4261	0.9499	0.9279	0.9381	0.9595
Neural Network 1 (24-x-1)	87.4093	0.6932	0.6999	0.6952	0.7110
Neural Network 2(24-x1-d1-x2-d2-1)	276.2794	0.6595	0.5997	0.6245	0.6387

Table 26 Snort IDS FPR experiment results

Dataset	Legitimate activity type	True negatives	False positives (Alerts)	FPR (%)
CTU-Normal-7	P2P D/Ls; Web browsing	471239	14284	3.031158287
CTU-Normal-12	P2P D/Ls; Web browsing	863374	14646	1.69636797
CTU-Normal-22	HTTPS Web browsing	2621640	15928	0.607558627
CTU-Normal-31	HTTPS Web browsing	1242661	1419	0.114190435
–	–	–	Average	1.36231883

An experiment was carried out using the Snort IDS due to its longstanding presence and usage in the world of network intrusion detection. The default rule sets that are to be utilised are therefore deemed to be more thorough, mature and well developed as a result. The default rule sets that are to be used are known as the 'Community Snort Rules' and are free to use, and open source, under the GNU general public licence version 2 [63]. The results are shown in Table 26.

It can be observed that there is a range of FPR values for each dataset i.e., from 0.11% to 3.03%. Additionally, the higher FPRs are present in the scenarios that include P2P usage (CTU-Normal-7 and CTU-Normal-12), despite these datasets containing a fraction of the number of packets in the pure web browsing traffic datasets (CTU-Normal-22 and CTU-Normal-31). Therefore, we can tentatively conclude that as more complex scenarios are included in the normal traffic, the FPR rate

Table 27 F1 score for bio-optimised hyperparameter improvement

Classifier	Default hyperparameters	Bio-optimised hyperparameters	Improvement
Decision Tree	0.961800	0.969803	0.008003
Random Forest	0.974219	0.977448	0.003229
Ada Boost	0.950865	0.958826	0.007961
Linear SVC	0.655630	0.740954	0.085324
K Neighbors	0.938194	0.940588	0.002394
Neural Network 1 (24-x-1)	0.695254	0.837873	0.142619
Neural Network 2 (24-x1-d1-x2-d2-1)	0.624558	0.827529	0.202971

on even known legitimate traffic will also increase, especially when used in a larger network with more active clients.

The bio-inspired hyperparameter optimisation was executed with a population size of 10 and a generational limit of 10 for the ML classifiers. In the case of two ANNs, a population and generational limit of 15 was chosen due to the much larger gene pools present and thus allowed more hyperparameter set evaluations to take place. Do refer to Table 27.

The highest performing classification after hyperparameter optimisation is performed by RF classifier, despite a minor improvement in the F1-score performance after optimisation. In the other classifiers, such as Linear SVM which initially scored very low, the GA optimisation has led to noticeable performance improvement. Additionally, despite the large search spaces present in the gene pools, ANNs resulted in the largest improvements, taking both their F1-score performances above 0.82, with the single hidden layer network having a higher score (0.8379) but lower overall improvement after GA hyperparameter optimisation.

13.1 Bio-optimised Model Validation

When the bio-optimised RF classifier hyperparameter sets are used to train the final models that are to be used in the product for live detection, it is necessary to evaluate their FPR performance against known legitimate data, as already conducted with Snort in Table 26. The FPR is measured using similar datasets as with Snort, and the results are shown in the Table 28.

In this case, only the 13 CTU-13 derived RF classifier models are considered, like the models from the ISOT and ISCX datasets can incur larger FPR's due to their more generalised and diverse behaviour datasets. But the clear advantage of our model over Snort IDS performance is clearly seen.

Table 28 The false positive rate comparison of the bio-optimised RF-GA on CTU-13 models against Snort's performance

Dataset	Legitimate activity type	FPR of Snort	FPR of optimised RF-GA on CTU-13 models
CTU-Normal-7	P2P Downloads; Web browsing	3.031158287	0
CTU-Normal-12	P2P Downloads; Web browsing	1.69636797	0
CTU-Normal-22	HTTPS Web browsing	0.607558627	0
CTU-Normal-31	HTTPS Web browsing	0.114190435	0
Average		1.36231883	0

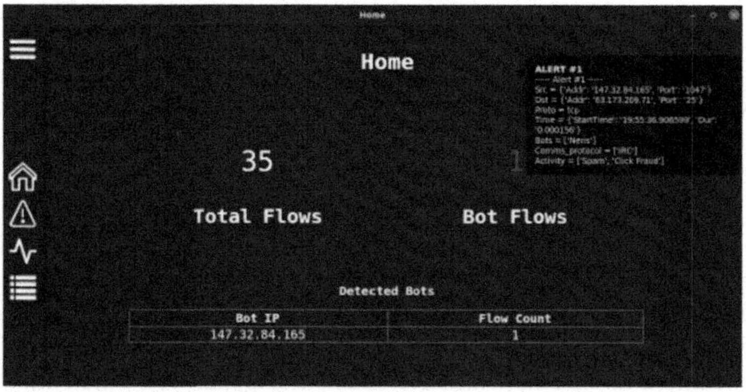

Fig. 10 The software output of botnet detection—regular and bot traffic

For the highest performing classifier after optimisation, i.e. the RF classifier with GA, the data is presented in Table 29. An average accuracy of 99.85% is achieved across all three datasets—CTU-13, ISOT and ISCX, which is a very good achievement. This also suggests that the RF optimisation with GA makes the classifier performance more consistent across different datasets.

Tables 30, 31 and 32 shows a comparison of our work with similar work. The papers mentioned within the table uses similar datasets. The data for our work shows individual dataset accuracy for CTU-13, ISOT and ISCX datasets. Comparing our work of optimised ML classification (RF-GA) where we had used flow-based data without IP addresses or port numbers, the accuracy and F1-score of our work are better than other work. Our work will also withstand bot evasion attacks like encryption, packet fragmentation etc. as it can only use flow-based data.

The software output when the botnet detection is executed, is shown in Figs. 10 and 11. The figures show the normal traffic flow and bot flow in text and activity graph, with alerts being generated when bot activity is detected.

Table 29 RF classifier's bio-optimised hyperparameters using GA and their performance with 10 generations and population size of 10

Dataset		Pre-Processes size (MB)	Botnet Class (%)	F1-Score (RF)	GA Fitness Chromosome	Precision (RF-GA)	Recall (RF-GA)	F1-Score (RF-GA)	Accuracy (RF-GA)	GA Improvement
CTU-13	1	629	1.45	0.9758	[187, 'entropy', 2, 2, 0.0, None]	0.9825	0.9710	0.9775	0.9987	0.001716027
	2	401	1.16	0.9884	[100, 'entropy', 2, 2, 0.0, None]	0.9924	0.9869	0.9897	0.9995	0.001315723
	3	1002	0.57	0.9991	[187, 'entropy', 3, 1, 0.0, 'balanced']	0.9998	0.9997	0.9994	0.9999	0.0002534
	4	251	0.23	0.9848	[36, 'entropy', 2, 1, 0.0, None]	0.9875	0.9847	0.9863	0.9998	0.001558873
	5	28.9	0.69	0.9692	[192, 'entropy', 2, 1, 0.0, 'balanced']	0.9899	0.9513	0.9781	0.9993	0.008989795
	6	124	0.83	0.9964	[70, 'entropy', 2, 1, 0.0 None]	0.9969	0.9985	0.9969	0.9998	0.000433455
	7	25.3	0.06	0.8513	[75, 'gini', 3, 1, 0.0, 'balanced']	0.9129	0.7998	0.8685	0.9997	0.017174908
	8	657	0.21	0.9727	[145, 'gini', 3, 2, 0.0, None]	0.9737	0.9742	0.9730	0.9997	0.000323613
	9	464	8.86	0.9760	[136, 'gini', 3, 2, 0.0, Noe]	0.9739	0.9738	0.9780	0.9928	0.002041924
	10	291	8.12	0.9992	[198, 'entropy', 4, 2, 0.0, 'balanced']	0.9995	0.9993	0.9993	0.9997	9.47989E-05
	11	23.4	7.61	0.9991	[100, 'gini', 2, 1, 0.0, None]	0.9998	0.9997	0.9991	0.9997	3.30176E-05
	12	72.5	0.67	0.9361	[169, 'gini', 2, 1, 0.0, None]	0.9653	0.9120	0.9453	0.9986	0.009255644
	13	428	2.08	0.9838	[100, 'entropy', 3, 1, 0.0, None]	0.9852	0.9845	0.9845	0.9987	0.000752687
ISOT		352	7.73	0.9887	[172, 'gini', 3, 1, 0.0, None]	0.9936	0.9857	0.9906	0.9973	0.001888855
ISCX		64.7	36.61	0.9921	[114, 'gini', 3, 1, 0.0, 'balance']	0.9931	0.9946	0.9947	0.9940	0.00261135
Average value		320.92	5.13	0.9742	NA	0.9831	0.9677	0.9774	0.9985	0.003229605

Table 30 Comparison: CTU-13 dataset

Author Name	Dataset	Technique	Performance achieved
Chen et al. [11]	CTU-13	RF	Detection Rate = 93.6%
Khan et al. [30]	CTU-13 + ISOT	DT	Accuracy = 98.7%
Sinha et al. [62]	CTU-13	LSTM	Accuracy = 96.2%, TPR = 94.6%, TNR = 96.3% (only for Scenarios 6, 7, 10, 11 & 12)
Hassan et al. [26]	CTU-13	Embedding and NN	Accuracy = 75.2% Precision = 76.6% Recall = 75.1% F1-Score = 74.8%
*Proposed Work	CTU-13	RF & GA	Accuracy = 99.9%, Precision = 98.14% Recall = 99.25% F1-Score = 97.5%

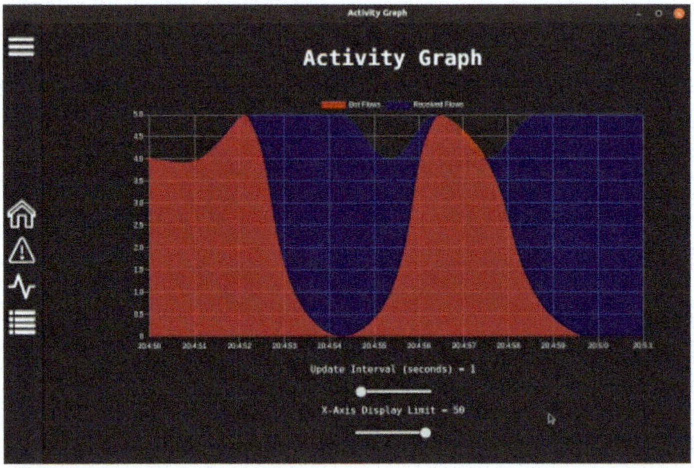

Fig. 11 The software output of botnet detection—activity graph

14 Future Work

Future work would include the inclusion of more datasets that include more diverse types of bot samples and malicious activities. Moreover, the datasets will be used to label any bot network behaviour from the bot sources that do not specifically define these behaviours individually. Due to this, the models used in the IDS only label detected botnet behaviour and provide an array of the range of possible characteristics that a bot sample exhibits.

The neural network models were found to be not as effective as initially expected, therefore, it is recommended that additional feature extraction techniques be used, or by using different types of neural networks with data that is processed in a different manner, such as a recurrent neural networks (RNNs) being applied with data that is processed in time windows, to exploit the spatial-temporal nature potentially exhibited in modern botnet network communications.

Also, an evaluation using a network flow time window can be conducted, with varying the time window length parameter for analysis of the flows. Finally, more bio-optimisation techniques can be explored for the optimisation of the classifier hyperparameters such as firefly, particle swarm optimization, ant colony etc. for comparison purposes. The optimisation times can be compared and it will also be possible to optimise the feature selection.

15 Conclusion

From the initial analysis of botnets and their operations, it is clear that they pose a large threat to networks globally. They are extremely heterogeneous and possess the ability to autonomously propagate and communicate with a C&C architecture to allow the mass coordination of actions and attacks. Additionally, it is shown that existing IDS are becoming increasingly ineffective to the ever-increasing and effective methods of botnet evasion techniques. Such systems, such as Snort, exhibit large FPRs, even in the presence of known legitimate network traffic. Additionally, the rule sets employed utilise signatures from the communication payload data of bots, which can be obfuscated in various ways to evade this direct payload signature-matching detection mechanism.

The ML methodology successfully enabled network flow features to be extracted and analysed from the publicly available representative, realistic datasets utilised in the evaluation. This methodology proves that botnet C&C communication detection can be facilitated by the characterisation of bot behavioural features using ML classification algorithm, and without the need for signature-based analysis, the network-based behaviour can be analysed. Bio-inspired hyperparameter optimisation was explored using Genetic Algorithm (GA) to allow more efficient convergence to optimal hyperparameters for each classifier. Overall, the RF classifier with GA achieved the highest average F1 score of 97.74% across all of the datasets and accuracy of

Table 31 Comparison: ISCX dataset

Author Name	Dataset	Technique	Performance achieved
Bijalwan et al. [9]	ISCX	KNN & DT	Accuracy = 96.41%
Alauthaman et al. [2]	ISCX + ISOT	DT & NN	Accuracy = 99.2%
Bijalwan [8]	ISCX	Adaboost & DT	Accuracy = 98.36% F1-Score = 98.54% Precision = 98.85% Recall = 98.23%
S-K et al. [64]	ISCX	K-means & RF	Accuracy = 99.97% Detection Rate = 98.94% False Alarm Rate = 3.80%
Alauthaman et al. [3]	ISCX + ISOT	Reinforcement learning	Accuracy = 98.3%
Hosseini et al. [28]	ISCX	CNN & LSTM	Accuracy = 99.0% Precision = 98.0% Recall = 100%
Hassan et al. [26]	ISCX	Embedding and NN	Accuracy = 80.9% Precision = 80.9% Recall = 80.9% F1-Score = 80.9%
*Proposed Work	ISCX	RF & GA	Accuracy = 99.4%, Precision = 99.87% Recall = 99.84% F1-Score = 99.5%

99.85% with FPR of 0% against known legitimate datasets when CTU-13 models were trained and serialised with the bio-optimised hyperparameters.

The network flow data is advantageous as it does not incur the privacy concerns that .pcap files present, as they do not include the raw payload data directly. Therefore, they can be publicly distributed more easily and allow further research to be conducted and cross-validated to methodologies in related work. Also, the tools were used to facilitate the feature extraction from raw .pcap files. The ML methodology is then utilised in the product to allow live network flow data feature extraction in the detector's sniffer module.

The product was successfully designed, implemented and tested against all relevant requirements to ensure that the implementation was sufficient. The GUI was developed to present the data to the user in various formats and implements dynamic data loading and pseudo-persistence between pages.

Acknowledgement This is an invited paper for 16th International Conference on Global Security, Safety & Sustainability (ICGS3-24), themed on "Cybersecurity and Human Capabilities through

Table 32 Comparison: ISOT dataset

Author Name	Dataset	Technique	Performance achieved
Pektas and Acarman [49]	ISOT	RF	F1-Score = 99.0%, Accuracy = 99.5%, Recall = 99.0%, Precision = 99.0%
Alauthaman et al. [3]	ISOT + ISCX	Reinforcement learning	Accuracy = 98.3%
Mai & Park [37]	ISOT	K-means	Detection Rate = 97.11%
Khan et al. [30]	ISOT + CTU-13	DT	Accuracy = 98.7%
Alauthaman et al. [2]	ISOT + ISCX	DT & NN	Accuracy = 99.2%
Hassan et al. [26]	ISOT	Embedding and NN	Accuracy = 99.9% Precision = 99.9% Recall = 99.9% F1-Score = 99.9%
Velasco-Mata et al. [71]	ISOT	Embedding and NN	Accuracy = 99.9% Precision = 99.9% Recall = 99.9% F1-Score = 99.9%
*Proposed Work	ISOT	RF & GA	Accuracy = 99.7%, Precision = 99.66% Recall = 98.79% F1-Score = 99.1%

Symbiotic Artificial Intelligence", from 25 to 27 November 2024. The YouTube link of the project and demo of the software developed: https://youtu.be/gNQjC91VtOI.

References

1. Acarali D, Rajarajan M, Komninos N (2018) Modelling Botnet propagation in networks with layered defences. In: 2018 International symposium on networks, computers and communications (ISNCC), Rome, pp 1–6. https://doi.org/10.1109/ISNCC.2018.8530934
2. Alauthaman M, Aslam N, Zhang L et al (2018) A P2P Botnet detection scheme based on decision tree and adaptive multilayer neural networks. Neural Comput Appl 29:991–1004. https://doi.org/10.1007/s00521-016-2564-5
3. Alauthman M, Aslam N, Al-kasassbeh M, Khan S, Al-Qerem A, Choo K-KR (2020) An efficient reinforcement learning-based Botnet detection approach. J Netw Comput Appl 150 (Online). https://doi.org/10.1016/j.jnca.2019.102479

4. Almashor M, Dehghantanha A (2024) Robust IoT Botnet detection framework resilient to gradient based adversarial attacks. SN Comput Sci 5(5):242. https://doi.org/10.1007/s42979-024-03242-0
5. Argus (2012) Argus—Auditing network activity (Online). Available at: https://qosient.com/argus/gettingstarted.shtml. Accessed 16 Mar 2024
6. Baker AR, Caswell B, Poor M (2004) Snort 2.1 intrusion detection, 2nd edn. Syngress Publishing
7. Beigi EB, Jazi HH, Stakhanova N, Ghorbani AA (2014) Towards effective feature selection in machine learning-based botnet detection approaches. In: 2014 IEEE conference on communications and network security, pp 247–255. https://doi.org/10.1109/CNS.2014.6997492
8. Bijalwan A (2020) Botnet forensic analysis using machine learning. Secur Commun Netw 2020:1–9. https://doi.org/10.1155/2020/9302318
9. Bijalwan A, Chand N, Pilli E, Rama Krishna C (2016) Botnet analysis using ensemble classifier. Perspect Sci 8:502–504. https://doi.org/10.1016/j.pisc.2016.05.008
10. Bullard C (2020) ra(1)—argus-client—Debian testing—Debian Manpages. *Manpages.debian.org* [Online]. Available: https://manpages.debian.org/testing/argus-client/ra.1.en.html. Accessed 28 Sept 2024
11. Chen R, Niu W, Zhang X, Zhuo Z, Lv F (2017) An effective conversation-based botnet detection method. Math Probl Eng 2017
12. CIC (2014) Canadian Institute for Cybersecurity, Botnet 2014—datasets—research—University of New Brunswick (Online). Available at: https://www.unb.ca/cic/datasets/botnet.html. Accessed 19 Mar 2024
13. Decision Trees (2019) The documentation titled "1.10. Decision Trees," (Online). Available: https://scikit-learn.org/stable/modules/tree.html#decision-trees. Accessed 25 Nov 2024
14. Distance Metric - Sklearn (2020) The documentation titled "sklearn.neighbors.DistanceMetric - scikit-learn 0.23.2 documentation". *Scikit-learn.org* (Online). Available: https://scikit-learn.org/stable/modules/generated/sklearn.neighbors.DistanceMetric.html
15. Dollah RFM, Faizal M, Arif F, Mas'ud MZ, Xin LK (2018) Machine learning for http botnet detection using classifier algorithms. J Telecommun Electron Comput Eng (JTEC) 10(1–7):27–30
16. Fatima A, Maurya R, Dutta MK, Burget R, Masek J (2019) Android malware detection using genetic algorithm based optimized feature selection and machine learning. In: 2019 42nd International conference on telecommunications and signal processing (TSP), pp 220–223. https://doi.org/10.1109/TSP.2019.8769039
17. Fogla P, Sharif MI, Perdisci R, Kolesnikov OM, Lee W (2006) Polymorphic blending attacks. In: Proceedings of the 15th USENIX security symposium. Vancouver, Canada, pp 241–256
18. Fowler KR (2015) Introduction to good development. In: Developing and managing embedded systems and products
19. Frank E, Hall M, Witten I (2016) The WEKA workbench—data mining: practical machine learning tools and techniques. Cs.waikato.ac.nz (Online0. Available: https://www.cs.waikato.ac.nz/ml/weka/Witten_et_al_2016_appendix.pdf. Accessed 20 Nov 2023
20. Garcia S, Grill M, Stiborek J, Zunino A (2014) An empirical comparison of botnet detection methods. Comput Secur 45:100–123. https://doi.org/10.1016/j.cose.2014.05.011
21. Gardiner J, Cova M, Nagaraja S (2014) Command & control: understanding, denying and detecting—a review of malware C2 techniques, detection and defences. University of Birmingham, Birmingham, UK
22. Gebhart G (2004) Worm propagation and countermeasures. SANS Institute
23. Godínez-Bautista A, Padierna L, Rojas-Domínguez A, Puga H, Carpio M (2018) Bio-inspired metaheuristics for hyper-parameter tuning of support vector machine classifiers. In: Fuzzy logic augmentation of neural and optimization algorithms: theoretical aspects and real applications, pp 115–130. https://doi.org/10.1007/978-3-319-71008-2_10
24. Gu G, Zhang J, Lee W (2008) Botsniffer: detecting botnet command and control channels in network traffic. In: Proceedings of the 15th annual network and distributed system security symposium (Online). https://corescholar.libraries.wright.edu/cse/7

25. Haddadi F, Phan D-T Zincir-Heywood AN (2016) How to choose from different botnet detection systems NOMS 2016–2016? IEEE/IFIP network operations and management symposium, Istanbul, pp 1079–1084. https://doi.org/10.1109/NOMS.2016.7502964
26. Hassan M, Haque M, Tozal M, Raghavan V, Agrawal R (2021) Intrusion detection using payload embeddings. IEEE Access
27. Holz T (2005) A short visit to the bot Zoo. IEEE Secur Privacy Mag 3(3):76–79. https://doi.org/10.1109/msp.2005.58
28. Hosseini S, Nezhad A, Seilani (2021) Botnet detection using negative selection algorithm, convolution neural network and classification methods. In: Evolving systems. Springer, pp 1–15
29. John GH, Langley P (1995) Estimating continuous distributions in Bayesian classifiers. In: Proceedings of the eleventh conference on uncertainty in artificial intelligence. Morgan Kaufmann Publishers Inc., San Francisco, CA, USA
30. Khan RU, Zhang X, Kumar R, Sharif A, Golilarz NA, Alazab M (2019) An adaptive multi-layer botnet detection technique using machine learning classifiers. Appl Sci 9(11). Art no. 2375
31. Khattak S, Ramay N, Khan K, Syed A, Khayam S (2014) A taxonomy of Botnet behavior, detection, and defense. IEEE Commun Surv Tutor 16(2):898–924. https://doi.org/10.1109/surv.2013.091213.00134
32. Lin K, Chen S, Hung J (2014) Botnet detection using support vector machines with artificial fish swarm algorithm. J Appl Math 2014:1–9. https://doi.org/10.1155/2014/986428
33. Liu Y, Sun J, Sun R, Wen Y (2010) Next generation internet traffic monitoring system based on netflow. In: 2010 International conference on intelligent system design and engineering application. Changsha, pp 1006–1009. https://doi.org/10.1109/ISDEA.2010.337
34. Li X, Wang L, Sung E (2008) AdaBoost with SVM-based component classifiers. Eng Appl Artif Intell 21(5):785–795. https://doi.org/10.1016/j.engappai.2007.07.001
35. Mahi A (2023) A comprehensive literature review on machine learning models application in detecting and preventing Botnet attacks on IoT devices (Online). Available: https://shorturl.at/R5U3q
36. Mahmoud M, Nir M, Matrawy A et al (2015) A survey on botnet architectures, detection and defences. IJ Netw Secur 17(3):272–289
37. Mai L, Park M (2016) A comparison of clustering algorithms for botnet detection based on network flow. In: 2016 Eighth International conference on ubiquitous and future networks (ICUFN). Vienna, pp 667–669. https://doi.org/10.1109/ICUFN.2016.7537117
38. Mannikar R, Di Troia F (2024) Enhancing Botnet detection in network security using profile hidden Markov models. Appl Sci 14(10):4019. https://doi.org/10.3390/app14104019
39. Mecklin J (2020) Current time—Bulletin of the atomic scientists. Bulletin of the Atomic Scientists (Online). Available: https://thebulletin.org/doomsday-clock/current-time/#full-statement. Accessed 06 Dec 2023
40. Moubayed A, Injadat M, Shami A (2020) Optimized random forest model for Botnet detection based on DNS queries. arXiv:2012.11326. Accessed 10 Jan 2024
41. Nearest Neighbors (2020) The documentation titled "1.6. Nearest Neighbors - scikit-learn 0.23.2 documentation". Scikit-learn.org (Online). Available: https://scikit-learn.org/stable/modules/neighbors.html#classification
42. Nguyen TT, Armitage G, Branch P (2022) A hybrid deep learning model for network anomaly detection using payload and flow features. IEEE Trans Netw Serv Manag 19(4):4406–4421
43. Nivargi V, Bhaowal M, Lee T (2006) Machine learning based botnet detection. CS-229 Final Project Report. Stanford University, USA
44. NSFOCUS (2023). Botnet Trends 2023 review and 2024 predictions (Online). Available: https://nsfocusglobal.com/company-overview/resources/botnet-trends-2023-review-and-2024-predictions/. Accessed 10 Nov 2024
45. Pandas (2014a) The documentation titled "DataFrame - Pandas 1.0.3 documentation". Pandas.pydata.org (Online). Available: https://pandas.pydata.org/pandas-docs/stable/reference/frame.html. Accessed 28 Mar 2024

46. Pandas (2014b) The documentation titled “Pandas.read_csv - pandas 1.0.3 documentation”. Pandas.pydata.org (Online). Available: https://pandas.pydata.org/pandas-docs/stable/reference/api/pandas.read_csv.html. Accessed 28 Mar 2024
47. Pandas (2014c) The documentation titled “pandas.DataFrame.loc - pandas 1.0.3 documentation” (Online). Available at: https://pandas.pydata.org/pandas-docs/stable/reference/api/pandas.DataFrame.loc.html. Accessed 27 Mar 2020
48. Pedregosa F, Varoquaux G, Gramfort A, Michel V, Thirion B, Grisel O, Blondel M, Prettenhofer P, Weiss R, Dubourg V et al (2011) Scikit-learn: machine learning in python. J Mach Learn Res 12:2825–2830
49. Pektas A, Acarman T (2017) Effective feature selection for botnet detection based on network flow analysis. In International conference automatics and informatics
50. Pektaş A, Acarman T (2017) Effective feature selection for botnet detection based on network flow analysis. In: International conference automatics and informatics
51. Pektaş A, Acarman T (2018) Botnet detection based on network flow summary and deep learning. Int J Netw Manage 28(6). Art no. e2039
52. Putman C, Abhishta, Nieuwenhuis LJM (2018) Business model of a Botnet. In: 26th Euromicro International conference on parallel, distributed and network-based processing (PDP). https://doi.org/10.1109/pdp2018.2018.00077
53. Qrator (2024) Q1 2024 DDoS attacks statistics and overview. Qrator.net. Online: https://qrator.net/blog/details/q1-2024-ddos-attacks-statistics-and-overview
54. Resende PAA, Drummond AC (2018) Adaptive anomaly-based intrusion detection system using genetic algorithm and profiling. Secur Privacy 1:e36 (Online). https://doi.org/10.1002/spy2.36
55. Roesch M (1999) Snort—Lightweight intrusion detection for networks. In: LISA ’99: Proceedings of the 13th USENIX conference on system administration. Seattle, Washington, pp 229–238
56. Saad S, Traore I, Ghorbani A, Sayed B, Zhao D, Lu W, Felix J, Hakimian P (2011) Detecting p2p botnets through network behavior analysis and machine learning. In: 2011 Ninth annual international conference on privacy, security and trust, pp 174–180. https://doi.org/10.1109/PST.2011.5971980
57. Scikit-learn (2019a) Scikit-learn: machine learning in Python - scikit-learn 0.23.2 documentation. Scikit-learn.org (Online). Available: https://scikit-learn.org/stable/index.html. Accessed 09 Nov 2024
58. Scikit-Learn (2019b) Scikit-Learn - 1.13. Feature selection (Online). Available at: https://scikit-learn.org/stable/modules/feature_selection.html#feature-selection. Accessed 2 Apr 2020
59. Scikit-Learn (2019c) Scikit-Learn - 1.11. Ensemble methods (Online). Available at: https://scikit-learn.org/stable/modules/ensemble.html#forest. Accessed 26 Mar 2024
60. Shah S, Issac B (2018) Performance comparison of intrusion detection systems and application of machine learning to Snort system. Future Gener Comput Syst 80:157–170
61. Sinha K, Viswanathan A, Bunn J (2019) Tracking temporal evolution of network activity for botnet detection. ArXiv.org. https://arxiv.org/abs/1908.03443
62. Sinha K, Viswanathan A, Bunn J (2019) Tracking temporal evolution of network activity for Botnet detection. arXiv:1908.03443. Accessed 10 Sep 2024
63. Snort (2020) The documentation titled “Snort Licence” (Online). Available at: https://www.snort.org/license. Accessed 4 Mar 2024
64. Soheily-Khah S, Marteau P, Bechet N (2018) Intrusion detection in network systems through hybrid supervised and unsupervised machine learning process: a case study on the ISCX dataset. In: 2018 1st International conference on data intelligence and security (ICDIS). https://doi.org/10.1109/icdis.2018.00043
65. Sommer R, Paxson V (2010) Outside the closed world: on using machine learning for network intrusion detection. In: 2010 IEEE symposium on security and privacy. Berkeley/Oakland, CA, pp 305–316. https://doi.org/10.1109/SP.2010.25
66. Song Y, Locasto ME, Stavrou A, Keromytis AD, Stolfo SJ (2007) On the infeasibility of modeling polymorphic shellcode∗. In: Proceedings of the 14th ACM conference on computer and communications security. New York, US, pp 541–551

67. Spamhaus (2019) Botnet Threat Report 2019. Spamhaus.com [Online] Spamhaus. Available at: https://www.spamhaus.com/custom-content/uploads/2020/04/2019-Botnet-Threat-Report-2019-LR.pdf. Accessed 06 December 2024
68. Support Vector Machines (2019) 1.4. Support Vector Machines - scikit-learn 0.23.2 documentation. Scikit-learn.org (Online). Available: https://scikit-learn.org/stable/modules/svm.html#
69. TensorFlow (2019) The documentation titled "TensorFlow". TensorFlow (Online). Available: https://www.tensorflow.org/. Accessed 29 Oct 2024
70. Vacca JR (2010) Network and system security. Syngress, Burlington, MA
71. Velasco-Mata J, Gonzalez-Castro V, Fernandez E, Alegre (2021) Efficient detection of botnet traffic by features selection and decision trees. IEEE Access 9:120567-120579
72. Wang P, Wu L, Aslam B, Zou CC (2009) A systematic study on Peer-to-Peer botnets. In: 2009 Proceedings of 18th International conference on computer communications and networks. San Francisco, CA, pp 1–8. https://doi.org/10.1109/ICCCN.2009.5235360

The manufacturer's authorised representative in the EU is Springer Nature Customer Service Centre GmbH, Europaplatz 3, 69115 Heidelberg, Germany. If you have any concerns regarding our products, please contact ProductSafety@springernature.com

Printed and bound by CPI Group (UK) Ltd, Croydon, CR0 4YY
30/09/2026
02225563-0019